Lithium-Ion Cells: Materials and Applications

Lithium-Ion Cells: Materials and Applications

Yury Koshtyal

Alexander Rumyantsev

Basel • Beijing • Wuhan • Barcelona • Belgrade • Novi Sad • Cluj • Manchester

Authors
Yury Koshtyal
PJSC "MMC "Norilsk Nickel",
Moscow, Russia
Laboratory of Lithium-Ion Technology,
Ioffe Institute,
Saint-Petersburg, Russia

Alexander Rumyantsev
Laboratory of Lithium-Ion Technology,
Ioffe Institute,
Saint-Petersburg, Russia

Editorial Office
MDPI AG
Grosspeteranlage 5
4052 Basel, Switzerland

For citation purposes, cite as indicated below:

Lastname, Firstname, Firstname Lastname, and Firstname Lastname. Year. *Book Title*. Series Title (optional). Basel: MDPI, Page Range.

ISBN 978-3-0365-8267-2 (Hbk)
ISBN 978-3-0365-8266-5 (PDF)
https://doi.org/10.3390/books978-3-0365-8266-5

Contents

List of Figures and Tables

Abbreviations

1050	Aluminium alloy
1070	Aluminium alloy
1085	Aluminium alloy
1100	Aluminium alloy
1235	Aluminium alloy
3003	Aluminium alloy
8021	Aluminium alloy
1N30	Aluminium alloy
3C	Computer, communication, and consumer electronics
AAM	Anode-active material
AC	Alternating current
AG	Artificial graphite
AND	$NC(CH_2)_4CN$, CAS № 111-69-3
BET	Brunauer, Emmett, and Teller theory
BEV	Battery electric vehicle
BMLMFP	Bio-modified LFP, contains several mass percents $[Ca_5(PO_4)_3(OH)_2]$
BMS	Battery management system
BOL	Beginning of life
BOM	Bill of materials
BP	Biphenyl, $C_{12}H_{10}$, CAS № 92-52-4
BPU	Battery protection unit
BS	$C_4H_8O_3S$, CAS № 1633-83-6
C1100	Copper
CAM	Cathode-active material
CAN	Controller Area Network
CB	Carbon black
CB (A)	Acetylene black
CC CV (CCCV)	Mode of charging constant current then constant voltage
CCS	Coating layer containing inorganic particles
CEI	Cathode–electrolyte interphase
CF	Carbon fibres
CHB	Chlorobenzene
CHB	$C_{12}H_{16}$, CAS № 827-52-1
CID	Current interruptive device
CMC (Na-CMC)	Sodium salt of carboxymethyl cellulose
CRC	$LiCoO_2$
CTC	Cell to chassis

DBDMB	$C_{16}H_{26}O_2$, CAS № 7323-63-9
DBP	Dibutyl phthalate absorption number
DC	Direct current
DCIR	Direct current internal resistance
DEC	Diethyl carbonate
DEPA	$C_8H_{17}O_5P$, CAS № 867-13-0
DFA	$C_7H_6F_2O$, CAS № 452-10-8
DMC	Dimethyl carbonate
DoD	Depth of discharge
DOE	Department of Energy
DTD	$C_2H_4O_4S$, CAS № 1072-53-3
EC	Ethylene carbonate
ED	Electro drive
EDLC	Electric double-layer capacitor
EDS	Energy-dispersive spectroscopy
EES	Electric energy storage
EGBE	$C_8H_{12}N_2O_2$, CAS № 3386-87-6
EMC	Ethyl methyl carbonate
EOL	End of life
EPA	Environmental Protection Agency
ES	$C_3H_6O_4S$, CAS № 1073-05-8
EsB	Energy-selective back-scattered
ESS	Energy storage system
F2EC	$C_3H_2F_2O_3$, CAS № 311810-76-1
FB	C_6H_5F, CAS № 462-06-6
FEC	Fluoroethylene carbonate
FEC	$C_3H_3FO_3$, CAS № 114435-02-8
G (fl.)	Flake graphite
G (sph.)	Spheroidised graphite
Gr	Graphite
Grph.	Graphene
H-NCM	High nickel content NCM
H-TP	$C_{18}H_{22}$, CAS № 61788-32-7
HBL	High-temperature binder layer
HC	Hard carbon
HDPE	High-density polyethylene
HE	High energy
HE	High-energy lithium-ion cells

HEV	Hybrid electric vehicle
HFP	Hexafluoropropylene
HOPG	Highly oriented pyrolytic graphite
HP	High power
HPPC	Hybrid pulse power characteristic
HRL	Heat resistance layer
HVAC)	Heating, ventilation, and air conditioning
IACS	International Annealed Copper Standard
ICE	Internal combustion engine
IoT	Internet of Things
IT	Information Technology
LAGP	$Li_{1+x}Al_xGe_{2-x}(PO_4)_3$
LATP	$Li_{1.5}Al_{0.5}Ti_{1.5}(PO_4)_3$
LCO	Lithium cobalt oxide – $LiCoO_2$
LFP (LFP/C)	Carbon-coated lithium ferro-phosphate – $LiFePO_4$
LGPS	$Li_{10}GeP_2S_{12}$
LIB	Lithium-ion cell with battery controller, or battery management system and battery case
LIBOB	Lithium bis(oxalato)-borate
LiBOB	$LiB(C_2O_4)_2$, CAS № 244761-29-3
LIBS	Laser-induced breakdown spectroscopy
LIC	Lithium-ion cell
LiDFBP	$C_4F_2LiO_8P$
LIFSI	$F_2LiNO_4S_2$, CAS № 171611-11-3
LIODFB	$LiBF_2(C_2O_4)$, CAS № 409071-16-5
$LiPO_2F_2$	$LiPO_2F_2$, CAS № 24389-25-1
LiTFSI	$Li[N(SO_2CF_3)_2]$, CAS № 90076-65-6
LLZO	$Li_7La_3Zr_2O_{12}$
LLZO-SM	$Li_{7-2x+y}Mg_xLa_{3-y}Sr_yZr_2O_{12}$, $(0.1 \leq x \leq 0.3, 0 \leq y \leq 0.5)$
LLZTO	$Li_{6.4}La_3Zr_{1.4}Ta_{0.6}O_{12}$
LMFP	Carbon-coated lithium manganese–ferro-phosphate—$LiMn_xFe_{(1-x)}PO_4$, $x < 1$
LMO	Lithium manganese oxide—$LiMn_2O_4$
LMO/C	Lithium manganese spinel/hard carbon or soft carbon
LMO/LTO	Lithium mixed oxide (NCM or NCA or LMO)/lithium titanate
LMR	Lithium-manganese-rich NCM
LNMO	High-voltage nickel manganese spinel
LNO	High-nickel-content NCM or NCA
LTO	Lithium titanate—$Li_4Ti_5O_{12}$

M.O.	Mixed oxides (NCM or/and NCA or/and LMO)
MCMB	Mesophase carbon microbeads
mHEV	Micro hybrid electric vehicle
mHEV	Mild hybrid electric vehicle
MLCC	Multilayer ceramic capacitors
MLCC-type	Lithium-ion cell with construction similar to multilayer ceramic capacitor
MMDS	$C_2H_4O_6S_2$, CAS № 99591-74-9
MPC	$C_8H_8O_3$, CAS № 13509-27-8
mSSLICs	Miniature solid-state lithium-ion cells
MWCNTs	Multi-wall carbon nanotubes
N201	Nickel alloy
Na-NiCl$_2$	Sodium–nickel–chloride
Na-S	Sodium–sulphur batteries
NC, NCO	Lithium nickel cobalt oxide
NCA	$Li_{1+x}Ni_eCo_fAl_gO_{2+d}$ $e + f + g = 1; 0.8 \leq e \leq 0.92, 0.015 \leq g \leq 0.05, x \leq 0.07$
nCA (nC)	Current numerically equal to the rated capacity
NCM (A:B:C)	$Li_{1+x}Ni_aCo_bMn_cO_{2+d}$ $0 \leq x \leq 0.05$, $a + b + c = 1$, designated as where $a = A/(A + B + C)$, $b = B/(A + B + C)$, $c = C/(A + B + C)$, $x \leq 0.07$
NCMA	Lithium nickel cobalt manganese aluminium oxide
NG (NG-core)	Carbon-coated natural graphite
NMC	see NCM
NMP	N-methyl-2-pyrrolidone
OAN	Oil absorption number
OLOs	Over-lithiated layered oxides
oTPh	$C_{18}H_{14}$, CAS № 84-15-1
PAN	Polyacrylonitrile
PbA	Lead–acid batteries
PbA/C	Lead–acid–carbon (hybrid cell)
PC	Propylene carbonate
PCB	Battery protection circuit board
PCLB	Pouch lithium ceramic battery
pcs	Pieces
PE	Polyethene
PEHM	Polyethene with high melting point
PELS	Polyethene with low melting point
PET	Polyethene terephthalate
PhBA	$C_6H_5B(OH)_2$, CAS № 98-80-6
PHEV	Plug-in hybrid electric vehicle
PHOS	Phosphazene

PMIA	Poly (m-finylene isophthalamide)
PMMA	Polymethyl methacrylate
PMS	$C_3H_6O_3S$, CAS № 16156-58-4
PP	polypropylene
PRS	$C_3H_4O_3S$, CAS № 21806-61-1
PS	$C_3H_6O_3S$, CAS № 1120-71-4
PT	Lithium-ion cells used for manufacturing of power tools batteries
PTC	Positive temperature coefficient
PTFE	Polytetrafluoroethylene
PVDF	Polyvinylidene fluoride
PVDF-CTFE	Poly(vinylidene fluoride-co-chlorotrifluoroethylene
R&D	Research and Development
Ref.	Reference
SAE	SAE International (formerly the Society of Automotive Engineers)
S_{BET}	Specific surface area determined according BET theory
SBR	Styrene-butadiene rubber
SC	Soft carbon
SD	Self-discharge
SEI	Solid-electrolyte interphase
SFL	Safety functional layer
SN(SCN)	$C_4H_4N_2$, CAS №. 110-61-2
SoC	State of charge
SOH	State of health
SRS	Ceramic coating on separator
SWNTs	Single-wall carbon nanotubes
T&D	Transmission and Distribution
TAB	$C_{11}H_{16}$, CAS № 2049-95-8
TEMPO	2,2,6,6-tetramethylpiperidine-1-oxyl radical
Tg	Glass-transition temperature
TMSB	$C_9H_{27}O_3BSi_3$, CAS № 4325-85-3
TMSPate	$C_9H_{27}O_4PSi_3$, CAS № 10497-05-9
TMSPite	$C_9H_{27}O_3PSi_3$, CAS № 1795-31-9
TTFP	$(CF_3CH_2O)_3P$, CAS № 370-69-4
UHMW, (U) HMW	Ultra-high-molecular-weight
UPS	Uninterruptible power supplies
V2F (VDF)	Vinylidene fluoride
VA	$C_4H_6O_2$, CAS № 108-05-4
VC	Vinylene carbonate
VC	$C_3H_2O_3$, CAS № 872-36-6
VDA	German Association of the Automotive Industry
VEC	$C_5H_6O_3$, CAS № 4427-96-7
VGCF	Vapour-grown carbon fibres
WDS	Wavelength-dispersive X-ray spectroscopy
xEV	Various types of electric vehicles

About the Authors

Yury Koshtyal

Yury Koshtyal received Bachelor's (2006), Master's (2007), and PhD degrees in Solid-State and Physical Chemistry (2011) at Saint Petersburg Institute of Technology (Technical University). The research was devoted to the structural and chemical transformations that occur during the synthesis of photoactive active materials by molecular layering (atomic layer deposition).

During his education period, he carried out several internships in Écoles des Mines d'Alès (IMT Mines Alès, France). In addition, he received scholarships from Languedoc Roussillon (France), the Government of France (Bourse de stage d'études), and the St. Petersburg Government (Russia).

Since 2013, he has worked in the lithium-ion technology laboratory at the Ioffe Institute (St. Petersburg, Russia) and conducted the research and development of anode and cathode materials for lithium-ion cells.

Since 2016, he has conducted research in the field of thin-film electrode materials obtained via atomic layer deposition for solid-state lithium-ion cells (Institute of Metallurgy of Mechanical Engineering and Transport, Peter the Great Saint Petersburg Polytechnic University).

In 2023, he joined PJSC "MMC "Norilsk Nickel" and participates in the R&D of high-nickel cathode materials for lithium-ion cells as well as high-energy lithium-ion and sodium-ion cells.

Alexander Rumyantsev

Alexander Rumyantsev received Bachelor's (2005), Master's (2007), and PhD degrees in Electrochemistry (2011) at Saint Petersburg Institute of Technology (Technical University), studying the electrochemical processes occurring in lithium-ion cells.

In 2011, he joined the Ioffe Institute and participated in the organisation of the lithium-ion technology laboratory. The field of research includes the electrochemical testing of promising materials (anodes, cathodes, electrolytes) for lithium-ion cells and lithium metal cells, lithium-ion cell technology development, and the conduction of lithium-ion cell and battery tests.

Preface

Many articles and books have been published on the topics designated in this volume. However, most of the literary sources are formatted as follows:

- They describe the results of scientific articles on the synthesis and study of perspective materials;
- They reveal circuit and design solutions for constructing control systems and manufacturing batteries;
- They are educational materials.

At the same time, a small number of publications include the following:

- A description of materials that are produced industrially and used in the LIC manufacturing process;
- A demonstration of the industrially produced LICs' energy and power parameters;
- An analysis of the characteristics of manufactured miniature lithium-ion cells, solid-state LICs and lithium metal cells, and all-solid-state cells.

Considering the popularity of the topics discussed, one can hope to find quite detailed information on the Internet. Indeed, modern search engines make it possible to find a sufficiently large number of relevant documents. However, while conducting such research, we encountered the following challenges:

- The data are somewhat fragmented, and their systematisation and structuring are required;
- The search results do not always meet search queries. For instance, the relevant to the topic data was found, but they didn't match the query;
- As the accumulated data grow, the search time for new information increases;
- The choice of search engine and location (different countries) affects the search results;
- The data are not indexed in search engines despite the correct keywords and website being requested;
- The information disappears due to website updates;
- The found data require processing; for example, many presentations show changes in the shape of the discharge curves depending on the discharge current strength. In addition, Ragone plots are necessary for a correct comparison and, therefore, the mathematical processing of presented results is required.

Thus, this book was written to systematise and structure information on industrially produced materials for LIC manufacturing and industrially produced and promising LICs (and lithium metal rechargeable cells) for various applications.

Yury Koshtyal and Alexander Rumyantsev
Authors
Saint Petersburg, Russia

Acknowledgements

The authors are grateful for assistance with the preparation of the manuscript and for providing support to Kudryashova L.N.; for proofreading and editing the original text to Kuznetsov V.P.; for translation of the 2nd, 4th, 6th, 7th, and 8th chapters to Sukhinin A.N.; for translation of the 3rd chapter to Solonitsyna A.P.; for translation of the 5th chapter to Association of Professional Translators (APP LLC); for fruitful discussions to Belyaev S.S., Voronzova E.V., Zhdanov V.V., Loginova M.M., Maksimov M. Yu., Makhonina E.V., Rykovanov A.S. and others; for help in preparing the manuscript for publication to Vishnyakov P.S.

1 Introduction

Lithium-ion cells (LICs) and batteries, due to their functional parameters such as specific energy, specific power, cycle life, and performance in broad temperature intervals, are in demand for various applications. Despite a long development history, new materials for LIC manufacturing and the development of batteries with improved characteristics are still intensively discussed topics in the scientific and technical literature.

1.1. Highlights

1. A review of commercially available materials and promising new materials offered by start-ups and leading manufacturers for the production of lithium-ion cells with improved characteristics was carried out (Chapter Materials for Lithium-Ion Cells).
2. Ragone plots for lithium-ion cells of different applications are given, as follows:

2.1 High-energy LICs for smartphones, LICs in 18650 and 21700 cases, and perspective LICs and lithium metal cells (including those with solid-state electrolyte) (Chapter Advanced High-Power Lithium-Ion Cells for Electric Hand Tools).

2.2 High-power LICs in 18650 and 21700 cases for power tools (Chapter Advanced High-Power Lithium-Ion Cells for Electric Hand Tools).

2.3 LICs for various hybrid and electric vehicles (Chapter Lithium-Ion Cells and Batteries for Hybrid, Electric Passenger Vehicles).

2.4 LICs for stationary energy storage systems (Chapter Lithium-Ion Electric Energy Storage for Stationary Applications).

2.5 Miniature LICs (0.16 mAh ÷ 1.22 Ah, in cylindrical, prismatic, button, coin cases), fully solid-state (bulk, MLCC-type, thin-film), and LICs for medical applications (Chapter Miniature Lithium-Ion Cells for the Internet of Things, Wearable Devices, and Medical Applications).

3. The coverage of the considered cells in each of the assessed areas (Chapters 3–7). More than 200 cells are mentioned in the book (see Appendix A).

1.2. Search Methodology

To increase the coverage of initial information, various search engines (Baidu, Google, Yahoo, etc.), publication bases (SAE), and file storage systems (docin.com, slideshare.net, wenku.baidu.com, etc.) were used. The search was carried out using keywords and pictures and from different locations. We also used the principle applied when looking for mushrooms in the forest: when helpful information was found, we looked around for more. The initial data for review included articles from battery-devoted websites and materials (booklets, reports) from specialised scientific and technical conferences (Europe, China, USA, South Korea, Japan). Also, the authors' personal experience in the development of power sources, laboratory tests, and scientific and technical articles (including from industrial companies) and monographs were considered. Thousands of documents were viewed, and more than 1300 references are cited in the present book.

1.3. What Information Can Be Found in the Book?

The book provides background information on the main characteristics of materials (active anode materials, active cathode materials), binders, thickeners, conductive additives, electrolytes (including functional additives), separators, current collectors, housings, and current leads, as well as their qualitative effect on the specific energy and capacity of manufactured cells (Chapter Materials for Lithium-Ion Cells). Methods for increasing the specific energy (Chapter Lithium-Ion Cells with High Specific Energy), power, and safety (Chapter Lithium-Ion Cells and Batteries for Hybrid, Electric Passenger Vehicles) of LICs are described that include choosing a design solution, materials, and manufacturing technology.

The book also provides background information (weight, sizes, and performance parameters) on the LICs and batteries of micro hybrid, mild hybrid, strong hybrid, and plug-in hybrid electric vehicles, as well as modules for making the stationary energy storage devices and cells mentioned in the Highlights. In addition, the effect of discharge current strength (power) on the change in specific energy (Ragone plots) of the series cell is shown for each of the considered applications. However, the authors did not have the opportunity to verify all of the information presented in practice. Moreover, data tend to become outdated; in this regard, we advise the reader to consider the data presented in the book as a first approximation. More detailed, accurate, and up-to-date data can be requested directly from the manufacturers.

1.4. How to Use This Book

Like many authors, we hope that our work will be in demand. However, we are also aware that not all information will be relevant for the reader. Nowadays, most books are published in electronic form, and the reader can use the search function to save time. However, given the possibility of presenting the same content in different words, search queries may sometimes not match the wording used in the book. In this regard, for navigation through the book, one can use the table of contents, the list of names of figures and tables, and indices (manufacturer-brand LIC—numbers of tables and/or figures; keywords).

To simplify the presentation of information, we decided to use abbreviations and designations. Abbreviations related to chemical substances (materials), words, or phrases are defined on first mention in the text and the "Abbreviations" section of the book. One can find brief designations (models) of cells and batteries in the captions of the figures or tables. They are also listed in alphabetical order in the Appendix (Table A.2) where the manufacturer's name and the number of tables and figures describing their properties can be found. The main characteristics of the cells and batteries are indicated in the table in which they are first mentioned.

Naturally, the information provided may not be detailed enough. In this case, one can look for references given in the text. In other words, the book can be used as a starting point (or map) for further information retrieval.

1.5. Who Might Benefit from This Book?

The authors hope that this book will help save time and deepen knowledge in the field of the characteristics of manufactured and promising lithium-ion cells and lithium metal cells for specialists in the development of various equipment, lithium-ion cells, and batteries; scientists conducting research in the field of creating new materials; marketers; students; and those simply interested in the topic.

2 Materials for Lithium-Ion Cells

Power sources enable the autonomous operation of electrical devices. Power sources can be constructed based on various electrochemical systems: lead–acid, nickel–cadmium, nickel–metal hydride, lithium-ion cells, etc.

Lithium-ion cells are in high demand for the manufacturing of batteries for electrical devices (smartphones, electric tools, hybrid and electric vehicles, uninterruptible power supplies, quadcopters, etc.) due to their high specific (relative to a unit of mass and/or volume) power and energy, cycle life, performance in a wide temperature range, price, and other parameters.

The lithium-ion cell electrochemical system can be presented [1] as follows (Equation (1)):

$$(+)[H_P]/[Li_x\ H_P]//Electrolyte//[LiH_N]/[H_N](-) \tag{1}$$

where

[HP] is a structure into which lithium can be reversibly introduced (or extracted from it) at high potentials of approximately +3 ÷ +5 V relative to Li^+/Li;
[HN] is a structure into which lithium can be reversibly introduced (or extracted from it) at low potentials below +2 V relative to Li^+/Li.

In such a case, for the sake of simplicity reactions with electrode-active materials during lithium-ion cell charge/discharge can be presented as follows:
Positive electrode (Equation (2)):

$$[LiH_P] \underset{\text{Discharge}}{\leftarrow} \overset{\text{Charge}}{\rightarrow} \left[Li_{(1-x)}H_P\right] + xLi^+ + xe^- \tag{2}$$

Negative electrode (Equation (3)):

$$\left[Li_{(1-x)}H_N\right] + xLi^+ + xe^- \underset{\text{Discharge}}{\leftarrow} \overset{\text{Charge}}{\rightarrow} [LiH_N] \tag{3}$$

Summary equation (Equation (4))

$$\left[Li_{(1-x)}H_N\right] + [LiH_P] \underset{\text{Discharge}}{\leftarrow} \overset{\text{Charge}}{\rightarrow} [LiH_N] + \left[Li_{(1-x)}H_P\right] \tag{4}$$

In fact, one should consider side processes occurring during charge/discharge. For instance, part of lithium doesn't intercalate into anode and participates in the formation of solid electrolyte interface. The LIC operation principle can be described in short as follows. When being charged, lithium ions leave the structure of the cathode-active material, acquiring a solvate shell. Then, while passing through electrolyte and separator pores, they approach the negative electrode, lose the solvate shell at the electrolyte/anode material interface, and enter the structure of the anode-active material (Figure 1). At the same time, electrons move via the external electrical circuit (from the active material through the conductive network to the current collector and tab) in the same direction (from the cathode to the anode).

During charging, the potential of the cathode-active material (positive electrode) increases and the potential of the negative electrode decreases (relative to Li^+/Li). The increasing LIC voltage at charging is the difference between the potentials of the positive and negative electrodes.

LICs are manufactured in a controlled atmosphere using high-tech equipment. The main manufacturing stages [3,4] are as follows (a sample layout of the plant with a list of equipment can be found in brochure [5], Figure 2). At the first stage, slurries of cathode and anode materials are prepared in vacuum mixers. Cathode slurry consists of cathode-active material (examples of applied materials are given in Table 1), a binder, a conductive additive, and a solvent (N-methyl-2-pyrrolidone, etc.). Anode slurry consists of anode-active material, a binder (and thickener), and a solvent (water or N-methyl-2-pyrrolidone, etc.). In addition, it may contain conductive additives (carbon black, graphites, carbon fibres, etc.) when polyvinylidene fluoride (PVDF) is used as a binder. The concentration of the electrode slurry components may be different along with the height of the vessel (sedimentation at the bottom, aggregation of particles in the middle zone, and clarification at the top is possible), which may result in the non-uniform application of the slurry. Thus, the following measures are taken to reduce non-uniformity depending on the applied technology [6–9]:

- Preliminary dry mixing of cathode-active material and conductive additives;
- Special devices for feeding the active-layer components into the mixer;
- Special vacuum planetary mixers or systems consisting of a series of different mixing devices (binder solution preparation—low shear force, adding of conductive additive and mixing at high shear force, operation of ball mills for homogenisation, pre-mixing of cathode material and slurry of conductive additives, mixing and adding the specified amount of binder and solvent);
- Longer mixing time;
- Shortening of time between production of slurry and its application;
- Various devices for active-layer transfer to the coating machine.

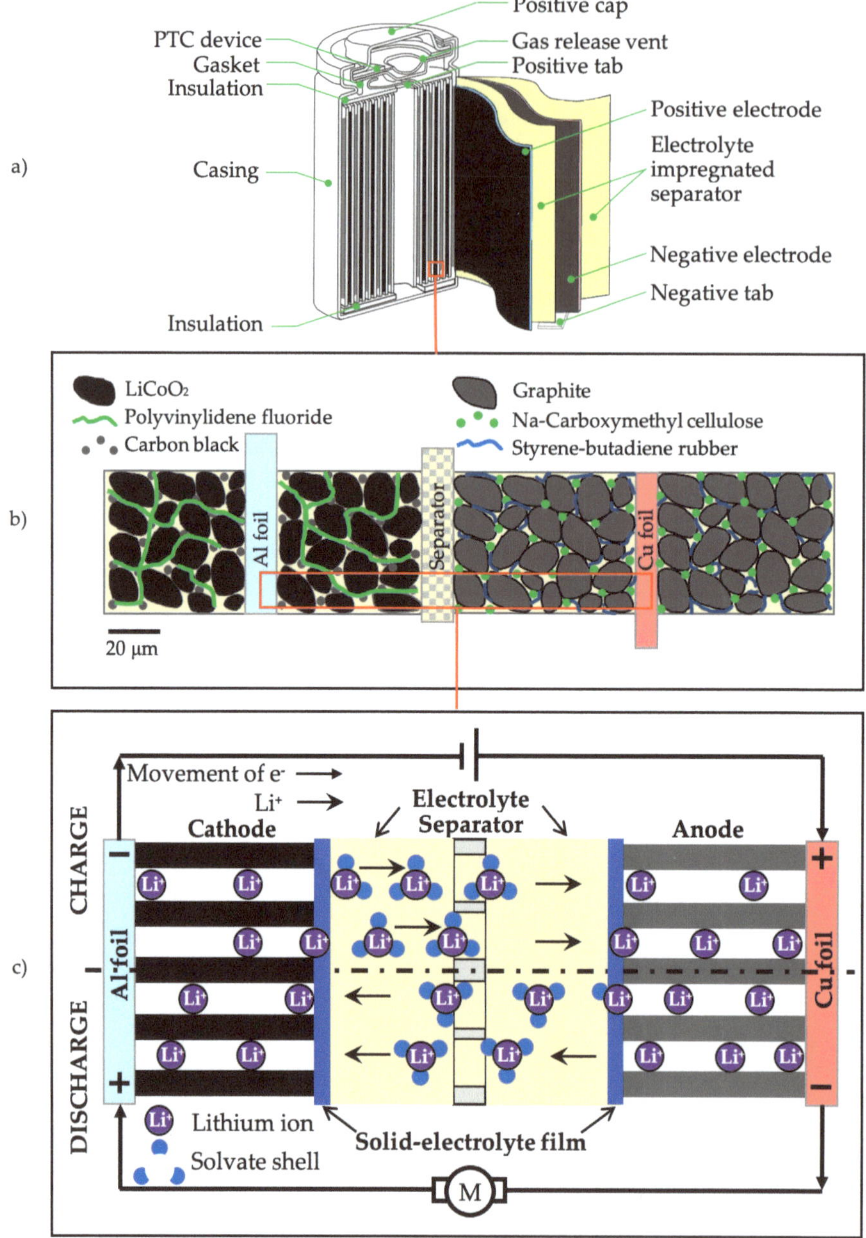

Figure 1. Block diagram of a lithium-ion cell (**a**) based on an image from Sanyo [2], electrode assembly (**b**), and a description of the operation principle during charging and discharging (**c**). Source: Figure by authors.

Table 1. List of materials for manufacture of LIC electrochemical system.

LIC Component	Type of Material	Material
Positive electrode	Cathode-active material	$LiCoO_2$ (LCO), $LiMn_2O_4$ (LMO), $LiFePO_4/C$ (LFP), $Li_{1+x}Ni_aCo_bMn_cO_{2+d}$ $0 < x < 0.05$, $a + b + c = 1$, designated as NCM (A:B:C), where $a = A/(A + B + C)$, $b = B/(A + B + C)$, $c = C/(A + B + C)$. NCM (1:1:1) = $Li_{1+x}Ni_{0.33}Co_{0.33}Mn_{0.33}O_{2+d}$, $Li_{1+x}Ni_eCo_fAl_gO_{2+d}$ $e + f + g = 1$; $0.8 < e > 0.92$, $0.015 < g < 0.05$
	Binder	Polyvinylidene fluoride (PVDF)
	Conductive additive	Carbon black, acetylene black, graphite, carbon fibres
	Aluminium foil	10 ÷ 25 μm, 3003, 1070, 1085 etc., may be coated with a functional layer
Negative electrode	Anode-active material	Artificial graphite (AG), carbon-coated natural graphite (NG-core), graphitising carbon material (soft carbon), non-graphitising carbon material (hard carbon). Mesophase carbon microbeads, MCMB (graphite or non-graphitising carbon), $Li_4Ti_5O_{12}$(/C), $Si(SiO_x)/C$ composite, Sn/Co/C
	Binder	Styrene-butadiene rubber (SBR), carboxymethyl cellulose (Na-CMC), latex, PVDF
	Conductive additive	ref. positive electrode
	Copper foil	6 ÷ 20 μm, electrolytic (electrodeposited) or rolled
Electrolyte		1 ÷ 1.5 M $LiPF_6$ solution of mixture with ethylene carbonate (EC), propylene carbonate (PC), dimethyl carbonate (DMC), ethyl methyl carbonate (EMC), etc. Functional additives: lithium salts and organic substances (vinylene carbonate (VC), fluoroethylene carbonate (FEC), etc.)
Separator		Porous films of polyethene (PE), polypropylene (PP), trilayer PP/PE/PP, which can be coated with particles of aluminium oxide, boehmite, PVDF film, etc.; cellulose separator, etc., with thickness in the range of 7 ÷ 40 μm depending on the application

Note: A 3D-carbon framework can be used as a binder for electrode manufacture according to the Neocarbonix^{TM} technology provided by Nanoramic Laboratories (FastCap). In proposed technology electrode paste is prepared from isopropyl alcohol (etc.) and carbon materials [10]. Source: Table by authors.

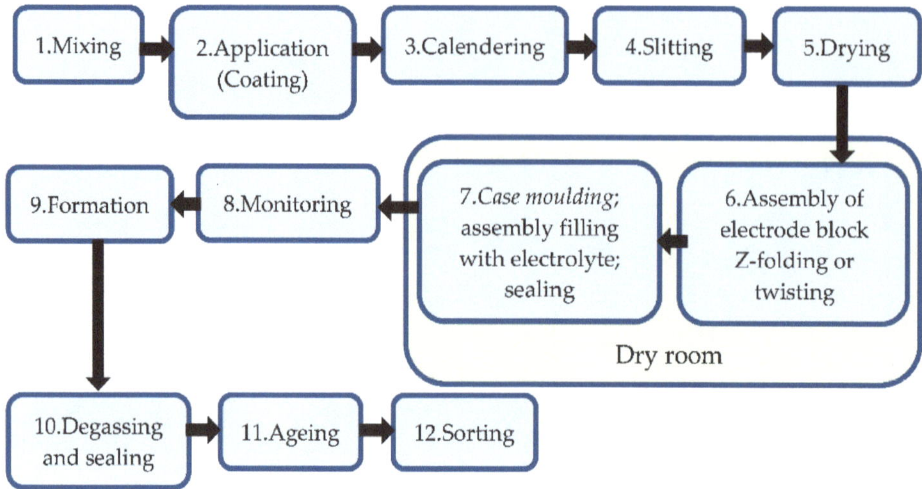

Figure 2. Manufacturing sequence for lithium-ion cells with laminated foil (pouch) cases and metal prismatic (cylindrical) cases. Source: Figure by authors.

At the second stage, a coating machine is used to apply cathode (anode) slurry [11] (under a controlled atmosphere [12]—argon, nitrogen, low moisture content—cathode materials with high content of nickel, manufacture of cells with lithium anode) to aluminium (copper) foil in series to one side and the other side (or to both sides in one pass of the foil [13,14]) as a layer (one strip, several strips, one strip with a break, several strips with a break [5]) with a preset thickness and application density expressed in mg/cm^2. The most widely spread coating application systems are as follows: a slot die coating system (a double slot die system) for industrial machines and a reverse commabar system (for pilot coating machines; the application rate reaches 15 m/minute [15]). The other systems for active layer application to foil are described in the presentation in [12]. The foil width may reach 650 mm. During coating, the foil motion speed may reach 70 $\div$ 80 m/min [11,15] when an industrial machine is used to apply the slurry.

Currently, there is a technology for applying an active layer with preset porosity and binder concentration in depth [16]. The electrode layer area adjoining the foil contains a greater amount of binder and has a denser structure. The external area of the layer has a greater porosity (it is less dense) and includes less binder. The electrode slurry (primarily anode) applied according to such technology has better adhesion to the current collector, and the charge current can be increased due to the formed porous structure of the electrode layer. Regardless of the method of active layer application, the drying rate is an important parameter. Fast drying can lead to a higher concentration of the binder in the upper part of the active layer [17–19], which reduces the adhesion of the active layer and increases electrode resistance.

Extrusion [20] is an alternative method for electrode manufacture. At the first stage, the dry mixing of an active material, conductive additive, and binder is carried out at high temperature and under reduced pressure. The produced homogeneous mixture is transferred to an extruder; a solvent, which may be a component of the electrolyte, is added, and the active layer is extruded onto a polymer film (polyethene terephthalate). After that, the active layer is transferred from the polymer film to the current collector during heating. Then, the calendering of the electrodes is carried out. The advantage of this technology is the exclusion of N-methyl-2-pyrrolidone from the manufacture of the electrode. The disadvantages are a substantially lower application rate, an extra stage of active-layer transfer from the polymer film to the current collector, and the difficulty in selecting proper conditions for producing an active layer that is uniform in terms of characteristics, which increases the manufacturing cost. It is possible that the process under development (the "HemKoop" project, Technical University of Braunschweig) that uses two extruders mounted in series will increase the demand for extrusion-based application technology. The technology under development makes it possible to apply the active layer directly onto the current collector without using a polymer film as an intermediate link [21].

At the third stage, electrode tape is calendered between rollers (if required, the rolls may have hot surfaces [22]) in order to:

- Ensure proper contact between the active material and conductive additive;
- Increase the adhesion of the active layer;
- Create the required density and porosity (later, during storage, drying, twisting, and filling with electrolyte, the porosity of the active layer increases, and it decreases during formation and cycling) [23]);
- Set electrode thickness.

At the fourth stage electrode manufacturing stage, the electrode tape is slit (or punched) [24]. The produced electrodes are dried (fifth stage) in a vacuum cabinet to remove traces of moisture [25]. At the sixth stage, the cells are assembled in a room with low moisture content since water can negatively affect the operation of the electrodes and partial hydrolysis of the electrolyte, and, ultimately, the cycle life and energy characteristics of the cell. Depending on the LIC design selected, the positive electrode, separator, and negative electrode are wound on a pin (stud) or a bushing (roll structure), or assembled as a stack separating the positive and negative electrodes with a separator layer (a Z-folder or a quicker manufacturing method is possible—an electrode assembly consisting of electrodes and separator sheets [26]). Both types of electrode assembly design have their advantages and disadvantages and are used by various manufacturers. For example, SDI uses prismatic cells with a roll structure [27], since this solution provides [28] better wetting of the entire electrochemical system with electrolyte. On the other hand, LG uses separate electrode cards isolated by a separator, as such a design is less prone to deformation during cyclic charge–discharge [29,30].

During the manufacture of LICs mounted in a soft case, tabs are welded to the electrode assembly using the ultrasonic welding method ((+) relatively thick (0.1 ÷ 0.7 mm) aluminium (for example, 1235, 1050) [31], (−) nickel (for example, N201) [32] or nickel-plated copper (for example, C1100) foil with a glued three-layer hot-melt adhesive film 72, 100 μm thick [33–35]). At the seventh stage, the electrode assembly is placed in a case stretched out from laminated foil [36–39] (pouch, thickness 68–153 μm, depth of stretched form 3 ÷ 8 mm [40]), and then a sheet of laminated foil (pouch) is welded (thermal welding) with a "pocket" left for electrolyte filling [41]. The electrolyte is filled, the pocket is sealed, and after inspection the cell is transferred for formation. Rolling through rollers to provide better electrolyte distribution within the laminated foil (pouch) case can be effectuated (the operation time is a few seconds).

During the manufacture of LICs mounted in prismatic, elliptical, or cylindrical cases (for the case manufacturing procedure, refer to [42]), the electrode assembly (block) is welded via the ultrasonic (or laser) method to the tabs (preliminarily welded to the LIC cover), and the welded assembly is placed inside the case [43]. The electrode assembly may be a roll [44] or several rolls 2—Lim50 [45], 4—SDI 60 Ah [46] (an increased number of rolls allows the reduction of the volume of non-occupied space inside the LIC prismatic case [47]), or consist of separated sheets of electrodes [48]. The cover is welded to the case by the laser. An electrolyte is filled through a special hole, and after inspection (the eighth stage), the assembly is transferred to the formation (the ninth stage). The formation includes several charge/discharge cycles (including those at high temperatures) at low currents (0.1 ÷ 0.5C, 20 ÷ 80% state of charge) according to the pre-set program [49,50]. At the tenth stage, degassing (during formation, a small amount of gas is released) and final sealing is carried out. At the eleventh stage, ageing is carried out: holding for several weeks at room temperature and high temperature with a state of charge of 80 ÷ 100%, measuring the open-circuit voltage, and then sorting by capacity. Finally, the cells are sorted taking into account their capacity. Before shipment to the customer, the given batch of cells is tested. For a more detailed and illustrative description of the LIC manufacturing procedure, refer to the brochure in [51].

The main characteristics of LICs are as follows:

- Specific energy (a characteristic of autonomous operation time);
- Power (permissible charge/discharge current, voltage, resistance);
- Cycle life (the main causes for the in-service drop in LIC performance and methods for detecting the causes are given in paper [52]);
- Performance in a wide temperature range.

The LIC's characteristics depend not only on the materials used at the manufacturing stage, but also on the manufacturing technology and design. For example, specific energy can be increased due to a more significant amount of active material in the volume and the mass of the LIC. The increment share of active materials can be achieved by increasing the thickness and density of the electrode-active layer and using thinner current collectors, a separator, a case, etc., at the manufacturing stage.

The specific power can be increased by reducing the active layer's thickness, increasing the active layer's porosity, increasing the amount of the conductive additive, etc. Usually, the structural and process solutions providing an increase in power cause a decrease in specific energy.

For more information about LICs, including types of designs, manufacturing technology, and the characteristics of the cells, refer to monograph [48]. The basic materials for LIC manufacture are briefly described in the sections below; for more information, refer to monograph [53].

2.1. *Cathode-Active Materials*

The primary cathode-active materials are lithium cobalt oxide (LCO), carbon-coated lithium iron phosphate (LFP), lithium manganese oxide spinel (LMO), lithium nickel cobalt aluminium oxide (NCA), and lithium nickel cobalt manganese oxide (NCM) with different mole ratios of nickel, cobalt, aluminium, and manganese. Table 2 summarizes the various cathode-active materials that differ in essential characteristics (Table 3) produced by major manufacturers. For example, there are several modifications of lithium cobalt oxide. The LC95 material has an increased tap density and can operate reversibly when charged to high potentials. The lithium cobalt oxide (LCO 983HA) produced by Pulead has a lower tap density, but retains stability at high levels of delithiation as well.

Beijing Easpring manufactures lithium cobalt oxide that can provide high discharge rates and high performance at high potentials relative to lithium. Materials of such a type allow high specific (volumetric) energy and power of LICs to be achieved, which can be in demand during the manufacture of power sources for quadcopters. Structural stability at high potentials is ensured by doping (Mn, Mg, Ti, Al [54]) and applying functional coatings (an artificial solid–electrolyte interphase (SEI) film).

The SEI film (sometimes referred to as CEI—cathode–electrolyte interphase) application on the surface of the cathode enables a reduction in the following:

- Resistance during interphase reactions (electrolyte–cathode material);
- Intensity of side reactions with electrolyte;
- Gas evolution.

Table 2. Basic characteristics of LIC cathode-active materials.

	Manufacturer	Brand	Type	d_{10}	d_{50}	d_{90}	S_{BET}	T_d	C_d	Ref.
					μm		m^2/g	g/cm^3	mAh/g	
LCO	Tianjin BM	HVC20 * (4.55 V)	CP, D	5.4	16.8	30.9	0.23	2.9	198.1	[55]
	Shanshan	LC95 (4.45 V)	CP, D	–	17.5	–	0.12	2.8	190	[56]
	Pulead	983HA (4.4 V)	CP	6.7	13.3	26.3	0.37	–	190	[57]
	Tianjin BM	BM450C	CP, P	3.6	6.5	11.4	0.45	2.4	176	[55]
	Easpring	–	CP, P	–	6	–	0.39	–	176	[58]
	Pulead	910H (4.35 V)	CP	3	8 ÷ 12	30	<0.5	–	165	[59]
	Shanshan	LC800E	D	–	17.5	–	0.15	2.9	165	[56]
	Shanshan	LC108R	P	–	6	–	0.41	2.35	161	[56]
LFP	Phostech	Ph P1	SE	–	3.4	–	–	<1.2	155	[60,61]
	Sud Chemie	P2	P	–	0.7	–	–	<0.7	155	[61,62]
	BASF	LFP-400	SE	–	10	–	–	1	156	[63,64]
	Hanwha	LFP-CNT	SE	2	6.6	–	12	0.6	>150	[65]
	Aleees	M12	SE	<1.5	4	<10	12	1	153	[66,67]
	Tatung	P13F	SE	1.5	3	5.5	5.5	0.95	153	[68]
LMO	Qianyun High Tech	QY-G02	SE	>6	<15	<35	0.6	2.15	110	[69]
		QY-X01	P	>6	<15	<32	0.65	2.25	108	[70]
NCM	3M	BC-618 (1:1:1)	D, SE	–	10.5	–	0.26	2.69	160	[71]
	3M	(1:1:1)	P	–	8.5	–	0.7	2.2	167	[71]
	Tianjin BM	HVT-6G	CP, P	3.5	5.5	8.5	0.58	1.96	162	[55,72]
	Daejung En. Mat.	(1:1:1)	P	<8	<12	<25	<0.4	>2.3	>160	
	BASF	(4:2:4)	SA	>4	8 ÷ 13	<28	<0.5	>2.0	163	[73]
	Tianjin BM	BMT708 (5:2:3)	P, SE	4	7.7 ÷ 9.5	<25	<0.7	>2.0	162	[74]
	Pulead	PU50D (5:2:3), 4.35V [S]	CP (P)	3.5	6	10.4	0.4	2.2	179 (3 ÷ 4.45 V)	[75,76]
	Henan Kelong	(6:2:2)	SE	–	12.6	–	<0.3		168.5	[77]
	L&F	NS-X15S 7:1.5:1.5	SE	–	15	–	–	>2.3	–	
	Xinxiang Tianli Energy	TLM 810 (8:1:1)	SE	>5	9 ÷ 12	–	<0.5	>2.3	–	
	–	CGH (80%)	SE	–	–	–	–	–	204	
	EcoproBM	CSG88	SE	–	–	–	–	–	≈220	[78]
	BTR	M8-S (4.35 V) [S]	CP	–	5	–	0.6	–	210	[79]

Table 2. *Cont.*

	Manufacturer	Brand	Type	d_{10}	d_{50}	d_{90}	S_{BET}	T_d	C_d	Ref.
					μm		m^2/g	g/cm^3	mAh/g	
NCA	BASF	HED 7050	P, SE	–	6.3	–	0.41	–	192	[80]
	BASF	HED 7150	SE	–	15.4	–	0.16	–	189	[80]
	BTR	N1-L (Ni-0.81)	SE	–	12	–	2.2	–	197	[81]
	BTR	N8-L (Ni-0.88)	SE	–	12	–	2.2	–	205	[81]
	BTR	N8-S (Ni-0.88) [S]	SE	–	6.5	–	–	–	195	[82]
	BTR	N9-C (Ni-0.92)	SE	–	12	–	0.5	–	215	[81]

Note: The materials can ensure increased cathode potential (CP), positive electrode density (D), power (P), safety (SA), and specific energy (SE) of a cell manufactured from them. Phostech=>Sud Chemie=>Clariant=>Johnson Matthey, *—lab scale, S—single crystal, C_d—discharge capacity, d_{10}, d_{50}, d_{90}—particle size distribution parameters, S_{BET}—specific surface area, T_d—tap density. Source: Authors compilation based on data from references mentioned in the table.

The artificial CEI also provides the increased stability of the cathode material structure during operation at high temperatures and potentials.

Along with the modifications remaining operable at high potentials, the LIC manufacturers can use LCOs operable at a potential below 4.35 V, for example, LC800E (high density) and LC108R (the particle shape and size ensure increased LIC power) and their equivalents.

Lithium iron phosphate is also available in several modifications. Phostech produced LFP with relatively large particles (d_{50} = 3.4 μm). Later, Sud Chemie proposed a method for producing smaller LFP particles, making it possible to manufacture LICs with a higher specific power. This type of material is sometimes designated as P2.

However, a small particle size requires increasing the proportion of binder in the electrode slurry. Alternatively, the formation of secondary ball-shaped particles from primary particles was proposed (P2S, Figure 3). In addition to BASF (LFP-400), Johnson Matthey and other manufacturers are also producing agglomerated iron phosphate. In addition, to improve characteristics of the LFP/C cathode, several approaches can be used:

- The $LiFePO_4$ structure may be distorted (Formosa Lithium Iron Oxide Corporation [83]);
- Doping with boron (Prayon [84]) and magnesium (Valence [85]) can be performed;
- Metal oxides are introduced (Aleees [66,67,86]);
- Conductive coatings (adding carbon nanofibers—Hanwha [65,87]) can be varied;
- The secondary particle shape, as per Hangzhou Changkai Energy Technology Co. [88] and Tatung [68], and the shape and size of primary particles [89] may be adjusted.

There are also some LFP-based materials in whose structure some iron atoms are substituted with manganese atoms (LMFP) [89–93], including bio-modified ones (BMLMFP, contains several mass percents $[Ca_5(PO_4)_3(OH)_2]$ [94,95]). Manganese added into the crystal structure changes the shape of a discharge curve (a step appears) and allows the increase in the average discharge voltage. Cells with LMFP cathode material are manufactured nowadays [91,96].

Table 3. Basic parameters of LIC cathode-active materials.

Parameter	Description	
Chemical composition	LCO, LFP, LMFP, LMO, NCM, NCA	
Main element distribution across a particle is uniform, or designed	Core–shell [78,97–100]	Gradient in terms of composition [101,102]
	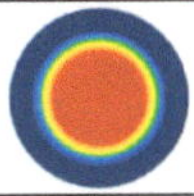	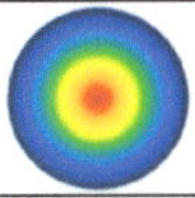
	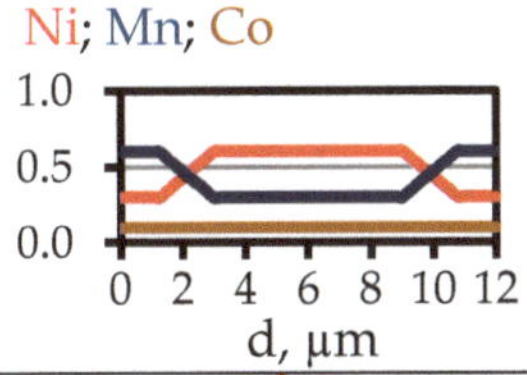	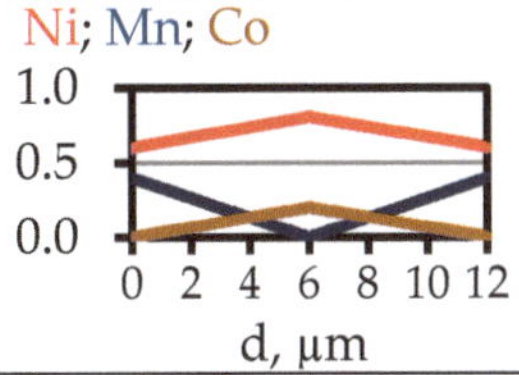
Dopants: Ti, Mg, Al, etc.	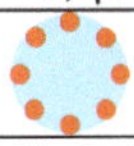 [80]	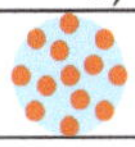 [103]
Functional coatings Electrically conductive coatings: Carbon [104] Nanotubes [65,87] Passivating coatings Polymeric [105] Inorganic [106] Ion-conducting coatings [105,106] Combined coatings [106]	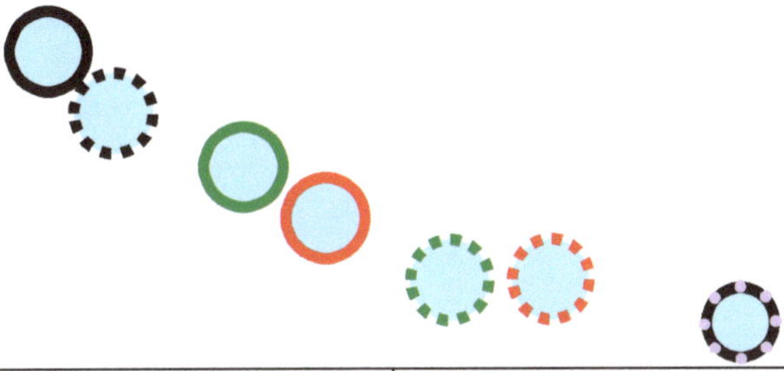	
Structure of particles	Solid [107]	Agglomerates [103]
	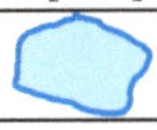	
Particle shape Cobble [57], Ball [108], Rod [68] and others [109–111]	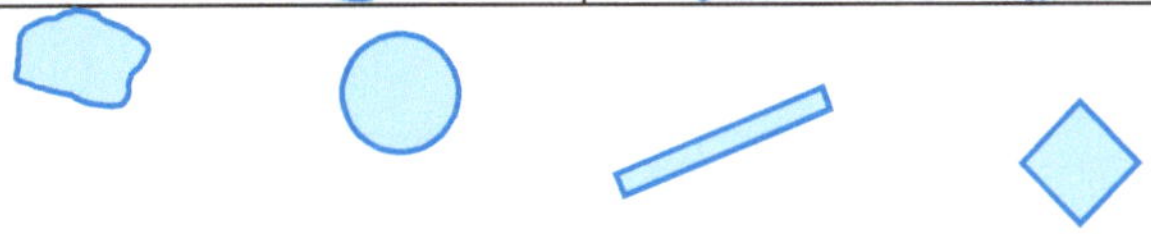	
Particle size	d_{50} (0.7 ÷< 20 µm), d99 < 45 µm	
Porosity of secondary particles) [44,108,112,113]	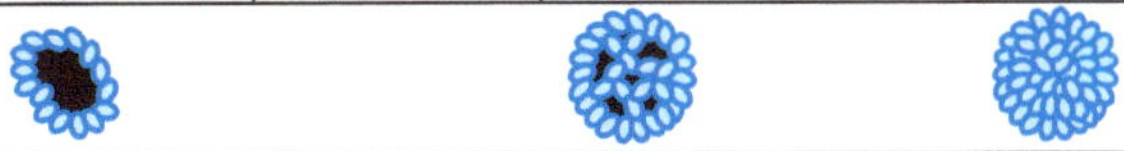	
Orientation of primary particles (crystallites) [108]		
Shape of primary particles [89,103]	 balls plates rods facetted particles	
Size of primary particles [89]	Less than 0.1 µm [104,114,115]—approx. 2.5 µm [103,116]	
Primary or secondary particles are coated [68,117]	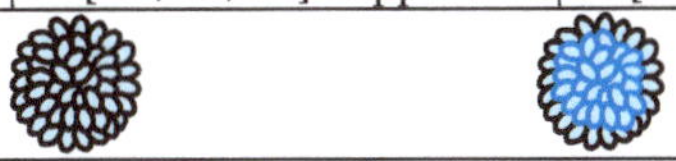	

Source: Table by authors.

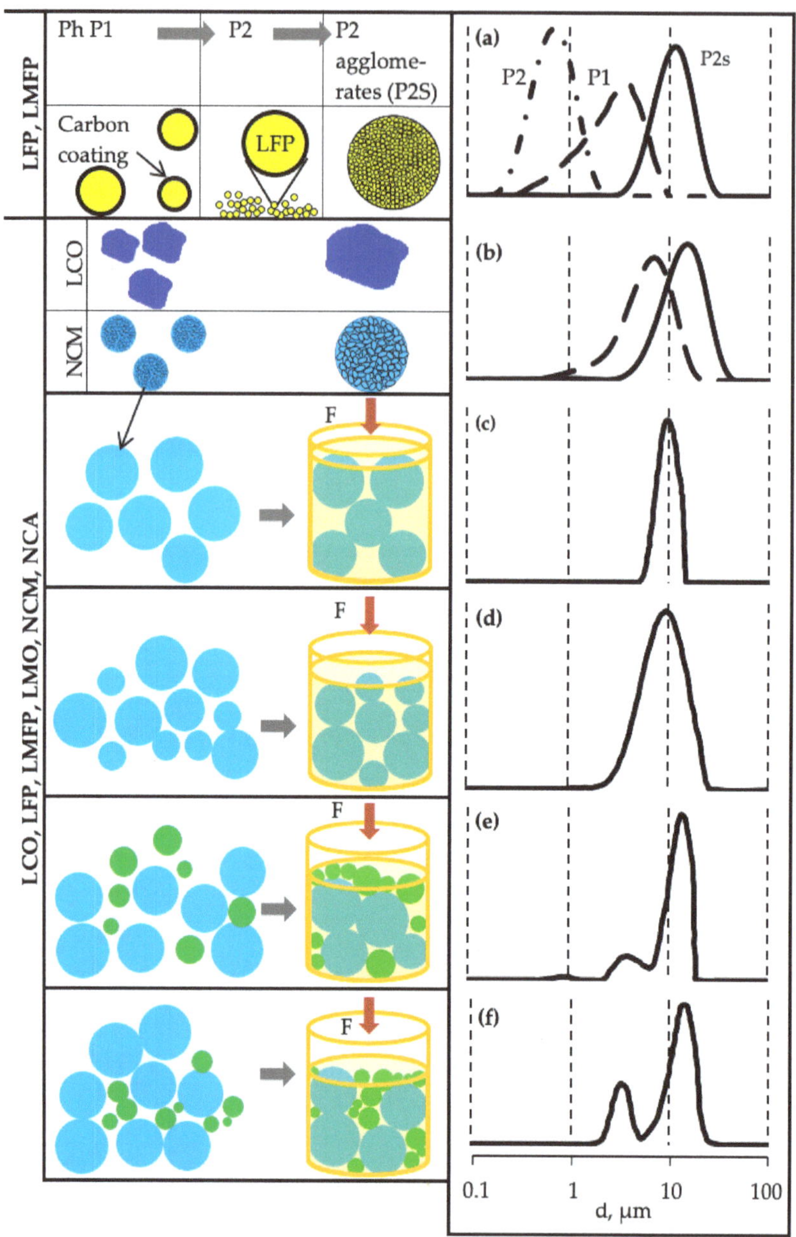

Figure 3. Shape of differential curves of particle size distribution: LFP (P1, P2, agglomerated P2), LCO, and NCM (NCA, LMO, LMR) (**a**–**d**) monomodal, (**c**)—narrow, (**d**)—broad, (**e**,**f**)—bimodal. Source: Figure by authors.

$LiMn_2O_4$ has a relatively low capacity (<120 mAh/g), but its crystal structure provides lithium intercalation/deintercalation in three directions, which increases the lithium diffusion rate and LIC power. The manufactured materials differ in particle size and shape [110,111,118], dopants (Al, Mg) stabilizing the structure [103], and tap density (the density of a pressed tablet) [118]. In addition, there is high-voltage nickel–manganese spinel ($LiNi_{0.5}Mn_{1.5}O_4$, LNMO) [119,120]. However, LNMO has not been widely used yet as there are no electrolytes capable of providing a sufficient cycle life at high potentials (the charge potential is approx. 5 V and the plateau is 4.7 V) of the cathode surface. Lithium nickel cobalt manganese oxides differ in chemical composition (content of transition metals and dopants). The most common ratios of Ni:Co:Mn are 1:1:1, 4:2:4, 5:2:3, 6:2:2, 8:1:1, but materials with other ratios are also available: 3.5:3.5:30 (LNCM-35 [121]), 6:1:3 (LNCM-60) [122], 7:1:2 [123], 8.3:1.2:0.5 [124], 8.3:0.6:1.1 [125], and 8.8:0.9:0.3 [126].

Besides Ni-rich cathode materials, there are lithium–manganese-rich cathode materials (LMR, or over-lithiated layered oxides—OLOs), which were introduced by Thackeray [127,128]. The general formula of Li-rich oxides is often written as $xLi_2MnO_3 \cdot (1 - x)LiMO_2$ ($M = Mn_aNi_bCo_c$, $a + b + c = 1$) and considered as a structure integrated from two layered oxides: Li_2MnO_3 and $LiMO_2$. Charging at high voltage activates Li_2MnO_3 and results in high discharge capacities, which may reach 250 mAh/g or even higher. There are few industrial-grade LMR materials [129,130]. Due to the high charge voltage and voltage fade during cycling, LMR is almost never used in the manufacture of lithium-ion cells. The results presented in the literature suggest that narrowing the operating voltage may solve the voltage fade problem and still provide relatively high capacities [131]. Due to the increased Mn and lower Ni and Co content, the cost of materials might be lower, enabling applications in industrial-grade lithium-ion cells.

Industrial-grade NCM cathode-active materials are produced by the lithiation of precursors with Li_2CO_3 or $LiOH \cdot H_2O$ (solid-state synthesis) [132,133]. The precursors are mixed hydroxides, or carbonates, or oxides of transition metals (and aluminium in case of NCA) are obtained by coprecipitation of transition metal sulphates by sodium hydroxide (or Na_2CO_3) in ammonia solution (complexing agent) [134]. Also, some new processes are still emerging. For instance, Nano One suggests using metals as initial reagents in their M2CAM (One-Pot process) to produce a precursor of cathode material [135].

A higher content of nickel increases capacity due to a more significant number of lithium ions being removed from the cathode material structure at a fixed potential [76] (when charged to a voltage of 4.2 V, more lithium is removed from the nickel-rich NCM cathode material than from $LiCoO_2$). However, a higher content of nickel negatively affects the cycle life [136], reduces thermal stability, and complicates the process of electrode manufacture (since they are more prone to hydrolysis) [76]. The review in [137] covers the problems and possible solutions for improving nickel-rich layered cathode materials (NCM, NCA).

A more significant share of cobalt (small amounts of cobalt in nickel-rich cathode materials do not stabilize the structure [138,139]) increases the stability of the structure, thus improving the cycle life. On the other hand, lithium cobalt oxide (hence, it can be assumed that NC, NCA, and NCM materials with a greater share of cobalt as well) demonstrates a high thermal effect at decomposition [140,141]. Cobalt is more expensive than nickel and manganese, which affects the cost of the cathode material. A higher manganese content improves safety (a minor exothermic effect during the decomposition of cathode materials with a high concentration of manganese is observed—NCM 424) and reduces cost (it is a less expensive transition metal than cobalt and nickel) [80]. However, manganese is less prone to form layered structures that can decrease the cathode material's capacity.

The structural stabilisation of nickel-rich layered cathode materials can be achieved by aluminium doping [138,139] (NCA materials). In industrially produced NCA materials, the molar fraction of nickel atoms in transition metals varies within the 0.8–0.92 range [81,126]. For example, the positive electrode of a high-power 21700 3 Ah LIC includes NCA with two types (small and big ones) of particles and an atomic content of transition metals: Ni—85.9 ÷ 87.5%, Co—10.6 ÷ 12.9%, and Al—1.2 ÷ 1.9% [142].

One of the advanced options for stabilising the structure of nickel-rich layered cathode materials is adding cobalt, manganese, and aluminium (quaternary materials) into the structure. According to the research results [143], secondary particles of the cathode material NCMA ($Li[Ni_{0.89}Co_{0.05}Mn_{0.05}Al_{0.01}]O_2$) were less destroyed compared to NCM ($Li[Ni_{0.90}Co_{0.05}Mn_{0.05}]O_2$) and NCA ($Li[Ni_{0.885}Co_{0.100}Al_{0.015}]O_2$) due to the more excellent mechanical stability and lower internal mechanical stress (at a high level of delithiation). In addition, the structural stability of the NCMA particles ensured a smaller amount of stress during the cycling growth of the surface wetted by electrolyte and, consequently, a lower rate of material destruction. As a result, a minor increase in charge transfer resistance and a longer cycle life (number of charge/discharge cycles) were achieved. Furthermore, the simultaneous presence of manganese and aluminium atoms in the cathode material structure increased the thermal stability and decreased the thermal effect at decomposition (improved safety).

LG [144] is one of the main consumers of NCMA material. The companies producing NCMA or preparing for NCMA series production are L&F [144], Cosmo AM&T [145], EcoproBM [146], Posco [147], and BTR [148,149]. The molar fraction of nickel in NCMA cathode materials being produced or under development ranges from 0.8 to 0.92 [148].

To combine high capacity and stability, it was recommended [101] that the concentration of transition metals should be varied in the secondary particles of cathode materials. A high concentration of nickel in the core centre ensures a high capacity. A lower nickel concentration and a higher concentration of manganese in the material layers close to the outer surface provide greater material stability (lower heat release during heating of the delithiated cathode material) and a longer cycle life at cycling. The electrolyte decomposition intensity decreases as well, which can be caused by both a lower concentration of nickel ions and a denser shell structure (the core and the shell may have different structures [150]). For more details on advances in the synthesis of materials with a core–shell structure, refer to the review in [151]. The manufacture of NCM materials with a core–shell structure (with a concentration gradient) is reported by Ecopro [98], 3M and Umicore [97], BASF [80], etc.

The performance of cathode materials (NCM, NCA) depends on the content of dopants. For example, BASF emphasises that a reduced amount of chromium dopant in mixed oxides can increase the cycle life [152]. On the other hand, some extra dopant atoms (the most referred to in the literature are Al, Mg, F, Ca, and Ga) can increase in ionic conductivity and facilitate the suppression of the following:

- Cation mixing (Li^{+}/Ni^{2+});
- Phase transformation (H2/H3 at high levels of delithiation, charge);
- Oxygen release [153].

The dopants can be evenly distributed in the volume or/and presented in the near-surface layers of the cathode material particles. For example, BASF reports a cathode material with aluminium dopant that is only introduced into the outer layers of secondary particles (aluminium diffuses from the Al_2O_3 coating) [80].

A combined variant has been applied by TIAX to increase the stability of lithium nickelate. According to the information presented in patents [154,155], a material core consists of lithiated nickel oxide doped with magnesium. The material shell includes lithium cobalt oxide. The results of material testing (CAM-7; the molar fraction of nickel in transition metals is 0.88 for modifications 7.3 and 7.4 [156]), including materials tested as a part of the LIC prototypes, are given in papers [156–164]. In addition, the major cathode material manufacturers BASF [165] and Johnson Matthey Battery Materials [166,167] have announced that they have acquired the rights to manufacture nickel-rich materials (CAM-7^{TM}, Gemx^{TM} under a licence from Tiax (Camx)).

GEMX-series cathode materials (gLNO^{TM}, gNMC^{TM}, gNCA^{TM}, gNCMA^{TM} [168,169]) have the following structure: a secondary particle consisting of cathode material grains whose outer boundaries are cobalt-rich. So, the outer surface of a secondary particle has a protective coating of cobalt-rich cathode material, and the grains inside of this particle are also protected by this coating (the material manufactured by Nano One has a similar structure [170]). When the secondary particle becomes broken (because of calendaring at elevated pressure or during cycling), a subsequent break is less likely than if secondary particles are used without a coating or if only a secondary particle is coated. Therefore, the side reactions (transference of transition metals into the electrolyte solution and further into the SEI anode, gas formation reactions accelerated by transition metals, phase transition from the layered to rock salt structure) between the electrolyte and the nickel-rich cathode material leading to a capacity decrease are less intensive. Also, the cathode materials (nickel-rich) produced according to GEMX manufacturing technology contain fewer LiOH and Li_2CO_3 substances. Their lower concentration results in an improvement in the material by diminishing the impact from side reactions—interactions with the electrolyte to form gas (CO_2) [168].

Various coatings can be applied to NCM materials to improve their functional properties:

- Carbon coating (increased electronic conductivity, Daejung Energy Materials);
- Coating with high ionic conductivity (CATL) [105], artificial SEI film produced by the polymerisation of organic molecules adsorbed on the surface [105];
- Reduced lithium concentration (due to application of a thin layer of cathode material precursor particles) [108]. Several functional coatings can be combined [171] to achieve better performance of the cathode material.

Along with capacity and average discharge voltage, the LICs' specific energy is also affected by material density [172], material energy density [103], and electrode manufacturing technology. Usually, cathode materials (NCM, NCA) are spherical particles (secondary particles) formed by smaller particles (primary particles) (Table 3). Cathode materials are industrially produced with different primary particle sizes [89], with secondary particles with the following characteristics:

- Various sizes (distribution);
- Hollow [44,112] (the thinner the layer ("shell"), the lower the internal resistance at low states of charge) [44];
- Porous [80];
- With crystallites oriented towards the outer surface of the secondary particles [108].

Single-crystal NCM and NCA (solid-like LCO, d_{50} sized $2 \div 6$ µm—N524T and PU50D, M8-S, N8-S—Table 2; due to small particle size, tap density may decrease down to 1.8 and below [173]) are industrially produced as well. Cathode materials with a small particle size, narrow distribution of secondary particles, and hollow and/or porous secondary particles make it possible to produce electrodes with high porosity, a large electrolyte-wetted surface and, consequently, high ionic conductivity (power). When the faces of the crystallites through which lithium passes during intercalation are oriented to the outer surface of the secondary particle, ionic conductivity increases during diffusion. Porous secondary particles (with a relatively high internal specific surface area) are more prone to phase transitions that reduce the performance of NCM and NCA (especially at high potentials and in the case of cathode materials with a high nickel concentration) [153]. The compacted structure of the secondary particles is due to the following:

- Increased size of primary particles;
- Larger particle size;

- Wide or bimodal distribution of secondary particles of the cathode material;

Allows increasing the amount of active material in an electrode (Figure 3) and reduction in the pressure during electrode rolling until the required active layer density is achieved [174].

The size and porous structure of secondary particles can be formed during precursor deposition. The structure of the secondary particles can further change depending on the chemical composition of the lithium precursor, the temperature, and the time modes of the lithiation reaction and subsequent annealing (cooling). The growth of primary particles and the degree of structure crystallinity increase as the annealing time increases. Single-crystal particles are less susceptible to phase transformations occurring at high potentials, along with increased electrode density. Therefore, increased cutoff voltage affects the structural change during the cyclic charge/discharge to a lesser extent.

Lithium cobalt oxide has a higher density than lithium nickel cobalt manganese oxide (Figure 4). With greater stability at high potentials, it is possible to use this material to manufacture cells with a high energy density (for portable electronic devices). The electrode density (the amount of active material per volume of the active layer) correlates with the thickness of a pressed tablet of the active material. It depends on the composition and structure of the active layer and pressure during electrode rolling. The pressed tablet density, in turn, depends on the applied pressure [55]. For example, when the pressure changed from 90 to 360 MPa, the density of an LCO tablet changed from 3.6 to 4.1 g/cm^3, and an NCM tablet density for high-power LICs varied from 3.1 to 3.25 g/cm^3 when the pressure rose from 110 to 190 MPa [55].

The proper specific energy and power of positive electrodes can be achieved when mixtures of cathode materials are used, for example, NCM:LMO—LG [175], NCA:NCM:LMO—SDI [142]; this has been reported for LICs whose positive electrodes include a mixture of cathode materials NCM:LMFP CALB [176,177], BYD [178]. Furthermore, if some of the NCM is replaced with LMFP in the electrode, it will reduce heat evolution during abusive external influences such as a cell puncture [176].

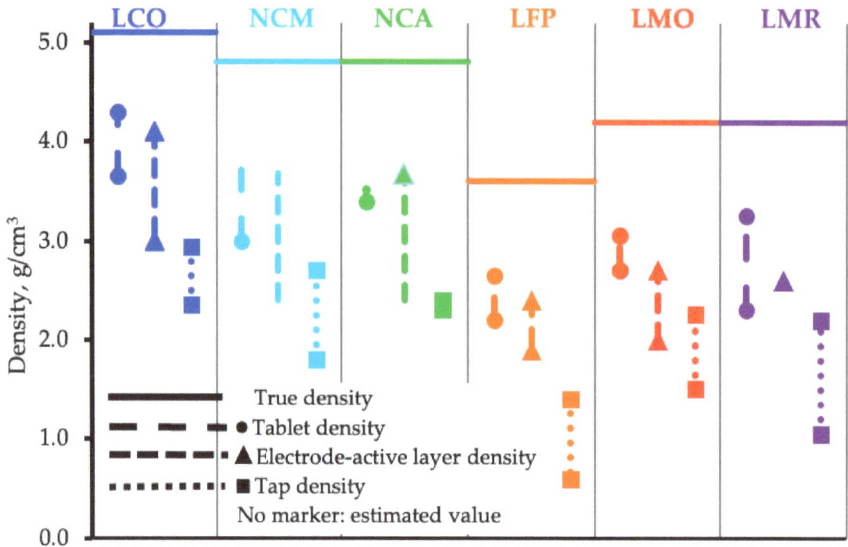

Figure 4. Density of cathode-active materials LCO [22,172], NCM [22,80,172], NCA [172], LFP [172,179,180], LMO [116,172,181,182], and LMR (depending on the composition) [106,129,172,183–185]. Source: Figure by authors.

2.2. *Anode-Active Materials*

The essential anode-active materials (Table 4) are natural graphite (NG), artificial graphite (AG), mesophase carbon (MCMB, including artificial graphite), graphitising carbon material with amorphous structure (soft carbon, SC), non-graphitising carbon material (hard carbon, HC), lithium titanate ($Li_4Ti_5O_{12}$), and silicon-containing composites (Si/C, Si/SiO_x/C).

The discharge capacity of graphites depends on the graphitisation degree (values Lc(002) and d(002) [186]). The amount of material with a crystal structure can be determined by the formula (3.44-d002)/0.0868 [187], d002 (interplanar distance in angstroms). The theoretical capacity of graphite is equal to 372 mAh/g (837 mAh/cm^3 [188], 741 mAh/cm^3, after swelling [189]).

Two modifications of graphite are available: hexagonal (2H, ABAB) and rhombohedral (3R, ABCABC). A share of the rhombohedral modification is calculated based on the ratio of line 101 intensities on X-ray diagrams $3R(\%) = I_{3R(101)}/(I_{3R(101)}/I_{2H(101)})$ [187]. An increased share of the 3R phase reduces the rate of lithium intercalation, leading to lithium plating on the surface; therefore, this fact is considered when selecting anode-active materials with a graphite structure.

Table 4. Basic characteristics of LIC anode-active materials.

	Manufacturer	Brand	Type	d_{10}	d_{50}	d_{90}	S_{BET}	T_d	C_d	Refs.
					μm		m^2/g	g/cm^3	mAh/g	
AG	Hitachi	MAGC	SE	–	20	–	1	1.1	345	[190]
	Hitachi	MAGE3	SE	–	22.5	–	3.9	0.91	361	[191]
	Shanshan	LK	SE	12.3	20.6	33.8	3.3	0.91	360	
	Shanshan	QCG-X	P	9.2	14.5	22.7	1.4	0.96	352	[192]
	Hitachi	SMG-A-PG-20A	SE	–	19.9	–	–	1.07	350	[191]
	Hitachi	SMG-A-PG-10A	P	–	14.2	–	3.3	1.05	349	[191]
	Shanshan	CP7-M	P	4.2	8.5	15.7	1.8	1.0	338	
	CSCC	UF1	P		1 ÷ 5		7	0.75	332	[193]
	Showa Denko	SCMG CF-C	P	2	6	13	–	–	330	[194]
	Shanshan	CAG-3MT	P	–	10.5	–	1.6	1.1	325	
MCMB	Shanshan	HCMS	SE	–	21	–	3.4	1.01	360	
	JFE	KMFC-GC	SE	–	23.3	–	0.68	–	354	[190]
	JFE	KMFC-GA	SE	–	24	–	0.41	–	341	[190]
	Shanshan	CMS-G15	P	8.6	15.6	26.3	1.2	1.42	324	
NG	Hitachi	MAGX-1	SE	–	20.4	–	4.4	0.76	363	[190]
	BTR	918	SE	–	<20	–	<4.5	<1.2	360	[195]
	Posco	PAS-C1	P	4.6	9	25	2.1	1.03	356	[196]
	BTR	MSG18	SE	–	<19	–	<2.7	<1.1	355	[197]
HC	Kureha	J	P	–	9	–	<7	–	450	[198]
	BTR	HC-1	P	–	<15	–	3 ÷ 7	<1.2	>400	[199]
	Kureha	S(F)	P	–	9	–	<8	–	300	[198]
	Kuraray	Kuranode	P	–	5	–	6	–	320	[200]

Table 4. *Cont.*

	Manufacturer	Brand	Type	d_{10}	d_{50}	d_{90}	S_{BET}	T_d	C_d	Refs.
					µm		m^2/g	g/cm^3	mAh/g	
SC	Posco	PSCAM400	P	–	7.5	–	–	0.9	400	[199]
	Posco	PSCAM280	P	–	10.5	–	–	1	280	
	BTR	SC-1A	P	–	4 ÷ 5	–	5	<0.8	260	
	Shanshan	SC-1	P	–	8 ÷ 14	–	–	>0.8	<245	
	Posco	PSCAM240	P	–	8.5	–	–	1	240	
LTO	BTR	LTO-1	P	<0.5	<1.5	<5	16	0.9	160	[201]
	BTR	LTO-S	SE	<3	6 ÷ 12	30	4	0.9	160	[201]
Si, SiO_x	Shanshan	Si-C-S-3	SE	–	8 ÷ 15	–	5 ÷ 7	0.4	500	[202]
	Hitachi	CRZ113R5 *	SE	–	6.4	–	2.3	–	413	[203]
	Hitachi	CRZ105 *	SE	–	10	–	1.0	–	401	[203]
	Hitachi	CRZ103 *	SE	–	10	–	1.9	–	400	[203]
	BTR	BSO-400	SE	<9	<19	<37	2 ÷ 5	<1.1	400	[204]
	BTR	S400	SE	<9	<19	<35	2 ÷ 5	<1.1	400	[205]

Note: This material is used for the manufacture of cells with high specific energy (SE) and high power (P), * 5% of silicon-containing material as a mixture with MAGE3. C_d—discharge capacity, d_{10}, d_{50}, d_{90}—particle size distribution parameters, S_{BET}—specific surface area, T_d—tap density. Source: Authors compilation based on data from references mentioned in the table.

Due to the low price and high degree of crystallinity resulting in high specific capacity, the natural graphite obtained after the spheroidisation process and covered with carbon, or another coating occupies a significant part of the anode material market.

Lithium intercalation into the graphite structure occurs along the edge planes (perpendicular to the basal planes). The presence of a significant number of edge planes within the structure of natural graphite leads to two adverse effects. First, excessive electron density on the edge planes interacts with electrolyte solvent molecules, leading to a growth of thicker SEI film and an increase in resistance during cycling [53,206]. Second, the possible intercalation of electrolyte components leads to the delamination of natural graphites [207,208]. Therefore, to reduce the negative impact of the above factors, natural graphite is spheroidised via mechanical processing. Then, a carbon coating (a sample layout of the modification unit is given in references [187,209,210]) or a polymeric coating (SMG A, Hitachi) is applied. The applied coatings can also cover the pores being active centres during reactions with electrolyte [206].

Heating at 1800 $\div$ 3000 °C graphitising carbon materials produces the artificial graphites used in LICs [211,212]. The artificial graphite-based anode-active materials have a turbostratic structure—a particle of material contains graphite crystallites oriented in different directions. Good quality artificial graphites ensure a longer cycle life and a higher charge rate (charge acceptance) than coated natural graphites.

The following functional characteristics of artificial graphites can be distinguished: the ability to provide the high specific energy of LICs (MAGE3, LK), high power (CAG3MT), high charge rate (QCG-X, SMG-A-PG-10A, graphite with the high specific surface area (UF1 CSSC (small particle size of 3 μm [193])), "activated", porous graphite [213]—a similar anode material probably used for the manufacture of Microvast LpCO^{TM} LICs [214–216]), high performance (and quick charging) at low temperatures (CP7-M, SCMG CF-C), low resistance at low states of charge (SCMG CF-C), and low swelling at charging (SMG-A-PG-20A, isotropic structure [217], low value of plane ratios 002/110 [208]).

For mesophase artificial graphites, the particle shape is close to spherical, and the crystallites' distribution in the particles depends on the production conditions. For more details on the structure and properties of this type of anode-active material, refer to publications [53,218–220].

The first carbon materials used for the manufacture of LICs were graphitising carbon (soft carbon, SC [221]) and non-graphitising carbon (hard carbon, HC) [222,223]. SC can transform into graphite at high temperatures, while HC cannot transform into graphite even at high temperatures [224]. The interplanar distance d002 decreases and the size of the coherent scattering regions (Lc) grows in the HC, SC, and graphite series (Table 5). The true density of HC and SC is less than that of graphite. Consequently, the density of the produced electrodes is lower. If the state of charge (lithiation) is high, the anode potential can reach the lithium potential, since intercalation and lithium adsorption in the pores occur [225]. During the subsequent discharge, a high irreversible capacity is observed (the effective capacity can vary in the range of 75 ÷ 87% of the capacity at charging [226]). The HC and SC charge and discharge curves are similar in shape. A higher value of the average charge potential allows the charging of the anode at higher currents and lower temperatures without metal lithium plating [187,227,228]. However, the low electrode density and high values of average discharge potential adversely affect the LICs' specific energy. Due to a possibly more significant share of the electrochemically active surface [229], the ionic conductivity of negative electrodes made based on amorphous carbons may become higher than those containing graphite as an anode-active material.

Table 5. Comparison of some characteristics of carbon anode-active materials [230].

Parameter	Non-Graphitising Carbon	Graphitising Carbon	Graphite
Structure			La Lc
d_{002}, nm	0.37 ÷ 0.38	0.34 ÷ 0.35	0.335 ÷ 0.34
Lc, nm	1.1 ÷ 1.2	2 ÷ 20	80
True density, g/cm^3	1.5 ÷ 1.6	2.0	2.2 (2.25)
Density of electrode-active layer, g/cm^3	0.9 ÷ 1.0	1.2	1.35 ÷ 1.9 [22,231]
Irreversible capacity, %	15	15	5 ÷ 10

Source: Authors compilation based on data from references mentioned in the table.

Since adsorbed moisture has a highly negative effect on the operation of HC, the electrodes are produced using N-methyl-2-pyrrolidone and PVDF (the exception is LBV-1001 Sumitomo Bakelite [232,233]). In addition, moisture adsorption can be decreased by applying coatings by depositing organic compound thermolysis products from the gas phase [234].

The electrolyte wetting of SC-based electrodes (at the same electrode density) is higher than for graphite-based ones [221]. If SC (30%) is added into the electrode slurry, the wetting of the electrode with electrolyte can be increased without a significant drop in the density of the graphite-based electrode [221]. For more details on carbon anode materials, refer to the presentation in [235].

Lithium titanate ($Li_4Ti_5O_{12}$, LTO) is an anode-active material with a spinel structure (true density—3.41 g/cm^3, theoretical capacity is 170 mAh/g, 580 mAh/cm^3 [188]). Intercalation/deintercalation in LTO occurs almost without changing the volume of the material (electrode) [236]. Therefore, this material has a long cycle life (a significant number of charge/discharge cycles) [237]. The decomposition of the main electrolyte components (ethylene carbonate and propylene carbonate) occurs at a potential of less than 1 V [206]. Since LTO has a high potential relative to lithium (approx. 1.55 V), then SEI film is not formed (unless special additives decomposing at high potentials on the anode are introduced into electrolyte). The high average charge potential and the spinel structure (high rate of lithium intercalation) allow LICs to be charged at room temperature with high currents without the risk of lithium plating. For example, it was demonstrated [238] that when charging with a current of 10C (up to 90% capacity), an LIC with an LTO anode can be charged in 5.6 minutes, while the charging time for an LIC with a negative electrode manufactured using HC and MCMB equals 7.3 and 12.8 minutes, respectively. The charge current limitation is caused by lithium plating in the form of dendrites on the anode, leading to microscopic short circuits.

The maximum peak charge current depends on the design of the LIC (electrodes). For example, it has been found that [239] LIC UF121285 (Panasonic, 5 Ah, cathode—NCM, anode—graphite), used for the manufacture of batteries for hybrid electric vehicles, can be charged with currents of 20C without a significant decrease in discharge capacity (50 cycles). The possibility of charging with high currents along with the methods to manufacture high-power LICs is likely provided by an excessive amount of the anode material. For example, for LICs with a graphite-based anode capable of being charged with high currents, the anode capacity for lithium is 1.5 times (or more) higher than the cathode capacity [112].

High-power LICs with a lithium-titanate-based anode SCiB^{TM} (Toshiba) can also be charged with a high current (20C). For example, a 10 Ah cell is charged with a current of 200 A to 80% capacity for 3 minutes [240]. The high charge rates permit using an electrochemical system with an LTO-based anode to manufacture power sources for devices belonging to the Internet of Things (IoT) category and wearable electronics [241].

The peak charge current at low temperatures is limited by the starting moment of lithium plating. It depends on the nature of the anode-active material and the manufacturing technology of the LICs and electrodes. For example, when charging is possible at a temperature of −20 °C, the peak charge current of LICs with graphite-based anodes varies from 0.05C [242] to 1C (Lishen, ≈80% of the rated value [243,244]). Therefore, when LICs with graphite-based anodes are charged at low temperatures, it shortens the cycle life.

LICs with LTO-based negative electrodes can be safely charged (without lithium plating on the anode) and discharged at a temperature of −20 °C (1C charge, 1C discharge—a capacity of ≈65% of the capacity output achieved during cycling at 25 °C) and, at the same time, have a rather long cycle life [240]. Representatives of Altairnano (Yinlong, Gree) reported on the possibility of charging LICs with a current of 0.1C at temperatures down to −50 °C [245]. The LTO-based cells can be equipped with positive electrodes, such as LMO (Enerdel [246], Leclanche [247], Lishen [244], Toshiba [248]) and NMC (Xalt (Kokam) [249]); some cells, such as NC (Leclanche [250, 251]) and LFP (CellTech [252], Liacon [250], Tianjin Institute [253,254]); or electrodes containing a mixture of NCM and LFP [255]. The shape of a discharge curve [216] (its "kinked" section) allows for the assumption that the LIC-positive electrode LpTO (Microvast) is also based on a mixture of cathode-active materials containing NCM.

To guarantee proper performance at low temperatures, it is essential to select the anode-active material and electrolyte capable of providing sufficient ionic conductivity, as well as the binder. Styrene-butadiene rubbers (SBRs) used as a binder for graphite-based electrodes have a relatively high glass transition temperature (-10 °C and above [256]). However, the rubbers may not guarantee the stability of the electrode structure at low temperatures, which will reduce the discharge capacity [257] and the cycle life of the LICs [258]. On the other hand, SBR with a higher glass transition temperature is less prone to deformation, which causes a less dense structure of the active layer, higher ionic conductivity, and better performance at low temperatures [259]. A binder with a low glass transition temperature enables the production of electrode-active layers with high density and improved performance at high temperatures. There are materials available on the market that include binders with high and low glass transition temperatures (Tg) and a core (high Tg)–shell (low Tg) structure [260].

A PVDF-based binder with a glass transition temperature of -40 °C can be used as an alternative to SBR-based binders [261]. However, PVDF has low electronic conductivity, so to increase the electrode conductivity (and increase power), additional conductive additives should be introduced, thus reducing the amount of active material and the specific energy of the LIC.

At high temperatures of 55 °C (and above), side reactions can occur on the LTO surface (especially at a high state of LIC charge), leading to gas formation, swelling of the case [224,262–265], and cycle life-shortening (6T SCiB^{TM} 64 Ah, 38 °C—4000, 50 °C—2000, 65 °C—500 [266]). Some methods for suppressing the gas formation and extending the cycle life of LICs with lithium-titanate-based anodes are described elsewhere [216,267–270].

LICs with LTO-based anodes have low specific energy (45 $\div$ 90 Wh/kg) due to the following factors:

-Relatively low values of tap density (the electrode density can reach 1.9 $\div$ 2.3 g/cm^3 [271] depending on a particle shape);
-Low discharge capacity (low specific energy of the negative electrode, Table 4);
-High potential relative to lithium (lower potential difference).

The maximum reported value of specific energy found for LICs of this type is 120 Wh/kg [272] (capacity is 30 Ah for a laminated foil (pouch) case). The main applications of LICs with LTO-based negative electrodes are electric buses, other electric vehicles, and stationary storage systems used for the storage and transmission of electric power [273].

To increase LICs' specific energy, a negative electrode based on graphite or a mixture of graphites (the mixture of graphites is required to achieve the optimum capacity and density of the active layer) is combined with a silicon-containing composite. The following silicon-containing composites are mentioned: SCN (Silicon Carbon Nanocomposite, Samsung [274], Figure 5), $SiO_x/Si/C$ (Hitachi Maxell [275,276], Shin-Etsu [277], Figure 5), and other materials listed in presentations from companies such as BTR [278,279], Lishen [243], Paraclete Energy [280], Shanshan [202], Yinlong [281], and Dr Xiao Chengwei [282].

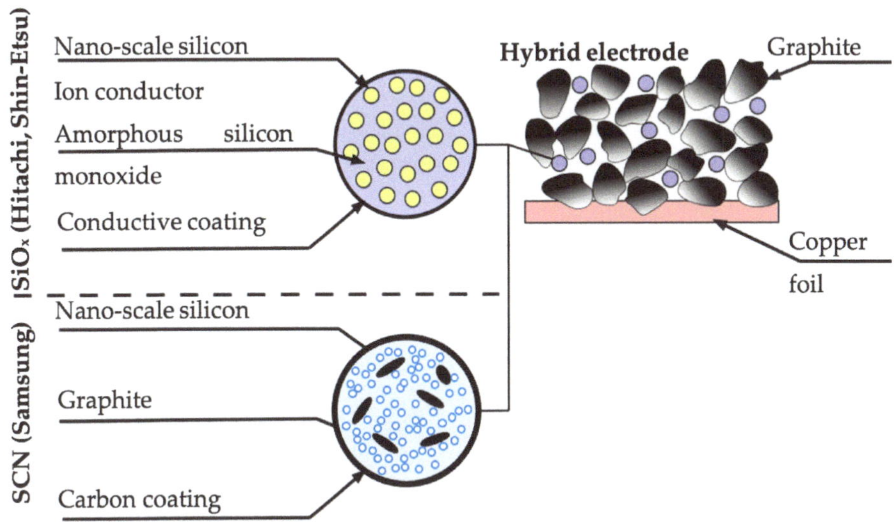

Figure 5. Conceptual models of $SiO_x/Si/C$ and silicon–carbon nanocomposite anode materials [274,283]. Source: Figure by authors.

According to the studies of the negative electrodes for a 18650 LIC with high specific energy using scanning electron microscopy and energy-dispersive X-ray spectroscopy, the size of the regions containing silicon and oxygen is less than 20 μm [106]. A mass fraction of silicon in the active layer is insignificant at just a few percent. The mass content is limited, since silicon and its compounds change their volume significantly when lithium is added/extracted to/from the structure. When a thin copper foil and a high density of active materials are used, an increase in the electrode volume can lead to mechanical stresses and the potential destruction of the electrodes [284].

Table 4 shows the characteristics of silicon-containing mixtures with minimum capacities. According to the report in [46], in 2015, the products developed by Shin-Etsu were demanded the most by major manufacturers of LICs with a high specific energy. Furthermore, the electrochemical characteristics of the electrodes based on a mixture of graphites and materials produced by Shin-Etsu are described in papers [277,285–287].

To improve capacity retention [288,289] during cycling, SiO-C composites can be pre-lithiated. The introduction of lithium into the SiO-C composite leads to the formation of Li_2SiO_3. This can be carried out by heating the SiO-C composite with lithium compounds using the electrochemical method or the redox method [290,291]. When cycling a cell with a SiO-C-based anode, lithium silicate Li_4SiO_4 formed during LIC charging is partially destroyed during LIC discharge (delithiation) in the initial cycles. On the contrary, the lithium silicate (Li_2SiO_3) contained in the pre-lithiated Li-SiO-C composite is more stable during initial charge/discharge cycles, which contributes to less of a decrease in material capacity during cycling [288].

During the manufacture of negative electrodes for Nexelion LICs (18650, 14430), Sony used graphite and a nanocomposite amorphous material SnCo/C [292,293] (sometimes referred to as SnCoTi) as titanium atoms were found during the study of the negative electrode for the first-generation LIC [294]. A conceptual model of the material structure is shown in Figure 6 (for original image see [295]).

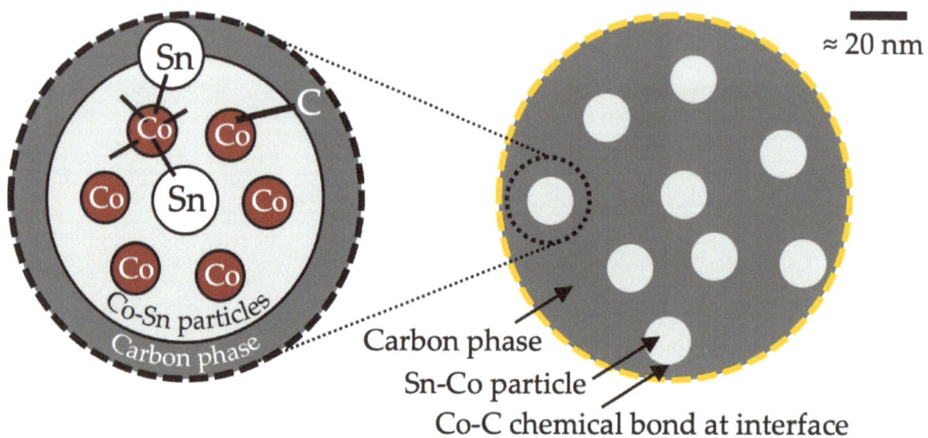

Figure 6. Schematic structure of SnCo/C composite anode material. Figure by authors.

The potential (relative to lithium) of the cathode-active and anode material depends on the chemical composition, structure, and degree of lithiation (delithiation). A voltage variation during the discharge of an LIC manufactured using active materials can be estimated based on the difference between the curves' characteristic of the respective materials (Figure 7). However, more accurate electrode potentials can only be determined when taking a measurement using a three-electrode cell, as an excessive amount of anode-active material is taken. The excessive amount is required to compensate for the irreversible capacity of the anode-active material (lithium ions are involved in forming an SEI film on the negative electrode [297]) and to guarantee that the negative electrode potential is higher than the lithium plating potential.

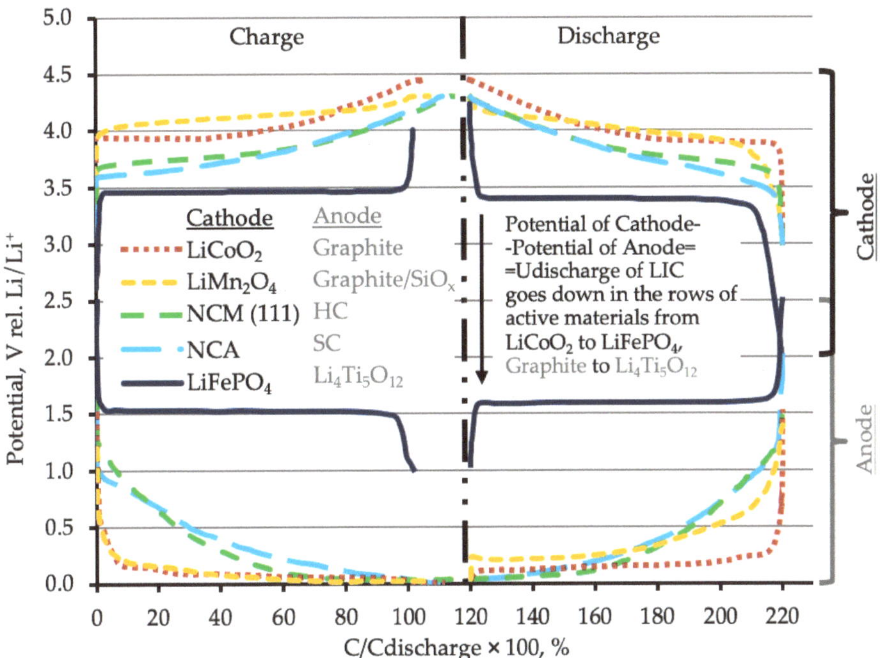

Figure 7. Charge and discharge curves of anode-active (graphite [206], graphite+SiO_x [277], HC [198], SC, $Li_4Ti_5O_{12}$ [201]) and cathode-active (LCO [58], LFP [62], LMO [103], NCA [296], NCM [296]) materials. Figure by authors.

2.3. Conductive Additives

The electronic conductivity of cathode materials varies from 10^{-1} (LFP/C) to 10^{-6} (LMO) S/cm. Due to potential hydrolysis and the deterioration of cathode material performance [298], a non-aqueous solution of polyvinylidene fluoride (N-methyl-2-pyrrolidone, etc.) is used for manufacturing LCO-, LMO-, NCM-, and NCA-based positive electrodes. PVDF is a dielectric that reduces electronic conductivity. Therefore, conductive additives are introduced into positive and negative electrodes (some of which are manufactured using aqueous binders) of high-power LICs and LICs with high specific energy to increase electronic conductivity. During the manufacture of high-power LICs, the electronic conductivity of the electrodes (primarily of the positive electrode) shall be higher than the ionic conductivity of the electrolyte so as not to limit the performance of the LICs [299].

For high-power LICs, it is necessary to increase the electronic conductivity on the surface of active materials, which can be achieved either by creating conductive coatings or by uniformly enveloping an active material particle with particles of a conductive additive [300]. Since the thickness of high-power LIC electrodes is relatively small, it is quite easy to achieve electronic conductivity through the electrode-active layer. With regard to LICs with high specific energy, on the contrary, conductivity near the particles is less critical due to low currents. Since the active layers of the electrode are relatively thick, it is necessary to create "conduction channels" through the entire electrode [300].

Along with increasing the electronic conductivity of electrode-active layer, conductive additives can do the following [301,302]:

- -Affect the preparation and drying rate of the active layer, etc.;
- -Affect the formation of a porous structure and electrode density (ionic conductivity);
- -Affect the active layer adhesion;
- -Provide heat dissipation from the electrode-active layer;
- -Provide electrical contact between the active layer and the current collector.

Various types of carbon black (Cabot LITX200, Denka Li-250 (50%), Imerys C65), conductive graphites (Imerys KS6L (for the cathode—isotropic spherical, artificial graphite), SFG6L (for the anode—anisotropic, flake, artificial graphite), etc.), and carbon fibres (Showa Denko, VGCF-H) are used as conductive additives (Table 6). In recent years, new commercial products have appeared: carbon blacks with a high specific surface area (for example, Li-S02 and Li-S03—Denka, S_{BET}—178 and 241 m^2/g, OAN—239 and 252 mL/100 g, respectively [303]), graphenes [304,305] with different morphologies (for production method, see [306]), single-wall [307] and multi-wall [308,309] carbon nanotubes, and porous carbon materials [310,311] (for example, Porocarb Lion 403 and Porocarb Lion 210 with internal porosity—30% and 20%, respectively).

During the manufacture of LIC-positive electrodes, the most demanded conductive additives are various grades of carbon black. The basic characteristics are specific surface area (particle size), oil absorption number (OAN), or dibutyl phthalate (DBP), the content of impurities. As the specific surface area increases (and the particle size decreases), the area of contact between the conductive additive and the active material increases as well. At the same time, with an increment in the quantity of the conductive additive, the dispersion of the electrode-active layer is more complicated, the intensity of side reactions grows, and the cycle life decreases [312].

Table 6. Electrically conductive additives for LIC electrodes.

Manufacturer	Type	Brand	S_{BET}, m^2/g	d, nm	OAN, mL/100 g	σ, S/cm	Fe, ppm	Reference
Cabot	CB	LITX50	50 ÷ 55	31	140 ÷ 150	–	<10	[300,301,313]
Cabot	CB	LITX200	125 ÷ 175	–	155 ÷ 165	–	<10	[300,301,314]
Cabot	CB	LITX300	180 ÷ 240	–	150 ÷ 180	–	<10	[304,315]
Cabot	CB	LITXHP	80 ÷ 120	–	200 ÷ 280	–		[316,317]
Cabot	Grph.	LITXG300	250 ÷ 350	<2000	–	–	<50	[304]
Cabot	Grph.	LITXG700	650 ÷ 750	<2000	–	–	<50	[304]
Denka	CB (A)	Li-100	68	35	–	5	2	[318]
Denka	CB (A)	Li-250	58	37	200 ^{D}	5	2	[318]
Denka	CB (A)	Li-400 (HS100)	39	48	177 ^{D}	7	2	[318]
Denka	CB (A)	Li-435 (FX35)	133	23	267 ^{D}	4	2	[303,318–321]
Imerys	CB	Super P	62	–	290	10	5	[302,322]
Imerys	CB	C45	45	35 ÷ 45	339	–	2	[322]
Imerys	CB	C65	62	30 ÷ 40	287	–	2	[322,323]
Imerys	G (sph.)	KS6	20	3400	114	1000	75	[322]
Imerys	G (sph.)	KS6L	20	3400	–	–	15	[322]
Imerys	G (fl.)	SFG6	17	3500	–	–	80	[322]
Imerys	G (fl.)	SFG6L	17	3500	–	–	25	[322]
Showa Denko	CF	VGCF-H	13	0.15 μm × 8 μm	–	1000	–	[324]
Cnano	SWNTs	FT2000	450	2 ÷ 4 nm × 500 μm	–	–	≤20 ^{F} wt.%	[325]
OCSiAl	SWNTs	TuBall	–	1.6 nm × 5 μm	–	–	<15 ^{F} wt.%	[307]
MNC	SWNTs	Meijo eDips	–	–	–	–	1 wt.%	[326–328]
Cabot	MW CNTs	Enermax 12	85 ÷ 110	30 ÷ 50 nm × 5 ÷ 12 μm	–	–	–	[329]
Cabot	MW CNTs	Enermax 61	230 ÷ 350	4 ÷ 16 nm × 5 ÷ 20 μm	–	–	–	[329]
Cnano	MW CNTs	LB117-54 St	–	d = 9 nm	–	–	15	[308]
Cnano	MW CNTs	FT9000	110 ÷ 250	d = 10 ÷ 25 nm × 10 μm	–	–	0.1 ÷ 5 ^{F} wt.%	[330]
Heraeus	Porous carbon	Porocarb Lion 403	40 ÷ 60	d_{90} ≤ 4	–	4	–	[310,311]
Heraeus	Porous carbon	Porocarb Lion 210	15 ÷ 25	d_{90} ≤ 8	–	10	–	[310,311]

Note: CB—carbon black, CB (A)—acetylene black, CF—carbon fibres, G (fl.)—flake graphite, G (sph.)—spheroidised graphite, Grph.—graphene, MWCNTs—multi-wall carbon nanotubes, SWNTs—single-wall carbon nanotubes, ^{D}—dibutyl phthalate absorption, ^{F}—total content of dopants, St—conductive additives dispersed in solvent. Source: Authors' compilation based on data from references mentioned in the table.

When the oil absorption number (OAN or DBP) is increased, the aggregation of the conductive additive structure (Figure 8) increases as well, which may cause the accumulation of a larger volume of electrolyte in the electrode [319]. Due to the rather large size of the agglomerates (up to 2 μm) formed from primary particles, high thermal and electrical conductivity are ensured, and the amount of conductive additive in the electrode can be reduced [331].

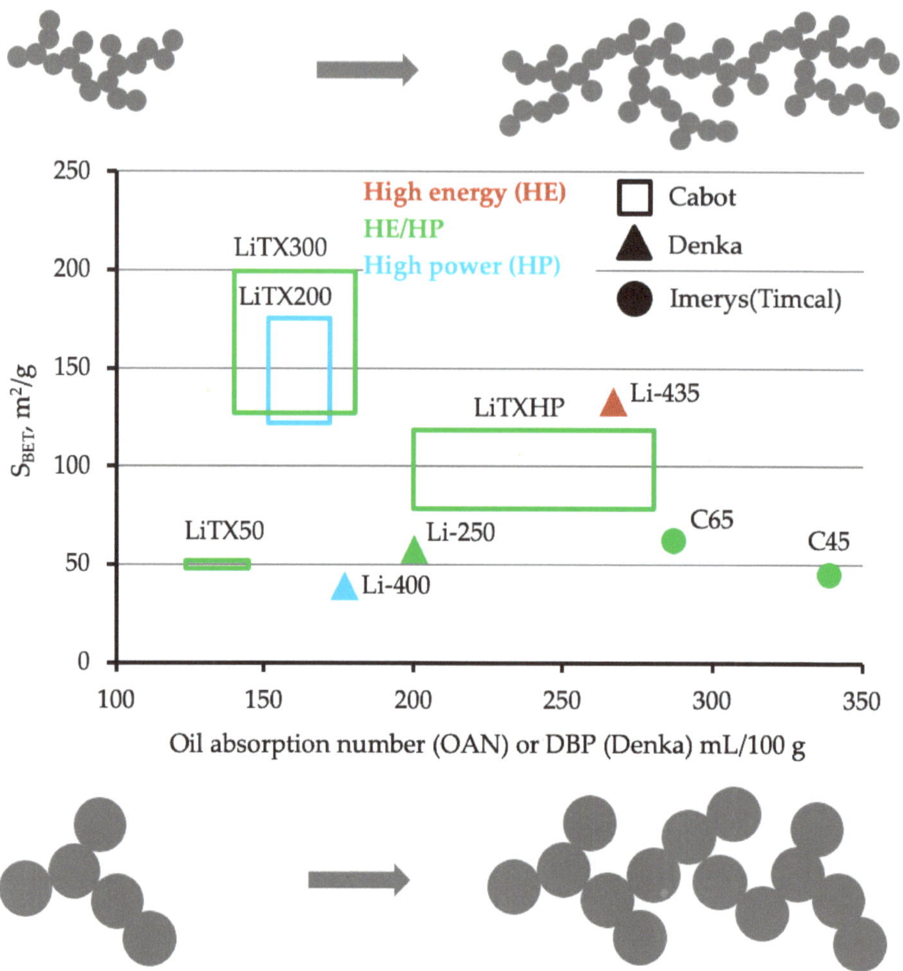

Figure 8. Layout of main conductive additives manufactured by Cabot, Denka, and Imerys in coordinate grid: specific surface area—oil absorption number or dibutyl phthalate. Source: Figure by authors.

The stability of the electrode-active layer over time and a uniform distribution of the conductive additive (with high specific surface area and developed structure) can be achieved by adding polyvinyl pyrrolidone [319] or other surfactants

(LAPONITE-RD (BYK) [332]). The distribution of conductive additives and a binder in the electrode-active layer can be evaluated based on scanning electron microscopy—energy-selective back-scattered (EsB), energy-dispersive spectroscopy (EDS), wavelength-dispersive X-ray spectroscopy (WDS), and laser-induced breakdown spectroscopy (LIBS) [333]. To achieve a high content of the conductive additive in the active layer (electrodes of high-power LICs), carbon black with large particles and a low aggregation structure of agglomerates can be used.

Impurities are dissolved at high potentials and are further deposited on the negative electrode, leading to the degradation of the SEI film (anode-active material) or dendrite formation and, consequently, potential microscopic short circuits. In addition, possible metal dopants in the form of large particles sized 20–30 μm whose thickness exceeds the thickness of the separator may also cause short circuits, negatively affecting the safety and cycle life of the LICs [319].

Conductive additives (Li-400, LiTX200) for manufacturing high-power LIC electrodes are characterised by a low oil adsorption number, allowing the increase of additives in the active layer. The specific energy can be increased by reducing the mass fraction of conductive additives. In this regard, it is possible to use conductive additives (Li-435) with a small particle size and a developed structure of agglomerates, which provide good electronic (bigger contact area) and ionic (more electrolyte is contained in the developed structure when wetting) conductivity at a low mass fraction in the electrode [321].

Manufacturers use various synthesis methods to produce carbon blacks that affect the characteristics of finished materials. For example, since there are no functional groups (primarily polar ones), the acetylene black made by Denka absorbs less moisture from the atmosphere and is not so prone to aggregation. On the other hand, the carbon black produced by Cabot has polar functional groups on its surface, and it absorbs more moisture from the atmosphere. However, the polar functional groups improve wetting with water and simplify carbon dispersion in aqueous binder solutions [306].

The primary particles of Denka acetylene black ($Lc = 2.5 \div 3.5$ nm, $C_0/2 = 0.351$ nm) are crystallised to a greater extent in comparison to Imerys carbon black (C45—$Lc = 2$ nm, $C_0/2 = 0.3575$ nm, C65—$Lc = 2$ nm, $C_0/2 = 0.3586$ nm [323]). A greater degree of carbon particle crystallinity provides better heat removal [331], a longer service life (number of cycles), and less gas formation [317].

Along with the said physical and chemical characteristics, conductive additives are estimated based on the following indicators:

-Minimum pressure (density) during electrode rolling necessary to ensure the preset electronic conductivity [312];
-Maintaining the preset electronic conductivity of electrodes at high depths of discharge [334];
-Minimum mass fraction in the electrode to ensure the proper performance parameters (discharge capacity and average voltage at the preset discharge current) [299,300];
-Stability at high potentials [300], etc.

The thermal and electrical conductivity of conductive graphites and carbon fibres (VGCF) are much higher than carbon blacks. However, due to the large particle size, the contact area with the active material particles is smaller. Therefore, they are used together with carbon black during the manufacture of positive electrodes. This method allows the positive effect of each additive on the LIC characteristics to be combined (Table 7). For example, to ensure the high density and electrical conductivity of an NCM-based electrode, the optimum ratios (with the total content of conductive additives within 2 wt.%) of carbon black C65 and conductive graphite KS6L are within (1:1) ÷ (1:3) [217,335].

Table 7. Effect of carbon black or graphite added to active layer on LIC parameters [302].

LIC Parameters	Carbon Nanotubes	Carbon Black	Graphite
Electrode electronic conductivity	Contact between particles and contact with current collector	Contact between particles and contact with current collector	Conductive paths through the entire electrode
Electrode ionic conductivity	Electrolyte absorption	Electrolyte absorption	Porosity setting
Cyclic stability	High due to good electronic conductivity of active layer	Flexible network, optimum formation of SEI film	Low rate of side reactions
Specific energy density	Less amount is used to obtain required electronic conductivity	Small amount required	Compressibility, low swelling
Electrode manufacture	Dispersions may not be stable in time. Special treatment (high shear rate mixing, surfactants) is required to obtain dispersion	Stable dispersions, compatibility with binder	Small amount of binder Low viscosity of slurry

Source: Authors' compilation based on data from reference mentioned in the caption of the table.

The specific surface area of conductive graphites is relatively small, and less binder is required to keep them within the electrode. Therefore, when the amount of binder and conductive additive(s) is fixed, and a mixture of graphite and carbon black is used, the adhesion (peeling force) of the active layer to the current collector is higher than when using carbon black alone [217,335].

To improve the characteristics of negative electrodes (graphite-based obtained with the use of aqueous binders), conductive graphites (SFG6) can be added into the active layer. Keep in mind that the specific surface area of conductive graphites is higher than that of graphites used as active materials, and it may become necessary to increase the binder mass fraction. Conductive graphites are involved in lithium intercalation/deintercalation processes. Due to the smaller particles, the conductive graphite fills the cavities between the active material particles, increasing the electrodes' density and LICs' specific energy. If conductive graphites are added, this also extends the cycle life (capacity retention during cycling) [217,335].

Graphenes and single-wall and multi-wall carbon nanotubes are among the most advanced conductive additives. Their main advantage is high conductivity at a lower mass fraction in the electrode-active layer. However, the large-scale implementation of these materials is limited by their price and complex dispersion (manufacturers offer carbon material dispersions in various solutions to make handling easier: for example, Cnano LB series [336] and OCSiAl—TUBALL [337]).

A combination of conductive additives may become an optimum solution. For example, according to the presentation in [321], Li 435 and carbon nanotubes improved the specific energy (up to 194.8 Wh/kg, 436 Wh/L) and power (a smaller reduction in average discharge voltage under a discharge current of 2C) of LIC in case 2614891. One of the reasons for the improvement in the characteristics may be the increase in electronic conductivity. Carbon nanotubes improve current collection from the particle surface, while acetylene black agglomerates are better at removing electrons through the active layer. The positive effect of the combined use of Enermax^{TM} carbon nanotubes and carbon black was described in reference [329]. Carbon nanotubes improve current collection through the active layer. Carbon black provides local current collection from a particle and keeps electrolyte within its porous structure. Thus, it enhances the wetting of particles with electrolyte and increases lithium ionic conductivity in the electrode, which is especially important for maintaining proper performance at low temperatures.

2.4. Current Collectors

Aluminium foils [338–341] of various grades (8021 (59%), 3003 (48.5%), 1070, 1100 (59.5%), 1N30 (59.8%), 1085 (61.5%), etc.) are used as positive electrode current collectors; the values of conductivity as percent IACS (International Annealed Copper Standard) [338,339,342]) and thickness ($10 \div 25$ μm) are given in brackets. The increased electrical conductivity and thickness of the current collector and the contact area (quality of the welded joint) of the current collector with a tab allow the internal resistance and heating of the LIC to be reduced [339,343]. To improve adhesion to the electrode slurry, the surface roughness of the foil can be increased by additional mechanical processing [339]. Also, perforated [43] and porous (appropriate for adding an electrode-active layer into the pore space—Fusporous [344–346]) aluminium foils are available on the market.

The improvement in LICs' functional characteristics may be achieved by using aluminium foils with conductive coatings containing various forms of carbon with a binder (0.5 ÷ 5 μm thick [43,207,347–353]), including nanotubes [354–356] and graphene [356]. In addition, nanowhiskers of aluminium carbide (without a binder) [357] and PTC coating [339,358,359] can be used as well. Along with increasing conductivity at the electrode slurry/current collector/carbon interface, carbon coatings improve adhesion between the electrode-active layer and the current collector and protect the surface from exposure to electrolyte components [360–362].

Aluminium foil (anode-active material—LTO), copper foils (carbon anode-active materials), and, in some cases, titanium foils (e.g., Zero-Volt^{TM} technology—Quallion [363], Saft [364]) are used as negative electrode current collectors. Foil materials are selected depending on their stability in the operating ranges of the potentials of anode-active materials [365,366]. The thickness of the applied foils varies from 6 ÷ 7 μm [367,368] (LICs with high specific energy) to 20 μm (high-power and large LICs). Most [369] of the applied copper foils are electrolytically produced [367,370]; however, rolled foils [371–373] retain a small market share. The electrolytic method makes it possible to produce foils of a greater width containing fewer dopants that have high conductivity (for example, NC-WS Furukawa—99% IACS [374]) and a tensile strength of approximately 320 MPa [375]. The characteristics of rolled foils (tough pitch copper foil) depend on the selected alloy. For example, Hitachi Cable used HCL02Z alloy (0.02% Zr, 99.98% Cu) to produce (roll) foil with a tensile strength of 450 MPa and conductivity of 97% IACS [372]). Even higher tensile strength (550 MPa) was achieved for foil made of Cu-0.12Sn (HS 1200, JX Nippon Mining & Metals) [376,377].

It is also possible to produce foils with higher strength through the electrolytic method, for example, NC-TSH [374] with a tensile strength of 490 MPa and conductivity of 95% IACS. Foils with higher tensile strength may be required for LICs with high specific energy, since the materials (Si, SiO_x) that significantly change the electrode's volume during charge/discharge [378] might be applied. To ensure work safety and proper cycle life, the applied copper foils must withstand the undergoing deformations [379].

Perforated copper foils [43,380] can reduce the mass fraction of the negative electrode current collectors in an LIC, but such foils have a lower tensile strength.

Along with conductivity and strength, the checked parameters include relative elongation (the allowable value increases with foil thickening [375]) and surface roughness (↓ surface roughness, ↑ wetting [381]); the special microstructure of grains in the volume guarantees plasticity in the case of tension–compression during charge–discharge [382] (recrystallisation can be performed to improve the mechanical properties of the foil [381]).

To improve the performance of copper current collectors, carbon coatings can be used, which perform the following functions [361]:

-Reduce electrical resistance at the foil/electrode slurry interface;
-Increase electrode slurry's adhesion to foil;
-Reduce copper corrosion when using slurries with a high pH.

2.5. Binders and Thickeners

Binders are used for the manufacture of LIC-positive and -negative electrodes. Also, binders can be a part of separator coatings [383,384], negative [383] and positive [379,383,385] electrode coatings, and conductive coatings on current collectors, and can also be used to increase electrolyte viscosity (PVDF/HFP [386,387]).

The following properties are considered when selecting a binder for electrodes:

-Stability within the operating (or broader) temperature range [258];
-Stability within the electrode-operating potential range [388,389];
-Adhesion of electrode slurry with the preset composition (it can be assessed based on the peel strength of electrodes with the given composition (ASTM D903) or when bent onto a rod [379,390];
-Viscosity of the solution with the preset concentration of binder [391];
-The possibility of uniform application (in the case of non-uniform application, a local excess of cathode-active material or a lack of anode-active material can lead to lithium plating during charging [379]);
-Easy handling (electrode slurry preparation);
-Price.

During the manufacture of positive electrodes, PVDF solutions in N-methyl-2-pyrrolidone (or other organic solvents) are mainly used as binders. Electrodes containing lithium iron (and manganese) phosphates can be produced [392,393] using both PVDF and water-soluble binders (Na-CMC/SBR).

Due to the low electronic conductivity of cathode materials, conductive additives (carbon black, graphite, etc.) are added into the active layer during the manufacture of positive electrodes (when $LiFePO_4/C$ is used). There are non-polar groups on the surface of conductive additives, which complicates the dispersing procedure (longer time and possible uneven distribution across the active layer may occur). Dispersal can be simplified and improved by using various surfactants (for example, LAPONITE-RD (BYK) — $[Na + 0.7[((Si_8Mg_{5.5}Li_{0.3})O_{20}(OH)_4] - 0.7]$ [394]), as well as TEMPO (2,2,6,6-tetramethylpiperidine-1-oxyl radical) oxidised cellulose nanofibres [395] (for example, CNF manufactured by DKS [396]), used together with Na-CMC [397,398]. The cross-sectional size of the fibre is several nanometres, and the length can reach several micrometres. There are $-COO^-$ groups and $-CH_2OH$ groups on the surface, which affect the dispersion and viscosity of the solution.

The advantages of aqueous binders are as follows:

- -High electrode drying rate (the water evaporation temperature is less than that of N-methyl-2-pyrrolidone);
- -Reduced cost of the production line equipment, as it is not necessary to capture and regenerate solvent (a sample of the feasibility study is given in paper [399]);
- -More uniform distribution of conductive additive at high electrode drying rates [392] (due to smaller SBR/CMC particles and lower water evaporation temperature);
- -A uniform film formed around the particles of cathode material, reducing the degree of electrolyte decomposition [385], etc.

The use of aqueous binders (CMC/SBR) for cathode-active materials (LCO, LMO, NCM) is limited by the following factors:

- -Deteriorated properties of active materials due to hydrolysis [298,400–402] and the formation of lithium carbonate and lithium hydroxide coatings [403,404] (the formation of a lithium carbonate coating is also possible when nickel-rich cathode materials are kept in the air, resulting in a drop in material energy and power [405]);
- -Water suspension of cathode materials has an alkaline pH of 10 ÷ 11 or more [401,406,407], leading to the partial dissolution of [408] aluminium foil oxide film (and aluminium foil) with the release of hydrogen gas, which reduces electrode slurry uniformity and affects the adhesion of the electrode-active layer;
- -Lower electrochemical stability compared to PVDF and its copolymers [389] at high potentials relative to lithium [257,388].

The following methods to reduce adverse reactions are reported: the formation of coatings on particles of cathode material (Li_2CO_3, [409], TiO_x [409]) and aluminium foil (carbon coatings [410,411]), pH reduction (stirring the suspension under a CO_2 atmosphere [407]), and adding acids to the electrode slurry (formic [406], polyacrylic [401], hydrochloric [412], and phosphoric acids [406,413]).

Aqueous binders can potentially be used for the manufacture of positive electrodes, for example, a water emulsion of PVDF-HFP particles—Kynarflex LBG (Arkema) [384,391,414,415], TRD202A (fluoro acrylic copolymer, JSR [388,404,409, 416], acrylonitrile copolymer (LA138 [417,418]), polyacrylate BA-310C [419,420], and materials (AN-19K, AN-15A) with a core–corona (polyacrylate) structure: Elebine (Senka) [421], Polymer L (Zeon) [379,385,422], Zeon AY-9391 [392], and Powerbinder (neutral pH when dissolved, no aluminium corrosion—Ielectrolyte [423,424]).

PVDF, applied as a binder for the manufacture of positive electrodes using organic solvents, is a powder or granules of a V2F homopolymer that differs in molecular weight, functional groups (Solvay Solef 5130, Arkema HSV1800, etc.), and degree of crystallinity [425]. The higher the binder's molecular weight, the longer the polymer chains, so less binder is required to maintain the proper adhesion. The functional groups can ensure the better dispersing ability of the binder and better affinity with the active material surface. For example, the polar functional groups distributed along the Solef 5130 polymer chain strengthen intramolecular bonds and bond to the cathode material surface, resulting in stronger adhesion [426]. When the degree of crystallinity is low, solvent and electrolyte penetrate more intensively into the polymer, which, on the one hand, leads to the swelling of the electrode-active layer (decreased electronic conductivity due to the loss of contact between conductive particles and active material). But, on the other hand, greater wetting of the active material particles by the electrolyte is provided (better ionic conductivity of the active layer). Apparently, the manufacture of electrodes requires a balance between ionic and electronic conductivities, and it is necessary to choose a binder with the optimal degree of crystallinity [427].

To prepare the binder, solid PVDF is dissolved in N-methyl-2-pyrrolidone or other organic solvents (acetone). Along with solid PVDF, Kureha produces ready-made solutions (the PVDF content varies from 5 to 13 wt.% and the viscosity ranges from 550 to 2100 mPa·s depending on the molecular weight of the polymer [53,428]) for preparing a slurry of electrode-active layer. In addition, there are some suspensions available on the market that contain a binder (PVDF) and a conductive additive, such as Toyo ink—LIOACCUM™ ONESHOT WANISU™ [429]. The manufacturer notes the advantages of this product as the possibility of producing suspensions with a high concentration of solid components, shorter preparation time of the electrode paste, reduced consumption of low-molecular polymers, and lower probability of coagulation of the conductive additive in the electrode. To reduce electrode resistance, a binder based on polymerised polyvinylidene fluoride and ionic liquid (PIOXCEL CB [430]) can be used.

Nickel-rich cathode materials may have an excessive amount of LiOH on their particles' surface, which increases the pH of the electrode slurry solution. If a homopolymer binder is used, an increase in electrode slurry pH may be accompanied by gelation. According to research carried out by Daikin, an electrode slurry prepared with a PVDF/PTFE copolymer binder (VT-475) was more resistant to gelation at higher pH. Furthermore, it was demonstrated that PVDF and PTFE copolymer as a binder allows the electrode density to be increased (due to a greater softness of VT-475 compared to PVDF) [431] and resistance to cracking is increased when twisting electrodes with an active layer density of 3.4 and 3.5 g/cm^3 [432].

When manufacturing negative electrodes using certain HC and SC anode materials, a PVDF-solution-based binder is also used. For some grades of graphites, [202] both PVDF and aqueous binders (or aqueous binders only) can be used.

The aqueous binders used for the manufacture of anodes based on graphite-active materials are styrene-butadiene rubber (SBR), acrylonitrile copolymer (LA133, etc.), PVDF water emulsion [427], etc.

The most common is anode-active layer manufacturing using SBR and Na-CMC as a binder and a thickener. Anode-active material slurry can be prepared by successively mixing the Na-CMC solution first with the conductive additive, and then with the active material. In both cases, mixing with Na-CMC is carried out at high speeds (about 600 rpm) to achieve a more uniform distribution across the volume. The rotation speed is reduced to 50 ÷ 300 rpm at the last stage, and a binder (SBR) [433] is added. Speed reduction is necessary to maintain the integrity of the binder particles (micelle structure). Inside the coil of the polymer (SBR particles), non-polar groups are presented, and the destruction of micelles will lead to an uneven distribution of the binder across the slurry [433]. Other options for slurry preparation are described in paper [434] (high molecular weight binder) and [435] (graphite/SiO_x/conductive additive, JSR binder).

The sodium salt of carboxymethyl cellulose is made from cellulose, sodium hydroxide, and monochloroacetic acid [436,437]. The basic parameters of Na-CMC are as follows:

-Particle size (the smaller particle size, the faster and easier the dispersion process, and the better the quality of the electrode [438]; it varies depending on the brand: SunRose, ≤50 μm; Akupure, ≤100 and ≤500 μm [439]);
-Molecular weight (the higher the molecular weight, the higher the viscosity of the solution at a fixed mass fraction [438,440]);
-Degree of polymerisation (the higher the degree of polymerisation, the higher the viscosity of the solution at a fixed mass fraction [436]);
-Degree of group substitution (the ratio of H/CH_2COONa in the OR groups; the higher the degree of substitution, the better CMC dissolves in water, and the lower the viscosity of the solution [438,440]);
-Viscosity of the solution at a fixed mass fraction (and rotational speed (usually 60 rpm) when determining it. At pH from 6 ÷ 9, the viscosity of Na-CMC aqueous solution is stable; when the rotation speed rises, the determined viscosity value decreases—shear thinning effect [437]);
-Maximum permissible moisture content (usually below 10 wt.%);
-Degree of purity (usually over 99.5% of basic substance, for BSH—min. 99% [441]) and various impurities [442], including NaCl [443];
-Water solution pH (5.5 ÷ 8.5 [443]).

CMC sodium salt simplifies the dispersion of the anode-active material (graphite and conductive additive) in an aqueous solution, increases the viscosity (serves as a thickener), stabilises the active layer, and increases the strength of the active layer (increased peeling force). Graphite is poorly wetted with water as it has non-polar functional groups on its surface. The CMC sodium salt composition contains both polar and non-polar parts of molecules. Thus, Na-CMC with a low molecular weight (higher distribution density across the volume) and a low degree of substitution (fewer polar groups) is required to improve the covering ability of graphite (carbon black). On the other hand, to enhance the distribution of non-polar carbon particles across the volume of the aqueous solvent, it is necessary to increase the degree of substitution; a greater molecular weight of the polymer is required to increase the viscosity of the solution. Therefore, there should be a balance between the characteristics, and only some grades of Na-CMC (Table 8) are appropriate for electrode manufacture.

When optimizing the electrode-active layer characteristics, manufacturers pay attention to the following parameters of Na-CMC:

- The small particle size allows the dissolution velocity to be increased and improves the applied electrode-active layer [438,444].
- Na-CMC with a high molecular weight:
    - Provides a more extended stability of the electrode paste over time [445];
    - Increases the resistance of the active layer to peeling [438];
    - Reduces the mass fraction of CMC in the electrode, and, consequently, increases the specific energy and reduces the electronic resistance of the negative electrode [434].

On the other hand, more time is needed to disperse Na-CMC with a higher molecular weight; as a result, the cost of electrode production will be higher. The water emulsion of styrene-butadiene rubber (approx. density of 1.5 g/cm^3 [446]) used as a binder contains $40 \div 51$ wt.% of polymer (Table 9). Styrene and butadiene monomers are randomly located in the polymer chain [433]. Polymer chains can form bonds with each other (cross-linking) and are twisted into coils of almost the same size (balls) [447]. There are polar and non-polar functional groups on the outer surface of the coils [447]. Polar groups improve the distribution of rubber particles in water (stabilisation of water emulsion). Non-polar groups provide bonds with the surface of graphite and conductive additive.

Table 8. Characteristics of the commercial sodium salt of carboxymethyl cellulose for the manufacture of LIC negative electrodes.

Brand	DS mol/C_6	Viscosity, 1% Solution mPa·s	pH	MW Thousands	Reference
CP Kelco (Cekol)					
30000	0.75 ÷ 0.85	2500 ÷ 3500	6.5 ÷ 8.0	–	[448]
Daicel					
2200	0.8	1500 ÷ 3000	5.5 ÷ 8.5	–	[443]
DKS (Cellogen)					
BSH-6	0.65 ÷ 0.75	3000 ÷ 4000	–	–	[441]
BSH-12	0.65 ÷ 0.75	6000 ÷ 8000	–	330 ÷ 380 *	
DOW					
10000PA	–	900 ÷ 1500	6.5 ÷ 8.0	–	[442,449]
30000PA	–	3000 ÷ 4000	6.5 ÷ 8.0	–	[449,450]
GL Chem					
GB-S01 (GB8-M20)	0.8 ÷ 0.9	800 ÷ 1000	–	300 ÷ 400	[451]
GB-S02 (GB8-H10)	0.8 ÷ 0.9	1500 ÷ 2500	–	500 ÷ 600	
GB-L01 (GB9-S10) **	0.8 ÷ 1.0	400 ÷ 800 (4%)	–	100	
Nippon Paper Industries (Sunrose)					
MAC200HC	0.85 ÷ 0.95	1000 ÷ 2500	6.0 ÷ 8.5	230 ÷ 325	[438,452]
MAC350HC	0.78 ÷ 0.88	2500 ÷ 5000	6.0 ÷ 8.5	300 ÷ 380	
MAC500LC	0.65 ÷ 0.75	3000 ÷ 5000	6.0 ÷ 8.5	300 ÷ 365	

Table 8. *Cont.*

Brand	DS mol/C_6	Viscosity, 1% Solution mPa·s	pH	MW Thousands	Reference
		Nouryon (AkuPure)			
2791	0.85 ÷ 1.0	1000 ÷ 2000	–	–	[439]
2795	0.85 ÷ 1.0	1500 ÷ 3000	–	–	
2885	0.7 ÷ 0.95	1500 ÷ 3000	–	–	
		Hunan Sentai Biotechnology			
BVH-ST3	≥0.9	800 ÷ 1600	6.0 ÷ 8.5	–	[453]
BVH-ST4	≥0.9	1600 ÷ 2500	6.0 ÷ 8.5	–	
BVH-ST5	≥0.9	2500 ÷ 4000	6.0 ÷ 8.5	–	

Note: *—1600 ÷ 1800 degree of polymerisation; **—for secondary cell membrane protection coating. Source: Authors' compilation based on data from mentioned references.

Table 9. Characteristics of commercial aqueous binders based on styrene-butadiene rubber for manufacture of LIC negative electrodes.

Brand	Tg, °C	D, μm	Solid Matter, %	pH	Viscosity mPa·s	Reference
			BASF			
21–11 ap			49 ÷ 51	6 ÷ 7	80 ÷ 400	[454]
Styrofan 7212	15	0.2	49 ÷ 51	7 ÷ 9	<200	[455]
			JSR			
TRD105A	−3	0.095	40	7.5	–	[456]
TRD104A	7	0.17	45	7.0	30	[256,456–458]
TRD2102	28	0.14	48	7.0	–	[256,459]
TRD102A	−10	0.09	48	7.0	–	[256,459]
TRD1002	−18	0.12	50	8.0	–	[256,459]
TRD2001	−5	0.17	48	7.5	–	[459–461]
			Nippon A&L			
AL 1002	40	0.15	48	6	–	[462]
AL 2001	7	0.22	48	5	–	
AL 3001	−15	0.17	48	7	–	
SN-307R	7	0.22	48	5	–	
			Zeon			
BM-480B	−15	≈0.1	40	–	–	[463,464]
BM-451B	10	≈0.15	40	8	<50	[385,465–467]
BM-430B	−37	–	40	8	5 ÷ 100	[463,468]
BM-400	−5	0.13	40	6	12	[53]

Source: Authors' compilation based on data from mentioned references.

Active-layer adhesion is maintained due to the following:

- The mechanical strength of the polymer;
- The strength of the contact between the polymer and active material (conductive additive and current collector);
- The distribution of the binder across the active layer, which depends, among other things, on the electrode manufacturing conditions (preparation and application of the active layer, drying).

The mechanical strength (and elasticity) of the binder depends on the ratio of the monomer composition and polymer structure (cross-linking, branching, chain length, etc.) and is estimated based on stress/strain curves [469]. The strength of the contact between the binding active material (conductive additive, current collector) depends on the composition and structure of the surface. The distribution across volume depends on the size of the binder particles, functional groups (the fewer hydrophilic groups, the more binder remains near the current collector after drying [470]) on the surface, and drying conditions.

A smaller binder size increases the number of contacts between the active material and the binder network penetrating the active layer. A higher drying temperature (rate) provides faster solvent removal and the transfer of some of the binder from the current collector to the outer part of the active layer, and uneven binder distribution across the thickness. The uneven binder distribution reduces the adhesion of the active layer to the current collector. On the contrary, a higher vacuum drying temperature (without drying in the air) can increase adhesion.

The characteristics binder manufacturers pay attention to are as follows:

-The temperature (glass transition temperature, Tg) at which the binder loses its elasticity, depending on the ratio of butadiene (Tg of butadiene polymer is −50 °C or lower) and styrene (Tg of styrene polymer is 100 °C) monomers [447]. Since the volume of the graphite-based negative electrode changes during charge/discharge, maintaining elasticity at low temperatures can improve the capacity retention during cycling as the integrity of the active layer is preserved.

-The rate of lithium-ion transfer from the electrolyte and their conductivity in the binder (reduced activation energy determined by measuring an impedance of coating on the highly oriented pyrolytic graphite (HOPG) electrode);

-The wetting with electrolyte (shorter absorption time and electrolyte droplet contact angle).

-The low migration rate at high drying rates (which is significant during the manufacture of an electrode with thick active layers, based on the studies using scanning electron microscopy and a distribution map for osmium previously introduced into the binder).

-The minor increase in thickness due to a small amount of swelling (resistance to electrolyte) or during cycling (measurement of mass and linear dimensions of the electrodes), which preserves electronic conductivity better (contact between particles and conductive additive) during operation [470].

-The maintenance of elasticity under cyclic loading [465] (the preservation of tensile strength after cyclic stretching the film of the binder wetted with electrolyte to a predetermined elongation value).

The last two characteristics and high tensile strength are important when active silicon-containing materials are used. The binder restrains the electrode volume increase (LIC mechanical deformations are reduced) [435] and preserves the structure of the active layer with a larger number of charge/discharge cycles (cycle life) [465]. The styrene-butadiene rubbers BM-451B [385,465–467], TRD105A [435,456], TRD302A [470,471], and S2910J-38 (SBR+hydrophylic polymer) [435,459], polyacrylates [279,421,472–476] including crosslinked polyacrylates, Alpha Aekyung [477,478], polymers with a semi-IPN structure [477–480], and polyimide/polyamide binders [379] etc. [481,482], can be used as binders for silicon-containing anodes.

For more information on binders, refer to papers [53,483] and reference [469].

2.6. Separators

The separator prevents electrical contact between the electrodes and, at the same time, provides lithium-ion passage.

The essential characteristics of separators include the following:

-Material (polyethene, polypropylene, cellulose);
-Thickness ↓ (specific energy ↑);
-Stability at high potentials ↑ (specific energy ↑, cycle life ↑);
-Gurley gas permeability ↑ (the integral characteristic depends on thickness, tortuosity of pores, pore distribution, etc.; power ↑);
-Average pore size (electrolyte retention, resistance);
-Pore size distribution (electrolyte distribution);
-Porosity ↓ (strength ↑; on the other hand, due to the decrement of the separator porosity with increased anode volume, the uneven distribution of lithium ions can lead to a local excess and a lack of lithium ions on the anode surface, resulting in a higher risk of dendrite formation);
-Wetting with electrolyte ↑ (power ↑);
-Degree of shrinkage at high temperatures ↓ (safety ↑, ensures separation of electrodes and prevents thermal runaway);
-Shutdown temperature ↓ (safety ↑, prevents thermal runaway, depends on separator structure and materials);
-Puncture strength ↑ (safety ↑);
-Tensile strength ↑ (easy handling during LIC manufacture ↑);
-Elastic modulus ↑ (probability of wrinkling during LIC manufacture ↓), etc.

Porous films with dielectric properties, mainly polyolefins such as polyethene (PE) ("wet" production method, two or three types of fibres are used [484]) and polypropylene (PP) ("dry" production method, stretching of PP film, multilayer PE/PP, PP/PE/PP) are used as separators. In addition, the use of cellulose separators (Toshiba SCiB™ [485]), which can have high porosity, is also reported. For example, the porosity of separators used by Nippon Kodoshi Corporation (NKK) [486,487], series TBL, varies within 63 ÷ 69%. However, unlike polyolefin separators, cellulose separators do not melt [488] and cannot shut down LICs mechanically. In addition, the potential moisture adsorption by cellulose separators (moisture is also contained in ceramic-coated separators [489–492]) can complicate LIC manufacture.

The porosity of polyethene separators varies from 30% (Tonen, 50 nm pore size [493]) to 53% (Asahi Hipore, 150 nm pore size [493]). The porosity of polypropylene separators is slightly higher and can reach 60% (Senior [489]). Small pore size and porosity can negatively affect LIC performance. During charging, the anode volume increases and the porosity of the separator decrease due to deformation, which can lead to an excessive amount of lithium ions in some anode areas and, consequently, to lithium plating [493]. On the other hand, when comparing the performance of prototypes assembled with separators differing in porosity, it was found that the capacity retention after 400 charge/discharge cycles was higher for LICs that had separators with tiny pores than for those that had separators with large pores. The observed effect is explained by the more uniform flow of lithium ions set by tiny pores [494].

Polyethene separators are more durable than polypropylene separators (the thickness can be reduced down to 8 μm (PE $\approx$ 7 μm, coating $\approx$ 1 μm S7Note, [495]), and they are better wetted with electrolyte [496]). On the other hand, their ionic resistance is higher, their oxidation resistance is lower (although this can be improved with a ceramic coating), and their shrinkage at high temperatures is more intensive (but separators can be enhanced with a ceramic coating) [497]. The maximum thickness of polyolefin separators used to manufacture LICs can reach $40 \div 45$ μm [498].

LIC operational safety can be enhanced by using three-layer separators: PP/PE/PP (Celgard [499]), PEHM/PELS/PEHM (Setela^{TM} Toray [175,494]; the melting point of PELS is below that of PEHM), or HDPE/(U) HMW-PE/HDPE (Evapore® Trilayer 5 μm, Brückner Maschinenbau (Siegsdorf, Germany) [500]). When the LIC temperature rises uncontrollably, polyethene melts (at approximately 120 °C). It covers the pores of the polypropylene layer (or polyethene with a high melting point), stopping LIC operation and reducing further heating.

The specific separator type (and brand) used during manufacture is selected based on LIC characteristics and testing results. As initial approximation, the choice of separator can be made on the base of dependency of Gurley air permeability vs. thickness. An example of a plot for Celgard separators is shown in Figure 9.

The improvement in the characteristics of LICs may be achieved by applying [501] ceramic coatings (ceramic particles + binder) to the following:

-The anode (SFL (safety functional layer)—SDI [502], HRL (heat resistance layer)—Panasonic [503]);
-The cathode [106,379,385,492];
-The separator [504] from one side (on the positive electrode side—protection from oxidation at high potentials; on the negative electrode side—the lesser possibility of anode thermal propagation [505]) or from both sides (the separator thickness and cost of the ceramic coating can be greatly increased) [505]).

Coatings' thickness can vary from 1 μm [495,504] to $7 \div 8$ μm [175]. Deposited on separator particles with a bimodal size distribution improve safety even at low application density (9 g/m^2) [142]. A reduced application density with maintained stability at high temperatures can be achieved using smaller inorganic particles (Al_2O_3) [506]. For example, some coatings have an application density of $0.5 \div 1$ g/m^2 (dry weight; water suspension with solid substance mass fraction of $20 \div 40\%$ is used upon application [14]). The most widely spread separator inorganic coatings are Al_2O_3 [175,383,498,507], AlOOH (boehmite) [175,498,507], TiO_2 [175,498], and $BaTiO_3$ [142].

For thermally resistant layer application, binders produced by the following manufacturers can be used:

-Zeon [508,509] (dissoluble in NMP acrylonitrile rubber BM-500B—binder [53], BM-720H—thickener);
-SDK (PNVA^{TM} GE191 [510–512]—water-soluble polymer N-vinylacetamide),
-Haodyne (SWA610 [513]), Solvay (PVDF aqueous latexes series XPH [514]);
-Arkema [414] (aqueous latexes or hard granules LBG, solvent-based LBG 8200).

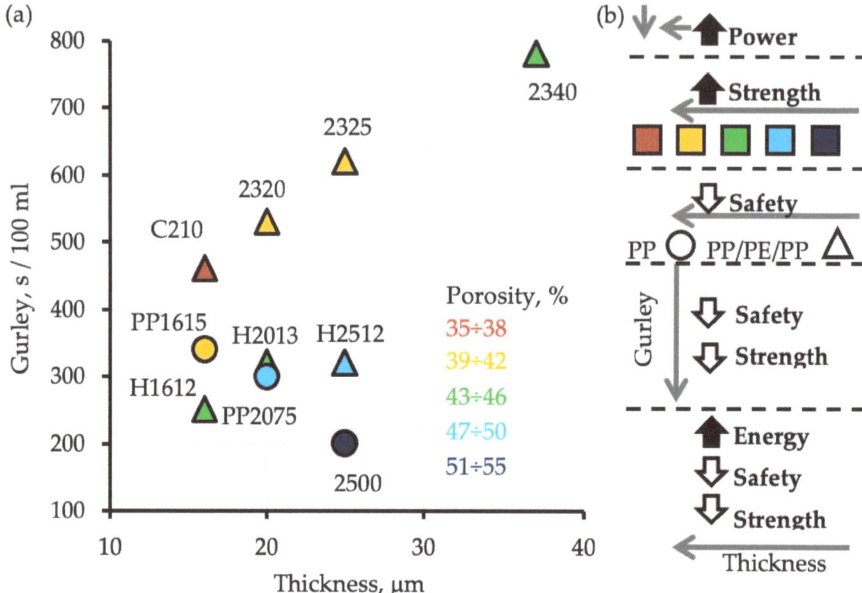

Figure 9. Air permeability, thickness, and porosity of some polypropylene and three-layer separators (Celgard) (**a**) and their qualitative effect (**b**) on power, specific energy, safety, and complexity of LIC manufacture. Source: Figure by authors.

Various functional additives can be introduced into the suspension of ceramic particles (substances improving wetting and dispersing ability, defoamers, etc. [394,414]) to enhance the quality of the applied layer. Ceramic particles can also be introduced into a portion of the separator volume so that two layers can be traced on the micrographs: one layer consists of polymer (21 ÷ 26 µm), and the other includes both polymer and ceramic particles (11 ÷ 12 µm) [504]. In addition, ceramic particles added into the external area of the polymer separator increase the thermal stability of the separator and its wetting with electrolyte [506].

Ceramic coatings are stable even at high temperatures and reduce the possibility of thermal runaway (preventing separator shrinkage) when an LIC is punctured (at the internal short circuit) [503]. Ceramic coatings can [492,497,505,515] block dendrite formation, stop or slow down separator oxidation (at high potentials), adsorb electrolyte decomposition products (HF), improve wetting, etc. Depending on the separator structure, coating, and temperature, the applied coating may partially break, remaining on the surface, or the ceramic particles may melt into the polymer separator [506].

The disadvantages of ceramic-coated separators are as follows:

- -Moisture absorption from the air (a typical value is usually expressed in ppm, 300 ppm [507], and more), which complicates LIC manufacture;
- -Increased resistance (at 100 kHz) [496].

The LIC performance is improved by applying the following functional polymeric coatings to the separator: aramid (PERVIO^{TM}, Sumitomo [498, 516]), meta-aramid [517,518], PVDF [257,384,414,519–523] (the structure of PVDF coating applied from aqueous solution and acetone is described elsewhere [519]), PVDF—hexafluoropropylene (PVDF-HFP [489,522,524]), PVDF-CTFE (poly(vinylidene fluoride-co-chlorotrifluoroethylene) [524], and polymethyl methacrylate (PMMA) [525].

Polymeric coatings perform the following functions:

- -Reduce the degree of shrinkage at high temperatures (Aramid, PVDF, PMMA);
- -Increase wetting with electrolyte (PVDF) [414];
- -Improve contact between the electrodes and the separator (PVDF coating; lower internal resistance due to a closer adherence of the electrodes to the separator [414]);
- -Lower the degree of deformation during cycling [526];
- -Simplify the manufacturing procedure during assembly, including stacking [526], to ensure the manufacture of a more compact roll of electrodes [526]);
- -Increase electrolyte viscosity (PVDF-HFP).

The manufacturers also offer separators with combined coatings: separator/ceramic coating/polymeric coating (PVDF) [489,507] or separator/ceramic particles+polymeric coating (PVDF [507,514], etc. [507]). This type of separator improves safety, but such separators are thicker, less permeable (Gurley), and can increase resistance [384].

Among the more advanced but still not widely used separators are those produced by the following companies:

- -Dream Weaver: the gold brand—combined nano- and microfibres of aramid, cellulose, and polyethene terephthalate (PET) [495,527,528]; the silver brand—fibres of acrylic copolymer and natural and modified cellulose [529,530];
- -Dupont [487] and Hinofiber [531]: polyimide separator;
- -Entek: three-layer polyethene separator with ceramic particles added into both of its external layers [506,532,533];
- -Freudenberg: polyester fibres reinforced with ceramic particles [534,535];

-Litarion (Electrovaya): the Separion brand [536]—ceramic particles added into the plexus of polyethene terephthalate fibres (equivalent of Freudenberg separators);
-Microvast: (poly (m-finylene isophthalamide)) PMIA separator [537];
-Mitsubishi Paper Mills: the Nanobase 0 brand, consisting of thin polyester fibres [538]; the Nanobase X brand, which is a separator made from polyester fibres with an applied ceramic coating [539,540];
-Optodot: the NPORE brand—ceramic membrane and a ceramic coating on the electrode [541];
-Toshiba: spin-coated electrospinning separator [542,543];
-Others [544].

Refer to works [53,499,505,545,546] for more details on the effect of separator characteristics on LIC performance.

2.7. Electrolytes

The electrolyte is a medium allowing for the movement of lithium ions (with a solvate shell). Manufacturers (for example, Capchem [547]) present electrolytes in catalogues classifying them according to LIC applicability with the following properties:

-A specific type of cathode material (LFP, NCM, LCO);
-A silicon-containing composite within the negative-electrode-active layer;
-High-density electrodes;
-Maintaining proper performance at high potentials;
-Maintaining proper performance at low temperatures;
-High power;
-Aggregate state: liquid, gel polymeric (polymerisation [387,548] at temperatures of $60 \div 70\ ^\circ C$), hard polymeric [328];
-In a cylindrical or prismatic case;
-Others.

The effect of the components' composition, ratio, and purity degree on the functional properties are described in papers [379,549–555]. Electrolyte production usually requires $2 \div 5$ solvents (with a purity of more than 99%) from the series of organic carbonates (ethylene carbonate, propylene carbonate, dimethyl carbonate, ethyl methyl carbonate, diethyl carbonate, etc.) and ethers (ethyl acetate, ethyl propionate, propyl propionate, propyl acetate, etc.). Fluorinated solvents (Solvay Energain^{TM} [556] and Daikin (an electrolyte including ethyl methyl carbonate, fluoroethylene carbonate, and fluoro-ether) in various proportions) [432]) can be used to improve performance at high potentials.

Lithium hexafluorophosphate is used as the main salt ($LiPF_6$ purity can reach 99.9% [557,558]), but other lithium salts can also be contained in the electrolyte [559]. The main salt concentration in the electrolyte can vary from 1 to 1.5 M [502] (usually

1 M). The electrolyte characteristics may be improved by adding several functional additives (purity > 99.5% [560–565]) with a concentration of a few percent. There are hundreds of functional additives whose effects on electrolyte properties have been studied. Table 10 (the symbols are interpreted in Table 11) gives the designations and CAS numbers of the additives presented in papers (reports, scientific and technical publications) produced by electrolyte manufacturers, LIC manufacturers, and companies dealing with reviews in the fields of LIC manufacture and marketing.

Table 10. Electrolyte functional additives for lithium-ion cells.

Designation	Formula	CAS No.	Concentration %/M	Function	Reference
AND	$NC(CH_2)_4CN$	111-69-3	1 ÷ 2	● ▼ ☀	[555,560]
BP	$C_{12}H_{10}$	92-52-4	0.1, 1 ÷ 3	● ▲ ■	[187]
BS	$C_4H_8O_3S$	1633-83-6	–	○ ☀	[187,566,567]
CHB	$C_{12}H_{16}$	827-52-1	2.6	● ■	[187,560,568]
DBDMB	$C_{16}H_{26}O_2$	7323-63-9	–	■ □	[549,569,570]
DEPA	$C_8H_{17}O_5P$	867-13-0	–	○ Si	[571]
DFA	$C_7H_6F_2O$	452-10-8	–	■	[175,572]
DTD	$C_2H_4O_4S$	1072-53-3	–	○ ☀	[566,573]
EGBE	$C_8H_{12}N_2O_2$	3386-87-6	0.5 ÷ 1	● ▼ ▼ ⚡	[574,575]
ES	$C_3H_6O_4S$	1073-05-8	–	○ ☀	[187,568,576]
FB	C_6H_5F	462-06-6	2.4 ÷ 7	Solvent	[574,577]
FEC	$C_3H_3FO_3$	114435-02-8	2.0 ÷ 7.0	○ Si ▼ ❄	[228,578,579]
F2EC	$C_3H_2F_2O_3$	311810-76-1	–	○ Si	[578]
H-TP	$C_{18}H_{22}$	61788-32-7	0.41	● ■	[53,577]
LiBOB	$LiB(C_2O_4)_2$	244761-29-3	0.5 ÷ 1 and above	○ □ ● ■ ☀	[502,549,566,580]
LiDFBP	$C_4F_2LiO_8P$	–	–	● ▼ ⚡	[502,581]
LIFSI	$F_2LiNO_4S_2$	171611-11-3	0.2M	▲ ❄	[573,582,583]
LIODFB	$LiBF_2(C_2O_4)$	409071-16-5	–	☀	[228]
$LiPO_2F_2$	$LiPO_2F_2$	24389-25-1	–	▲ ❄	[502,573]
LiTFSI	$Li[N(SO_2CF_3)_2]$	90076-65-6	2–5	▼ ○ ⬟ ☀ 🛡	[584]
MMDS	$C_2H_4O_6S_2$	99591-74-9	–	● ▼ ○	[566,585]
MPC	$C_8H_8O_3$	13509-27-8	1.6	💧	[53,142]
oTPh	$C_{18}H_{14}$	84-15-1	0.1 ÷ 0.2	● ■	[187]
PhBA	$C_6H_5B(OH)_2$	98-80-6	0.5	○	[586]

Table 10. *Cont.*

Designation	Formula	CAS No.	Concentration %/M	Function	Reference
PHOS	Phospha-zenes	–	–	⊘	[187,587]
PMS	$C_3H_6O_3S$	16156-58-4	–	◍	[560,588]
PRS	$C_3H_4O_3S$	21806-61-1	–	◍●▼✹ϟ	[568,589]
PS	$C_3H_6O_3S$	1120-71-4	1.3 ÷ 3.2	◍Δ●▼❄	[187,560,568,574]
SN(SCN)	$C_4H_4N_2$	110-61-2	0.7 ÷ 2.8	●▼▽ϟ	[568,574]
TAB	$C_{11}H_{16}$	2049-95-8	–	■	[560,578]
TMSB	$C_9H_{27}O_3BSi_3$	4325-85-3	–	●ϟ	[502]
TMSPate	$C_9H_{27}O_4PSi_3$	10497-05-9	0.5	◍●✹	[590,591]
TMSPite	$C_9H_{27}O_3PSi_3$	1795-31-9	–	●▽HF	[502,592]
TTFP	$(CF_3CH_2O)_3P$	370-69-4	–	◍●⊘	[106,593]
VA	$C_4H_6O_2$	108-05-4	–	◍	[187]
VC	$C_3H_2O_3$	872-36-6	0.3 ÷ 2.3	◍	[187]
VEC	$C_5H_6O_3$	4427-96-7	–	◍✹	[568]

Source: Authors' compilation based on data from mentioned references.

The introduction of functional additives [594] into the electrolyte can improve LIC performance. For example, organic additives have a potential range in which their structure is stable. When this potential range is exceeded, reduction (anode) or oxidation (cathode) occurs with the formation [379,595] of SEI film on the anode (SEI) or cathode (sometimes referred to as CEI; it may not be uniform, as deposition mainly occurs on active centres [596]). The SEI film on the anode can protect the anode from the penetration of solution components into the graphite structure, stabilise charged anode-active material, increase the intercalation rate, etc. In addition, SEI film formation on the cathode prevents the dissolution of metals, electrolyte oxidation, etc.

Overcharge protection can be ensured by using functional additives that behave as follows at high potentials:

-Polymerise, forming a relatively thick film (its thickness and functional characteristics depend on the concentration [187,572]) that prevents the movement of lithium ions (BP, CHB, etc.).

-Oxidise, with the formation of gases (for example, LiBOB—CO_2 [597]) that increase the internal pressure and activate the current interruptive device (CID, cylindrical LICs).

-Cause a strongly exothermic reaction (TAB [578]) leading to separator melting followed by isolation of the anode and cathode.

-Reversing (cathode)/reducing (anode) oxidation—redox additive [569,598].

Table 11. Interpretation of symbols given in Table 10.

	Sign	Function
Anode	◯	Forms a coating on the anode (SEI)
	Si	Improves performance of Si-containing materials
	△	Coating with high ionic conductivity
	▽	Prevents metal plating on the anode
	▣	Lithium graphite stabilisation
Cathode	●	Forms a coating on the cathode (CEI)
	▲	Coating with high ionic conductivity
	▼	Reduces dissolution of metals
	■	Overcharge protection
	□	Additive redox
Electrolyte	▲	Ionic conductivity increase
	▼	Reduced electrolyte decomposition degree
	⬢	Reduced electrolyte decomposition degree at float charge
	○	Reduced gas formation
Electrolyte	💧	Better anode wetting
	⊘	Reduced flammability
	HF	Hydrofluoric acid scavenging
LIC	☀	Improved performance (storage, ↓ increased resistance at cycling) at high temperatures
	❄	Improved performance at low temperatures
	⚡	Improved performance at high voltages
	🛡	Improved safety at high temperatures

Source: Table by authors.

Functional additives may improve electrolyte properties in the following manner: reduce flammability (non-flammable or low-flammable substances), increase ionic conductivity (salts with excellent solubility within the operating temperature range), and increase viscosity (VDF HFP polymers [257]) (Tables 10 and 11). Some characteristics can be improved due to the use of functional additives, but the additives can deteriorate others, and it becomes necessary to achieve the optimal composition. For example, polymers added into electrolyte reduce ionic conductivity, but Sony [386] managed to develop a gel polymer electrolyte with a high ionic conductivity of 10 mS/cm (at room temperature, the ionic conductivity of LIC electrolytes varies within 6.5 ÷ 12 mS/cm). Electrolyte components that improve combustion resistance can reduce ionic conductivity and increase the electrolyte's viscosity and density, which negatively affects LIC power [599].

The electrolyte requirements depend on the purpose of the lithium-ion cell. However, the high specific energy of LICs is provided by the electrodes' design and active materials, and the electrolyte ensures the proper performance of the electrochemical system (Table 12).

Table 12. Electrolyte requirements for LICs with high specific energy.

Electrode	Problems	Remedies
High-potential cathode	Electrolyte oxidation Metal ions transition into solution	Selection of solvents maintaining stability within the operating potential range Introduction of additives reducing the dissolution of cathode-active material
Cathode containing nickel-rich active materials	Electrolyte decomposition accelerated by nickel ions Metal ions transition into solution	Introduction of additives reducing the dissolution of cathode-active material
High-density anode active layer	Electrode hard-wetting Lithium plating	Low-viscosity solvents Optimised composition of solvents Selection of additives
Anode including silicon-containing nanocomposite	SEI film destruction caused by major volume changes during charge/discharge	Selection of effective additives forming an elastic SEI film Higher concentration of additives forming SEI film

Source: Table by authors.

The LICs used to manufacture batteries for electric vehicles have basic parameters such as safety, cycle life at cycling (durability), and performance in a wide temperature range (Table 13). The electrolyte used affects these performance characteristics.

For more information on electrolyte properties depending on electrolyte composition, refer to the special reviews in [549,594,598,600–602].

Table 13. Requirements for electrolytes used in LICs of electric vehicles batteries.

Parameter	Requirements for Electrolyte	Remedies
Safety	Safety in case of overcharge	Introduction of functional additives (Table 10)
	Non-flammable or low-flammable at high temperatures	
	Improved thermal stability of SEI film	
	Gel polymer technology	Addition of polymer
Cycle life at cycling and durability	Proper viscosity and conductivity	Selection of solvents and their ratios, addition of high-soluble lithium salts
	SEI film stability	Functional additives
	HF retention and gas evolution prevention	
Performance in a wide temperature range	Ionic conductivity in a wide temperature range should be sufficient to maintain the required LIC power	Combination of solvents to maintain proper performance in a wide temperature range
		Lithium salt is stable and capable of dissociation in a wide temperature range, solvate shell with a low energy barrier to destruction (selection of solvents)
	Low resistance of SEI film	Functional additives

Source: Table by authors.

2.8. Conclusions

This section briefly describes the manufacturing technology and materials of a lithium-ion cell electrochemical block which, along with the design, determine the characteristics of the LIC under development. In the following sections, lithium-ion cells for various applications are considered.

The technical requirements for lithium-ion cell and lithium-ion battery (LIB) characteristics are imposed considering the specifications of the product powered by them. From the moment of introduction (1991) until 2018, the largest market share (expressed in Wh) was occupied by LIBs for mobile devices and portable electronics (Figure 10). This segment includes the following:

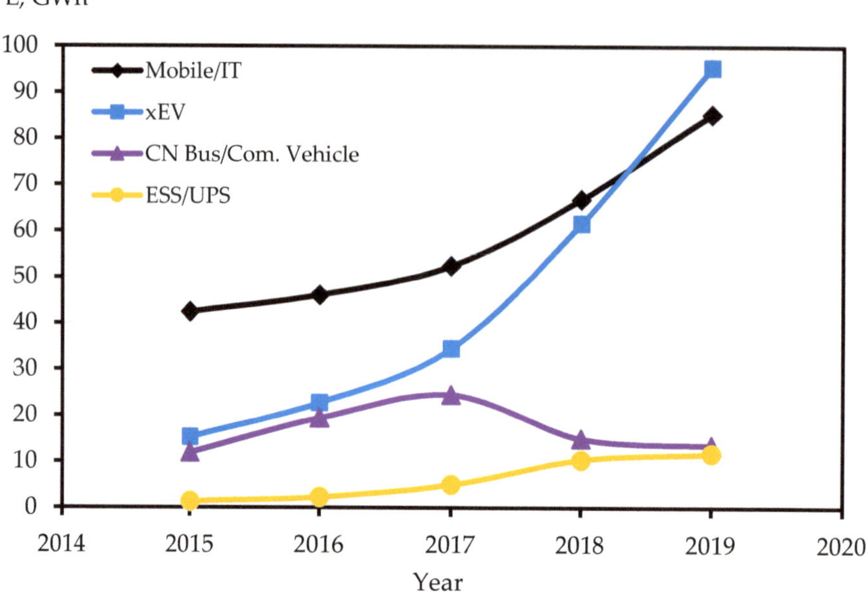

Figure 10. Evolution of LIBs market for the 2015–2019 period (initial data were taken from [574]). Source: Figure by authors.

-LICs with high energy density (Chapter Lithium-Ion Cells with High Specific Energy) used in batteries for tablets, laptops, smartphones, and cells in the 18650 (21700) case.

-High-drain cells in the 18650 or 21700 case (Chapter Advanced High-Power Lithium-Ion Cells for Electric Hand Tools), used for powering electric tools, vapers, and other applications.

-LICs for portable power banks (the capacity varies in the 0.5 ÷ 15 Ah range).

-LICs for wearable devices, batteries, etc. In batteries for wearable devices (headphones, smartwatches, etc.), high-energy LICs are mainly used. However, due to their smaller dimensions [603] (and smaller proportion of active materials) and, in some cases, the need to provide fast charging (more porous electrodes), their energy density (specific energy) is lower than that of LICs used to power smartphones and tablets. Information about some LICs used to power wearable devices can be found in the presentation in [604], and in Chapter Miniature Lithium-Ion Cells for the Internet of Things, Wearable Devices, and Medical Applications.

-LICs for the manufacture of batteries for electric bicycles, scooters, etc.—cylindrical (18650 with a capacity of up to 3.2 Ah [498] and 21700 with a capacity of up to 4.5 Ah [605]) and prismatic, including those with a cathode based on lithiated iron phosphate [606]. Toshiba's lithium-titanate cells are also mentioned [607]. Examples of designs and a description of the market structure of batteries for electric bicycles and scooters can be found elsewhere [358,608,609].

Since 2018, the demand for batteries for passenger hybrid (micro hybrid, mild hybrid, strong hybrid, plug-in hybrid) and electric vehicles has begun to dominate. Therefore, the descriptions of cells and batteries of this market segment are given in Chapter Advanced High-Power Lithium-Ion Cells for Electric Hand Tools (18650 and 21700 cells with high specific energy) and Chapter Lithium-Ion Cells and Batteries for Hybrid, Electric Passenger Vehicles.

A smaller market share is occupied by batteries for buses and commercial vehicles. The main differences between the batteries used in passenger cars are the less strict requirements for specific energy and higher requirements for safety. Since bigger batteries are used, LFP/C, LFP/Gr, LMO/LTO, and NMC/LTO cells are mainly applied in this type of battery [610]. However, some other cells are used; for example, Mercedes announced the use of LFP/Li solid-state batteries for eCitaro buses [611].

A detailed description of the designs, characteristics, and market of lithium-ion batteries for electric buses can be found in references [612–614].

Due to the development of energy production from renewable energy sources and the necessity of the efficiency and profitability increment of grids, the demand for stationary energy storage is growing. Although LIBs for stationary applications occupy the smallest share of the lithium-ion battery market, they have also grown, reaching 11.5 GWh in 2019 [574]. Information about the design of various lithium-ion energy storage devices, including modules and batteries, can be found in Chapter Lithium-Ion Electric Energy Storage for Stationary Applications.

In the Conclusions, a comparison is made of the specific energy and power of LICs for various applications.

References

1. Glaize, C.; Genies, S. *Lithium Batteries and Other Electrochemical Storage Systems*; John Wiley & Sons, Inc.: Hoboken, NJ, USA, 2013.
2. Chen, H.; Shen, J.L. A degradation-based sorting method for lithium-ion battery reuse. *PLoS ONE* **2017**, *12*, e0185922. [CrossRef]
3. Lithium Battery Manufacturing (Electropaedia). Available online: https://www.mpoweruk.com/battery_manufacturing.htm (accessed on 29 October 2018).
4. LTO Lithium Titanate Batteries by Tiankang™. Available online: https://www.youtube.com/watch?v=YE55Db2M9Yw (accessed on 12 March 2018).

5. Wang, F. Lithium-ion battery intelligent product line integrated solution (Lead). In Proceedings of the Battery Japan 2019 (Exhibition), Tokyo, Japan, 27 February–1 March 2019; p. 69.
6. Stadler, B. Fully Continuous Mixing of LIB Electrode Slurries. In Proceedings of the NAATBatt 2016 Annual Meeting & Conference, Indian Wells, CA, USA, 29 February–3 March 2016; p. 23.
7. Ken, N. Preparation of Electrode Slurries (Buhler). Available online: https://wenku.baidu.com/view/de5097500b1c59eef8c7b4b4 (accessed on 3 December 2019).
8. Ohata, T. Relationship between Mixing Processes, Slurry Dispersion Properties, and Electrode Performance (Primix). In Proceedings of the Advanced Automotive Battery Conference-Europe, Strasbourg, France, 25 June 2013; p. 23.
9. Masi, A. NETZSCH Helios fast and efficient dispersion system for Li-ion Battery Electrode coatings. In Proceedings of the 2nd Virtual Battery Exhibition, Virtual, 27 April–3 May 2021; p. 16.
10. Cao, B.; Chen, J.; Yu, T.; Yan, J.; Du, T.; Wagner, J.; Brambilla, N. NeocarbonixTM Electrodes for Li-Ion Batteries: Increased $/Kwh Savings and Scalability. *ECS Meeting Abstracts* **2021**, *MA2021-02*, 346. [CrossRef]
11. 科路得产业化软包全电池制作工艺大讲坛第三讲 (Pouch battery production technology forum, 科路得产业(company name), Lection №3). Available online: https://mp.weixin.qq.com/s?__biz=MzIyMzIwOTMzNA==&mid=2649829045&idx=1&sn=57ede028d9ae1edf1485bc6990f8a8b8&chksm=f0243cacc753b5ba2332edcc075b1b6a8c4260975b1142e40cfc81b0d241d1b38b9123bda0ec&scene=21#wechat_redirect (accessed on 14 December 2020).
12. Kolbusch, T. Innovative lab & pilot coating concepts for Li-ion and solid state batteries (Coatema). In Proceedings of the 2nd Virtual Battery Exhibition, Virtual, 27 April–3 May 2021; p. 48.
13. B&W MEGTEC Awarded Contract to Supply Coating Equipment to Maker of Lithium-Ion Batteries. Available online: https://www.babcock.com/news/megtec-awarded-contract-to-supply-coating-equipment (accessed on 6 April 2019).
14. Glawe, A. High efficient electrode and separator membrane coating lines for lithium-ion batteries. In Proceedings of the 2nd Virtual Battery Exhibition, Virtual, 27 April–3 May 2021; p. 36.
15. Mayer, J.; Michalowski, P.; Kwade, A. From Coating to Cell Stacking—Current challenges of economic electrode/cell production. In Proceedings of the 2nd Virtual Battery Exhibition, Virtual, 27 April–3 May 2021; p. 16.
16. 朱高稳 (Zhu, G.). 双层高速涂布机-EV电池之重器 (Double-layer high-speed coating machine—the most important tool for EV batteries), (Company name –Katop). In Proceedings of the "2017' 第二届动力电池应用国际峰会暨第三届中国电池行业智能制造研讨会 (Second International Summit on Power Battery Applications and the Third China Battery Industry Intelligent Manufacturing Seminar), Beijing, China, 16–17 November 2017; p. 15.

17. Muller, M.; Pfaffmann, L.; Jaiser, S.; Baunach, M.; Trouillet, V.; Scheiba, F.; Scharfer, P.; Schabel, W.; Bauer, W. Investigation of binder distribution in graphite anodes for lithium-ion batteries. *J. Power Sources* **2017**, *340*, 1–5. [CrossRef]
18. Jaiser, S.; Muller, M.; Baunach, M.; Bauer, W.; Scharfer, P.; Schabel, W. Investigation of film solidification and binder migration during drying of Li-Ion battery anodes. *J. Power Sources* **2016**, *318*, 210–219. [CrossRef]
19. 科路得产业化软包全电池制作工艺大讲坛第五讲 (Pouch battery production technology forum, 科路得产业 (company name), Lection №5). Available online: https://mp.weixin.qq.com/s?__biz=MzIyMzIwOTMzNA==&mid=2649829078&idx=1&sn=4a735c0c997493e60a3c5ef97dda99f1&chksm=f0243ccfc753b5d9c34f8e1ace508f1c25899f326cbba3b84873ee1391318c92a9d431af3836&scene=21#wechat_redirect (accessed on 14 December 2020).
20. Schaefel, T. GAIA Li-Ion Batteries: Evolution or Revolution? Available online: http://www.extraenergy.org/files/TimSch%E4fer-GAIA.pdf (accessed on 20 November 2019).
21. Gottschalk, L.; Mayer, J.; Kwade, A. Raw material/Mixing: Production of LIB slurries—Transfer from batch to continuous processes. In Proceedings of the 2nd Virtual Battery Exhibition, Virtual, 27 April–3 May 2021; p. 13.
22. 科路得产业化软包全电池制作工艺大讲坛第六讲 (Pouch battery production technology forum, 科路得产业 (company name), Lection №6). Available online: https://mp.weixin.qq.com/s?__biz=MzIyMzIwOTMzNA==&mid=2649829093&idx=1&sn=fe837f45fa902a3d6ac27386f691bb22&chksm=f0243cfcc753b5eac925bdc6d0293e9fa28070e27e78eea8b1cf43b175213fb756b5a980dacb&scene=21#wechat_redirect (accessed on 14 December 2020).
23. Zhang, J.; Ramadass, P.; Fang, W. Internal Short, Thermal Behavior and Porous Electrode. In Proceedings of the 2013 China Forum on Technical & Market Development of Positive & Negative Materials for Advanced LIB, Tianjin, China, 8–10 April 2013; p. 49.
24. 科路得产业化软包全电池制作工艺大讲坛第七讲 (Pouch battery production technology forum, 科路得产业 (company name), Lection №7). Available online: https://mp.weixin.qq.com/s?__biz=MzIyMzIwOTMzNA==&mid=2649829114&idx=1&sn=2dc47740c768efe98fe10071eb13991a&chksm=f0243ce3c753b5f511989c868f021f4ecc44eb9665011ba93795cba7e3a055a3c7a7db476de4&scene=21#wechat_redirect (accessed on 14 December 2020).
25. 报告22-田宏国-动力锂电池全产业链关键环节水分问题的系统解决方案 (Report №22—Tian Hongguo—Systematic solutions to moisture problems in key links of the entire power lithium battery industry chain) (Time High-Tech). Available online: http://www.docin.com/p-2068642689.html (accessed on 3 December 2019).
26. Woerner, M. Factory of the future Battery processed by OMRON. In Proceedings of the 2nd Virtual Battery Exhibition, Virtual, 27 April–3 May 2021; p. 21.
27. Analysis Of Samsung SDI Power Battery Technology For Lithium Battery Giant. Available online: http://www.optimumnanolithiumbattery.com/news/analysis-of-samsung-sdi-power-battery-technolo-4920947.html (accessed on 6 August 2018).
28. ESS Batteries by Samsung SDI (May 2017). Available online: https://www.samsungsdi.com/upload/ess_brochure/201705SamsungSDI_ESS_EN.pdf (accessed on 6 August 2018).

29. Advanced Batteries for Energy Storage. Available online: https://www.lgchem.com/upload/file/product/LGChem_Catalog_Global_2018.pdf (accessed on 14 November 2024).
30. LG Chem ESS Cell. Available online: https://www.youtube.com/watch?v=CU0RGPnwYYA (accessed on 12 March 2019).
31. 8 mm Width Aluminum Tab as Positive Terminal for Pouch Cell, 50pcs/Box—EQ-PLiB-ATC8. Available online: http://www.mtixtl.com/8mmwidthAluminumTabasPositiveTerminalforPolymerLi-ionBattery50pc.aspx (accessed on 6 August 2018).
32. 8mm Width Nickel Tab with Adhesive Polymer Tap as Negative Terminal for Pouch Cell, 50pcs/Box—EQ-PLiB-NTA8. Available online: http://www.mtixtl.com/8mmWidthNickelTabasNegativeTerminalforPolymerLi-ionBattery50pcs.aspx (accessed on 6 August 2018).
33. Hot Melt Adhesive (Polymer Tape) for Heat Sealing Pouch Cell Tabs (100 mL x 4mm W x 0.1 mm Thickness) EQ-PLIB-HMA4. Available online: http://www.mtixtl.com/HotMeltAdhesivePolymerTapeforHeatSealingPouchCellTabs100mL.aspx (accessed on 6 August 2018).
34. DNP 极耳胶 (DNP Polar ear glue). Available online: http://www.shmetal.cn/en/product-31071-28038-70188.html (accessed on 14 November 2024)..
35. 洋一 望月 (Youichi Mochizuki); 正隆 奥下(Masataka Okushita). Battery Tab and Lithium Ion Battery Using the Same. JP5292914B2, 12 5 2008.
36. 2013年锂电池铝塑包装膜市场供需状况 (Lithium battery aluminum plastic packaging film (pouch) market supply and demand situation) (ITRI IEK). Available online: https://wenku.baidu.com/view/3a4017d251e79b8968022684 (accessed on 6 August 2018).
37. Hubei King Plastic Aluminum Laminated Film Expertise. Available online: https://www.google.com/url?sa=t&source=web&rct=j&opi=89978449&url=https://pub-mediabox-storage.rxweb-prd.com/exhibitor/document/exh-45f78af2-cddc-4b1d-8652-96c7a41f011a/3190cf25-ab79-4ee7-82de-2d1ed4565ac8.pdf&ved=2ahUKEwjXtdyJjceJAxUjExAIHZIKJMwQFnoECA0QAQ&usg=AOvVaw3-T7YReYPMOwbHW7miS5Wb (accessed on 14 November 2024).
38. Aluminum-plastic Composite Film for Lithium Battery. Available online: http://www.sy-ec.com/a/en/Products/Plastic_composite_membrane__the_pion/20121103462.html (accessed on 16 April 2021).
39. Pouch films. Available online: https://www.szkejing.com/category-883.html (accessed on 8 May 2024).
40. SJP-68 Aluminum Laminated Film for Ion Batteries. Available online: https://www.alibaba.com/product-detail/SJP-68-Aluminum-laminated-film-for_62456604402.html?spm=a2700.details.0.0.4d5a7e86Tlyokp (accessed on 26 January 2022).
41. 墨染科路得产业化软包全电池制作工艺大讲坛第八讲 (Pouch battery production technology forum, 科路得产业(company name), Lection №8). Available online: http://www.360doc.com/content/19/0717/08/65038027_849283814.shtml (accessed on 14 December 2020).
42. Roever, M. Moving cell case production onto the fast lane. In Proceedings of the VDMA Virtual Battery Exhibition 2021, Virtual, 27 April–3 May 2021; p. 23.
43. World Smart Energy Week 2014 東京現場直擊系列報導二(Tokyo World Smart Energy Week coverage report, part 2). Available online: http://www.materialsnet.com.tw/material/DocView_MaterialNews.aspx?id=11697 (accessed on 7 November 2018).

44. Nagai, H.; Morita, M.; Satoh, K. Development of the Li-ion Battery Cell for Hybrid Vehicle. SAE Technical Paper 2016-01-1207. In Proceedings of the SAE 2016 World Congress and Exhibition, Detroit, MI, USA, 12–14 April 2016; p. 8. [CrossRef]
45. Topics. *GS Yuasa Technical Report 2016*; GS Yuasa Corporation: Kyoto, Japan, 2016; Volume 2, pp. 37–51.
46. *LIB材料市場速報 (Materials Market News) (15Q1)*; B3 Corporation: Tokyo, Japan, 2015; p. 30. Available online: https://www.docin.com/p-1282371064.html?docfrom=rrela (accessed on 22 May 2024).
47. Topics. *GS Yuasa Technical Report 2016*; GS Yuasa Corporation: Kyoto, Japan, 2016; Volume 1, pp. 37–50.
48. Dahn, J.; Ehrlich, G.M. Lithium-Ion batteries. In *Linden's Handbook of Batteries;* 4th ed.; Reddy, T.B., Ed.; McGraw-Hill: New York, NY, USA, 2011; p. 1457.
49. Mao, C.Y.; An, S.J.; Meyer, H.M.; Li, J.L.; Wood, M.; Ruther, R.E.; Wood, D.L. Balancing formation time and electrochemical performance of high energy lithium-ion batteries. *J. Power Sources* **2018**, *402*, 107–115. [CrossRef]
50. An, S.J.; Li, J.L.; Du, Z.J.; Daniel, C.; Wood, D.L. Fast formation cycling for lithium ion batteries. *J. Power Sources* **2017**, *342*, 846–852. [CrossRef]
51. Heimes, H.H.; Kampker, A.; Lienemann, C.; Locke, M.; Offermanns, C.; Michaelis, S.; Rahimzei, E. Lithium-Ion Cell Production Process. Available online: https://www.pem.rwth-aachen.de/global/show_document.asp?id=aaaaaaaaabdqbtk (accessed on 2 May 2021).
52. Birkl, C.R.; Roberts, M.R.; McTurk, E.; Bruce, P.G.; Howey, D.A. Degradation diagnostics for lithium ion cells. *J. Power Sources* **2017**, *341*, 373–386. [CrossRef]
53. *Lithium-Ion Batteries;* Springer: New York, NY, USA, 2009; p. 452.
54. Lu, M. Future Trends and Key Issues in the Global Lithium-ion Batteries Market and Related Technologies. In Proceedings of the 13th China International Battery Fair (CIBF 2018), Shenzhen, China, 22–24 May 2018; p. 30.
55. 天津巴莫公司-酸LCO品介 (Tianjin B&M Technology Co Ltd. Lithium cobalt oxide (LCO) product introduction). Available online: https://wenku.baidu.com/view/68eb1dbcd4bbfd0a79563c1ec5da50e2524dd1fc (accessed on 10 December 2020).
56. Xu, L. 高性能正被材料的市揚這用及友展(Application on the market and development of high-performance cathode materials). Available online: http://www.docin.com/p-1136007242.html (accessed on 2 August 2016).
57. Pulead LCO-983HA. Available online: http://www.pulead.com.cn/product.php?id=88 (accessed on 18 January 2017).
58. Technology, B.E.M. High-power LCO, mainly used in UAV, Drone, Electric toy and E-Cigarette batteries with high-rate requirement. In Proceedings of the China International Battery Fair (CIBF 2018), Shenzhen, China, 22–24 May 2018; p. 6.
59. Pulead LCO-910H. Available online: http://www.pulead.com.cn/product.php?id=87 (accessed on 18 January 2017).
60. Phostech Lithium. Innovative Material. Available online: http://www.yumpu.com/en/document/view/43911190/brochure-phostech-v2-phostech-lithium-inc/7 (accessed on 30 January 2017).

61. SC-LFP Presentation for Customer. Available online: http://wenku.baidu.com/view/ec72b73b3968011ca3009148.html (accessed on 30 January 2017).
62. Chen, T. 磷酸亚铁锂正极材料的发展 (Development of lithium iron phosphate cathode materials) Life Power P2. In Proceedings of the BL Energy Storage, Chengdu, China, 18 March 2013; p. 23.
63. HED™ LFP-400 Lithium Iron Phosphate Cathode Materials. Available online: http://product-finder.basf.com/group/corporate/product-finder/en/literature-document:/Brand+HED-Product+Data+Sheet--HED+LFP-English.pdf (accessed on 30 August 2019).
64. 巴斯夫 (BASF) 2013_Presentation_-_Luna_-_Tianjin_8-04-2013. Available online: http://wenku.baidu.com/view/097257e4551810a6f524868f (accessed on 20 February 2017).
65. Han, K. Hanwha's Active Materials for LIB (LFP-CNT composite). In Proceedings of the China Industrial Association of Power Sources (CIAPS), Shanghai, China; 2013.
66. Lithium Metal Phosphate. Available online: https://www.aleees.com/en/products-services/cam-material/ (accessed on 14 November 2024).
67. Aleees 磷酸铁锂正极材料 (Lithium Iron Phosphate Cathode Material). Available online: http://www.fentijs.com/index.php?m=yp&c=com_index&a=show&modelid=14&catid=890&id=1310&userid=454&page=1 (accessed on 14 November 2024).
68. Ke, C.-Y. Development of $LiFePO_4$/C cathode materials with improving the high temperature and C-rate performances. In Proceedings of the International Conference on Olivines for Rechargeable Batteries, Montreal, QC, Canada, 25–28 May 2014; p. 26.
69. QY-G02 $LiMn_2O_4$ (Qianyun High Tech). Available online: https://www.qianyun-tech.com/Product.aspx?funId=85&typeId=0&id=3413 (accessed on 17 May 2024).
70. QY-X01 $LiMn_2O_4$ (Qianyun High Tech). Available online: https://www.qianyun-tech.com/Product.aspx?funId=85&typeId=0&id=3190 (accessed on 17 May 2024).
71. 3M Marches Electroniques. Available online: http://www.docin.com/p-206678983.html (accessed on 17 February 2017).
72. 产品中心 世界级高端锂电材料供应商 (Product Center World-class supplier of high-end lithium battery materials). Available online: http://www.bamo-tech.com/pro.aspx?AboutCateId=85&CateId=85 (accessed on 10 December 2020).
73. HEDTM NCMs Superior Quality Cathode Materials. Available online: http://aerospace.basf.com/common/pdfs/BASF_HED_NCMs_Aerospace_Data_Sheet_sfs.pdf (accessed on 30 August 2019).
74. Wu, M. Development and production of high performance & high quality cathode materials for lithium-ion batteries (Tianjin B&M Science and Technology Joint-Stock Co. Ltd.). In Proceedings of the China Li-Ion Battery Positive, Negative Materials Technology and Market Development Forum, Tianjin, China, 8–10 April 2013; p. 36.
75. Leaflet of Qinghai Taifeng PULEAD Lithium-Energy Technology Co., LTD (580HA, 986, 988M, P800, PU50D, PU57D, PU60E, PU80E). In Proceedings of the 13th China International Battery Fair (CIBF2018), Shenzhen, China, 22–24 May 2018.
76. Guo, Y. The Chinese EV Market and LIB Cathode Supply Chain (Pulead). In Proceedings of the 37th International Battery Seminar&Exhibit, Vitual, 28 July 2020; p. 30.

77. Henan Kelong KL206 (NCM 622). Available online: http://www.hnkl.com.cn/eng/product/81_14 (accessed on 14 November 2024).
78. Investor Relations 2019 Think Tomorrow Technology For Future 세계 1위 양극소재 기업 에코프로비엠 (Ecopro BM, the world's No. 1 cathode material company). Available online: https://file.irgo.co.kr/data/BOARD/ATTACH_PDF/eb3a1cb1719b2c5bdd3bb290aff222b4.pdf (accessed on 18 January 2021).
79. NCM单晶品 易加工、高压实、高电压、高容量、长寿命、存储性能优良 (Single crystal products are easy to process, high compaction, high voltage, high capacity, long life, and excellent storage performance). Available online: https://www.btrchina.com/product/91.html (accessed on 16 January 2021).
80. Fetcenko, M. 适用于小型移动、固定以及交通运输的金属氢化物镍电池技术进展 (Nickel Metal Hydride Batteries for Portable, Stationary and Transportation Application). In Proceedings of the 12th China International Battery Fair (CIBF2016), Shenzhen, China, 22–26 May 2016; p. 37.
81. Materials, B.N.E. High Nickel NCA. In Proceedings of the China International Battery Fair (CIBF2018), Shenzhen, China, 22–24 May 2018; p. 1.
82. NCA单晶品 高容量、高安全性 (Single crystal product, high capacity, high safety). Available online: https://www.btrchina.com/product/89.html (accessed on 16 January 2021).
83. Formosa 磷酸铁锂正极材料 (Lithium Iron Phosphate Cathode Material). Available online: http://www.fentijs.com/index.php?m=yp&c=com_index&a=show&modelid=14&catid=890&id=1309&userid=454&page=1 (accessed on 14 November 2024).
84. Press Release CP-2012-16-R. Umicore and Prayon Join Forces to Develop and Produce Phosphate-Based Cathode Materials for Lithium-Ion Batteries. 15 May 2012. Available online: https://www.umicore.com/storage/migrate/20120515PrayonUmicoreEN.pdf (accessed on 14 November 2024).
85. Valence U1-12RT 12.8v 37Ah TESTED Lithium Iron Magnesium Phosphate Battery. Available online: https://goodsunsolarpanels.myshopify.com/products/valence-u1-12rt-12-8v-37ah-tested-lithium-iron-magnesium-phosphate-battery?srsltid=AfmBOoqygw6gXuJqZ11oloWNdbl9Ce5YBaXdjnP232kDDV2bGiTB2rl8 (accessed on 14 November 2024).
86. Wu, P.-J. Metal-oxide Enhanced LFP Material with High Energy Density for Lithium-Ion Rechargeable Battery. In Proceedings of the AABC 2009, Long Beach, CA, USA, 9–11 June 2009; p. 9.
87. Park, S. In Proceedings of the 이차전지/전기자동차 기술개발동향과 시장분석 및 사업화 전략 세미나—한국산업기술협회 (Secondary battery/electric vehicle technology development trends, market analysis and commercialization strategy seminar—Korea Industrial Technology Association), Seoul, Republic of Korea, 4 September 2013; p. 47.
88. Xu, Z.; Han, S.; Tao, L.Q.; He, L.Y.; Yi, Z.L.; Huang, Y.F.; Ye, L.; Gu, T. Which LIB Cathode Will Dominate for the EVs and ESS—From the Industry Point of View? In Proceedings of the 232nd Ecs Meeting, National Harbour, MD, USA, 1 October 2017; p. 28.

89. Liang, G. New development on lithium metal phosphate cathode materials. In Proceedings of the 2013 China Forum on Technical & Market Development of Positive & Negative Materials for Advanced LIB, Tianjin, China, 8–10 April 2013; p. 22.
90. Dow Energy Materials launches new manganese iron phosphate material for improved battery performance; 10–15% higher energy density. Available online: https://www.greencarcongress.com/2012/11/dem-20121122.html (accessed on 14 May 2024).
91. Kon, K.; Nemoto, M.; Nakahata, R.; Yamashita, H.; Ohgami, T.; Abe, H.; Kanamura, K. リン酸マンガン鉄リチウムの リチウムイオン二次電池用正極特性 (Characteristics of $LiMn_{1-x}Fe_xPO_4$ for lithium ion battery). *FB テクニカルニュース (Technical News))* **2017**, *73*, 20–25.
92. World Smart Energy Week 2012 東京現場直擊系列報導(二) (Tokyo live coverage series (Part 2)). Available online: https://www.materialsnet.com.tw/DocView.aspx?id=10099 (accessed on 6 November 2018).
93. Huang, H.-T. Lithium Manganese Iron Phosphate: The Next-Generation Olivine Cathode Material for Li-ion Batteries (HCM). In Proceedings of the 13th China International Battery Fair (CIBF 2018), Shenzhen, China, 22–24 May 2018; p. 26.
94. Upreti, S. Bio-mineralized cathode and anode materials for electrochemical cell. 9,799,883B2, 9 June 2014.
95. Magnis Resources Limited (Investor Presentation). Available online: https://www.asx.com.au/asxpdf/20180226/pdf/43rxtsdt6f1ppx.pdf (accessed on 6 April 2019).
96. Bernard, P. Saft's Advanced & Beyond Lithium-Ion Technologies for Mobility Applications. In Proceedings of the Advanced Automotive Battery Conference (AABC 2021 Europe), Virtual, 20 January 2021; p. 22.
97. Singh, J. Advanced High Energy Li-Ion Cell for PHEV and EV Applications (ES210). Washington, DC, USA, 10 June 2015; p. 17.
98. Ecopro Catalogue. Available online: http://www.doc88.com/p-515799292034.html (accessed on 17 February 2017).
99. Lu, Z.; Eberman, K.; Mi, N. 3M Lithium-Ion battery Material and Solution for EV and Energy Storage Application. In Proceedings of the 11th China International Battery Fair (CIBF 2014), Shenzhen, China, 18 June 2014; p. 25.
100. Sun, Y.-K.; Kim, D.-H.; Yoon, C.S.; Myung, S.-T.; Prakash, J.; Amine, K. A Novel Cathode Material with a Concentration-Gradient for High-Energy and Safe Lithium-Ion Batteries. *Adv. Funct. Mater.* **2010**, *20*, 485–491. [CrossRef]
101. Sun, Y.K.; Chen, Z.H.; Noh, H.J.; Lee, D.J.; Jung, H.G.; Ren, Y.; Wang, S.; Yoon, C.S.; Myung, S.T.; Amine, K. Nanostructured high-energy cathode materials for advanced lithium batteries. *Nat. Mater.* **2012**, *11*, 942–947. [CrossRef]
102. Amine, K.; Koenig, G.; Chen, Z.; Belharouak, I.; Sun, Y.K. New High Energy Gradient Concentration Cathode Material. Available online: https://www.energy.gov/sites/prod/files/2014/03/f10/es016_amine_2012_o.pdf (accessed on 30 August 2019).
103. (KETI), Rechargeable Battery-LIB-Cathode (in Korean). In Proceedings of the SNE Research, Pangyo-ro, Republic of Korea, 23 March 2013; p. 49.

104. Gauthier, M. Phostech's advanced C-$LiFePO_4$ cathode for EV-PHEV-HEV: Challenges and Opportunities. In Proceedings of the PHEV'09 Plug-in Hybrid and Electric Vehicles, Montreal, QC, Canada, 28–30 September 2009; p. 38.
105. Lu, H.-L. Current Situation Regarding xEV-Batteries & Materials in the Chinese Market and Future Outlook (ITRI IEK). In Proceedings of the 7th Korea Advanced Battery Conference 2016, Coex Seoul, Korea, 29–30 September 2016; p. 40.
106. Lu (Mark), H.-L. Commercial Technology & Product Development Trends of Cathode & Anode Materials for LIB in 2015. In Proceedings of the Second International Forum on Cathode & Anode Materials for Advanced Batteries, Hangzhou, China, 22 April 2015; p. 55.
107. Lithium Cobalt Oxide $LiCoO_2$. Available online: http://www.nichia.co.jp/ru/product/battery_lic.html (accessed on 2 August 2016).
108. 高性能动力三元材料技术开发进展 谭欣欣 (Advances in high-performance power ternary materials technology development, Hunan Shanshan Energy Technology co. Ltd.). In Proceedings of the 2016'第五届中国电池市场年会, 暨第一届动力电池应用国际峰会, 第二届中国电池行业智能制造研讨会 (The 5th China Battery Market Annual Conference, the 1st International Power Battery Application Summit, and the 2nd China Battery Industry Intelligent Manufacturing Seminar) (CBEA2016), Beijing, China, 14–15 November 2016; p. 35.
109. Burrell, A. Addressing the Voltage Fade Issue with Li-Mn Rich Oxide Cathode Materials (Es161). In Proceedings of the 2013 Vehicle Technologies Program Annual Merit Review and Peer Evaluation Meeting, Washington, DC, USA, 13–17 May 2013; p. 31.
110. 中科院宁波材料所-新型锰基正极材料研究 (Ningbo Institute of Materials, Chinese Academy of Sciences—Research on new manganese-based cathode materials)—Development of new type Mn-based cathode materials（LMO，LMP，OLO. In Proceedings of the 2013 China Forum on Technical & Market Development of Positive & Negative Materials for Advanced LIB, Tianjin, China, 8–10 April 2013; p. 26.
111. Lee, S.; Hu, H.S.; Xia, Y.G.; Xiao, F.; Liu, Z.P. Morphology-control preparation and electrochemical performance of Mn-spinel cathode materials (in Chinese). *Chin. Sci. Bull.* **2013**, *58*, 3350–3356. [CrossRef]
112. Nagai, H. Lithium-Ion Secondary Battery. US8945768B2, 6 May 2011.
113. 古滨诗. 巴斯夫正极材料的 发展和市场应用 (BASF's cathode materials Development and market application). In Proceedings of the 3rd, Advanced Batteries for xEV/ESS Conference, Wuhan, China, 11–13 November 2015; p. 20.
114. Bishop, P. An Overview of Nanomaterials for Energy Applications. In Proceedings of the EuroNanoForum 2015, Riga, Latvia, 10–12 June 2015; p. 28.
115. Liang, G. Highlights on Phostech's Recent Business Development and Advanced Life Power (R) Grade. Available online: https://www.yumpu.com/en/document/read/34821972/highlights-on-phostechs-recent-business-development-and (accessed on 14 November 2019).
116. Kweon, H.J. Iljin LMO cathode materials. In Proceedings of the SNE Research, Seoul, Republic of Korea, 27 March 2013; p. 32.

117. Campbell, S.A.; Talaie, E.; Esmaeilirad, A. Stabilized High Nickel NMC Cathode Materials for Improved Battery Performance. US20200373560A1, 5 May 2020.
118. ZTM (Xinxiang Zhongtian New Energy Technology Co., Ltd.). Available online: http://www.xxztgy.com/product (accessed on 20 March 2021).
119. 5 V High Voltage Cathode Material, Lithium Nickel Manganese Oxide, Basic Physical and Chemical Indicators. Available online: http://www.nem-cn.com/products/157 (accessed on 17 April 2019).
120. Wise, R. BASF Advanced Battery Materials for xEV and High-Performance Consumer Products. In Proceedings of the International Conference on the Frontier of Advanced Batteries, CIBF 2014, Shenzhen, China, 21 June 2014; p. 19.
121. Zhang, X.; Zhang, Y.; Liu, J.; Yan, Z.; Chen, J. Syntheses, challenges and modifications of single-crystal cathodes for lithium-ion battery. *J. Energy Chem.* **2021**, *63*, 217–229. [CrossRef].
122. LNCM-60 (Qingdao LNCM Co., LTD). In Proceedings of the China International Battery Fair (CIBF2018), Shenzhen, China, 22–24 May 2018.
123. Shahan, Z. LG Chem Has Begun Mass Production Of NCM712 Batteries In Poland. Available online: https://cleantechnica.com/2020/06/11/lg-chem-began-mass-production-of-ncm712-batteries-in-poland-in-q1/ (accessed on 18 January 2021).
124. Ternary Cathode Material. Available online: https://en.gem.com.cn/Products/info.aspx?itemid=5821 (accessed on 14 November 2024).
125. LL8313. Available online: https://en.brunp.com.cn/services-products/material-business?tp=28 (accessed on 14 November 2024).
126. Chen, Y.; Song, S.; Zhang, X.; Liu, Y. The challenges, solutions and development of high energy Ni-rich NCM/NCA LiB cathode materials. *J. Phys. Conf. Ser.* **2019**, *1347*, 012012. [CrossRef]
127. Johnson, C.S.; Kim, J.S.; Lefief, C.; Li, N.; Vaughey, J.T.; Thackeray, M.M. The significance of the Li_2MnO_3 component in 'composite' xLi_2MnO_3 center dot $(1-x)LiMn_{0.5}Ni_{0.5}O_2$ electrodes. *Electrochem. Commun.* **2004**, *6*, 1085–1091. [CrossRef]
128. Thackeray, M.M.; Kang, S.H.; Johnson, C.S.; Vaughey, J.T.; Benedek, R.; Hackney, S.A. Li_2MnO_3-stabilized $LiMO_2$ (M = Mn, Ni, Co) electrodes for lithium-ion batteries. *J. Mater. Chem.* **2007**, *17*, 3112–3125. [CrossRef]
129. 5V锰酸锂 (Lithium Manganese Oxide). Available online: http://www.qianyun-tech.com/en/Product.aspx?funId=151 (accessed on 30 January 2021).
130. 新型高电压层状三元正极材料 New Grade High Voltage Layered Three Elements Cathode Material-OLO. Available online: http://www.btrchina.com/product/53.html (accessed on 31 January 2017).
131. Burrell, A. Voltage Fade, an ABR Deep Dive Project: Status and Outcomes (Es161). In Proceedings of the DOE Annual Merit Review, Washington, DC, USA, 16–20 June 2014; p. 23.
132. How It's Made—NCM Cathode Materials for Lithium Batteries. Available online: https://www.youtube.com/watch?v=fXxlVioQrZs (accessed on 15 February 2022).

133. The Mass Production of the Cobalt-Free Cathode Material of SVOLT Officially Starts. Available online: https://www.youtube.com/watch?v=c-IKU5bf_SQ (accessed on 15 February 2022).
134. Li-Ion Battery Cathode Manufacture in Australia a Scene Setting Project. Available online: https://fbicrc.com.au/wp-content/uploads/2020/07/Li-ion-Battery-Cathode-Manufacturing-in-Aust-1.pdf (accessed on 15 February 2022).
135. Nano One's M2CAM. Available online: https://www.youtube.com/watch?v=4i1T6s_NdAQ (accessed on 16 February 2021).
136. Levasseur, S. Insights into NMC degradation processes for high energy systems: How far can we push? (Umicore). In Proceedings of the Advanced Automotive & Industrial Battery Conference Europe, AABC2017, Mainz, Germany, 31 January 2017; p. 26.
137. Julien, C.M.; Mauger, A. NCA, NCM811, and the Route to Ni-Richer Lithium-Ion Batteries. *Energies* **2020**, *13*, 6363. [CrossRef]
138. Dahn, J.; Li, H.; Zhang, N.; Liu, A.; Cormier, M. An Unavoidable Challenge for Ni-Rich Positive Electrode Materials for Li-Ion Batteries. In Proceedings of the 37th International Battery Seminar&Exhibit, Vitual, 28 July 2020; p. 27.
139. Li, H.; Cormier, M.; Zhang, N.; Inglis, J.; Li, J.; Dahn, J.R. Is Cobalt Needed in Ni-Rich Positive Electrode Materials for Lithium Ion Batteries? *J. Electrochem. Soc.* **2019**, *166*, A429–A439. [CrossRef]
140. Roth, P.E.; Doughty, D.H. In Proceedings of the Advanced automotive battery and ultracapacitor conference (AABC-06), Baltimore, MD, USA, 15–19 May 2006.
141. Doughty, D.H. *Vehicle Battery Safety Roadmap Guidance. Subcontract Report NREL/SR-5400-54404*; National Renewable Energy Laboratory: Golden, CO, USA, 2012; p. 131.
142. Lu, H.-l. The Development Status & Trends of Main Materials for the WW Power LIB. In Proceedings of the 2016'第五届中国电池市场年会, 暨第一届动力电池应用国际峰会, 第二届中国电池行业智能制造研讨会 (The 5th China Battery Market Annual Conference, the 1st International Power Battery Application Summit, and the 2nd China Battery Industry Intelligent Manufacturing Seminar) (CBEA), Beijing, China, 14–15 November 2016; p. 37.
143. Kim, U.-H.; Kuo, L.-Y.; Kaghazchi, P.; Yoon, C.S.; Sun, Y.-K. Quaternary Layered Ni-Rich NCMA Cathode for Lithium-Ion Batteries. *ACS Energy Lett.* **2019**, *4*, 576–582. [CrossRef]
144. POSCO Chemical, which Goes through L&F Cathode Material LG, to Tesla, Should also Pay Attention. Available online: https://www.world-today-news.com/%EC%98%81%EC%83%81%81-posco-chemical-which-goes-through-lf-cathode-material-lg-to-tesla-should-also-pay-attention/ (accessed on 18 January 2021).
145. Lee, S. Cosmo AM&T Developing High-Nickel Batteries with Samsung, LG. Available online: http://www.thelec.net/news/articleView.html?idxno=413 (accessed on 18 January 2021).
146. Lee, S. Eco Pro BM Develops Next-Gen NCMA Cathode. Available online: http://www.thelec.net/news/articleView.html?idxno=777 (accessed on 18 January 2021).
147. 포스코 2차전지 핵심, 양극재High~! High니켈 (The core of POSCO's secondary batteries, cathode material High-Nickel). Available online: http://product.posco.com/homepage/product/kor/jsp/news/s91w4000120v.jsp?SEQ=396 (accessed on 18 January 2021).

148. Yan, W.; Jia, X.; Yang, S.; Huang, Y.; Yang, Y.; Yuan, G. Synthesis of Single Crystal $LiNi_{0.92}Co_{0.06}Mn_{0.01}Al_{0.01}O_2$ Cathode Materials with Superior Electrochemical Performance for Lithium Ion Batteries. *J. Electrochem. Soc.* **2020**, *167*, 120514. [CrossRef]
149. Song, X.; Liu, G.; Yue, H.; Luo, L.; Yang, S.; Huang, Y.; Wang, C. A novel low-cobalt long-life $LiNi_{0.88}Co_{0.06}Mn_{0.03}Al_{0.03}O_2$ cathode material for lithium ion batteries. *Chem. Eng. J.* **2021**, *407*, 126301. [CrossRef]
150. Kim, U.-H.; Ryu, H.-H.; Kim, J.-H.; Mücke, R.; Kaghazchi, P.; Yoon, C.S.; Sun, Y.-K. Microstructure-Controlled Ni-Rich Cathode Material by Microscale Compositional Partition for Next-Generation Electric Vehicles. *Adv. Energy Mater.* **2019**, *9*, 1803902. [CrossRef]
151. Hou, P.; Zhang, H.; Zi, Z.; Zhang, L.; Xu, X. Core–shell and concentration-gradient cathodes prepared via co-precipitation reaction for advanced lithium-ion batteries. *J. Mater. Chem. A* **2017**, *5*, 4254–4279. [CrossRef]
152. Chintawar, P. Automotive grade NCM cathodes for Li ion batteries (BASF). In Proceedings of the AABC 2011, Pasadena, CA, USA, 24–28 January 2011; p. 24.
153. Lu, Y.; Zhang, Y.; Zhang, Q.; Cheng, F.; Chen, J. Recent advances in Ni-rich layered oxide particle materials for lithium-ion batteries. *Particuology* **2020**, *53*, 1–11. [CrossRef]
154. Onnerud, P.R.; Shi, J.J.; Dalton, S.L.; Lampe-Onnerud, C. Lithium Metal Oxide Materials and Methods of Synthesis and Use. 7381496, 20 May 2004.
155. Ofer, D.; Pullen, A.W.; Sriramulu, S. Polycrystalline Metal Oxide, Methods of Manufacture Thereof, and Articles Comprising the Same. 9391317 B2, 25 September 2015.
156. Rempel, J.; Pullen, A.; Kaplan, D.; Barnett, B.; Sriramulu, S. High-Nickel Cathode/ Graphite Anode Cells for Diverse DoD Applications (24-1). In Proceedings of the 48th Power Sources Conference, Marriott Tech Center Hotel, Denver, CO, USA, 11–14 June 2018; pp. 404–407.
157. Allen, J.L.; Delp, S.A.; Jow, T.R. *Evaluation of TIAX High Energy CAM-7/Graphite Lithium-Ion Batteries at High and Low Temperatures*; Army Research Laboratory: Adelphi, MD, USA, 2014; p. 20.
158. Rempel, J. High Energy High Power Battery Exceeding PHEV-40 Requirements (ES209). In Proceedings of the 2014 DOE VTP Merit Review, Washington, DC, USA, 18 June 2014; p. 21.
159. Rempel, J. High Energy High Power Battery Exceeding PHEV-40 Requirements (ES209). In Proceedings of the 2015 DOE VTP Merit Review, Washington, DC, USA, 10 June 2015; p. 29.
160. Rempel, J. High Energy High Power Battery Exceeding PHEV-40 Requirements (ES209). In Proceedings of the 2016 DOE VTP Merit Review, Washington, DC, USA, 7 June 2016; p. 25.
161. Rempel, J. Materials Development for High Energy High Power Battery Exceeding PHEV-40 Requirements (ES260). In Proceedings of the 2015 DOE VTP Merit Review, Washington, DC, USA, 10 June 2015; p. 31.

162. Rempel, J.; Ofer, D.; Pullen, A.; Kaplan, D.; Barnett, B.; Sriramulu, S. High Energy Li-ion Cells with CAM-7® Cathode and Si-based Anodes (24-2). In Proceedings of the 48th Power Sources Conference, Marriott Tech Center Hotel, Denver, CO, USA, 11–14 June 2018; pp. 408–411.
163. Ofer, D.; Kaplan, D.; Yang, C.; Dalton-Castor, S.; McCoy, C.; Kotwal, T.; Rutberg, R.; Barnett, B.; Sriramulu, S. CAM-7®/LTO Lithium-Ion Technology for Robust, High-Power Vehicle Batteries (33-2). In Proceedings of the 48th Power Sources Conference, Marriott Tech Center Hotel, Denver, CO, USA, 11–14 June 2018; pp. 561–564.
164. Ofer, D.; Menard, M.; McCoy, C.; Kaplan, D.; Yang, C.; Barnett, B.; Sriramulu, S. CAM-7®/LTO Lithium-Ion Technology for Structurally Integrated Vehicle Batteries (33-1). In Proceedings of the 48th Power Sources Conference, Marriott Tech Center Hotel, Denver, CO, USA, 11–14 June 2018; pp. 557–560.
165. Joint News Release. 12 April 2016. Available online: https://www.basf.com/global/en/media/news-releases/2016/04/p-16-177 (accessed on 14 November 2024).
166. Johnson Matthey Obtains License for the GEMXTM Advanced Battery Material Platform from CAMX Power. 21 November 2018. Available online: https://matthey.com/documents/161599/162236/JM-obtains-license-for-GEMX.pdf/f79a061f-4ea3-c9eb-7912-e804ee7623ba?t=1650968128837 (accessed on 14 November 2024).
167. Johnson Matthey Obtains License for the GEMX™ Advanced Battery Material Platform from CAMX Power. Available online: http://www.thebatteryshow.eu/resources/news-and-editorial/news-container/2018/12/03/johnson-matthey-obtains-license-for-the-gemx%E2%84%A2-advanced-battery-material-platform-from-camx-power/ (accessed on 12 March 2019).
168. Sahin, K. GEMX Cathode Platform, Applications (gLNO, gNMC, gNCA), and Extensions. In Proceedings of the Advanced Automotive Battery Conference (AABC USA), Virtual, 3 November 2020; p. 27.
169. Sahin, K. Crystallite Surface Engineering in Polycrystalline High Nickel Cathode Materials. In Proceedings of the Advanced Automotive Battery Conference (AABC 2021 Europe), Virtual, 19 January 2021; p. 22.
170. Nano One's Coated Single Crystal Cathode Materials Explained. Available online: https://www.youtube.com/watch?v=Z73dTNTkgpA&feature=emb_logo (accessed on 19 January 2021).
171. 陈彦彬 (Chen Yanbin); 刘亚飞 (Liu Yafei); 张学全 (Zhang Xuequan). 高能量密度动力锂电正极三元材料的研究进展 (Research progress on high energy density power lithium battery cathode ternary materials) (Easpring). In Proceedings of the 2016'第五届中国电池市场年会, 暨第一届动力电池应用国际峰会, 第二届中国电池行业智能制造研讨会 (The 5th China Battery Market Annual Conference, the 1st International Power Battery Application Summit, and the 2nd China Battery Industry Intelligent Manufacturing Seminar) (CBEA), Beijing, China, 14 November 2016; p. 56.
172. Gao, J.-K. Positive material requirement of EDV Li-ion batteries from LFP to LMO/NCM to Li-rich Mn based material, etc. In Proceedings of the China Forum on LEMD, Tianjin, China, 8–10 April 2013; p. 40.

173. QY-901G（单晶型 (Single crystal form)). Available online: http://www.qianyun-tech.com/Product.aspx?funId=115&typeId=674&id=3183 (accessed on 30 January 2021).
174. Collins, N. Rising to the Technical Challenges of Automotive LiB Applications with Johnson Matthey. In Proceedings of the Advanced Automotive Battery Conference (AABC 2021 Europe), Virtual, 20 January 2021; p. 7.
175. Lu, H.-S. Commercialized Technology and Future Market Development Trends of Electrolyte & Separator for LIB. In Proceedings of the Third International Forum on Electrolyte & Separator Materials for Advanced Batteries, Nanjing, China, 18–19 October 2018; p. 22.
176. Development and Application of CALB Olivine-Phosphate Batteries. Available online: https://www.eiseverywhere.com/file_uploads/fe4ca0e2cc1e74c003c8cfe731dc05e2_O11.03_Fangfang_Pan.pdf (accessed on 30 January 2017).
177. Yaxhou, X. Latest technical development and application of CALB NCM LFMP battery. In Proceedings of the 11th China International Battery Fair 2014 for Energy/Green Power, Shenzhen, China, 18–20 June 2014; p. 17.
178. Lu, H.-l. An Update on the Chinese xEV Market, and a Technical Comparison of its Batteries (ITRI IEK). In Proceedings of the World Mobility Summit 2016, Munich, Germany, 17–19 October 2016.
179. Hao, R. Discoveries and Understanding of Materials and Cell Chemistry for xEV Batteries. In Proceedings of the Advanced Automotive Battery Conference (AABC 2021 Europe), Virtual, 20 January 2021; p. 21.
180. LFP系列 高容量、高压实(Series high capacity, high compaction) (BTR). Available online: https://www.btrchina.com/product/88.html (accessed on 30 January 2021).
181. $LiMn_2O_4$ QY-103G 高比容量 长循环寿命(High specific capacity and long cycle life). Available online: http://www.qianyun-tech.com/Product.aspx?funId=85&typeId=0&id=3496 (accessed on 30 January 2021).
182. $LiMn_2O_4$ Qy-101. Available online: http://www.qianyun-tech.com/Product.aspx?funId=85&typeId=0&id=3001 (accessed on 30 January 2021).
183. Jansen, A.N. Cell Fabrication Facility Team Production and Research Activities (Es030). In Proceedings of the 2013 U.S. DOE Vehicle Technologies Program Annual Merit Review and Peer Evaluation Meeting, Washington, DC, USA, 14 May 2013; p. 28.
184. Zheng, J.M.; Myeong, S.J.; Cho, W.R.; Yan, P.F.; Xiao, J.; Wang, C.M.; Cho, J.; Zhang, J.G. Li- and Mn-Rich Cathode Materials: Challenges to Commercialization. *Adv. Energy Mater.* **2017**, *7*, 25. [CrossRef]
185. Mn-Based Multi-Element Composite Cathode Material Series (FL-1). Available online: https://www.btrchina.com/en/product/146.html (accessed on 8 February 2021).
186. 「先進・革新蓄電池材料評価技術開発(第 2 期)」 (中間評価)分科会 資料 (Advanced and innovative storage battery material evaluation technology development (2nd period)" (interim evaluation) subcommittee materials) 7–1 (NEDO). Available online: https://www.nedo.go.jp/content/100927228.pdf (accessed on 14 March 2021).
187. Yoshio, M.; Gunawardhana, N. Modern graphite, electrolyte and its additives (Saga University). In Proceedings of the Shenzhen International Li-ion Battery Summit, Shenzhen, China, 3–6 December 2011; p. 91.

188. Harada, Y.; Ise, K.; Takami, N. チタンニオブ酸化物負極による高容量化で超急速充電が可能な次世代リチウムイオン二次電池 SCiB™ (Next-Generation SCiB™ Lithium-Ion Rechargeable Battery with High-Capacity Titanium-Niobium-Oxide Anode Capable of Ultrahigh-Speed Charging). Technology Administration & Planning Office, Corporate Technology Planning Division, Toshiba Corporation, 1-1, Shibaura 1-chome, Minato-ku, Tokyo 105-8001, Japan. *Toshiba Rev.* **2018**, *73*, 4–8.
189. Pan, L. Challenges of Wearable Batteries. In Proceedings of the 37th Annual International Battery Seminar & Exhibition, Virtual, 29–30 July 2020; p. 37.
190. IIT LIB Related Study Program 07–08 (September 2007), Chapter 3 Power Tool Market Trends/LIB cells, materials Market Bulletin (07Q3). 2007, pp. 56–91. Available online: https://wenku.baidu.com/view/d8a2d02d7375a417866f8f91?_wkts_=1715775728737&needWelcomeRecommand=1 (accessed on 15 May 2024).
191. Hitachi Chemical Anode Materials (SCM Hypnergy Material Tech. Co., Ltd.). Available online: http://www.scmbattery.com/en/product_show.php?id=100 (accessed on 30 July 2018).
192. 人造石墨 (Artificial graphite). Available online: https://www.shanshantech.com/artificial-graphite (accessed on 14 November 2024).
193. China steel chemical company brochure (Battery Japan 2019). Available online: https://app.box.com/s/ll3naz3x2ca6aq3t9rsh42arj0rtobw5 (accessed on 20 May 2021).
194. SCMG (Structure Controlled Micro Graphite) Showa Denko. In Proceedings of the China International Battery Fair (CIBF2018), Shenzhen, China, 22–24 May, 2018; p. 1.
195. 高端天然石墨918系列产品 High Grade Natural Graphite 918 Series (BTR). Available online: http://www.btrchina.com/product/36.html (accessed on 16 February 2018).
196. 제품. Available online: http://www.poscochemtech.com/kr/sub.do?MENU_SEQ=131&PAGE_SEQ=74&LANG=ko_KR (accessed on 30 August 2019).
197. 天然石墨MSG、518系列产品, BTR (MSG18, 518, MSG-H). Available online: http://www.btrchina.com/product/38.html (accessed on 16 February 2018).
198. High-performance Anode Material. CARBOTRON® P. *Kureha Battery Materials Japan Co., Ltd.* Available online: http://www.kureha.co.jp/development/story/pdf/catalog_hc_eg_20120924.pdf (accessed on 2 June 2016).
199. SC/HC Series Products (BTR). Available online: http://www.btrchina.com/product/44.html (accessed on 30 July 2018).
200. Anode Material for Lithium-Ion Battery Kuranode™ Biohardcarbon. Available online: http://www.kuraray-c.co.jp/KURANODE/en.html (accessed on 3 June 2019).
201. 纳米钛酸锂系列(Nano lithium titanate series) LTO Series. Available online: http://www.btrchina.com/product/47.html (accessed on 2 June 2016).
202. Tech, S. Progress on Next-generation anode materials for LIB (Shanshan Tech). In Proceedings of the China Li-Ion Battery, Positive, Negative Materials Technology and Market Development forum, Tianjin, China, 8–10 April 2013; p. 49.
203. 日立化成氧化硅素（SiO）技术优势 (Hitachi Chemical's technical advantages of silica oxide (SiO)). Available online: http://www.scmbattery.com/cn/product_show.php?id=100 (accessed on 30 July 2018).

204. New SiO Based Anode Material (BTR). Available online: http://www.btrchina.com/product/46.html (accessed on 30 July 2018).
205. New Type High Capacity Si Based Composite Materials (BTR). Available online: http://www.btrchina.com/product/45.html (accessed on 30 July 2018).
206. Nishida, T. The Limitations of Anode Material and Future Development Direction (Hitachi Chemical). In Proceedings of the Asia-Pacific Lithium Battery Congress 2014, Shenzhen, China, 26–28 March 2014; p. 38.
207. Showa Denko Advanced Battery Materials. Available online: http://www.sdk.co.jp/assets/files/english/rd/gijutsu2015_09_e.pdf (accessed on 14 June 2023).
208. 陈韦志博士(Dr. Chen Weizhi); 颜瑞宾博士 (Yan Ruibin). Mesophase Graphite Materials for Power Li-Ion Batteries (CSSC). In Proceedings of the 年第3届中国（武汉）锂电新能源产业国际高峰论坛（ABEC 2015，锂电"达沃斯"）(The 3rd China (Wuhan) Lithium Battery New Energy Industry International Summit Forum (ABEC 2015, Lithium Battery "Davos"), Wuhan, China, 11–13 November 2015; p. 33.
209. Yoshio, M.; Wang, H.Y.; Fukuda, K.; Hara, Y.; Adachi, Y. Effect of carbon coating on electrochemical performance of treated natural graphite as lithium-ion battery anode material. *J. Electrochem. Soc.* **2000**, *147*, 1245–1250. [CrossRef]
210. Yoshio, M.; Yoshitaki, H. Safety issue of graphite anode in lib for EV (HEV). In Proceedings of the IBA 2013 meeting, Barcelona, Spain, 10–15 March 2013; p. 27.
211. Heimes, H.H.; Kampker, A.; von Hemdt, A.; Kreisköther, K.D.; Michaelis, S.; Rahimzei, E. Manufacturing of Lithium-Ion Battery Cell Components. Available online: https://www.pem.rwth-aachen.de/global/show_document.asp?id=aaaaaaaaaffpnvh (accessed on 2 May 2021).
212. Sharova, V.; Wolff, P.; Konersmann, B.; Ferstl, F.; Stanek, R.; Hackmann, M. Evaluation of Lithium-Ion Battery Cell Value Chain. Available online: https://www.econstor.eu/bitstream/10419/217243/1/hbs-fofoe-wp-168-2020.pdf (accessed on 15 February 2022).
213. Zheng, Z.; Deng, T.; Nie, Y.; Zhou, X. Graphite Negative Material for Lithium-Ionbattery, Method for Preparing the Same and Lithium-On Battery. US9281521 B2, 12 April 2013.
214. Zheng, Zh. Why Fast Charging Technology Is Prefered in EV, PHEV, and HEV? Available online: https://slidesplayer.com/slide/11647867/ (accessed on 14 November 2024).
215. Mattis, W.L. The Development of Safe, Fast Charging and Superb Battery Cycle Life Technology For xEV Applications. In Proceedings of the 2nd HEV Market & Advanced Battery Technology development Seminar, Hangzhou, China, 14 October 2016; p. 29.
216. Microvast Solutions for HEV and PHEV. Available online: https://wenku.baidu.com/view/df931c2d9b6648d7c1c74689.html (accessed on 19 May 2019).
217. Spahr, M.E. Carbonaceous Anode Materials and Conductive Additives as Key To Higher Performance Lithium-ion Batteries. In Proceedings of the 7th International Automotive Battery Conference Europe, Mainz, Germany, 31 January 2017; p. 25.
218. Auguie, D.; Oberlin, M.; Oberlin, A.; Hyvernat, P. Microtexture of mesophase spheres as studied by high-resolution conventional transmission electron-microscopy (CTEM). *Carbon* **1980**, *18*, 337–346. [CrossRef]

219. Yen, J.-P.; Lee, C.-M. Development of advanced anode materials for power Li-ion batteries (China Steel Chemical Corporation). In Proceedings of the 2nd International Forum on Positive & Negative Materials, Hangzhou, China, 22–24 April 2018; p. 23.
220. Chen, Y.-C. CSCC's Recent Development in Mesophase Graphite. In Proceedings of the 12th China International Battery Fair (CIBF2016), Shenzhen, China, 24–26 May 2016; p. 32.
221. Lee, J.-J. "Soft Carbon" as a LIB anode material for xEV application. In Proceedings of the Korea Advanced Battery Conference, Seoul, Republic of Korea, 29 May 2013; p. 24.
222. Ulmann, P.A. Carbon-Based Negative Electrode Active Materials for Lithium-Ion Batteries Past, Present and Trends towards the Future. In Proceedings of the MAT4BAT Summer School at EIGSI, La Rochelle, France, 2 June 2015; p. 15.
223. Kanno, R. 構造からみたリチウム電池電極材料 (Structural Aspects of Materials for Lithium Battery Electrodes). *GS Yuasa Tech. Rep.* **2006**, *7*, 1–11.
224. Wachtler, M. Anode Materials (ZSW). In Proceedings of the Li-ion batteries Lecture Winter Term 2016/17, Ulm, Germany, 7–21 November 2016; p. 108.
225. Mochida, I.; Ku, C.H.; Korai, Y. Anode performance and insertion mechanism of hard carbons prepared from synthetic isotropic pitches. *Carbon* **2001**, *39*, 399–410. [CrossRef]
226. 新型材料 (New Materials). Available online: https://www.shanshantech.com/new-type-materials (accessed on 14 November 2024).
227. Imoto, H. The development of non-graphitezable carbon "Carbotron P" for automotive lithium-ion batteries (KMBJ). In Proceedings of the Advanced automotive battery conference Europe (AABC), Strasbourg, France, 24–28 June 2013; p. 19.
228. Gao, J. Progress of xEv LIB battery technology and its requirement on electrolyte development. In Proceedings of the 中国锂电池电解质、隔膜材料技术与 市场发展论坛 (China Lithium Battery Electrolyte and Separator Material Technology and Market Development Forum), Guangzhou, China, 14–16 August 2013; p. 39.
229. Pfaffmann, L.; Birkenmaier, C.; Muller, M.; Bauer, W.; Mitsch, T.; Feinauer, J.; Kramer, Y.; Scheiba, F.; Hintennach, A.; Schleid, T.; et al. Investigation of the electrochemically active surface area and lithium diffusion in graphite anodes by a novel OsO4 staining method. *J. Power Sources* **2016**, *307*, 762–771. [CrossRef]
230. 吴羽化工硬碳负极材料 (Chemical Hard Carbon Anode Material)CARBOTRON P. Available online: http://www.achatestrade.com/product/276705278 (accessed on 7 August 2018).
231. 低膨胀、长循环天然石墨(Low expansion, long cycle natural graphite). Available online: https://www.btrchina.com/product/79.html (accessed on 2 February 2021).
232. The Advantage of Hard Carbon for Cold Cranking& Rapid Charging Requirement (Sumitomo Bakelite). In Proceedings of the Second International Forum on Cathode & Anode Materials for Advanced Batteries, Hangzhou, China, 22–24 April 2015; p. 23.
233. Saito, T.; Fujiwara, H.; Abe, Y.; Kumagai, S. Hard Carbon/SiO_x Composite Active Material Prepared from Phenolic Resin and Rice Husk for Li-ion Battery Negative Electrode. *Int. J. Soc. Mater. Eng. Resour.* **2018**, *23*, 142–146. [CrossRef]

234. Izawa, T. Development of High Performance Carbon Anode Material (Kuraray). In Proceedings of the Advanced Automotive & Industrial Battery Conference Europe, AABC2017, Mainz, Germany, 31 January 2017; p. 30.
235. Li-ion電池負極 (battery negative electrode) (I). Available online: http://carbon.cm.kyushu-u.ac.jp/handout/emef/20151021.pdf (accessed on 6 November 2018).
236. Blumrich, P. Opportunity Charging EV Concept for Urban Transport and Advantage of Toshiba's LTO Technology (SCiB™). In Proceedings of the Advanced Automotive Battery Conference (AABC), Mainz, Germany, 1 February 2017; p. 21.
237. Ziebarth, B.; Klinsmann, M.; Eckl, T.; Elsasser, C. Lithium diffusion in the spinel phase $Li_4Ti_5O_{12}$ and in the rocksalt phase $Li_7Ti_5O_{12}$ of lithium titanate from first principles. *Phys. Rev. B* **2014**, *89*, 7. [CrossRef]
238. Zhang, N. Application Research on Lithium-ion Power Battery for Electric Vehicle (Lishen). In Proceedings of the China-Korea Lithium Battery Workshop, Shenzhen, China, 4–6 June 2009; p. 57.
239. Yang, X.; Miller, T. Fast Charging Lithium-Ion Batteries. *SAE Tech. Pap.* **2017**, 2017-2001-1204. [CrossRef]
240. Characteristics of SCiB™ cells. Available online: http://www.scib.jp/en/product/cell.htm (accessed on 15 May 2019).
241. SLBとは (About "SLB"). Available online: https://www.nichicon.co.jp/products/slb/about/ (accessed on 14 November 2024).
242. Li, T. Coslight Li-ion pouch power cell solution. In Proceedings of the 34th International Battery Seminar & Exhibit 2017, Ft. Lauderdale, FL, USA, 21 March 2017; p. 25.
243. Zhang, N. "开放、协同、共享" 力神动力电池创新发展(Open, collaborative, and shared" Lishen Power Battery Innovation and Development) (Lishen). In Proceedings of the 2016 年第一届动力电池应用国际峰会 (The First International Summit on Power Battery Applications), Beijing, China, 14 November 2016; p. 38.
244. Zhang, N. Lishen High Power Battery Technology for Start-stop. In Proceedings of the 2nd HEV Market & Advanced Battery Technology Development Seminar, Hangzhou, China, 13–14 October 2016; p. 30.
245. Manev, V.; Coleman, M. Large Temperature Range PHEV and Start-Stop Li4Ti5O12 Batteries. In Proceedings of the HEV Market & Advanced Battery Technology Development Seminar, Beijing, China, 18 April 2014; p. 30.
246. 高功率钛酸锂电池 在HEV与PHEV上的应用—钛酸锂负极的性能和应用前景 柳俊 (Application of high-power lithium titanate batteries in HEV and PHEV—Performance and application prospects of lithium titanate anode, Liu Jun) (Enerdel). In Proceedings of the 2017 第一届钛酸锂电池产业链专题讨论会 (The first lithium titanate battery industry chain symposium) (LBIF17), Wuxi, China, 7 July 2017; p. 28.
247. Khare, N. Battery Storage for a Stable Microgrid (Leclanche). In Proceedings of the IEEE Industrial Electronics Society (Swiss Chapter) Workshop on "Trends in Microgrid Applications", Zurich, Switzerland, 27 March 2018; p. 22.

248. Takami, N.; Kosugi, S.; Honda, K. New SCiB[TM] High-Safety Rechargeable Battery for HEV Application. Technology Administration & Planning Office, Corporate Technology Planning Division, Toshiba Corporation, 1-1, Shibaura 1-chome, Minato-ku, Tokyo 105-8001, Japan. 東芝レビュ *(Toshiba Rev.)* **2008**, *63*, 54–57.
249. Manev, V.; Dahlberg, K.; Gopu, S.; Cochran, S. 12V Start-Stop and 48V Mild Hybrid LMO-LTO Batteries (Xalt). In Proceedings of the 35th International Battery Seminar & Exhibit, Ft. Lauderdale, FL, USA, 28 March 2018; p. 20.
250. STALLION Project: "Safety testing approaches for large Lithium-ion battery systems"(1st of October 2012 to 31st of March 2015). Available online: https://cordis.europa.eu/docs/results/308/308800/final1-final-report-v3.pdf (accessed on 19 May 2019).
251. LecCell 30Ah High Energy Lithium Nickel Cobalt Oxide (NCO)—Titanate (LTO) System. Available online: http://blizzard.cs.uwaterloo.ca/iss4e/wp-content/uploads/2019/04/leccell-30ah-high-energy.pdf (accessed on 14 June 2023).
252. Our Technologies. Available online: https://celltechsolutions.fi/products/our-technologies/ (accessed on 14 November 2024).
253. Zhiyuan, T.; Sun, L.; Wang, Q.; Ling, G. Cylindrical Single-Piece Lithium-Ion Battery of 400Ah and Its Preparation Method. US9502736B2, 22 November 2016.
254. Tang, Z. 唐. 圆柱形超大容量单体 400Ah 钛酸锂/磷酸铁锂电池的产业化 (Industrialization of cylindrical ultra-large capacity single 400Ah lithium titanate/lithium iron phosphate battery) (Tianjin Institute). In Proceedings of the 2017第一届钛酸锂电池产业链专题讨论会 (The first lithium titanate battery industry chain symposium) (LBIF17), Wuxi, China, 7 July 2017; p. 77.
255. Kokam Lithium Polymer Cells Brochure. Available online: http://kokam.com/wp-content/uploads/2016/03/SLPB-Cell-Brochure.pdf (accessed on 28 February 2018).
256. Anode SBR Blinder for Lithium Ion Batteries (JSR, TRD102A, TRD104A, TRD2102). Available online: http://www.scmbattery.com/en/product_show.php?id=99 (accessed on 8 August 2018).
257. Smart Energy Week 2017 日本現場報導 系列一(Japan Live Report Series 1). Available online: https://www.materialsnet.com.tw/DocView.aspx?id=24918 (accessed on 7 November 2018).
258. Yen, J.P.; Chang, C.C.; Lin, Y.R.; Shen, S.T.; Hong, J.L. Effects of Styrene-Butadiene Rubber/Carboxymethylcellulose (SBR/CMC) and Polyvinylidene Difluoride (PVDF) Binders on Low Temperature Lithium-Ion Batteries. *J. Electrochem. Soc.* **2013**, *160*, A1811–A1818. [CrossRef]
259. Lin, J. BAK xEV Battery R&D and Industrialization. In Proceedings of the 2014 U.S.-China Electric Vehicle and Battery Technology Workshop, Seattle, WA, USA, 18 August 2014; p. 53.
260. WQA-438, WPA-409, WRA-268 Binder (Gredmann). Available online: https://wenku.baidu.com/view/d54f1b759a6648d7c1c708a1284ac850ad0204e1 (accessed on 18 April 2019).
261. Solef PVDF Typical Properties EN-229548. Available online: https://www.solvay.cn/en/binaries/Solef-PVDF-Typical-Properties_EN-229548.pdf (accessed on 8 August 2018).

262. Qin, Y.; Chen, Z.; Belharouak, I.; Amine, K. Mechanism of LTO Gassing and potential solutions (es112). In Proceedings of the 2011 DOE Hydrogen and Fuel Cells Program, and Vehicle Technologies Program Annual Merit Review and Peer Evaluation, Washington, DC, USA, 9–13 May 2011; p. 18.
263. Wu, K.; Yang, J.; Zhang, Y.; Wang, C.Y.; Wang, D.Y. Investigation on $Li_4Ti_5O_{12}$ batteries developed for hybrid electric vehicle. *J. Appl. Electrochem.* **2012**, *42*, 989–995. [CrossRef]
264. Belharouak, I.; Koenig, G.M.; Tan, T.; Yumoto, H.; Ota, N.; Amine, K. Performance Degradation and Gassing of $Li_4Ti_5O_{12}/LiMn_2O_4$ Lithium-Ion Cells. *J. Electrochem. Soc.* **2012**, *159*, A1165–A1170. [CrossRef]
265. Bernhard, R.; Meini, S.; Gasteiger, H.A. On-Line Electrochemical Mass Spectrometry Investigations on the Gassing Behavior of $Li_4Ti_5O_{12}$ Electrodes and Its Origins. *J. Electrochem. Soc.* **2014**, *161*, A497–A505. [CrossRef]
266. Yasuda, T.; Koiwa, K.; Soda, M. SCiB™ Technology and Its Potential Military Applications. In Proceedings of the 48th Power Sources Conference, Denver, CO, USA, 11–14 June 2018; pp. 174–177.
267. He, Y.B.; Li, B.H.; Liu, M.; Zhang, C.; Lv, W.; Yang, C.; Li, J.; Du, H.D.; Zhang, B.A.; Yang, Q.H.; et al. Gassing in $Li_4Ti_5O_{12}$-based batteries and its remedy. *Sci. Rep.* **2012**, *2*, 9. [CrossRef]
268. Lin, J. Electrolyte research for LTO system (BAK). In Proceedings of the 2013 China Forum On LIB Electrolyte and Separator Technology and Market Development, Guangzhou, China, 15 August 2013; p. 32.
269. Yamate, S.; Kozono, S.; Ohkubo, K.; Katayama, Y.; Nukuda, T.; Murata, T. チタン酸リチウム負極に有機・無機二成分表面被膜 を電気化学的に形成したリチウムイオン電池の 高温寿命性能 (High Temperature Life Performance for Lithium-ion Battery Using Lithium Titanium Oxide Negative Electrode with Electrochemically Formed Surface Film Comprising Organic-Inorganic Binary Constituents). *GS Yuasa Tech. Rep.* **2009**, *6*, 27–31.
270. Tong, T.; Groesbeck, C. 10 Minute Lto Ultrafast Charge Public Transit Ev Bus Fleet Operational Data—Analysis of 240,000 km, 6 Bus Fleet Shows Viable Solution. *World Electr. Veh. J.* **2012**, *5*, 261–268. [CrossRef]
271. 任建国. 高性能微纳结构钛酸锂材料的开发和性能研究 (BTR) (Ren Jianguo. Development and performance research of high-performance micro-nanostructure lithium titanate materials). In Proceedings of the First LTO Battery Industrial Forum (LBIF2017), Wuxi, China, 6–7 July 2017; p. 22.
272. 王驰伟. 钛酸锂电池技术开发进展 (JEVE, 天津市捷威动力工业有限公司) (Wang Chiwei. Development progress of lithium titanate battery technology (JEVE, Tianjin Jiewei Power Industry Co., Ltd.). In Proceedings of the First LTO Battery Industrial Forum (LBIF2017), Wuxi, China, 6–7 July 2017; p. 20.
273. Lu, H.L. Recent Status & Future Outlook of World LTO Anode Material and Related LIBs (ITRI IEK). In Proceedings of the First LTO Battery Industrial Forum (LBIF2017), Wuxi, China, 6–7 July 2017; p. 27.
274. Lee, H. The future of new energy vehicles and automotive batteries (Samsung). In Proceedings of the 13th China International Battery Fair (CIBF2018), Shenzhen, China, 22–24 May 2018; p. 19.

275. Nagai, R.; Kita, F.; Yamada, M.; Katayama, H. Development of Highly Reliable High-capacity Batteries for Mobile Devices and Small- to Medium-sized Batteries for Industrial Applications. *Hitachi Rev.* **2011**, *60*, 28–32.
276. 日立化成 氧化硅 (Hitachi Chemical Silicon Oxide) SiO. Available online: http://www.achatestrade.com/product/277847378 (accessed on 27 October 2018).
277. Park, J.; Park, S.S.; Won, Y.S. In situ XRD study of the structural changes of graphite anodes mixed with SiOx during lithium insertion and extraction in lithium-ion batteries. *Electrochim. Acta* **2013**, *107*, 467–472. [CrossRef]
278. 岳敏 (Yue Min); 闫慧青 (Yan Huiqing). 快充类负极材料研究探讨 (Research on Fast Charged Anode Materials, BTR). In Proceedings of the 2016'第五届中国电池市场年会, 暨第一届动力电池应用国际峰会, 第二届中国电池行业智能制造研讨会 (The 5th China Battery Market Annual Conference, the 1st International Power Battery Application Summit, and the 2nd China Battery Industry Intelligent Manufacturing Seminar) (CBEA), Beijing, China, 14–15 November 2016; p. 30.
279. 新型负极材料的发展现状及LIB体系影响 (Development status of new anode materials and impact on LIB system) (BTR). Available online: http://www.lidianshijie.com/jishu/201712/11/3679.html (accessed on 19 May 2019).
280. SILO Silicon™ Silicon Anode Material. Available online: https://www.paracleteenergy.com/silicon-anode-material/ (accessed on 14 November 2024).
281. 动力电池的创新与发展 (Innovation and development of power batteries) (Yinlong). In Proceedings of the First LTO Battery Industrial Forum (LBIF2017), Wuxi, China, 6 July 2017; p. 41.
282. Xiao, C. 中国车用动力电池发展现状及趋势 (Development status and trends of vehicle power batteries in China). In Proceedings of the 13th China International Battery Fair (CIBF2018), Shenzhen, China, 22 May 2018; p. 36.
283. Laminated Lithium Ion Battery (Hitachi Maxell). Available online: https://wenku.baidu.com/view/a9a6c04fe518964bcf847c42.html?from=search (accessed on 20 July 2018).
284. IIT LIB-Related Study Program 10–11 (August 2010). p. 23. Available online: https://wenku.baidu.com/view/4edc5bed0975f46527d3e14e.html?_wkts_=1715778356718&needWelcomeRecommand=1 (accessed on 15 May 2024).
285. Yuan, Q.F.; Zhao, F.G.; Zhao, Y.M.; Liang, Z.Y.; Yan, D.L. Evaluation and performance improvement of Si/SiOx/C based composite as anode material for lithium ion batteries. *Electrochim. Acta* **2014**, *115*, 16–21. [CrossRef]
286. Yuan, Q.F.; Zhao, F.G.; Zhao, Y.M.; Liang, Z.Y.; Yan, D.L. Reason analysis for Graphite-Si/SiOx/C composite anode cycle fading and cycle improvement with PI binder. *J. Solid State Electrochem.* **2014**, *18*, 2167–2174. [CrossRef]
287. Yoshida, S.; Okubo, T.; Masuo, Y.; Oba, Y.; Shibata, D.; Haruta, M.; Doi, T.; Inaba, M. High Rate Charge and Discharge Characteristics of Graphite/SiOx Composite Electrodes. *Electrochemistry* **2017**, *85*, 403–408. [CrossRef]
288. Hirose, T.; Kohta, T.; Matsuno, T.; Osawa, Y.; Furuya, M.; Sakai, R.; Matsui, C.; Koide, H. Comparison of the Structure and Phase Changes of Carbon-Coated SiO and Li-Doped Carbon-Coated SiO During Repeated Charge–Discharge Cycling. *J. Electrochem. Soc.* **2020**, *167*, 120523. [CrossRef]

289. Reynier, Y.; Vincens, C.; Leys, C.; Amestoy, B.; Mayousse, E.; Chavillon, B.; Blanc, L.; Gutel, E.; Porcher, W.; Hirose, T.; et al. Practical implementation of Li doped SiO in high energy density 21700 cell. *J. Power Sources* **2020**, *450*, 10. [CrossRef]
290. Kamo, H.; Hirose, T.; Nishiura, N.F. Method for Producing Negative Electrode Active Material for Nonaqueous Electrolyte Secondary Batteries and Method for Producing Negative Electrode for Nonaqueous Electrolyte Secondary Batteries. EP3444877A1, 03 April 2017.
291. Hirose, T.; Kamo, H. Negative Electrode Active Material, Mixed Negative Electrode Active Material, and Method for Producing Negative Electrode Active Material. EP3467913A1, 10 April 2017.
292. Xin, F.X.; Whittingham, M.S. Challenges and Development of Tin-Based Anode with High Volumetric Capacity for Li-Ion Batteries. *Electrochem. Energy Rev.* **2020**, *3*, 643–655. [CrossRef]
293. Todd, A.D.W.; Ferguson, P.P.; Fleischauer, M.D.; Dahn, J.R. Tin-based materials as negative electrodes for Li-ion batteries: Combinatorial approaches and mechanical methods. *Int. J. Energy Res.* **2010**, *34*, 535–555. [CrossRef]
294. Wolfenstine, J.; Allen, J.L.; Read, J.; Foster, D. *Chemistry and Structure of Sony's Nexelion Li-ion Electrode Materials (ARL-TN-0257)*; USA Army Research Laboratory: Adelphi, MD, USA, 2006; p. 12.
295. Kubota, T. Solvent-Dependent Solid Electrolyte Interphases on Nongraphite Electrodes. In Proceedings of the International Conference on the Frontier of Advanced Batteries (CIBF2014), Shenzhen, China, 18–20 June 2014; p. 31.
296. *차세대 양극재 기술동향 및 전망 (Next-generation cathode material technology trends and prospects)*; 2013.03.27/SNE Research; Battery Material Biz Division ECOPRO CO., Ltd.: Seoul, Republic of Korea, 2013; p. 33.
297. Zhang, J.; Fang, W. Kinetics of the EDV Li-ion Cells on Electrode Thickness, Porosity and Tortuosity (Asahi Kasei). In Proceedings of the International Battery Seminar, Fort Lauderdale, FL, USA, 21–22 March 2017; p. 35.
298. Abraham, D. Revealing Aging Mechanisms in Lithium-Ion Cells. In Proceedings of the International Battery Seminar, Fort Lauderdale, FL, USA, 22 March 2017; p. 19.
299. Lu, W.; Su, X.; Jansen, A.; Dees, D. High Rate Lithium Ion Battery Development. In Proceedings of the 34th Annual International Battery Seminar &Exhibit, Fort Lauderdale, FL, USA, 20–23 March 2017; p. 20.
300. Lei, H.; Blizanac, B.; Atanassova, P.; DuPaskier, A.; Oljaca, M. Carbon materials to Advance Li-ion and Lead-acid Battery Performance. In Proceedings of the CIBF 2014, International Conference on the Frontier of Advanced Batteries, Shenzhen, China, 20–22 June 2014; p. 29.
301. Lej, H.; Blizanac, B.; Koehlert, K.; Atanassova, P.; Oljaca, M.; Romney, G. Carbon Additives for Advanced Batteries. In Proceedings of the 10th China International Battery Fair (CIBF2012), Shenzhen, China, 20–22 June 2012; p. 25.
302. Spahr, M.E.; Bing, Y. Carbon conductive additives for lithium-ion batteries. In Proceedings of the CIBF2005, Shenzhen, China, 3 April 2005; p. 42.

303. Yoda, A. Newly Developed Denka Acetylene Black to Replace Carbon-Nanotubes in xEV LiB Cells? In Proceedings of the Advanced Automotive Battery Conference (AABC 2021 Europe), Virtual, 20 January 2021; p. 32.
304. Lei, H.; DuPasquier, A.; Korchev, A.; Laxton, P.; Atanassova, P.; Oljaca, M.; Smith, G. New carbon Materials for Advance Batteries (Cabot). In Proceedings of the CIBF2016 International Conference on the Frontier of Advanced Batteries, Shenzhen, China, 24–26 May 2016.
305. Li, R.Q. Grapphene Mass Production and Application Products at SuperC. In Proceedings of the 12th International Conference on the Frontier of Advanced Batteries, CIBF2016, Shenzhen, China, 24–26 May 2016; p. 21.
306. Lei, H.; Bliznak, B.; DuPasquier, A.; Moeser, G.; Oljaca, M. Electrode Conductive Carbon Additives for Li-ion Battery Advancement (Cabot). In Proceedings of the second International Forum on Positive & Negative Materials for Advanced Batteries, Hangzhou, China, 22–24 April 2015; p. 23.
307. Graphene Nanotubes. Available online: https://ocsial.com/nanotubes/ (accessed on 14 November 2024).
308. Mao, O. New Generation of Carbon Nanotube and Graphene Materials for Li-Ion Battery Applications (Cnano). In Proceedings of the 12th China International Battery Fair (CIBF 2016), Shenzhen, China, 24–26 May 2016; p. 37.
309. Mao, O. Study and development of conductive additive for the Si based anode materials for lithium ion batteries (Cnano). In Proceedings of the 13th China International Battery Fair (CIBF 2018), Shenzhen, China, 22–24 May 2018; p. 31.
310. Hantel, M. POROCARB(R)—The Synthetic Carbon Performance Additive—A Cathode View. In Proceedings of the Korea Advanced Battery Conference 2022 (KABC 2022), Korea Institute of Science and Technology Center, Seoul, Republic of Korea, 20–21 September 2022; p. 20.
311. Porocarb® Products. Available online: https://www.northerngraphite.com/porocarb-products/ (accessed on 14 November 2024).
312. Lei, H.; Blizanac, B.; DuPasquier, A.; Koehlert, K.; Ojaca, M.; Romney, G. Li-ion Battery Positive Electrode Additives for Performance Advancement. In Proceedings of the 2013 China Forum on Technical & Market Development of Positive & Negative Materials for Advanced LIB, Tianjin, China, 8–10 April 2013; p. 22.
313. LITXTM 50 Conductive Additive. Available online: www.cabotcorp.jp/~/media/files/product-datasheets/datasheet-litx-50pdf.pdf (accessed on 1 March 2018).
314. LITXTM 200 Conductive Additive. Available online: www.cabotcorp.jp/~/media/files/product-datasheets/datasheet-litx-200pdf.pdf (accessed on 26 February 2018).
315. Litx® 300 CONDUCTIVE ADDITIVE. Available online: https://www.cabotcorp.com.br/~/media/files/product-datasheets/datasheet-litx-300pdf.pdf (accessed on 14 November 2024).
316. LITX® HP Conductive Additive (Cabot). Available online: www.cabotcorp.com.br/~/media/files/product-datasheets/datasheet-litx-hppdf.pdf (accessed on 14 November 2024).

317. Lei, H.; Laxton, P.; Korchev, A.; DuPasquier, A.; Shen, M.; Oljaca, M. Performance Carbon Conductive Additives for Li-ion Battery Advancement. In Proceedings of the 13th China International Battery Fair (CIBF 2018), Shenzhen, China, 22–24 May 2018; p. 23.
318. DENKA BLACK Li のグレード (grade). Available online: http://www.denka.co.jp/scm/product/scm/detail_003523.html (accessed on 26 February 2018).
319. 电化乙肤黑产品介绍 (Denka black Li, CIBF 2018). Available online: https://app.box.com/s/x3abnmjquxdme932xpnzf6y13bcwbdw6 (accessed on 20 May 2021).
320. DENKA 锂电池导电剂 乙炔黑 (Lithium battery conductive agent acetylene black) HS-100、SB50L. Available online: http://www.cnpowdertech.com/web-show-0-890-1639-454-1.html (accessed on 20 August 2018).
321. DENKA 锂电池导电剂 乙炔黑 (Lithium Battery Conductive Agent Acetylene Black) Li-400,Li-250. Available online: https://www.achatestrade.com/productinfo/782551.html (accessed on 14 November 2024).
322. C-Nergy Leaflet (Timcal, Imerys). Available online: https://mtikorea.co.kr/web/smh/pdf/LeafletC-NERGYC65_45.pdf (accessed on 14 November 2024).
323. Spahr, M.; Zuercher, S.; Rodlert-Bacilieri, M.; Mornaghini, F.; Gruenberger, T. High-Conductive Carbon Black with Low Viscosity. WO 2017/005921 A1, 8 July 2017.
324. VGCF-H (Showa Denko). Available online: www.sdkc.com/documents/VGCF-H.pdf (accessed on 26 February 2018).
325. FT2000系列粉末. Available online: http://www.cnanotechnology.com/h-pd-45.html (accessed on 20 March 2021).
326. Mass Production Technology for Single-Wall Carbon Nanotubes. Available online: https://www.aist.go.jp/aist_e/list/latest_research/2014/20140417/20140417.html (accessed on 17 April 2019).
327. Single Walled Carbon Nano Tube. Available online: https://www.meijo-nano.com/en/products (accessed on 14 November 2024).
328. MEIJO eDIPS. Available online: https://download.wezhan.cn/contents/sitefiles2057/10287992/files/541168..pdf?response-content-disposition=inline%3Bfilename%2A%3Dutf-8%27%27IMG.pdf&response-content-type=application%2Fpdf&auth_key=1730969825-3dc91b02d6554b5287caab19a7e8637c-0-e8a0804e3116687f39ff09d3f84f478e (accessed on 14 November 2024).
329. Kechagia, P. Application of Advanced Carbon Materials in Lithium-Ion Battery Electrodes. In Proceedings of the Advanced Automotive Battery Conference (AABC 2021 Europe), Virtual, 19 January 2021; p. 15.
330. FT9000系列粉末 (Series powder). Available online: http://www.cnanotechnology.com/h-pd-42.html (accessed on 20 March 2021).
331. Introduction of Denka Black (Denki Kagaku Kogyo K.K.). Available online: http://utsrus.com/index.php?file_id=10768&Itemid=7&option=com_virtuemart&page=shop.getfile&product_id=5833 (accessed on 14 October 2018).

332. Additives for Li-Ion Batteries. Available online: https://www.byk.com/en/service/downloads/technical-brochures?tx_rbdownload_brochuredownloads%5Baction%5D=download&tx_rbdownload_brochuredownloads%5Bcontroller%5D=Brochure&tx_rbdownload_brochuredownloads%5Bitem%5D=259&type=658352&cHash=59def4847c56046d969ce298e00e3b5a (accessed on 14 November 2024).
333. Bauer, W. Investigation of the Additive Distribution in Electrodes for Lithium-Ion Batteries. In Proceedings of the MRS Fall Meeting and Exhibite, Boston, MA, USA, 25–30 November 2018; p. 25.
334. High Performance Materials for Advanced Lithium-Ion Batteries (Cabot). Available online: http://www.cabotcorp.com/~/media/files/brochures/specialty-carbon-blacks/brochure-high-performance-materials-for-advanced-lithium-ion-batteries.pdf (accessed on 14 October 2018).
335. Zürcher, S. Carbon Materials for Li-Ion Battery: Features and Benefits. In Proceedings of the Frühjahrstagung des Arbeitskreises Kohlenstoffe, Meitingen, Germany, 26 April 2016; p. 20.
336. Carbon Nanotubes Powder, Carbon Nanotubes and Graphene Conductive paste (Cnano). Available online: http://www.cnanotechnology.com/h-col-103.html (accessed on 20 March 2021).
337. Product Catalog 01.2020 (OCSiAl). Available online: https://ocsial.com/media/file/2020/06/22/CATALOG_ENG_200203.pdf (accessed on 21 March 2021).
338. Ashizawa, K.; Yamamoto, K. リチウムイオン電池用アルミニウム箔 (Aluminum Foil for Lithium-Ion Battery). *Furukawa-Sky Rev.* **2009**, *5*, 1–6.
339. Motoi, T.; Tanaka, H.; Katoh, O. リチウムイオン電池集電体用アルミニウム箔 (Aluminum Foil for Lithium-ion Battery Current Collector). *UACJ Tech. Rep.* **2014**, *1*, 125–130.
340. Aluminum Foil Brochure of Henan Mingtai Al. Inductrial Co. Ltd. (CIBF2018). Available online: https://app.box.com/s/0oiy21mtee2gethqg58up3718ho7whq5 (accessed on 20 May 2021).
341. Battery-Grade Aluminum Foil Cathode Foil Materials for Li-Ion Battery & Capacitor Manufacturers (Targray). Available online: https://www.targray.com/li-ion-battery/foils/aluminum (accessed on 14 October 2018).
342. UACJ Aluminum Foil. Available online: https://ufo.uacj-group.com/products/foil.html (accessed on 14 October 2018).
343. Zhang, N. Progress of Lishen High Power Li Ion Battery for HEV Application. In Proceedings of the International Seminar on HEV Market and Advanced Battery Technology Development, Beijing, China, 18 April 2014; p. 46.
344. リチウムイオン二次電池集電体「ファスポーラスR」を開発 Development of Lithium-ion Secondary Battery Current Collector, "FUSPOROUS". Available online: https://www.uacj.co.jp/review/furukawasky/009/pdf/09_abst16.pdf (accessed on 14 October 2018).
345. Nemoto, M.; Kubota, M.; Abe, H.; Tanaka, Y.; Kanamura, K. リチウム二次電池用多孔質集電体正極の開発(2) Development of the Cathode with Porous Current Collector for Lithium Secondary Batteries (2). *FB* テクニカルニュース **2015**, *71*, 16–21.

346. Kubota, M.; Nemoto, M.; Abe, H.; Tanaka, Y.; Kanamura, K. リチウム二次電池用多孔質集電体正極の開発 Development of the Cathode with Porous Current Collector for Lithium Secondary Batteries. *FB テクニカルニュース (Technical news)* **2014**, *70*, 28–32.
347. Functional Power Collector, Primer Coated Foil (Nippon Graphite Industries). Available online: https://app.box.com/s/g6yoz5a5l2g7j6buiwoxixo1z28hvsf8 (accessed on 20 May 2021).
348. Mornaghini, F.F.C.; Cericola, D.; Ulmann, P.; Hucke, T.; Spahr, M.E. Carbon conductive additives for electrodes in electrochemical energy storage devices. In Proceedings of the IBA Meeting 2013, Barcelona, Spain, 6–11 October 2013; p. 21.
349. Tomozawa, H. Highly Conducting Carbon-Coated Current Collector "SDX" for large Li-ion Batteries. In Proceedings of the Advanced automotive batteries (AABC Europe), Mainz, Germany, 31 January 2017; p. 25.
350. Ranafoil TM (Toyo Aluminium K.K.). Available online: http://www.toyal.co.jp/eng/products/haku/hk_ranafoil.html (accessed on 15 October 2018).
351. Carbon-Coat Foil for Electrode Collector of Lithium-Ion Batteries and Electrolytic Double Layer Capacitors (Nippon foil MFG Co., Ltd.). Available online: https://wenku.baidu.com/view/025a1b8b680203d8ce2f24ba.html?from=search (accessed on 26 October 2018).
352. 電池用カーボンコート箔 SHX (Carbon coated foil for batteries SHX). Available online: http://www.shohoku.co.jp/wp-content/uploads/2019/08/263317899550abd075a57e3e217ddb5b.pdf (accessed on 17 June 2024).
353. Takeda, A.; Nakamura, T.; Yokouchi, H.; Tomozawa, H. Resistance Reduction Effect by SDX® in Lithium-Ion Batteries. In Proceedings of the 232nd ECS Meeting, National Harbor, MD, USA, 1–5 October 2017.
354. Product Catalog January 2020. Available online: https://ocsial.com/media/file/2020/06/22/CATALOG_ENG_200203.pdf (accessed on 14 November 2024).
355. Nanotube-Coated Foil by OCSiAl Increases Energy Density and Prolongs Battery Life. Available online: https://ocsial.com/news/-nanotube-coated-foil-by-ocsial-increases-energy-density-and-prolongs-battery-life-/ (accessed on 14 November 2024).
356. Foiltec leaflet (Battery Japan 2019). Available online: https://app.box.com/s/o4kmo2w03tcr1ewxpxd4uzxdyx7avr2s (accessed on 20 May 2021).
357. Toyal-Carbo. Available online: http://www.toyal.co.jp/eng/products/haku/hk_tc.html (accessed on 16 October 2018).
358. Lu, M. Key Factors of the Power Battery Development in 2013：E-Motorcycle & EV. In Proceedings of the 2013 EV Market Trends Seminar (The 3rd International Taiwan Electric Vehicle Show), Taipei, 12 April 2013; p. 46.
359. CIAPS. 化学电源前沿技术概述2015 (Introduction to Frontline Technology for Chemical Power). Available online: https://wenku.baidu.com/view/76210e1ab80d6c85ec3a87c24028915f804d840d.html?from=search (accessed on 30 October 2018).
360. Primed Current-Collectors for Lithium-Ion Batteries (Armor). Available online: http://naatbatt.org/wp-content/uploads/2016/03/35-ARMOR.pdf (accessed on 16 October 2018).
361. Primed Current Collectors for Batteries (Armor Ensafe, CIBF2018). Available online: https://app.box.com/s/sp8tid6bykf2ngl06r19qgj5v8f77k25 (accessed on 20 May 2021).

362. Functional Coatings for Lithium-Ion Batteries. Available online: https://vonardenne.com/fileadmin/user_upload/brochures/SPA_Batteries/SPA_Batteries_eng.pdf (accessed on 14 November 2024).
363. Clarke, A.; Schrantz, K.; Macsalka, J. Benefits of Quallion's Zero-VoltTM Technology. In Proceedings of the 1st Medical Battery Conference, Dusseldorf, Germany, 16–17 November 2017; p. 22.
364. Ma, C.; Borthomieu, Y. A New Design Approach to Li-Ion Cells for LEO Applications. In Proceedings of the 48th Power Sources Conference, Marriott Tech Center Hotel, Denver, CO, USA, 11–14 June 2018; pp. 235–237.
365. Myung, S.T.; Hitoshi, Y.; Sun, Y.K. Electrochemical behavior and passivation of current collectors in lithium-ion batteries. *J. Mater. Chem.* **2011**, *21*, 9891–9911. [CrossRef]
366. Hendricks, C.E.; Mansour, A.N.; Fuentevilla, D.A.; Ko, J.K.; Waller, G.H. Copper Dissolution Investigations in Large-Format Lithium-ion Cells. In Proceedings of the 48th Power Sources Conference, Marriott Tech Center Hotel, Denver, CO, USA, 11–14 June 2018; pp. 54–57.
367. KCFT Leaflet 2019. Available online: https://app.box.com/s/zlqw3lrhzw8qh1of7ykengiutzev0fv0 (accessed on 21 May 2021).
368. Nippon Denkai Ltd., Products for Lithium-Ion Battery. Available online: https://www.nippon-denkai.co.jp/english/product/ (accessed on 19 January 2021).
369. IIT LIB-Related Study Program 10–11 (August 2010), Chapter 5 LIB Materials Market Bulletin (10Q3). Available online: https://wenku.baidu.com/view/4edc5bed0975f46527d3e14e.html (accessed on 26 October 2018).
370. Kondo, K.; Akolkar, R.N.; Barkley, D.P.; Yokoi, M. *Copper Electrodeposition for Nanofabrication of Electronics Devices*; Springer: New York, NY, USA, 2014; p. 282.
371. UACJ Copper Foil. Available online: https://ufo.uacj-group.com/en/products/copper.html (accessed on 19 October 2018).
372. Hitachi Metal Copper Foil. Available online: http://www.hitachi-cable.com/products/news/20100105.html (accessed on 19 October 2018).
373. Rolled Copper Foil for Lithium Ion Battery (Hitachi). Available online: https://www.marklines.com/statics/topSuppliers/img/exhibit/jsae2010/images/hitachigroup_P1070444.jpg (accessed on 30 August 2019).
374. Yoshida, S.; Sikiya, S.; Mitose, K. Development of the High Tensile Strength and the High Thermal Stability Copper Foil NC-TSH. *Furukawa Electr. Rev.* **2018**, *49*, 67–69.
375. NC- Foil Thickness 5–20 μm (The both Surfaces Smooth Copper Foil, Furukawa). Available online: http://www.furukawa.co.jp/foil/en/product/lithum/nc-ws.html (accessed on 20 October 2018).
376. High-Performance Copper Materials for Connectors, FPCs, and So on (JX Nippon Mining & Metals). Available online: https://www.youtube.com/watch?v=CM7Nn0Tp_GU&feature=emb_logo (accessed on 7 February 2021).
377. HS1200—High Strength Copper Foil for Battery. Available online: https://www.jx-nmm.com/english/products/alloy/rolling_foil/hs1200.html (accessed on 14 November 2024).
378. Takahashi, H.; Aragaki, M.; Hikami, T. The quantitative evaluation of anode thickness change for lithium-ion batteries. *Furukawa Rev.* **2015**, *46*, 30–36.

379. World Smart Energy Week 2013 東京現場直擊系列報導三 (Tokyo Live Coverage Series Report Three). Available online: https://www.materialsnet.com.tw/material/DocView_MaterialNews.aspx?id=10933 (accessed on 6 November 2018).
380. Perforated Metal Foil for Rechargeable Batteries. Available online: https://www.fukuda-kyoto.co.jp/uploads/pdf/Punching_Sheet_Leaflet.pdf (accessed on 7 November 2018).
381. LS Mtron Copper Foil. Available online: https://lsmtron.co.kr/pdf/Copper%20Foil.pdf (accessed on 26 October 2018).
382. 古河电工 负极集流体铜箔 (Furukawa Electric Copper foil for negative electrode current collector). Available online: http://www.achatestrade.com/Copy%20of%20JFE%20Mineral%20%E4%BA%BA%E9%80%A0%E7%9F%B3%E5%A2%A8%E8%B4%9F%E6%9E%81%E6%9D%90%E6%96%99 (accessed on 28 October 2018).
383. Kremer, K. Development of water-borne slurry with nano-sized ceramic particles for functional layer. In Proceedings of the Advanced automotive battery conference Europe (AABC), Strasbourg, France, 26–28 June 2013; p. 31.
384. Lithium Ion Battery PVDF Electrode Binders & Separator Coatings (Arkema). Available online: https://www.extremematerials-arkema.com/export/sites/technicalpolymers/.content/medias/downloads/brochures/kynar-brochures/2017-new-kynar-battery-brochure-optimized.pdf (accessed on 8 November 2018).
385. Wang, J. 电池粘结剂的技术动向及对电池性能的影响 (Technical trends of battery binders and their impact on battery performance) (Zeon). In Proceedings of the 2nd HEV Market & Advanced Battery Technology development Seminar, Hangzhou, China, 13–14 October 2016; p. 35.
386. 西, 美 (Nishi M.). 高分子ゲルを電解質とするリチウムイオン二次電池. (Lithium-ion secondary battery using polymer gel as electrolyte). 高分子 *Polymer* **2005**, *54*, 870–873. [CrossRef]
387. Jian, L. Electrolyte Development for High Energy Density Li-ion Batteries (BAK). In Proceedings of the 2nd International Forum on Electrolyte and Separator for Advanced Batteries, Shenzhen, China, 11–13 November 2015; p. 34.
388. Ugawa, S.; Masuda, K.; Kajiwara, I. リチウムイオン電池用水系バインダーWater-Based Binder for LIB Application. *JSR Tech. Rev.* **2014**, *121*, 10–15.
389. Solef® PVDF for Binders to Improve Battery Performance. Available online: https://www.solvay.cn/en/binaries/Solef-PVDF-for-Li-Ion-Batteries_EN.pdf-220706.pdf (accessed on 20 October 2018).
390. Li, J.; Daniel, C.; Mohanty, D.; Wood III, D.L. Thick Low-Cost, High-Power Lithium-Ion Electrodes via Aqueous Processing (Es164). In Proceedings of the DOE Annual Merit Review, Washington, DC, USA, 8 June 2016; p. 22.
391. Kynar® and Kynar Flex®: Polyvinylidene Fluoride (PvdF) resins For Batteries (Arkema). Available online: https://espanol.kynar.com/export/sites/kynar-latam/.content/medias/downloads/literature/kynar-pvdf-resins-for-battery.pdf (accessed on 22 October 2018).
392. Starke, B.; Seidlmayer, S.; Jankowski, S.; Dolotko, O.; Gilles, R.; Pettinger, K.H. Influence of Particle Morphologies of $LiFePO_4$ on Water- and Solvent-Based Processing and Electrochemical Properties. *Sustainability* **2017**, *9*, 888. [CrossRef]

393. World Smart Energy Week 2014 東京現場直擊系列報導三 (Tokyo live coverage series three). Available online: https://www.materialsnet.com.tw/material/DocView_MaterialNews.aspx?id=11698 (accessed on 7 November 2018).
394. LAPONITE-RD Rheology Additive Based on Synthetic Phyllosilicate for Aqueous Systems to Improve the Rheological Properties in the Low Shear Range. Available online: https://www.byk.com/en/products/additive-guide/laponite-rd (accessed on 14 November 2024).
395. Isogai, A.; Saito, T.; Fukuzumi, H. TEMPO-oxidized cellulose nanofibers. *Nanoscale* **2011**, *3*, 71–85. [CrossRef]
396. 第一工业制药 锂电池纳米级分散剂 (Daiichi Industrial Pharmaceutical Lithium Battery Nanoscale Dispersant) CNF (RHEOCRYSTA 1-2SX). Available online: https://www.achatestrade.com/productinfo/782743.html (accessed on 15 May 2024).
397. 平成 29 年度セルロースナノファイバー活用製品の 性能評価事業委託業務 (セルロースナノファイバーを適用したアイドリ ングストップ車用リチウムイオン電池の実用化に 向けた課題抽出) (FY2017 performance evaluation project for products utilizing cellulose nanofibers (Identification of issues for the practical application of lithium-ion batteries for idling-stop vehicles using cellulose nanofibers). Available online: https://www.env.go.jp/earth/ondanka/cnf/mat36_dksH29.pdf (accessed on 22 January 2021).
398. 平成28年度セルロースナノファイバー活用製品 の性能評価事業委託業務 (セルロースナノファイバーを適用したアイドリ ングストップ車用リチウムイオン電池の実用化に向けた課題抽出) (FY2016 performance evaluation project for products utilizing cellulose nanofibers (Identification of issues for the practical application of lithium-ion batteries for idling-stop vehicles using cellulose nanofibers). Available online: https://www.env.go.jp/earth/ondanka/cnf/mat16.pdf (accessed on 22 January 2021).
399. Wood, D.L.; Quass, J.D.; Li, J.L.; Ahmed, S.; Ventola, D.; Daniel, C. Technical and economic analysis of solvent-based lithium-ion electrode drying with water and NMP. *Dry. Technol.* **2018**, *36*, 234–244. [CrossRef]
400. Dong, L. *Impact of Moisture on the Properties of $LiNi_xCo_yMn_{1-x-y}O_2$ (NMC) Cathodes for Lithium-Ion Batteries*; The Ohio State University: Columbus, OH, USA, 2018.
401. Hawley, W.B.; Parejiya, A.; Bai, Y.C.; Meyer, H.M.; Wood, D.L.; Li, J.L. Lithium and transition metal dissolution due to aqueous processing in lithium-ion battery cathode active materials. *J. Power Sources* **2020**, *466*, 10. [CrossRef]
402. Wood, M.; Li, J.L.; Ruther, R.E.; Du, Z.J.; Self, E.C.; Meyer, H.M.; Daniel, C.; Belharouak, I.; Wood, D.L. Chemical stability and long-term cell performance of low-cobalt, Ni-Rich cathodes prepared by aqueous processing for high-energy Li-Ion batteries. *Energy Storage Mater.* **2020**, *24*, 188–197. [CrossRef]
403. Zhang, X.Y.; Jiang, W.J.; Zhu, X.P.; Mauger, A.; Qilu; Julien, C.M. Aging of $LiNi_{1/3}Mn_{1/3}Co_{1/3}O_2$ cathode material upon exposure to H_2O. *J. Power Sources* **2011**, *196*, 5102–5108. [CrossRef]
404. Cetinel, F.A.; Bauer, W. Processing of water-based $LiNi_{1/3}Mn_{1/3}Co_{1/3}O_2$ pastes for manufacturing lithium ion battery cathodes. *Bull. Mater. Sci.* **2014**, *37*, 1685–1690. [CrossRef]

405. Zhuang, G.V.; Chen, G.Y.; Shim, J.; Song, X.Y.; Ross, P.N.; Richardson, T.J. Li_2CO_3 in $LiNi_{0.8}Co_{0.15}Al_{0.05}O_2$ cathodes and its effects on capacity and power. *J. Power Sources* **2004**, *134*, 293–297. [CrossRef]
406. Loeffler, N.; Kim, G.T.; Mueller, F.; Diemant, T.; Kim, J.K.; Behm, R.J.; Passerini, S. In Situ Coating of $LiNi_{0.33}Mn_{0.33}Co_{0.33}O_2$ Particles to Enable Aqueous Electrode Processing. *Chemsuschem* **2016**, *9*, 1112–1117. [CrossRef] [PubMed]
407. Kimura, K.; Sakamoto, T.; Mukai, T.; Ikeuchi, Y.; Yamashita, N.; Onishi, K.; Asami, K.; Yanagida, M. Improvement of the Cyclability and Coulombic Efficiency of Li-Ion Batteries Using Li $Ni_{0.8}Co_{0.15}Al_{0.05}O_2$ Cathode Containing an Aqueous Binder with Pressurized CO_2 Gas Treatment. *J. Electrochem. Soc.* **2018**, *165*, A16–A20. [CrossRef]
408. Aluminum E-pH (Pourbaix) Diagram. Available online: https://corrosion-doctors.org/Corrosion-Thermodynamics/Potential-pH-diagram-aluminum.htm (accessed on 7 February 2021).
409. Watanabe, T.; Hirai, K.; Ando, F.; Kurosumi, S.; Ugawa, S.; Lee, H.; Irii, Y.; Maki, F.; Gunji, T.; Wu, J.F.; et al. Surface double coating of a $LiNi_aCobAl_{1-a-b}O_2$ (a > 0.85) cathode with TiO_x and Li_2CO_3 to apply a water-based hybrid polymer binder to Li-ion batteries. *Rsc Adv.* **2020**, *10*, 13642–13654. [CrossRef]
410. Doberdo, I.; Loffler, N.; Laszczynski, N.; Cericola, D.; Penazzi, N.; Bodoardo, S.; Kim, G.T.; Passerini, S. Enabling aqueous binders for lithium battery cathodes—Carbon coating of aluminum current collector. *J. Power Sources* **2014**, *248*, 1000–1006. [CrossRef]
411. Loeffler, N.; von Zamory, J.; Laszczynski, N.; Doberdo, I.; Kim, G.T.; Passerini, S. Performance of $LiNi_{1/3}Mn_{1/3}Co_{1/3}O_2$/graphite batteries based on aqueous binder. *J. Power Sources* **2014**, *248*, 915–922. [CrossRef]
412. Li, C.C.; Lee, J.T.; Tung, Y.L.; Yang, C.R. Effects of pH on the dispersion and cell performance of $LiCoO_2$ cathodes based on the aqueous process. *J. Mater. Sci.* **2007**, *42*, 5773–5777. [CrossRef]
413. Kukay, A.; Sahore, R.; Parejiya, A.; Hawley, W.B.; Li, J.L.; Wood, D.L. Aqueous Ni-rich-cathode dispersions processed with phosphoric acid for lithium-ion batteries with ultra-thick electrodes. *J. Colloid Interface Sci.* **2021**, *581*, 635–643. [CrossRef]
414. Park, S. High-Performance Separator Coatings for Lithium-Ion Batteries. Available online: https://page.arkema.com/rs/253-HSZ-754/images/20191112-Separator%20Coating-WEBINAR-DOWNLOAD.pdf (accessed on 20 November 2019).
415. Huzar-EA, M. How to build BETTER and More SUSTAINABLE Batteries. Available online: https://page.arkema.com/rs/253-HSZ-754/images/20191016-how-to-build-better-sustainable-batteries-WEBINAR-DOWNLOAD.pdf (accessed on 20 November 2019).
416. 水性正极、隔膜涂层粘结剂（日本JSR) (Water-based cathode and separator coating binder) Japan JSR. Available online: http://www.scmbattery.com/cn/product_show.php?id=129 (accessed on 26 October 2018).
417. Chengdu Indigo Power Sources Binder Brochure (CIBF2018). Available online: https://app.box.com/s/rwox53pwhjqd6j5q04apefbgh44q5tvd (accessed on 20 May 2021).
418. Products: LA132, LA133. Available online: http://www.cd-ydl.com/en/index.php?go=product/show-3.html (accessed on 14 November 2024).

419. BA-310C Plus (Blue Ocean & Black Stone). Available online: http://www.bobstech.com/product/ba-310c-plus-water-based-binder/?portfolioCats=29 (accessed on 15 May 2019).
420. Cathode Binder. A Water-Based Polyacrylate Binder for Effectively Improving Compact Density and Efficiently Improving Cell Durability (Blue Ocean & Black Stone). Available online: https://app.box.com/s/pkiyqzi0kfpx1d9my0h4jd5ieswnxuem (accessed on 20 May 2021).
421. Improved both Anode and Cathode Characteristics through the Use of Core-Corona Type Nanoparticles (Elebine, Senka corp.). Available online: https://app.box.com/s/ddz6n8y9fuo5d8bwta82nc3abrdz236k (accessed on 20 May 2021).
422. World Smart Energy Week 2016 東京現場直擊 系列報導一 (Tokyo Live Report Series 1). Available online: https://www.materialsnet.com.tw/DocView.aspx?id=24126 (accessed on 7 November 2018).
423. Powerbinder Leaflet (Ielectrolyte). Available online: https://ielectrolyte.net/wp-content/uploads/2018/04/f868cb50a57bff814d00ea44783d59f9.pdf (accessed on 17 April 2019).
424. iElectrolyte Kansai Japan. Available online: https://ielectrolyte.net/en/wp-content/uploads/2018/04/Water-based-process-for-cathode.pdf (accessed on 15 May 2024).
425. Battery Solutions with Kynar® PVDF Lithium-Ion Focus (Arkema). Available online: http://www.extremematerials-arkema.com/export/sites/technicalpolymers/.content/medias/downloads/market-presentations/Arkema-Lithium-Ion-Battery-Market-Presentation.pdf (accessed on 26 October 2018).
426. Solef® PVDF for Li-Ion Batteries. Available online: https://www.solvay.jp/ja/binaries/Solef-PVDF-for-Li-Ion-Batteries_EN.pdf-220706.pdf (accessed on 23 January 2021).
427. Toigo, C.; Singh, M.; Gmeiner, B.; Biso, M.; Pettinger, K.-H. A Method to Measure the Swelling of Water-Soluble PVDF Binder System and Its Electrochemical Performance for Lithium Ion Batteries. *J. Electrochem. Soc.* **2020**, *167*, 020154. [CrossRef]
428. High Performance Binder for Electrode Kureha KF Polymer. Available online: https://www.kureha.co.jp/en/business/material/pdf/KFpolymer_BD_en.pdf (accessed on 14 November 2024).
429. Toyocolor Supplies Battery Electrode Material for New Toyota Prius. Available online: https://www.prlog.org/12516609-toyocolor-supplies-battery-electrode-material-for-new-toyota-prius.html (accessed on 14 November 2024).
430. High Conductivity Binder for High Voltage type Electrode Application - Ion Conductive Binder PIOXCEL. Available online: https://www.piotrek-il.co.jp/productsindex3059.html (accessed on 14 November 2024).
431. Shinko, N. TUBALL™ and PVdF Modifier VT-475 for High Density Electrodes (Nobuhiro Shinka, DAIKIN Industries). Available online: https://www.youtube.com/watch?v=zQ55hZ3F3Gw (accessed on 22 January 2021).
432. Sunstrom, J.; Falzone, A.; Gilmore, M.; Hendershot, R.; Meserole, C.; Sandoval, A.; Grumbles, E.; Costa, M.; Haque, A. Pushing the Energy Limits of Lithium Ion Batteries through Fluorinated Materials. *SAE Tech. Pap.* **2019**, 2019-2001-0595. [CrossRef]

433. 科路得产业化软包全电池制作工艺大讲坛第二讲 (Lecture 2 of Kelode Industrialized Soft Pack Full Battery Manufacturing Technology Forum). Available online: https://mp.weixin.qq.com/s?__biz=MzIyMzIwOTMzNA==&mid=2649829030&idx=1&sn=8872e88ad743b9a29b5afcb891022ae1&chksm=f0243cbfc753b5a92fa052a49cdc7073d54381305a42a9cb52a7ab70b5136a7f81f314a98694&scene=21#wechat_redirect (accessed on 14 December 2020).
434. LiBの性能を向上する高粘度CMC 第一工業製薬 社報 No.574 拓人2015秋 (High viscosity CMC that improves LiB performance Daiichi Kogyo Seiyaku Company Newsletter No.574 Takuto 2015 Autumn). Available online: https://www.dks-web.co.jp/catalog_pdf/574_1.pdf (accessed on 29 October 2018).
435. Asai, Y.; Kurosumi, S.; Masuda, K.; Honda, T.; Ugawa, S.; Lee, H.; Yamashita, T. Development of binder for Si anode and evaluation of electrode expansion. *JSR Tech. Rev.* **2018**, *125*, 7–11.
436. CMC Book. Available online: http://www.bisi.cz/cmsres.axd/get/cms$7CVwRhc3USVqgzxkKF96gI$2BChNrXcTq$2BOUz0Xj7EmggLlJILTc$2BnjT05VW4kCumkdM (accessed on 14 December 2020).
437. 迟蓝kyegjxlb42. 锂离子电池 CMC-Na 使用手册 Lithium-ion battery CMC-Na User Manual. Available online: http://www.360doc.com/content/19/0124/09/61956981_810962169.shtml[11.12.2020 (accessed on 14 December 2020).
438. Sunrose Mac (Nippon Paper Industries). Available online: https://www.nipponpapergroup.com/english/products/mt_pdf/AABCpamphlet.pdf (accessed on 4 March 2020).
439. AkuPure Carboxymethyl Cellulose (CMC) for lithium ion battery anodes. Available online: https://celluloseethers.nouryon.com/siteassets/applications-lithium-ion-battery/leaflet-nouryon-leaflet-akupure-en-2-1909046.pdf (accessed on 14 December 2020).
440. DAICEL FINECHEM会社绍介资料-CMC (Introduction to CMC). Available online: https://wenku.baidu.com/view/a4f5f37da31614791711cc7931b765ce04087a6a (accessed on 10 December 2020).
441. CELLOGEN Carboxymethyl Cellulose. Available online: http://www.dalian-diligence.com/images/diyi.pdf (accessed on 14 December 2020).
442. 陶氏化学 锂电池增稠剂 (Dow Chemical Lithium Battery Thickener) WAL CRT 30000PA, WAL CRT 10000PA. Available online: https://www.achatestrade.com/productinfo/782636.html?templateId=1133605 (accessed on 14 November 2024).
443. Daicelmiraizu, CMC for Lithium-Ion Battery. Available online: https://www.daicelmiraizu.com/en/products/wsp_electron.html (accessed on 12 October 2020).
444. Sodium Carboxymethylcellulose SUNROSE® MAC Series. Available online: https://www.nipponpapergroup.com/english/products/mt_pdf/%E8%8B%B1%E6%96%87%EF%BE%8E%EF%BD%B0%EF%BE%91%EF%BE%8D%EF%BE%9F%EF%BD%B0%EF%BD%BC%EF%BE%9E%EF%BE%8D%EF%BE%9F%EF%BD%B0%EF%BD%BC%EF%BE%9E%E6%8E%B2%E8%BC%89%E7%94%A8%EF%BC%9AMAC%EF%BD%BC%EF%BE%98%EF%BD%B0%EF%BD%BD%EF%BE%9E.pdf (accessed on 14 November 2024).

445. リチウムイオン電池用CMC—第一工業製薬 第一工業製薬社報No.564拓人2013春 (CMC for lithium-ion batteries—Daiichi Kogyo Seiyaku Daiichi Kogyo Seiyaku Newsletter No.564 Takuto 2013 Spring). Available online: https://www.dks-web.co.jp/catalog_pdf/564_3.pdf (accessed on 29 October 2018).
446. Styrene-Butadiene Rubber (SBR) Binder for Li-ion Battery Anode 500 g/bottle—EQ-Lib-SBR. Available online: https://www.mtixtl.com/Styrene-ButadieneRubberSBRbinderforLi-ionBatteryAnode260g/bottle.aspx (accessed on 26 March 2021).
447. 關於SBR粘結劑扯點你不知道的，工藝和研發的都看看 (What you don't know about SBR adhesives, take a look at the process and R&D). Available online: https://www.luoow.com/dc_tw/108833752 (accessed on 14 December 2020).
448. 斯比可 CEKOL 30000 CMC 池稠 (CP Kelco CEKOL 30000 CMC Lithium Battery Thickener) . Available online: https://www.achatestrade.com/productinfo/782671.html (accessed on 14 November 2024).
449. 美国DOWCMC系列产品性能 (American DOWCMC series product performance). Available online: http://www.scmbattery.com/cn/product_show.php?id=114 (accessed on 14 December 2020).
450. WAL CRT 30000PA. Available online: https://download.wezhan.cn/contents/sitefiles2057/10287992/files/540928..pdf?response-content-disposition=inline%3Bfilename%2A%3Dutf-8%27%27WAL%2BCRT%2B30000%2BPA%25e8%25a7%2584%25e6%25a0%25bc%25e4%25b9%25a6%25ef%25bc%2588%25e4%25b8%25ad%25e6%2596%2587%25ef%25bc%2589.pdf&response-content-type=application%2Fpdf&auth_key=1730971019-2c6201ae1a744974a733c8e80a4ba3cc-0-b4d144c026e873065601a8c6cdc6e2f4 (accessed on 14 November 2024).
451. MicellCMC®. Available online: http://www.glchem.co.kr/MICELLCMC (accessed on 14 November 2024).
452. CMC 日本製紙 MAC系列 負極粘結劑 甲基纖維素鈉 鋰電 超級電 (CMC Nippon Paper MAC series negative electrode binder sodium carboxymethylcellulose lithium battery super battery). Available online: https://world.taobao.com/item/564908834198.htm?spm=a21wu.12321156-tw.0.0.ba92767af6H74p (accessed on 15 December 2020).
453. 电池级 CMC (Battery-Grade CMC). Available online: https://www.xtsentai.com/products_details_1/15.html (accessed on 14 November 2024).
454. 巴斯夫 锂电池粘稠剂 (BASF Lithium Battery Adhesive) Binder 21-11 ap. Available online: https://www.achatestrade.com/productinfo/782682.html (accessed on 14 November 2024).
455. Technical Information Styrofan 7212 (Basf). Available online: http://www.hailichem.com/en/product/sbr-latex-7212.html?file=40be76712faf32934c71395fceba789d (accessed on 15 December 2020).
456. 负极SBR粘结剂（日本JSR, TRD104A, TRD105A, TRD202A）(Negative SBR binder (Japanese JSR, TRD104A, TRD105A, TRD202A). Available online: http://www.scmbattery.com/cn/product_show.php?id=99 (accessed on 8 August 2018).
457. JSR TRD104A. Available online: http://www.achatestrade.com/product/276813918 (accessed on 28 October 2018).

458. JSR104A SBR丁苯橡膠乳液 超級電容粘結劑 鋰電池粘結劑 原裝進口 (SBR styrene-butadiene rubber emulsion, super capacitor binder, lithium battery binder, originally imported.). Available online: https://world.taobao.com/item/564995955693.htm?spm=a21wu.12321156-tw.0.0.7370dd3bQNRwen (accessed on 15 December 2020).
459. JSR Brochure (CIBF2018). Available online: https://app.box.com/s/npd83ia8f8g6t7zyp3m91i8gtntyf3fe (accessed on 20 May 2021).
460. JSR TRD2001, TRD102A. Available online: https://wenku.baidu.com/view/263c60d0240c844769eaee58.html (accessed on 28 October 2018).
461. 高端水性バインダーSBR(日本JSR) (High-end water-based binder SBR (Japan JSR). Available online: http://www.scmbattery.com/jp/product_show.php?id=116 (accessed on 15 December 2020).
462. NIPPON A&L 池粘合（水性） (Lithium Battery Binder (Water-Based)) SBR SN-307R. Available online: https://www.achatestrade.com/productinfo/782674.html (accessed on 14 November 2024).
463. Electrode for Electrochemical Element. WO2013039131A1, 14 September 2011.
464. Lee, Y.; Choi, J.; Ryou, M.H.; Lee, M. Polymeric Materials for Lithium-ion Batteries (Separators and Binders). *Polym. Sci. Technol.* **2013**, *24*, 603–611.
465. Binders for Lithium-Ion Rechargeable Batteries (Zeon). Available online: http://www.zeon.co.jp/business_e/enterprise/imagelec/battery.html (accessed on 29 October 2018).
466. Zeon binder brochure (CIBF2018). Available online: https://app.box.com/s/9lfglo3xq4hzf3mc79moi8rb3c822y01 (accessed on 20 May 2021).
467. Specifications for BM-451B. Available online: https://wenku.baidu.com/view/5cc1175586c24028915f804d2b160b4e767f81e5 (accessed on 15 December 2020).
468. 日本瑞翁Zeon SBR BM-430B 丁苯橡膠乳液 超級電容 鋰電池粘結劑 (Zeon SBR BM-430B styrene-butadiene rubber emulsion supercapacitor lithium battery binder). Available online: https://world.taobao.com/item/564841661597.htm?spm=a21wu.12321156-tw.0.0.75f57235CXjOs7 (accessed on 15 December 2020).
469. 这是我看过锂电池粘结剂最全面的一篇文章! (This is the most comprehensive article I have read on lithium battery binders). Available online: http://www.360doc.com/content/19/0604/17/57990096_840381370.shtml (accessed on 14 December 2020).
470. 高性能粘结剂在电池中的应用研究. (Research on the application of high-performance binders in batteries). Available online: http://www.360doc.com/content/19/0604/07/61956981_840259431.shtml (accessed on 15 December 2020).
471. Deheryan, S.; Kurosumi, S. JSR TRD302A Aqueous Binder for Si Containing LIB Anodes. In Proceedings of the Lithium Battery Chemistry Symposium 2019: Held at AABC Europe, Strasbourg, France, 27–31 January 2019; p. 346.
472. Introduction of Osaka Soda Functional Battery Materials (Battery Japan 2019 leaflet). Available online: https://app.box.com/s/4rdcyhre4u45axhyilg9posxc4a4tr2n (accessed on 20 May 2021).
473. Anode Binder. A Water-Based Polyacrylate Binder for Effectively Suppressing Electrode Expansion and Efficiently Improving Cell Durability (Blue Ocean & Black Stone). Available online: https://app.box.com/s/bo4p1hmwj0rjtsixqzkym8s0533pmmy2 (accessed on 14 June 2023).

474. BA-290S (Blue Ocean & Black Stone). Available online: http://www.bobstech.com/product/ba-290s-paa-binder-for-silicon-based-anode/?portfolioCats=28 (accessed on 15 May 2019).
475. Technical Data Sheet ETERSOL 1730. Available online: https://www.eternal-group.com/FileUpload/ProductEN/_ETERSOL%201730%20TDS.pdf (accessed on 15 December 2020).
476. Sumitomo Seika Chemicals Co., Ltd. Aquacharge (aqueous binder). Available online: https://www.sumitomoseika.co.jp/en/product/detail.php?id=41 (accessed on 19 January 2021).
477. LIB Materials (Aekyung Chemical). Available online: http://www.akc.co.kr/en/product/list.do?bcode_id=BCE9&mcode_id=BCE9_1&scode_id=BCE9_1_1 (accessed on 29 October 2018).
478. 爱敬化学 硅负极用PAA胶 (Aekyung Chemical PAA glue for silicon negative electrodes) (Aekyung Chemical). Available online: http://www.fentijs.com/index.php?m=yp&c=com_index&a=show&modelid=14&catid=1540&id=25200&userid=454&page=1 (accessed on 15 May 2024).
479. Semi-IPN Structure. Available online: http://www.akc.co.kr/en/product/list.do?bcode_id=BCE9&mcode_id=BCE9_1&scode_id=BCE9_1_2 (accessed on 29 October 2018).
480. Kim, N.S.; Chung, B.J.; Kim, S.J.; Choi, K.S. A Semi-IPN Polymeric Binder for High Performance Silicone Negative Electrodes in Lithium Ion Batteries. *ECS Meet. Abstr.* **2014**, MA2014-02, 315. [CrossRef]
481. Binder for Silicon Based Anode (Daxin). Available online: http://www.daxinmat.com/?sn=844&lang=en-US&c=235&s=307&p=308 (accessed on 17 April 2019).
482. Lithium Ion Batteries Resin for Anode Binders (Nippon Kayaku). Available online: https://wenku.baidu.com/view/93ceb764001ca300a6c30c22590102020640f258 (accessed on 17 April 2019).
483. Истомина, А.С.; Бушкова, О.В. Полимерные связующие для электродов литиевых аккумуляторов Часть 1. Поливинилиденфторид, его производные И другие коммерциализованные материалы. Электрохимическая энергетика **2020**, *20*, 115–131, (Istomina, A.S.; Bushkova, O.V. Polymer binders for lithium battery electrodes Part 1. Polyvinylidene fluoride, its derivatives and other commercialized materials. Electrochemical energy 2020, V. 20, p. 115-131. [CrossRef]
484. Yoshino, A. 3-D Imaging of separator pore structure and Li + diffusion behavior. In Proceedings of the IBA 2013 Meeting, Barcelona, Spain, 10–15 March 2013; p. 18.
485. 簡析2012年鋰電池隔離膜技術發展現況 (ITRI IEK) (Brief analysis of the current development status of lithium battery isolation film technology in 2012). Available online: https://wenku.baidu.com/view/8d74d959f01dc281e53af0b3.html (accessed on 7 November 2018).
486. 锂离子动力电池隔膜 (Lithium-ion power battery separator)/Cellulion Cellulose separator for Li-ion battery. Available online: http://www.scmbattery.com/cn/product_show.php?id=124 (accessed on 1 November 2018).
487. Morin, B.; Hennessy, J.; Arora, P. Developments in nonwovens as specialists membranes in batteries and supercapacitors. In *Advances in Technical Nonwovens*; Kellie, G., Ed.; Woodhead Publishing (Elsevier): Sawston, Cambridge, UK, 2016; pp. 311–339.

488. What Is the Melting Point of Cellulose? (Quora). Available online: https://www.quora.com/What-is-the-melting-point-of-cellulose (accessed on 1 November 2018).
489. Senior Product Manual (CIBF 2018). Available online: https://app.box.com/s/8ktzau8zdnrqpdq5y6vggusbt4mcyf50 (accessed on 20 May 2021).
490. Celgard's Family of Separator Technology Solutions (CIBF2018). Available online: https://app.box.com/s/ruav18pq9goykn4fsodj8rcbjynh0tao (accessed on 20 May 2021).
491. Smart Energy Week 2018日本現場報導系列(一) (Japan's current market share (1)). Available online: https://www.materialsnet.com.tw/DocView.aspx?id=32886 (accessed on 7 November 2018).
492. Oh, K.; Wang, J. Functional materials coated on separator or electrode for lithium ion batteries (Zeon). In Proceedings of the Third International Forum on Electrolyte & Separator Materials for Advanced Batteries, Nanjing, China, 19 October 2017; p. 27.
493. Xin, L. 动力电池隔膜微观结构与性能的关系 (Discussion on relation between the properties of the power battery separator with the microstructure (Tianjin DGM Tech Co., Ltd.). In Proceedings of the 2nd International Forum on Electrolyte and Separator for Advanced Batteries, Shenzhen, China, 11–13 November 2015; p. 29.
494. リチウムイオン2次電池用セパレータ開発動向 (東レ株式会社, Setela) (Development trends of separators for lithium-ion secondary batteries (Toray Industries, Inc., Setela)). TRC News 01.05. 2017, pp. 1–6. Available online: https://www.toray-research.co.jp/service/trcnews/pdf/201705-01.pdf (accessed on 16 May 2024).
495. Morin, B. 3rd Generation Separators: Using Thermally Stable Separators to Turn the Aluminum Current Collector into a Fuse (Dream Weaver). In Proceedings of the 34th Annual International Battery Seminar and Exhibit 2017, Fort Lauderdale, FL, USA, 20–23 March 2017; pp. 27, 135.
496. Xi, S. BYD电动车锂离子电池技术发展与应用及其对隔膜的发展需求 (Development and application of lithium-ion battery technology for electric vehicles and its development needs for separators). In Proceedings of the 中国锂电池电解质、隔膜材料技术与市场发展论坛 (the China Lithium Battery Electrolyte and Separator Material Technology and Market Development Forum), Guangzhou, China, 14–16 August 2013; p. 25.
497. Cardillo, S. Технология изготовления сепаратора для литий-ионных батарей (Celgard) (Separator manufacturing technology for lithium-ion batteries). In Proceedings of the 27-я Международная специализированная выставка "Автономные источники тока" Interbat2018 (27th International specialized exhibition "Autonomous Power Sources), Moscow, Russia, 21–23 March 2018; p. 21.
498. Lu (Mark), H.-L. Present Technology & Market Development Trends of Electrolyte & Separator for LIB. In Proceedings of the 201511 The Second International Forum on Electrolyte & Separator for Advanced Batteries, Shenzhen, China, 11 November 2015; p. 33.
499. Arora, P.; Zhang, Z.M. Battery separators. *Chem. Rev.* **2004**, *104*, 4419–4462. [CrossRef]
500. Maschinenbau, B. Evapore® Trilayer 5 μm. In Proceedings of the 2nd Virtual Battery Exhibition, Virtual, 27 April–3 May 2021; p. 1.
501. Premium Coating Separator (Enerever). Available online: https://app.box.com/s/i7ygsdtubx6igwzznzxsmcs2vhyfdjd6 (accessed on 20 May 2021).

502. SNE Research. 배터리4대주요소재시장동향및개발전략방향 (Battery. 4 Major Materials Market Trends and Development Strategy Directions). In Proceedings of the 2016 7th Korea Advanced Battery Conference (KABC2016), Coex, Seoul, Republic of Korea, 29 September 2016; p. 50.
503. Lu, M. Present Technology & Market Development Trends of Electrolyte & Separator for LIB. In Proceedings of the 中国锂电池电解质、隔膜材料技术与 市场发展论坛 (China Lithium Battery Electrolyte and Separator Material Technology and Market Development Forum), Guangzhou, China, 14–16 August 2013; p. 48.
504. Nguyen, T.T.D.; Abada, S.; Lecocq, A.; Bernard, J.; Petit, M.; Marlair, G.; Grugeon, S.; Laruelle, S. Understanding the Thermal Runaway of Ni-Rich Lithium-Ion Batteries. *World Electr. Veh. J.* **2019**, *10*, 79. [CrossRef]
505. Zhang, J.; Ramadass, P.; Fang, W. Separator for Li-ion. In Proceedings of the 中国锂电池电解质、隔膜材料技术与 市场发展论坛 (China Lithium Battery Electrolyte and Separator Material Technology and Market Development Forum), Guangzhou, China, 14–16 August 2013; p. 44.
506. Pekala, R. Ceramic-Modified Separators Designed to Push the Limits of Cell Performance. In Proceedings of the 38th Annual International Battery Seminar & Exhibition, Virtual, 9 March 2021; p. 26.
507. Functional Separator Professional Manufacturer (Dinho). Available online: https://app.box.com/s/bodhi8wir9s8g23zj34w7s5qnrwfkw7i (accessed on 20 May 2021).
508. Nakajima, J.; Fujino, A.; Ikuta, S.; Hayashi, T.; Fukumoto, Y.; Kasamatsu, S. Lithium Ion Secondary Battery. EP1667255A1, 13 September 2004.
509. Fujikawa, M.; Suzuki, K.; Inoue, K.; Shimada, M. Cylindrical Lithium Battery Resistant to Breakage of the Porous Heat Resistant Layer. US7419743B2, 4 April 2006.
510. Poly-N-vinylacetamide (Showa Denko). Available online: http://www.sdk.co.jp/english/products/126/132/13830.html (accessed on 28 October 2018).
511. New Technology Insights Heat-Resistant Coating Binder PNVATM GE191 Series. Available online: https://www.sdk.co.jp/innovation/english/points/pnva.html (accessed on 24 January 2021).
512. R&D Story Heat-Resistant Coating Binder PNVATM GE191 Series. Available online: https://www.sdk.co.jp/innovation/english/story/pnva.html (accessed on 24 January 2021).
513. Functional Binders for Coated Separators (SWA610). Available online: https://en.haodyne.com/productdetail/product002.html (accessed on 20 March 2021).
514. Mathivet, T.; Baert, T. Solvay's New Developments in Electrolyte Additives and Solef PVDF Binders. In Proceedings of the Advanced Automotive Battery Conference Europe (AABC), Mainz, Germany, 30 January–2 February 2017; p. 23.
515. Zhang, J.; McCallum, I.; Fang, W. Li-ion Safety and Ceramic Coated Separator (Celgard). In Proceedings of the 2nd International Forum on Electrolyte and Separator for Advanced Batteries, Shenzhen, China, 11–13 November 2015; p. 26.
516. Sumitomo Chemical to more than double PERVIO Li-ion separator production capacity; supplier to Panasonic, Tesla. Available online: https://www.greencarcongress.com/2015/06/20150611-sumitomo.html (accessed on 2 November 2018).

517. Separators for Lithium-ion Batteries (Teijin Lielsort). Available online: https://www.teijin.com/rd/technology/separator/ (accessed on 15 May 2024).
518. Teijin's Coating Technology. Available online: https://teijin-lielsort.com/en/technology/ (accessed on 14 November 2024).
519. Chuanming, G. Cangzhou Mingzhu Wet Separator & Coating Technics Development. In Proceedings of the 2nd International Forum on Electrolyte and Separator for Advanced Batteries, Shenzhen, China, 11–13 November 2015; p. 56.
520. Chuanming, G. 隔膜涂布技术的应用与展望 (Application and prospects of separator coating technology) (Cangzhou Mingzhu). In Proceedings of the Third International Forum on Electrolyte & Separator Materials for Advanced Batteries, Nanjing, China, 19 October 2017; p. 30.
521. About "LIELSORT" (Teijin). Available online: https://www.teijin-china.com/focus/lielsort/products/ (accessed on 8 November 2018).
522. Solvay Solution for Li-Ion Battery. Available online: https://wenku.baidu.com/view/620dc4546bd97f192279e951.html (accessed on 8 November 2018).
523. Lithium Ion Battery Focuc Battery Solutions with Kynar® PVDF (Arkema). Available online: https://www.extremematerials-arkema.com/export/sites/technicalpolymers/.content/medias/downloads/brochures/kynar-brochures/2017-kynar-battery-brochure.pdf (accessed on 8 November 2018).
524. Solef® PVDF for Flexible Battery Separators. Available online: https://www.solvay.jp/ja/binaries/Solef-PVDF-Flexible-Separators-for-Batteries_EN-220705.pdf (accessed on 24 January 2021).
525. 复合耐高温热粘性涂层隔膜 (Composite high temperature resistant thermal adhesive coating separator) PMMA (Dinho). Available online: http://dinhotech.com/241646fb-2205-fd8d-2563-884913cf350e/af23b7a2-f989-cea3-8c7f-de384f75d1fc.shtml (accessed on 15 May 2019).
526. Hough, L. Advances in Electrolyte Ingredients & Specialty Polymers for Li-Ion Batteries. In Proceedings of the Advanced Automotive Battery Conference (AABC USA), Virtual, 5 November 2020; p. 17.
527. Dreamweaver Gold MSDS. Available online: https://www.dreamweaverintl.com/uploads/5/7/8/8/57886015/msds-_dreamweaver_gold.pdf (accessed on 3 November 2018).
528. Dreamweaver GoldTM: When Safety is the Primary Concern. Available online: https://www.dreamweaverintl.com/uploads/5/7/8/8/57886015/dwi_brochure-_gold_170419.pdf (accessed on 3 November 2018).
529. MSDS Silver Separator (Dreamweaver). Available online: https://www.dreamweaverintl.com/uploads/5/7/8/8/57886015/msds_silver.pdf (accessed on 7 November 2018).
530. DWI Brochure Silver. Available online: https://www.dreamweaverintl.com/uploads/5/7/8/8/57886015/dwi_brochure_silver_170306.pdf (accessed on 7 November 2018).
531. Hinanofiber, PI180. Available online: https://www.hinanofiber.com/productinfo/1277697.html (accessed on 20 March 2021).

532. Wood, W.; Lee, D.; Frenzel, J.; Wandera, D.; Spitz, D.; Wimer, A.; Waterhouse, R.; Pekala, R.W. Structure-Property Relationships of Ceramic-Modified Separators (Entek). In Proceedings of the 33rd Annual International Battery Seminar & Exhibit, Fort Lauderdale, FL, USA, 23 March 2016; p. 33.
533. Waterhouse, R.; Emanuel, J.; Frenzel, J.; Lee, D.; Peddini, S.; Patil, Y.; Fraser-Bell, G.; Pekala, R.W. Structure, properties, and performance of inorganic-filled separators (Entek). In Proceedings of the 29th International Battery Seminar & Exhibit, Fort Lauderdale, FL, USA, 12–15 March 2012; p. 30.
534. Ceramic Sheet Separators for Lithium Ion Batteries (Freudenberg). Available online: https://separators.freudenberg-pm.com/Products (accessed on 2 November 2018).
535. Loeble, M. Freudenberg Performance materials The Safety Separator. In Proceedings of the 13th China International Battery Fair (CIBF 2018), Shenzhen, China, 22–24 May 2018.
536. Zschech, D. CERIO ® Technologie für Elektro- und Hybridfahrzeuge (LiTec). In Proceedings of the DRIVE-E Akademie, Erlangen, Germany, 9 March 2010; p. 62.
537. Mattis, W.L. Nonflammable, Fast Charging and Superb Battery Cycle Life Battery Technology For Automotive Applications (Microvast). In Proceedings of the 18th International Meeting on Lithium Batteries (IMLB), Chicago, IL, USA, 19–24 June 2016; p. 39.
538. Non-Woven Substrate for Battery Separator NanoBase0 (NB0) Mitsubishi Paper Mills Limited. Available online: https://www.mpm.co.jp/bs/bs-en/pdf/nb0.pdf (accessed on 14 November 2024).
539. Non-Woven Separator with Ceramic Coating NanoBaseX (NBX, For LIB) Mitsubishi Paper Mills Limited. Available online: https://www.mpm.co.jp/bs/bs-en/pdf/nbx.pdf(accessed on 14 November 2024).
540. Ceramic Coated Non Woven Separator "NanobaseX" characteristics and its benefits (Mitsubishi Paper Mills). Available online: https://wenku.baidu.com/view/fd3ea37533687e21af45a9a6.html?sxts=1541191604995 (accessed on 2 November 2018).
541. Optodot Boehmite Separator Patent Portfolio. Available online: http://optodot.com/lithium-battery-technology (accessed on 2 November 2018).
542. New Separator-Free Lithium-Ion Rechargeable Battery Developed by Toshiba. Available online: https://www.toshiba.co.jp/about/press/2018_06/pr0401.htm (accessed on 26 June 2020).
543. エレクトロスピニング法によるナノファイバー膜の高速形成技術 (High-speed formation technology for nanofiber membranes using electrospinning method). Technology Administration & Planning Office, Corporate Technology Planning Division, Toshiba Corporation, 1-1, Shibaura 1-chome, Minato-ku, Tokyo 105-8001, Japan. *Toshiba Rev. (Front. Res. Dev.)* **2017**, *72*, 74–75.
544. Nestler, T.; Schmid, R.; Munchgesang, W.; Bazhenov, V.; Schilm, J.; Leisegang, T.; Meyer, D.C. Separators—Technology Review: Ceramic based Separators for Secondary Batteries. In Proceedings of the 1st International Freiberg Conference on Electrochemical Storage Materials, TU Bergakademie Freiberg, Freiberg, Germany, 3–4 June 2013; pp. 155–184.

545. Santhanagopalan, S.; Zhang, Z.M. Separators for Lithium-Ion Batteries. In *Lithium-Ion Batteries: Advanced Materials and Technologies*; Yuan, X., Liu, H., Zhang, J., Eds.; Green Chemistry and Chemical Engineering; Crc Press-Taylor & Francis Group: Boca Raton, FL, USA, 2011; pp. 197–251.
546. Lee, H.; Yanilmaz, M.; Toprakci, O.; Fu, K.; Zhang, X.W. A review of recent developments in membrane separators for rechargeable lithium-ion batteries. *Energy Environ. Sci.* **2014**, *7*, 3857–3886. [CrossRef]
547. Electrolytes for Lithium-ion batteries (Capchem). Available online: http://en.capchem.com/project/lithium-ion-batteries (accessed on 19 February 2018).
548. 第一工业制药 锂电池用凝胶聚合物电解质 (Dai Ichi Kogyo Seiyaku Gel polymer electrolyte for lithium batteries) ELEXCEL ACG-127 (DKS). Available online: http://www.achatestrade.com/product/276795429 (accessed on 16 May 2024).
549. Wachtler, M. Li-ion Batteries Lecture Winter Term 2016/17 Electrolytes. 6 February 2017; 61.
550. Electrolyte Solutions and Flame Retardant (Targray). Available online: https://www.slideshare.net/mobile/Targray-Technology/electrolyte-solution-for-lithiumion-battery-manufacturing (accessed on 28 October 2018).
551. Ding, M.S.; Xu, K.; Zhang, S.S.; Amine, K.; Henriksen, G.L.; Jow, T.R. Change of conductivity with salt content, solvent composition, and temperature for electrolytes of LiPF6 in ethylene carbonate-ethyl methyl carbonate. *J. Electrochem. Soc.* **2001**, *148*, A1196–A1204. [CrossRef]
552. Logan, E.R.; Tonita, E.M.; Gering, K.L.; Li, J.; Ma, X.W.; Beaulieu, L.Y.; Dahn, J.R. A Study of the Physical Properties of Li-Ion Battery Electrolytes Containing Esters. *J. Electrochem. Soc.* **2018**, *165*, A21–A30. [CrossRef]
553. Ohkubo, K.; Sukino, K.; Yamate, S.; Kozono, S.; Katayama, Y.; Nukuda, T. Enhancement of Low Temperature Power Performance for Lithium-ion Cells with Lithium Titanium Oxide Negative Electrode by New Functional Electrolyte: Application of Carboxylic Acid Ester as Solvent and Organic Titanium Compound as Additive. *GS Yuasa Tech. Rep.* **2009**, *6*, 14–19.
554. Yaakov, D.; Gofer, Y.; Aurbach, D.; Halalay, I.C. On the Study of Electrolyte Solutions for Li-Ion Batteries That Can Work Over a Wide Temperature Range. *J. Electrochem. Soc.* **2010**, *157*, A1383–A1391. [CrossRef]
555. Lingli, K. Advance of R&D of high voltage electrolyte system (Lishen). In Proceedings of the 2nd International Forum on Electrolyte and Separator for Advanced Batteries, Shenzhen, China, 11–13 November 2015; p. 27.
556. High Performance Materials for Batteries (Solvay). Available online: https://www.solvay.com/sites/g/files/srpend221/files/2018-10/High-Performance-Materials-for-Batteries_EN-v1.7_0_0.pdf (accessed on 6 June 2019).
557. SDS LiPF6 (Stella Chemifa). Available online: https://www.stella-chemifa.co.jp/products/files/031102.pdf (accessed on 8 November 2018).
558. Lithium Hexafluorophosphate (JGHITEC). Available online: http://jghitec.cn/cn/products/hexafluorophosphate/ (accessed on 19 March 2021).

559. Synthesis and application technology of new electrolyte salts (Tinci). In Proceedings of the 2013 China Forum On LIB Electrolyte and Separator Technology and Market Development, Guangzhou, China, 14 August 2013; p. 53.
560. Abe, K. Electrolyte Reagents for Lithium-Ion Batteries (Advanced Energy Materials R&D Center, Chemicals Company, UBE Industries, Ltd.). Available online: https://www.sigmaaldrich.com/technical-documents/articles/technology-spotlights/electrolyte-reagents-for-lithium-ion-batteries.html (accessed on 5 September 2018).
561. Products. Available online: http://www.hicomer.com/en/page-47842.html (accessed on 14 November 2024).
562. Hopax Started the Expansion of Its Production Center for Additives for Lithium-Ion Batteries and Electroplating. Available online: https://www.hopaxfc.com/en/blog/hopax-expansion-production-center-for-additives-for-lithium-ion-batteries-and-electroplating (accessed on 14 November 2024).
563. Shijiazhuang Suntec-Chem Co., Ltd. Lithium-Ion Battery Electrolyte Additives 1. Available online: http://www.suntechem.com/product/5/ (accessed on 19 January 2021).
564. Shijiazhuang Suntec-chem Co., Ltd. Lithium-Ion Battery Electrolyte Additives 2. Available online: http://www.suntechem.com/product/5/#c_product_list-15561015348126805-2 (accessed on 19 January 2021).
565. Lithium Battery Electrolyte Additives. Available online: https://www.szcheerchem.com/li-ion-battery-electrolyte-additives/ (accessed on 14 November 2024).
566. Nelson, K. Studies of the Effects of Electrolyte Additives on the Performance of Lithium-Ion Batteries. Master's thesis, Dalhousie University, Halifax, NS, Canada, 2014.
567. Mei, Y.; Shijiazhuangsun-tec Chemical Co., Ltd. Electrolyte additives, CIBF2018 Brochure. Available online: https://app.box.com/s/xrpigz93ktmsrck00nai1lds8nnu8fkc (accessed on 20 May 2021).
568. Wang, F.; Gan, C.; Yuan, X. Industrial progress of nonaqueous liquid electrolytes for lithium-ion batteries. 储能科学与技术 *(Energy Storage Sci. Technol.)* **2016**, *5*, 1–8.
569. Zhang, L.; Zhang, Z.; Khalil, A. Redox Shuttle Additives for Lithium-Ion Battery. In *Lithium Ion Batteries—New Developments*; Belharouak, I., Ed.; InTech: Rijeka, Croatia, 2012; pp. 173–188.
570. Lamanna, W.M.; Bulinski, M.; Jiang, J.; Magnuson, D.; Pham, P.; Triemert, M.; Dahn, J.; Wang, R.; Moshurchak, L.; Garsuch, R. Chemical Redox Shuttle—Design for High Voltage. In Proceedings of the 214th ECS Meeting (Pacific Rim Meeting on Electrochemical and Solid-State Science Meeting), Honolulu, HI, USA, 12–17 October 2008; p. 1278.
571. 锂电最新报告—IIT 11Q1 10LIB Market. Available online: https://wenku.baidu.com/view/8e556d2b915f804d2b16c19f.html (accessed on 6 September 2018).
572. Belov, D.; Yang, M.-H. Ineffectiveness of electrolyte additives for overcharge protection in Li-ion battery. *ECS Trans.* **2007**, *6*, 29–44. [CrossRef]
573. Huang, D. 三元启停电池电解液的研究进展 (Development strategy of electrolyte used in Li ion battery with NMC for start-stop system). In Proceedings of the 2nd HEV Market & Advanced Battery Technology development Seminar, Hanzhou, China, 13–14 October 2016; p. 18.

574. Takeshita, H. Latest LIB Market Trends (The Automotive and Power Storage Markets Move Into Full-scale Growth. In Proceedings of the 10th Int'l Rechargeable Battery Expo Battery Japan 2019, Tokyo, Japan, 28 February 2019; pp. 1–6.
575. Hong, P.B.; Xu, M.Q.; Chen, D.R.; Chen, X.Q.; Xing, L.D.; Huang, Q.M.; Li, W.S. Enhancing Electrochemical Performance of High Voltage (4.5 V) Graphite/$LiNi_{0.5}Co_{0.2}Mn_{0.3}O_2$ Cell by Tailoring Cathode Interface. *J. Electrochem. Soc.* **2017**, *164*, A137–A144. [CrossRef]
576. Wrodnigg, G.H.; Besenhard, J.O.; Winter, M. Ethylene sulfite as electrolyte additive for lithium-ion cells with graphitic anodes. *J. Electrochem. Soc.* **1999**, *146*, 470–472. [CrossRef]
577. IIT E07Q3 3PT & Material Market. Available online: https://wenku.baidu.com/view/d8a2d02d7375a417866f8f91.html (accessed on 6 September 2018).
578. High Performance Materials for Batteries (Solvay). Available online: https://www.solvay.cn/en/binaries/High-Performance-Materials-for-Batteries_EN-220573.pdf (accessed on 6 September 2018).
579. Yang, C.-W.; Wu, F.; Wu, B.-r.; Ren, Y.-H.; Yao, J.-W. 含FEC电解液的锂离子电池低温性能研究 (Low-Temperature Performance of Li-Ion Battery with Fluoroethylene Carbonate Electrolyte). 电化学 *(Electrochemistry)* **2011**, *17*, 63–66.
580. Zhang, S.S.; Xu, K.; Jow, T.R. An improved electrolyte for the $LiFePO_4$ cathode working in a wide temperature range. *J. Power Sources* **2006**, *159*, 702–707. [CrossRef]
581. Han, J.G.; Park, I.; Cha, J.; Park, S.; Myeong, S.; Cho, W.; Kim, S.S.; Hong, S.Y.; Cho, J.; Choi, N.S. Interfacial Architectures Derived by Lithium Difluoro(bisoxalato) Phosphate for Lithium-Rich Cathodes with Superior Cycling Stability and Rate Capability. *ChemElectroChem* **2017**, *4*, 56–65. [CrossRef]
582. Electrolyte for Lithium-Ion Batteries: IONEL™ LF-101 Lithium bis(fluorosulfonyl)imide (LiFSI). Available online: https://www.shokubai.co.jp/en/products/detail/lifsi/ (accessed on 14 November 2024).
583. World Smart Energy Week 2012 東京現場直擊系列報導(三) (Tokyo live coverage series (3)). Available online: https://www.materialsnet.com.tw/DocView.aspx?id=10120 (accessed on 6 November 2018).
584. 3M™ Battery Electrolyte HQ-115 (Technical Data). Available online: https://multimedia.3m.com/mws/media/829379O/3mtm-battery-electrolyte-hq-115.pdf (accessed on 6 September 2018).
585. 1,5,2,4-Dioxadithiane 2,2,4,4-tetraoxide (MMDS). Available online: http://en.suntechem.com/product_detail/10.html (accessed on 14 November 2024).
586. Kishimoto, A.; Nakagawa, H.; Inamasu, T.; Yoshida, H. Investigation of Electrolyte Composition for Lithium Ion Batteries Operated at High Voltage with Ni,Co,Mn-based Positive Electrode. *GS Yuasa Tech. Rep.* **2015**, *12*, 6–11.
587. ホスファゼン [電池添加剤]/Phosphazenes [Battery Additives]. Available online: https://www.tcichemicals.com/eshop/ja/jp/category_index/12749 (accessed on 6 September 2018).

588. Abe, K.; Colera, M.; Shimamoto, K.; Kondo, M.; Miyoshi, K. Functional Electrolytes Recent Advances in Development of Additives for Impedance Reduction. In Proceedings of the International Battery Association meeting 2013, Barcelona, Spain, 10–15 March 2013; p. 22.
589. Song, W.F.; Hong, B.; Hong, S.; Lai, Y.Q.; Li, J.; Liu, Y.X. Effect of prop-1-ene-1,3-sultone on the Performances of Lithium Cobalt Oxide/Graphite Battery Operating Over a Wide Temperature Range. *Int. J. Electrochem. Sci.* **2017**, *12*, 10749–10762. [CrossRef]
590. Kong, L.; Su, Z.; Gao, J.; Sun, J. The mechanism study of TMSP additive in the lithium ion battery. In Proceedings of the 15th International Meeting on Lithium Batteries—IMLB 2010, Montreal, QC, Canada, 27 June–2 July 2010; p. A196.
591. Ren, Y.J.; Wang, M.Z.; Wang, J.L.; Cui, Y.L. Tris (trimethylsilyl) Phosphate as Electrolyte Additive for Lithium—Ion Batteries with Graphite Anode at Elevated Temperature. *Int. J. Electrochem. Sci.* **2018**, *13*, 664–674. [CrossRef]
592. Han, Y.K.; Yoo, J.; Yim, T. Why is tris(trimethylsilyl) phosphite effective as an additive for high-voltage lithium-ion batteries? *J. Mater. Chem. A* **2015**, *3*, 10900–10909. [CrossRef]
593. He, M.N.; Su, C.C.; Peebles, C.; Feng, Z.X.; Connell, J.G.; Liao, C.; Wang, Y.; Shkrob, I.A.; Zhang, Z.C. Mechanistic Insight in the Function of Phosphite Additives for Protection of $LiNi_{0.5}Co_{0.2}Mn_{0.3}O_2$ Cathode in High Voltage Li-Ion Cells. *Acs Appl. Mater. Interfaces* **2016**, *8*, 11450–11458. [CrossRef]
594. Abe, K. Nonaqueous Electrolytes and Advances in Additives. In *Electrolytes for Lithium and Lithium-Ion Batteries, Jow, T.R., Xu, K., Borodin, O., Ue, M., Eds.*; Springer: New York, NY, USA, 2014; pp. 167–207.
595. Tao, Y.; Yamate, S.; Ozaki, T.; Inamasu, T.; Yoshida, H.; Okuyama, R. リチウムイオン二次電池用グラファイト負極上の SEI 被膜の成長過程 (Growth Process of SEI Film on Graphite Negative Electrode for Lithium-ion Secondary Battery). *GS Yuasa Tech. Rep.* **2013**, *2*, 8–15.
596. Abe, K.; Ushigoe, Y.; Yoshitake, H.; Yoshio, M. Functional electrolytes: Novel type additives for cathode materials, providing high cycleability performance. *J. Power Sources* **2006**, *153*, 328–335. [CrossRef]
597. Zhang, S.S. An unique lithium salt for the improved electrolyte of Li-ion battery. *Electrochem. Commun.* **2006**, *8*, 1423–1428. [CrossRef]
598. Ярмоленко, О.В.; Юдина, А.В.; Игнатова, А.А. Современное состояние и перспективы развития жидких электролитных систем для литий-ионных аккумуляторов. Электрохимическая энергетика **2016**, *16*, 155–195, (Yarmolenko, O.V.; Yudina, A.V.; Ignatova, A.A. Current state and prospects for the development of liquid electrolyte systems for lithium-ion batteries. Electrochemical energy **2016**, 16, 155–195). [CrossRef]
599. Shibata, Y.; Sukino, K.; Tabuchi, T.; Inamasu, T.; Okuyama, R. Development of Higher Safety 12 Ah-class Lithium-ion Cells with High Performance Using Flame-resistant Electrolyte Containing Fluorinated Phosphate. *GS Yuasa Tech. Rep.* **2010**, *7*, 8–13.
600. Xu, K. Nonaqueous liquid electrolytes for lithium-based rechargeable batteries. *Chem. Rev.* **2004**, *104*, 4303–4417. [CrossRef]

601. Xu, K. Electrolytes and Interphases in Li-Ion Batteries and Beyond. *Chem. Rev.* **2014**, *114*, 11503–11618. [CrossRef]
602. Jow, T.R.; Xu, K.; Borodin, O.; Ue, M. *Electrolytes for Lithium and Lithium-Ion Batteries*; Springer Science+Business Media: New York, NY, USA, 2014; Volume 58, p. 476.
603. Applications Smart Wearable. Available online: https://www.atlbattery.com/en/details.html?product=wearables (accessed on 22 May 2019).
604. Lu, H.S. 2015年电池应用市场不同发展风貌与未来三年前景-IT与储能应用(研讨会简报) (Different development styles of the battery application market in 2015 and prospects for the next three years—IT and energy storage applications (symposium briefing)) (ITRI IEK). In Proceedings of the 轉型與創新-2014年電池市場回顧與2015年展望研討會, 台大醫院國際會議中心 (Transformation and Innovation—2014 Battery Market Review and 2015 Outlook Seminar, National Taiwan University Hospital International Conference Center), Taipei, Taiwan, 9 December 2014; p. 45.
605. Yang, M.-H. What's The Next Standard LIB cell for LEVs and EVs Applications? (Hitech energy). In Proceedings of the Advanced Automotive and Industrial Battery Conference (AABC2016, Europe), Mainz, Germany, 25–28 January 2016; p. 27.
606. $LiFePO_4$ Batteries and Battery Chargers. Available online: https://electricscooterparts.com/lifepo4batteries.html (accessed on 12 November 2018).
607. CES 2009: Schwinn's Tailwind Bike Uses Exclusive Toshiba Battery Tech (Video). Available online: https://www.treehugger.com/clean-technology/ces-2009-schwinns-tailwind-bike-uses-exclusive-toshiba-battery-tech-video.html (accessed on 14 June 2023).
608. 第12章E-Motorcycle和xEV巴士、卡车的市场动向（13Q1）(Chapter 12 Market Trends of E-Motorcycle and xEV Buses and Trucks (13Q1)) (B3). Available online: https://wenku.baidu.com/view/d00c136552d380eb62946dd8.html (accessed on 13 November 2018).
609. Lu, H.S. 2015年電池應用市場的不同發展風貌與未來三年前景-動力應用 (Different development styles of the battery application market in 2015 and prospects for the next three years—Power Applications). In Proceedings of the 轉型與創新-2014年電池市場回顧與2015年展望研討會, 台大醫院國際會議中心 (Transformation and Innovation—2014 Battery Market Review and 2015 Outlook Seminar, National Taiwan University Hospital International Conference Center), Taipei, Taiwan, 9 December 2014; p. 79.
610. Li-ion Batteries for Electric Buses 2018–2028 Technology, Market Trends, Forecasts and Key Players. Available online: https://www.idtechex.com/research/reports/li-ion-batteries-for-electric-buses-2018-2028-000595.asp (accessed on 9 November 2018).
611. Solid-State batteries are already being installed in city buses: Update. Available online: https://www.greencarreports.com/news/1132780_solid-state-batteries-are-already-being-installed-in-city-buses-update (accessed on 14 June 2023).
612. Lu, H.L. 電動大客車國際發展趨勢與展望 (International development trends and prospects of electric buses) (ITRI IEK). In Proceedings of the 公路公共運輸推動成果分享未來展望研討會 (Road Public Transport Promotion Achievements Sharing and Future Outlook Seminar) (MOT), Taipei, Taiwan, 20 April 2018; p. 39.

613. 宇通E7纯电动培训-6、7米- 宇通E7纯电动城市客车原理及操作手册 (Yutong E7 pure electric training—6, 7 meters—Yutong E7 pure electric city bus principle and operation manual) (Yutong). Available online: https://wenku.baidu.com/view/ca569790d5d8d15abe23482fb4daa58da0111c35.html?from=search (accessed on 7 November 2018).
614. Electric Buses in Cities. Driving Towards Cleaner Air and Lower CO_2 (Bloomberg New Energy Finance). Available online: https://c40-production-images.s3.amazonaws.com/other_uploads/images/1726_BNEF_C40_Electric_buses_in_cities_FINAL_APPROVED_(2).original.pdf?1523363881 (accessed on 22 May 2019).

3 Lithium-Ion Cells with High Specific Energy

3.1. Lithium-Ion Cells in Prismatic Cases

Users' requirements for portable electronic devices include fast response and small volume/weight, and long-term autonomous operation. These characteristics may be essentially improved by using lithium-ion cells (LICs) with high specific energy.

The lithium-ion battery (LIB, lithium-ion cell + battery controller, or battery management system and battery case) was used as a cell phone power supply for the first time by the Sony company in 1991 [1]. Despite a long history of evolution, even in recent years, a significant increase in specific energy has been observed, which can be exemplified by the specific energy of the LIBs produced by one of the leading manufacturers (see Figure 11). The plot below shows that the specific energy depends on the active materials used in fabrication and the form factor. For example, prismatic cells fabricated using plastic cases with large volume and weight have lower energy density than gel polymer LICs with cases made of laminated foil (pouches).

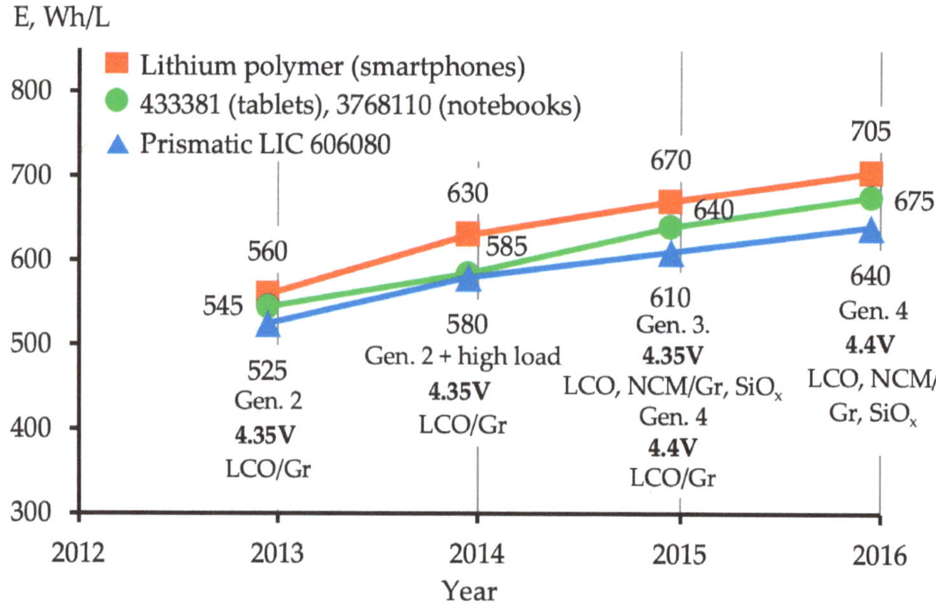

Figure 11. Energy density (Wh/L) for LICs of different form factors produced by one of the leading manufacturers (the values have been determined in discharging with a current of 0.2 C to 2.75 V). Initial data were taken from [2]. Source: Figure by authors.

Lithium-ion cells in laminated foil (pouch) cases [3] are often called gel polymer or polymer cells (batteries) because polymers are added to the electrolyte to thicken it. However, not all manufacturers add polymers to the electrolyte with further polymerisation when laminated foil (pouch) cases are used [3]. Among such polymers, PVDF-HFP is mentioned most often. In addition, some companies (e.g., Sony) add, besides the polymer, fine particles of inorganic materials for the sake of ensuring greater safety in emergencies [4].

The variations in the energy density of gel polymer cells during the last 12 years can be exemplified by those used as power supplies in Samsung S smartphones. From 2010 to 2022, the energy density of these LICs increased from 435 to 720 Wh/L (Table 14). The energy density growth is the result of improvements in materials and technology and cell volume growth.

As an active cathode material, lithiated cobalt oxide is used, whose density of the positive-electrode-active layer may reach 3.7 g/cm^3 [5] or higher. For example, the density of the active layer based on a mixture of lithiated cobalt oxide and the lithiated nickel cobalt manganese oxide may reach 4.0 g/cm^3 [6].

One of the trends in increasing the specific energy of gel polymer LICs is the rise in the charge (and, hence, average discharge) voltage. The cutoff charge voltages for LIBs used in state-of-the-art Samsung and Apple smartphones are 4.47 V [7] (BG991ABY), 4.45 V [8] (S3110), 4.43 V [9] (BG980ABY), 4.4 V (Table 14), and 4.35 V [10]. The necessary performance of the lithiated cathode material can be achieved in a wide range of potentials by stabilising the structure with dopants and modifying the surface with various functional coatings [11]. Coatings may be applied onto the active layer or the electrode during fabrication, or during the formation process when the electrolyte's functional additives interact with the surface of the active material. The functional coatings ensure the following:

- A decrease in resistance during interphase reactions (electrolyte—cathode material);
- A reduction in electrolyte side reaction intensity;
- The suppression of gas release in the process of LIC operation.

In addition, the stability of the cathode material structure at high temperatures and potentials can be improved.

Graphite-type anode materials are mainly used for manufacturing LICs with high specific energy. The main advantages of graphite are as follows:

- High reversible capacity (355 ÷ 360 mAh/g for artificial graphites obtained at a processing temperature of about 3000 °C [11]);
- Ability to provide high electrode density (a mix of graphites with particles of different sizes and shapes, artificial and natural graphite [6]);
- Low swelling at lithium intercalation/deintercalation [12].

Besides graphite, it is possible to add a nanocomposite containing silicon to the active layer of a high-specific-energy LIC negative electrode (SCN (Silicon Carbon Nanocomposite, Samsung [12]), SiO_x/Si/C (Hitachi Maxell [13], Shin-Etsu [14])).

Table 14. Characteristics of LIBs used to power Samsung S smartphones.

Phone Model	Year	Marking/ Size, mm Weight *	C_{rated}, Ah	U_{max} / U_{avrg} V	E_{min} Wh	E_{typ} Wh	E_{min} Wh/L	E_{typ} Wh/L	E_{min} Wh/kg	E_{typ} Wh/kg	Ref.
S22	2022	BS901ABY 496167	3.59	4.47 / 3.88	13.92	14.35	695	717	–	–	[15]
S21	2021	BG991ABY 5.2 × 60 × 69(66) m = 52.90 g	3.88	4.47 / 3.88	15.06	15.52	700	721	284	293	[7]
S20	2020	BG980ABY PGF536070A	3.88	4.43 / 3.86	14.98	15.44	673	694	–	–	[9]
S10+	2019	BG975ABU 5.3 × 47 × 91(87) m = 55.75 g	4.10	4.4 / 3.85	15.4	15.79	679	697	276	283	[16]
S10	2019	BG973ABU 5.3 × 44 × 83(79) m = 46.20 g	3.30	4.4 / 3.85	12.71	13.09	657	676	275	283	[17]
S9	2018	BG960ABA PGF564177HT	3.00	4.4 / 3.85	11.55	–	653	–	–	–	[18]
S8	2017	BG950ABA PGF544183H	3.00	4.4 / 3.85	11.55	–	629	–	–	–	[19]
S7	2016	BG930ABA PGF494088H	3.00	4.4 / 3.85	11.55	–	670	–	–	–	[20]
S6	2015	BG920ABE PGF394595	2.55	4.4 / 3.85	9.82	–	589	–	–	–	[21]
S5	2014	BG900BBE 5.5 × 42.3 × 84	2.80	4.4 / 3.85	10.78	–	552	–	–	–	[22]
S4	2013	EB-B600 5.3 × 57 × 63	2.60	4.35 / 3.8	9.88	–	519	–	–	–	[23]
S3	2012	EB-l1g6 5.4 × 50.5 × 63	2.10	4.35 / 3.8	7.98	–	464	–	–	–	[24]

Table 14. *Cont.*

Phone Model	Year	Marking/ Size, mm Weight *	C_{rated}, Ah	U_{max} U_{avrg} V	E_{min}	E_{typ}	E_{min}	E_{typ}	E_{min}	E_{typ}	Ref.
					Wh		Wh/L		Wh/kg		
S2	2011	EB-1A2GBU 5 × 46 × 59	1.65	4.2 3.7	6.11	–	450	–	–	–	[25]
S1	2010	EB57515ZVU ≈5 × 50 × 51	1.50	4.2 3.7	5.55	–	435	–	–	–	[26]

Note: C_{rated}—rated capacity, U_{max}—maximum charge voltage, U_{avrg}—average discharge voltage, E_{min}—minimal energy, E_{typ}—typical energy. * Weight of the battery with glue layer and without protective polymer sheets. Specific energy of smartphones S6–S9, S20, and S22 was calculated based on their overall dimensions indicated in LIB marking. According to presentation [5], the specific energy of LIBs in S8/S9 is 688 Wh/L and 278 Wh/kg, close to the characteristics of LIBs in iPhone X(L) (695 Wh/L, 275 Wh/kg, 511064L) and iPhone X(H) (687 Wh/L, 272 Wh/kg, S3110). Source: Authors' compilation based on data from references cited in the table.

However, the silicon weight fraction in the active layer is typically insignificant: only a few percent. The silicon-containing materials' weight fraction is limited in electrodes due to their volume change when lithium is introduced into or extracted from the structure. If thin copper foil is used, the increase in the electrode volume can result in mechanical strains and possible electrode damage [27].

According to a report [28], in 2015, Shin-Etsu products were the most demanded silicon-containing composites for fabricating negative electrodes.

The increase in LICs' specific energy may be ensured not only by using materials with high specific energies, but also by increasing their content in terms of LIC volume and weight. The increment in the volume of active materials may be achieved by reducing the fraction of conductive additive(s) and increasing the electrode thickness and density. These techniques lead to the growth of intrinsic resistance, which affects the maximum charge/discharge current. High-specific-energy LICs are not charged/discharged with high currents because of the possibility of severe heating.

Consider the characteristics of some LICs fabricated using lithiated cobalt oxide and graphite as the active materials. The elevation of discharge current leads to a lower mean discharge voltage, capacity, and specific cell energy (Figure 12a). When the discharge temperature decreases (Figure 12b), a similar dependence is observed. The necessary dependencies can be plotted to form a concept of variation in specific energy and power of the batteries used in smartphones (Figure 13). The energy and power characteristics presented in this section are approximate due to certain factors:

- In references devoted to cell development, the weight of the cell is specified. However, the smartphone battery weight includes the weights of the cell, battery management system, and battery connection cable. Therefore, it is necessary to understand whether the value applies to the battery or only to a cell when making a comparison.
- The LIC characteristics may scatter from sample to sample in one batch.
- The techniques used, the quality of curve digitising (performed with the use of Graphic for iPad [29] and GetData software [30]), and further calculation may also affect the obtained values.

The most up-to-date prismatic batteries' specific energy (discharge current C/20) reaches 275 ÷ 300 Wh/kg. The energy density (block of electrodes in laminated foil free of the battery controller) may reach 770 Wh/L. An L-shaped battery's specific energy and energy density (ICP5/110/64) [31] are slightly lower, namely, 250 Wh/kg and 690 Wh/L, respectively. An increase in the discharge current results in a reduction in specific energy. The performed measurements showed that the BG991ABY battery's discharge current increase to 1.5C is accompanied by a decrease in specific energy and energy density to 270 Wh/kg and 660 Wh/L, correspondingly.

Since customers are interested in optimising the weight–size characteristics of the device as a whole (smartphone, iPad, ultrabook, and others), the desired effect can also be achieved through the battery pack arrangement and shape. For instance, the power supplies for the smartphones Apple iPhone X [32] and iPhone XSmax [10] consist of two LICs. The efficiency of the device's internal space utilisation may be increased by using prismatic LICs with an additional step (step design) or LICs of a complex shape (hexagonal, smart clock battery) [33].

The performance characteristics of portable electronic devices (and other products) may be improved by increasing the charge rate. However, the acceleration of charge requires an increase in the electrode porosity and a decrease in thickness, and so on, which negatively affects the energy density, cost, etc. [34,35]. The LICs of some models [11] (case 425882, ATL) may be charged by currents of 1.5C (680 Wh/L, 0 ÷ 75%—30 min), 3C (550 Wh/L, 0 ÷ 75%—20 min), or 5C (460 Wh/L, 0 ÷ 75%—10 min [11]).

Among high-power LICs with high specific energies, there are those used to provide the performance of unmanned aerial vehicles (quadcopters and others). For instance, the nominal specific energy of the 6860C5 cell (ATL) reaches 235 Wh/kg (≈520 Wh/L). After 500 cycles of charge (1C)/discharge (5C) in the range of 3.0 ÷ 4.35 V at a temperature of 40 °C, 10% of the discharge capacity decrease and an 8% increase in cell thickness (swelling) are observed [36]. For the battery pack assembled using four LICs installed in series (4S1P), the manufacturer guarantees a lower cycle life (capacity of more than 90% of the nominal one after 200 operating cycles). The LIC case swelling may be caused by the formation of cavities in the positive-electrode-active layer due to the partial loss of contact with the current collector [37] and an increase in the thickness of the anode-active layer [38,39]. The case swelling may also be caused by the formation of various gases (hydrogen, first of all) within the LIC bulk [37]. It was found out that reduction in the cutoff discharge voltage from 3.0 V to 2.5 V leads to the intense generation of hydrogen within the bulk of LICs for portable electronic devices (in pouch cases) [3]. The hydrogen is generated at large discharge depths.

High-specific-energy LICs in pouch cases are also produced using lithiated nickel cobalt manganese oxide (NCM). Despite their high mass-specific energy, they seem to have relatively low energy densities (LG 59Ah, E ≈ 260 (≈470) Wh/kg (Wh/L) [40–44]; therefore, they are not used as power supplies for portable electronic devices. Among recent achievements (2019), particular attention should be paid to LICs with a specific energy and energy density of ≈300 Wh/kg and ≈700 Wh/L developed by the CATL company [45], which will be implemented in the production of automobile battery packs.

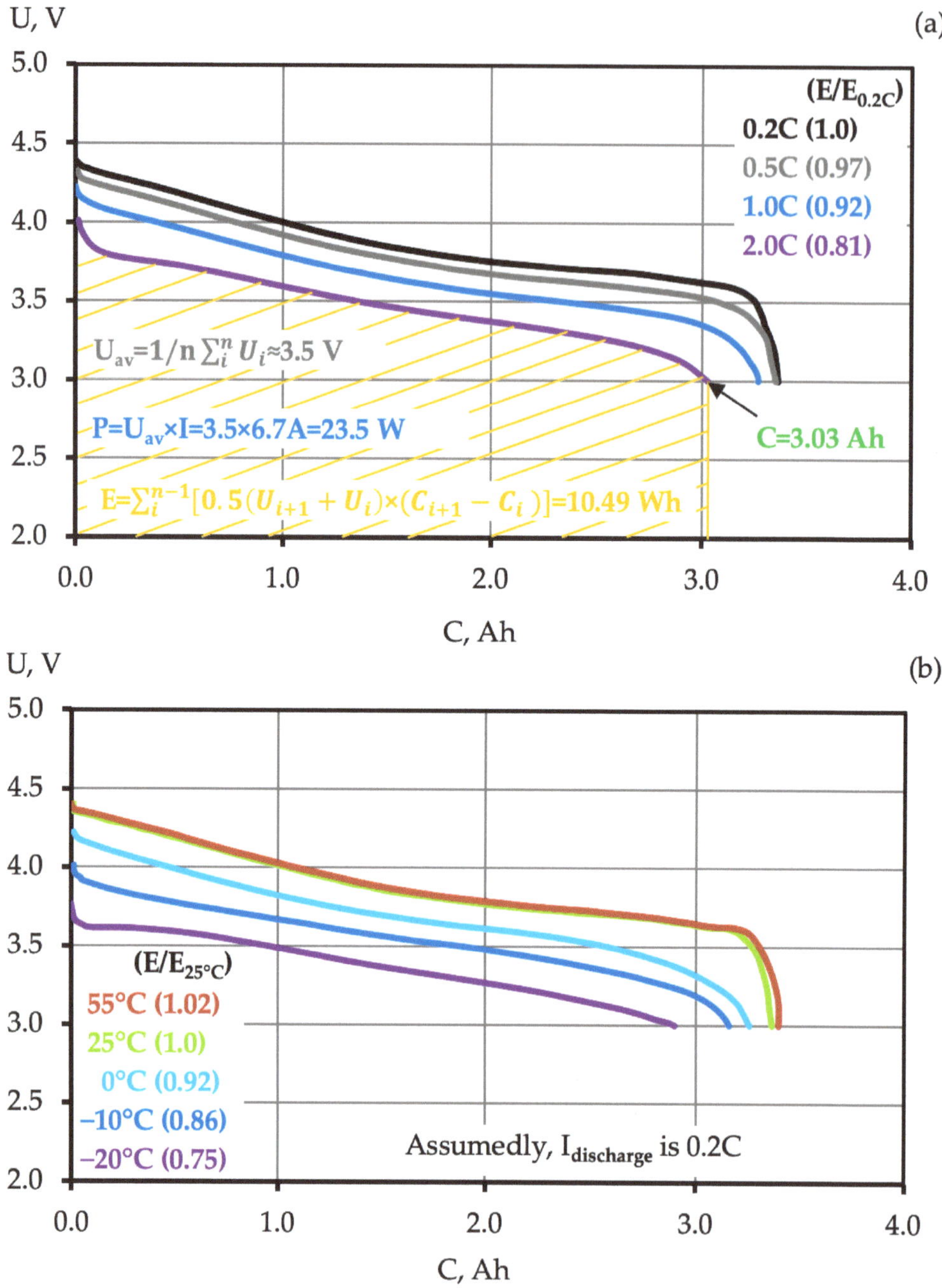

Figure 12. Effect of discharge current (**a**) and temperature (**b**) on the shapes of discharge curves of a high-specific-energy lithium polymer LIC (based on data from [38]). Source: Figure by authors.

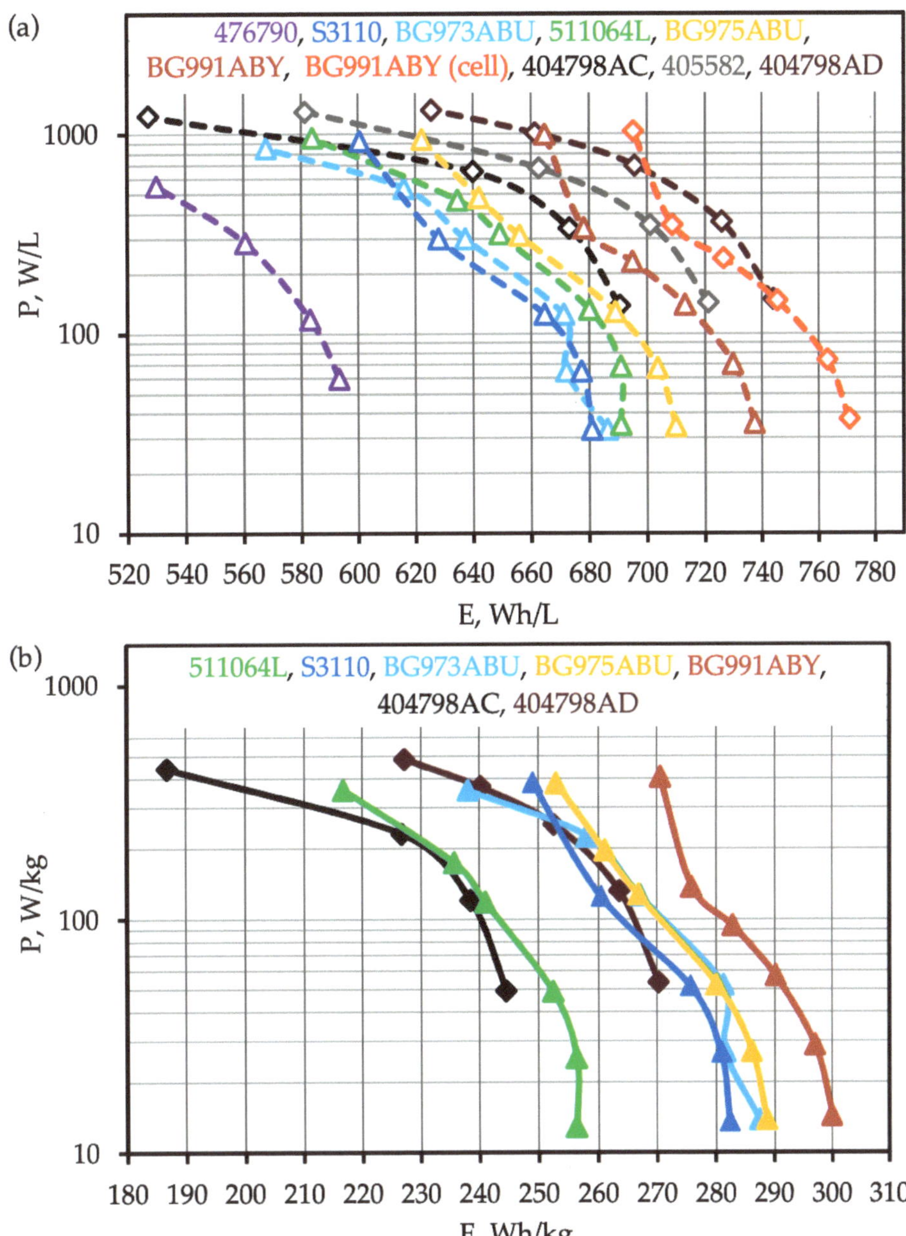

Figure 13. Energy (Wh/L, Wh/kg) vs. power (W/L, W/kg) of high-specific-energy lithium-ion batteries and cells 476790 [2], S3110* [8], BG973ABU* [17], 511064L* [31], BG975ABU* [16], BG991ABY* [7], 404798AC [46], 405582 [38], and 404798AD [46] (*data was taken from results of tests conducted in Ioffe Institute). Energy and power related to volume (**a**) and weight (**b**) of the cells (batteries), correspondingly. Source: Figure by authors.

3.2. Lithium-Ion Cells in Cylindrical Cases

Lithium-ion cells in cylindrical cases (14430, 14500, 14650, 18650, 21700, etc.) are used for battery packs in notebooks, photo-cameras, external batteries, electric vehicles, etc. Cylindrical LICs have a record-high energy density (Table 15). Nevertheless, in fabricating battery packs, it should be considered that being close-packed in a prismatic case, cylindrical LICs occupy only about 80% of the space (depending on their number and style of packing).

Table 15 lists the characteristics of the LICs of the two most frequently mentioned form factors, 18650 and 21700, produced using different pairs of active anode and cathode materials. The main specific feature of LIC 18650WH1 is the introduction into the negative-electrode-active layer of the nanocomposite amorphous material SnCo/C (sometimes referred to as SnCoTi, since the presence of titanium atoms was revealed in analysing the negative electrode of the first-generation LIC Nexelion [47]). The addition of the tin–cobalt alloy provides the 14430 and 18650 LICs with a higher capacity, namely, 0.9 Ah [48–52] and 3.5 Ah (Table 15), respectively. However, because of a lower average discharge voltage compared with LIC 18650 MJ1 (and some analogues), the total energy of 18650 WH1 is lower and equals 11.7 Wh. Sony designed this cell for battery packs for internally produced notebooks. Unfortunately, we failed to find its test results in publications.

Table 15. Characteristics of high-specific-energy LICs in cylindrical cases.

Trademark		18650Wh1, Nexelion, Sony (4.3 ÷ 2.0 V)	18650E1 (4.35 V), LG	18650MJ1, LG	21700M50, LG
Year		2011	2012	2014	2016
Cathode		LCO	LCO	NCM with high Ni content	
Anode		Graphite + SnCo/C	Graphite *	Graphite + SiO_x	
Overall dimensions, mm		d18; h65	d18.4; h65.05	d18.5; h65.2	d21.1; h70.15
Weight, g		53.5	<49	<49	<69
R AC (1 kHz), mOhm		–	<70 (PTC)	<40	<25
Nominal Voltage, V		≈3.5	3.75	3.635	3.635
Max. Direct current, A		–	4.65	10 (7 [53])	7.275
Energy, Wh		11.5	12	12.7	18.2
Specific	Wh/kg	226	245	255	263
energy	Wh/L	723 **	694	725	742
Ref.		[54]	[55]	[56]	[57]

Note: LCO is lithiated cobalt oxide, NCM is lithiated mixed nickel cobalt manganese oxide. The values specified in the table are only for the reference and should be refined for each LIC. d—diameter (mm), h—height (mm), PTC is the membrane with a positive temperature coefficient, * assumably, graphite is used as the anode material. A similar Samsung-produced cell, 18650 32 A, is fabricated using high-voltage lithiated cobalt oxide and artificial graphite. ** is declared by the manufacturer. Source: Authors' compilation based on data from references cited in the table.

In producing LIC 18650E1, high-voltage lithiated cobalt oxide is used as the active cathode material, enabling a cutoff charge voltage of 4.35 V. When the cutoff voltage increases, the discharge capacity increases (Figure 14). The effect of the discharge current on the form of the discharge curve of LIC 18650E1 (cathode LCO) and MJ1 (cathode High-Ni NCM) is given in Figure 15. At higher discharge currents (7A), the discharge curve of 18650E1 is lower than for MJ1.

Based on the mutual closeness of the average discharge voltage of LIC MJ1 and M50, it is possible to conclude that almost the same active materials might be used for their fabrication. Therefore, the increase in the LICs' overall dimensions from 18650 (MJ1) to 21700 (M50) allows the LIC Watt-hour cost to be decreased without modifying the production procedure [58]. A detailed comparison of 18650 and 21700 cells is given in papers [59,60].

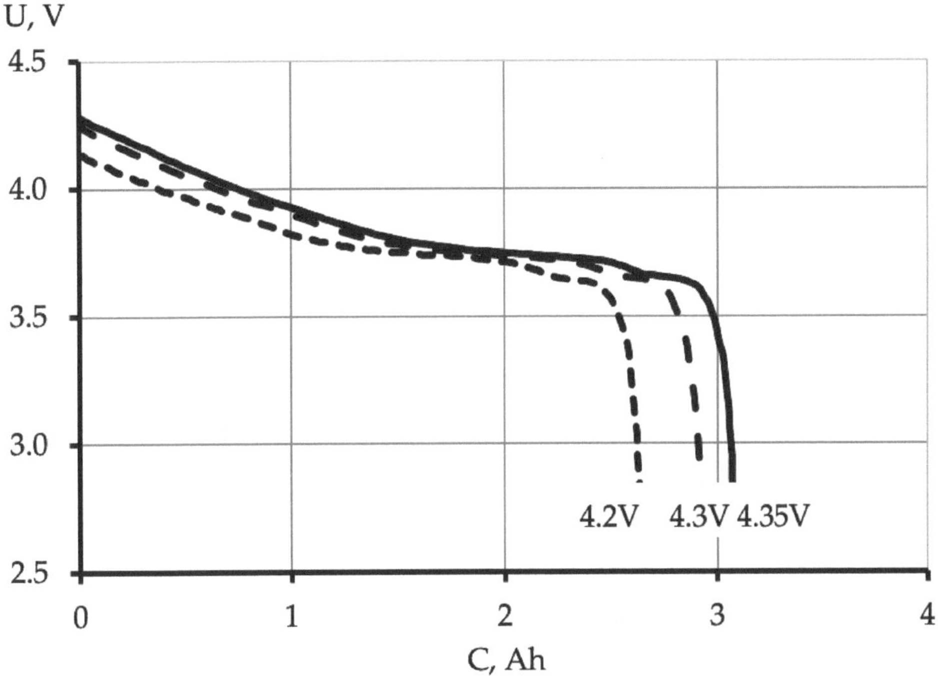

Figure 14. Effect of the cutoff charge voltage on the shape of LIC 18650 E1 discharge curves (based on data presented in reference [61]). Source: Figure by authors.

To assess the current level of the achieved characteristics, data taken from booklets and test results for 18650 LIC produced by leading manufacturers were generalised. For instance, the mathematical processing of the discharge curves gave the total energy (area under the discharge curve), maximal discharge capacity, and average discharge voltage (Table 16). Taking into account the measurement error of the studied LICs, it is possible to assume that the analysed LIC models with cathodes based on nickel-containing lithiated oxide have close characteristics. In contrast, E1 (18650) LICs, with cathodes based on lithiated cobalt oxide, have lower specific energy.

During discharge, the LIC intrinsic temperature increases (Figure 16a). The case wall temperature at the end of the discharge is proportional to the discharge current (Figure 16b) and, evidently, to the capacity (for LICs of the GA, 35E, and VC7 series).

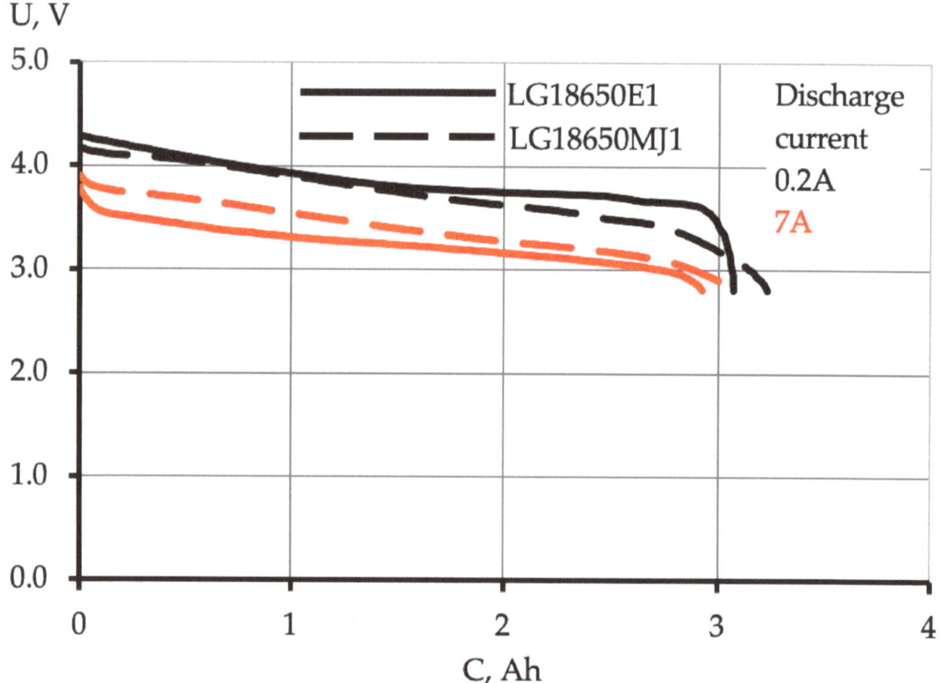

Figure 15. Discharge curves of LIC LG18650 E1 and LG18650 MJ1 obtained at discharge currents of 0.2 A and 7 A (based on data presented in references [53,61]). Source: Figure by authors.

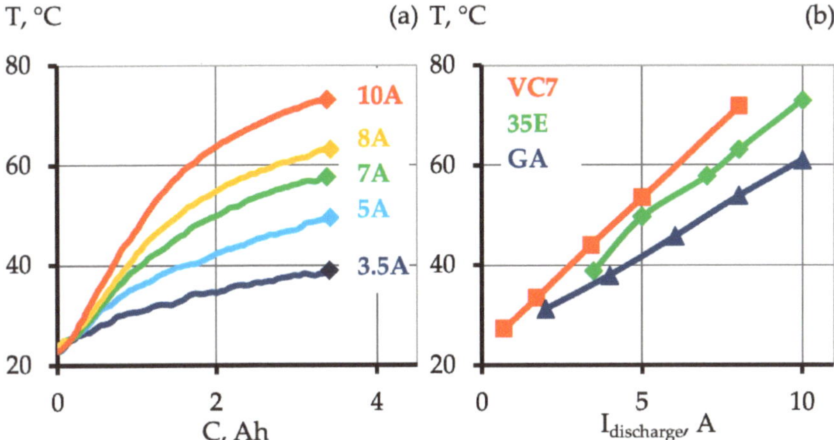

Figure 16. Variations in the LIC SDI 18650 35E case wall temperature in discharging with currents of 3.5A ÷ 10A (**a**) and case wall temperature of LIC Sony 18650 VC7 [65], SDI 18650 35E [66], and Panasonic 18650 GA [67] at the end of discharging vs. the discharge current (the initial sources for plots were taken from references given in square brackets) (**b**). Source: Figure by authors.

The effect of the discharge temperature (when charging is performed at room temperature) on the shape of discharging curves and variations in the case wall temperature has been demonstrated by the example of LIC VC7 (Figure 17). A reduction in the discharge temperature results in a decrease in the discharge capacity and average discharge voltage. Since high-specific-energy LICs have relatively high resistance, a minimum may be observed in the discharge curves (U vs. C) at low temperatures at the beginning of discharge. High internal resistance promotes LIC heating. The heating of LICs leads to a reduction in the internal resistance, increasing the discharge voltage and capacity. For instance, in the case of VC7 LIC cooled to −20 °C, the wall temperature at the end of the discharge can rise to 30 °C. Discharging VC7 with a current of 1.4C at higher temperatures also increases the wall temperature. The wall temperature of VC7 preheated to 60 °C may reach 80 °C at the end of discharge.

Based on the calculated specific powers and specific energies, Ragone diagrams (Figure 18) were plotted for the LICs in case 18650 considered in this section. The specific energy ranges from 240 to 256 Wh/kg (680 ÷ 720 Wh/L) at low discharge currents (0.2 A ÷ 0.68 A). When the discharge current increases to 7 A, the specific energy decreases to 190 ÷ 220 Wh/kg (550 ÷ 610 Wh/L), while power ranging from 465 to 540 W/kg is introduced into the active anode layer structure.

Table 16. Energy, discharge capacity, and average discharge voltage of lithium-ion cells in 18650 cases—LG E1 [61], LG MJ1 [53], Sanyo GA [62,63], Samsung 35E [64], and Sony VC7 [65], determined from the discharge curves.

	Mark	I, A U, V	0.2 Wh	1	3	5	7	10
				% of the value at the discharge current of 0.2 A				
Energy	E1	4.2 ÷ 2.8	9.93	95.7	90.6	86.0	78.4	
	E1	4.3 ÷ 2.8	11.20	95.4	89.6	85.5	77.9	
	E1	4.35 ÷ 2.8	11.82	96.4	90.7	86.4	80.2	
	MJ1	4.2 ÷ 2.8	11.94	97.7	93.8	90.9	87.6	
	GA	4.2 ÷ 2.8	12.29	96.5	92.6	90.0	87.5	83.9
	35E	4.2 ÷ 2.8	12.50	97.1	92.6	89.1	86.5	82.6
	VC7	4.2 ÷ 2.5	12.83	96.9	93.8	91.2	87.6	
			Ah	% of the value at the discharge current of 0.2 A				
Capacity	E1	4.2 ÷ 2.8	2.64	98.6	98.1	98.1	94.1	
	E1	4.3 ÷ 2.8	2.92	98.0	97.6	97.9	93.8	
	E1	4.35 ÷ 2.8	3.07	99.0	98.1	97.8	95.2	
	MJ1	4.2 ÷ 2.8	3.23	99.2	98.0	97.1	95.5	
	GA	4.2 ÷ 2.8	3.35	97.5	96.1	95.4	94.3	92.6
	35E	4.2 ÷ 2.8	3.40	98.6	97.1	96.0	95.0	93.1
	VC7	4.2 ÷ 2.5	3.54	99.0	98.1	97.2	96.0	
			V	% of the value at the discharge current of 0.2 A				
Aver. discharge voltage	E1	4.2 ÷ 2.8	3.83	97.1	91.6	86.9	82.8	
	E1	4.3 ÷ 2.8	3.86	97.4	92.0	87.6	83.2	
	E1	4.35 ÷ 2.8	3.87	97.4	92.5	88.6	84.5	
	MJ1	4.2 ÷ 2.8	3.73	98.4	95.7	93.3	91.7	
	GA	4.2 ÷ 2.8	3.70	98.9	96.2	94.3	92.7	90.5
	35E	4.2 ÷ 2.8	3.71	98.4	95.1	92.7	91.1	88.7
	VC7	4.2 ÷ 2.5	3.64	98.4	95.9	94.0	91.5	
Discharge current of LIC VC7, A			0.68	1.7	3.4	5	8	

Source: Table by authors.

The cycle life of different high-energy lithium-ion cells is presented in Figure 19. The increase in cycling temperature to 45 °C leads to a minor decrement in the cycle life. The subsequent elevation in temperature during the cycling test might lead to a decrease in capacity retention.

As a result, the structure of the cathode material is more stable in the case of cell charging to a lower potential. For instance, if the cutoff voltage is reduced in charging from 4.2 V to 4.0 V, the LIC 26F cycle life increases from 1000 to 2700 cycles (Figure 20). Suppose that NCM (especially with a high content of nickel) is used as the cathode. In that case, the cycle life increase is caused by a decrease in the probability of phase transition (H2/H3) occurring at relatively high delithiation extents and causing particle cracking [68–70].

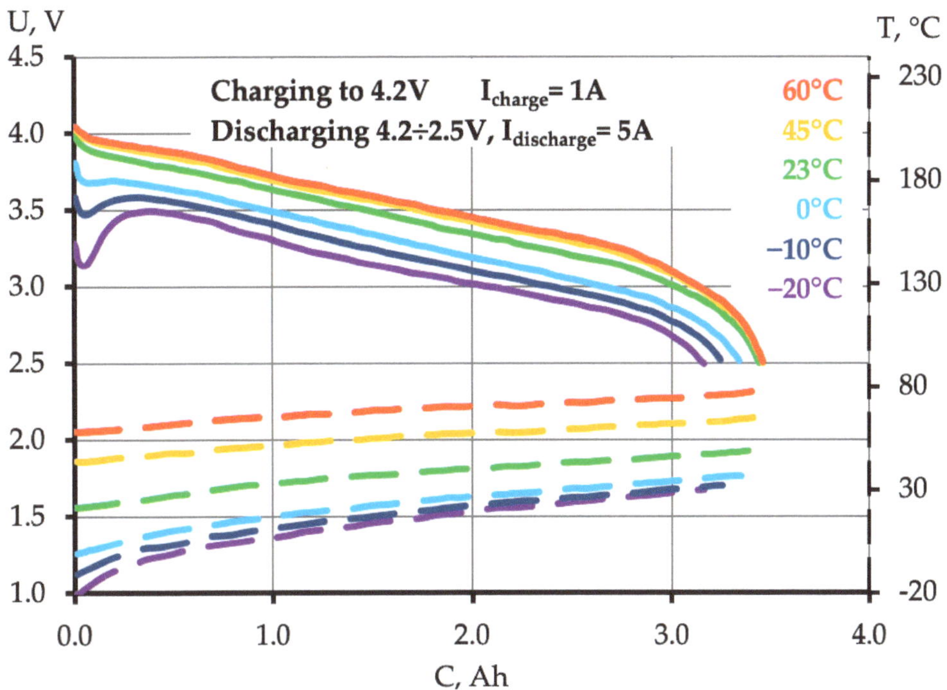

Figure 17. Effect of LIC Sony 18650 VC7 initial temperature on the LIC case's discharge capacity and LIC case's temperature variations in discharging with a current of 5 A (the initial data were taken from [65]). Source: Figure by authors.

To enhance the safety of high-energy LICs in cylindrical cases, a ceramic coating in the volume of the cell is provided (Sony, LG—coating on the separator (LG—SRS), or, in the case of SDI 32 A, coating on the anode (SFL)). LIC 18650 E1 also has a film with a positive temperature coefficient (PTC). When the temperature increases, the conductivity decreases; thus, the electric insulation of LICs occurs (Figure 21). The design of the cylindrical LICs 18650 and 21700 may stipulate other safety-ensuring automatically actuated components [71]. For instance, lithium bis(oxalate)borate added to the electrolyte oxidises at a higher LIC cathode potential (4.5 V and higher) and releases carbon dioxide; the internal pressure increases, which actuates the current interruptive device mounted on the lid (top) of the LIC [71]. Examples of the case lid designs for LIC 18650B, GA, 35E, MJ1, and VC7 are shown elsewhere [72]. A sharp rise in pressure may cause gas release through the "valves" mounted in the top and bottom parts of the case [72].

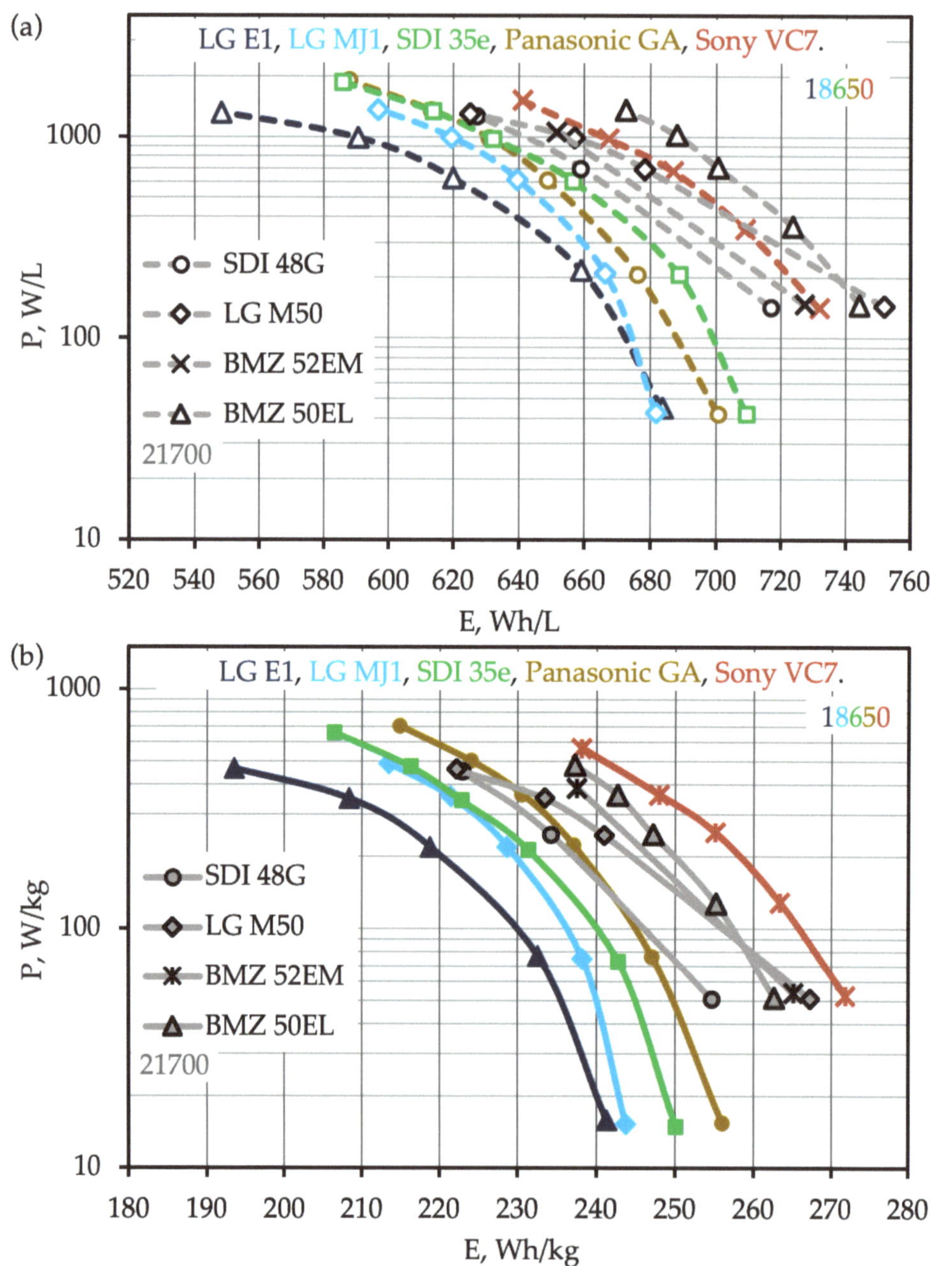

Figure 18. Energy (Wh/L, Wh/kg) vs. power (W/L, W/kg) for high-specific-energy cylindrical LICs 18650 (LG E1 [61], LG MJ1 [53], SDI 35e [64], Panasonic GA [62], Sony VC7 [65]) and 21700 (SDI 48G [73], LG M50 [73], BMZ 52EM, and BMZ 50EL [74] (the initial data were taken from references given in square brackets). Energy and power related to volume (**a**) and weight (**b**) of the cells (batteries), correspondingly. Source: Figure by authors.

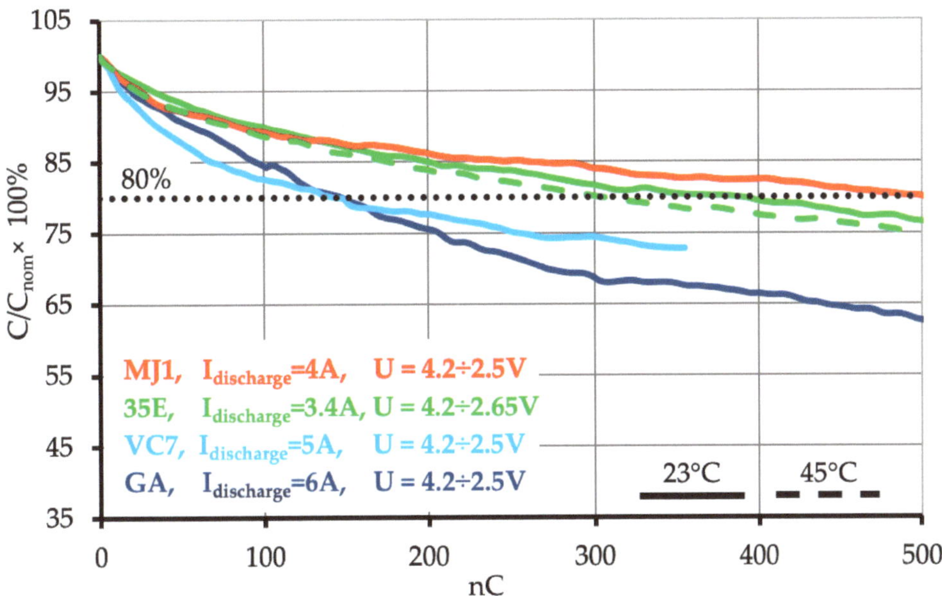

Figure 19. Effect of the number of charging (with currents of a few decimals of C) and discharging cycles on the capacity normalised to the initial one for various LICs: MJ1 [75], 35E [66], VC7 [65], and GA [67] (initial data were taken from references given in square brackets). Source: Figure by authors.

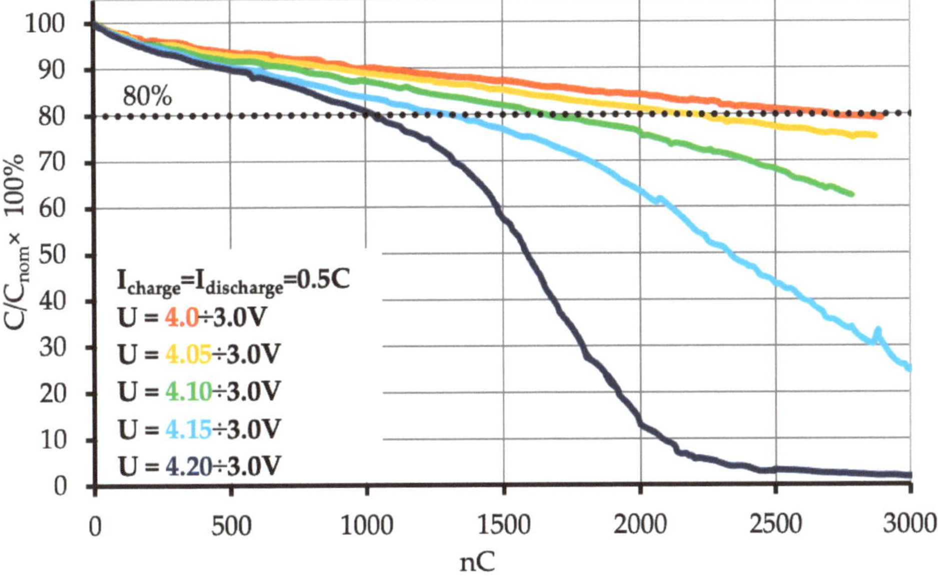

Figure 20. Effect of the upper charging limit of LIC SDI 18650 26F on the cycle life (initial data were taken from reference [71]). Source: Figure by authors.

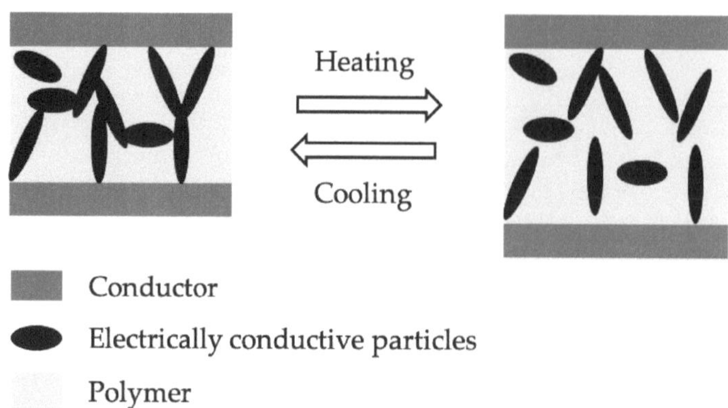

Figure 21. Positive temperature coefficient disc (or layer) operating principle. Source: Figure by authors.

In addition to the internal protection, an external one may be used. For instance, to improve the safety of the LIB power supply, the following are possible:

- Equip the battery with a chip (battery protection circuit board (PCB) [76]), protecting it against overcharging and high charging/discharging currents;
- Install a battery management system;
- Set LICs into a heat-resistant form [72,77–79] or non-flammable cooling fluid [80–82].

3.3. Methods for Increasing Lithium-Ion Cells' Specific Energy

The specific energy of a lithium-ion cell depends on a significant number of parameters (Table 17) related to both case parts and the electrochemical unit.

Case shape: Cylindrical cells are typically characterised by high energy density since, if electrodes are wound into a cylinder, the structure has almost no unoccupied space. On the other hand, it is necessary to consider that the batteries are typically prismatic, and to estimate battery energy density it is necessary to multiply the energy density of cylindrical cells by the fraction of the cylinders' volume in the total battery volume since cavities remain between the cylinders, even in close-packed arrangements.

The material the case is made from, as well as its density and thickness, affect the item weight and specific energy. Minimal thickness and density are characteristic of laminated foil. However, foil provides insufficiently reliable protection of the electrochemical system against mechanical damage. When the battery pack comprises an LIC in a pouch case, additional reinforcing frames may be needed, which will reduce the battery's specific energy. Moreover, the employment of the laminated foil (pouch) case implies using gel polymer electrolyte, since this diminishes the probability of electrolyte leakage through the terminal seals [3]. Thinning of the cases (plastic, aluminium, steel) reduces the case part fraction of the cell weight, but diminishes durability.

Table 17. Methods for increasing LIC specific energy.

	Method	Comments	W	C	U
Casing	Case shape (prism < ellipse < cylinder)	Even when cylindrical and elliptic cells are closely arranged, voids remain within the battery pack space	⇧		
	Case material (steel < aluminium < plastic < laminated foil (pouch))	*Case strength*, polymer introduction into the electrolyte for LICs in pouch cases	⇧		
	Case wall thickness	*Case strength*	⇧		
	Terminal weight	*Power* decrease may occur	⇧		
	A share of unoccupied space (a few electrode rolls instead of one in a prismatic deep-depth case)	**A complication of the manufacturing procedure**	⇧		
	Refusing additional structural parts ensures advanced safety (plates preventing puncturing, insulating tape for parts of electrodes, etc.)	*Safety in possible triggering of abnormal situations* (puncturing, short-circuit, deformation, etc.) may decrease	⇧		
	Case depth or height for LICs of the coin or button type	An **increase in the volume** of the power-supplied item is possible	⇧		
Separator	*Separator thickness*	**A complication of the manufacturing procedure** due to the reduction in the separator strength, decrease of *safety* in the possible triggering of abnormal situations	⇧		
	Using separators with functional coatings on the cathode side to improve the high-potential resistance	**Cost**, the *share of active materials*	⇩	⇧	⇧
Electrolyte	*Electrolyte share*	*Cycle life and power* decrease may occur	⇧		
	Using electrolytes of low *density* and low *viscosity*	The possibility of using **denser electrode-active layers**, the necessity of ensuring the case tightness or subsequent polymerisation of the electrolyte	⇧		
	Introduction to the electrolyte functional additives ensuring operation at high potentials, serviceability with cathodes with high content of nickel and anodes containing Si-C, SiO_x-C and SnCo-C composites	**Cost**	⇧	⇧	⇧

Table 17. *Cont.*

	Method	Comments	W	C	U
Electrodes	*Current collector thickness*	Potential increase in **internal resistance** and reduction in the *current collector strength*	⇧		
	Number and/or *area (width) of the current collector terminals*	An increase in the **internal resistance** is possible	⇧		
	Active layer thickness on the current collector	It is necessary to ensure electron conductivity through the active layer thickness The binder should provide the active layer stability in cycling (binder, cycle life)	⇧		
	The density of the active layer applied (*porosity*). It may be achieved by increasing the **calendaring pressure** using mixtures of active materials	*Power*. *Cycle life* may decrease due to possible degradation of the active layer particles and risk of dendrite growth (negative electrode); *the capacity* may decrease due to insufficient electrolyte wetting of the active layer particles and extracting them from the electrochemical reaction	⇧		
	Using a binder of **high molecular weight** (**peel strength**, at the same content)	Reduction in the *electrode's electron conductivity* => reduction in *specific power* is possible	⇧		
Positive electrode	Applying special cathode coatings reduces *separator oxidation*	**A complication of the manufacturing procedure, cost**	⇩	⇧	⇧
Electrode-active materials	**The tap density** of active materials	See Electrodes section in the table, **Density of the applied active layer** (*porosity*).	⇧		
	The true density of active materials	**The density of the applied active layer**	⇧		
	The density of particles (secondary ones)	**The density of the applied active layer**, *Power*	⇧		
	Size of particles (secondary ones)	**The density of the applied active layer**, *Power*	⇧		
	Bimodal or wide size distribution of particles	*Power*, and see **Electrodes section** in the table, **density of the applied active layer** (**porosity**).	⇧		
	Active materials with high **crystallinity** and a low *share of inactive phases*	**Cost**	⇧	⇧	
	Active material mixtures	An increase in the **active layer density** (e.g., LCO/NCM, AG/NG)	⇧		

Table 17. *Cont.*

	Method	Comments	w	C	U
Electrode auxiliary materials	**Aggregate structure of conductive additive agglomerates** (S_{BET}, **OAN**), reduction in the *conductive additive share*, fabrication of thicker electrode-active layers	**A complication of dispersion and application of the active electrode material**	↑		
(+) Electrode-active materials	Active materials with better **high-potential resistance** (impurities, special coatings)	**Cost**, LCO, NCM		↑	↑
	Layered cathode materials with an elevated nickel content in the structure	*Structural stability of the cathode-active material*		↑	
(+) Electrode auxiliary materials	*Reduction in the share of the conductive additive and binder*	The use of conductive additives and binders with improved characteristics, optimisation of the manufacturing procedure		↑	↑
(−) Electrode-active materials	*Anode-active material with a low average charge/discharge potential relative to* Li^+*/Li*	A risk of lithium deposition at high charge currents or reduced temperatures		↑	↑
	Isotropic-structure graphites, *low plane-to-plane 002/110 ratio* (low swelling during charging)	**Cost**	↑		
	Improvement **of the carbon materials crystallinity**	The *average discharge potential* decreases relative to Li^+/Li, **the difference between potentials** increases			↑
	High-capacity non-graphitisable carbon materials	High **irreversibility of the first cycle**, the probable necessity of using anhydrous solvents in fabricating the electrode slurry.		↑	↓
	Using graphite mixed with Si/C, SiO_x/C, SnCo(Ti)/C	Introduction of special additives ensuring elasticity and ionic conductivity of the SEI film. The design should allow the volume of the used negative electrodes to be changed. Elastic binder. Reduction in the LIC voltage (the relative-to-lithium potential is higher than that of graphite)		↑	↓
(−) Electrode auxiliary materials	Using conductive graphite particles as a conductive additive	Lithium intercalation/deintercalation is possible, but with higher **irreversibility** due to large surface area (small particle size)	↑		
	Using aqueous binders	Lower *electrode stability* at reduced temperatures	↑		
	Using **binders with higher elasticity**	**Cost**, but it becomes possible to use materials with high capacity and more extensive volume changes with lithium introduction or removal		↑	

Note: Colour indicates the **increment** (**bold**green) or *decrement* (*italic*blue) of parameter; w—weight part of active materials, C—capacity, U—voltage. For instance, *Case wall thickness* => *Case strength* => w↑—decrement of cell wall thickness leads to weaking of case strength and increment of specific energy due to decrease of weight inactive components (case parts) and increase of active materials in the weight of the cell. Source: Table by authors.

To improve the durability (and decrease the probability of puncturing), metal plates (copper, SDI) are introduced into the structure of the lithium-ion cell. However, this safety enhancement method reduces the specific energy.

The choice of arrangement and design of the electrode block affects the utilisation extent of the case's inner space and, hence, the electrode fraction of the LIC volume and specific energy. The case space utilisation decreases in the following order: cylindrical, elliptic, prismatic. In a laminated foil (pouch) prismatic case, the electrodes may be arranged as a card stack (LG) or a roll. Since both versions of the design are in high demand, one may assume that the space utilisation extent is approximately the same for both designs of shallow-depth laminated foil cases. In the aluminium (steel) cases used with prismatic LICs, the electrodes are typically arranged as one or several rolls. For deep-depth cases, the use of several small rolls of electrodes instead of one large electrode roll may reduce the share of unoccupied space and increase the specific energy [83]. On the other hand, the specific energy of small-sized LICs (e.g., for portable electronic devices) diminishes with decreasing thickness (prismatic case) and diameter (cylindrical case) because of the decrease in the fraction of active materials in the cell composition.

High-power cells need terminals with large cross-section areas that enable the passage of high discharge/charge currents. Some industrial applications need LICs with glass seal leads. In both cases, the casing part weight increases and reduces the specific energy, but augments the power or safety and performance at negative temperatures.

The mass fraction of electrolytes and separators in cells with high specific energy is significantly lower than in high-power cells. An example of the separator thickness effect on the increase in specific energy is given elsewhere [84]. The thickness reduction decreases the separator strength, the manufacturing process becomes more complicated, and emergency-situation safety becomes inhibited.

The dependence of the electrolyte fraction for cells of various specific energies can be found in the presentation in [85]. The decrease in the electrolyte fraction negatively affects the power and cycle life, since a lithium deficit may occur due to irreversible reactions taking place in cyclic charging/discharging.

Modern industry can produce current collectors that are thinner than those used in LIC fabrication. Nevertheless, they are not used because of their low strength, which appears to be insufficient for the coating procedure used in the electrode manufacturing process. Furthermore, since the anode swells during charging and shrinks during discharging, current collectors should be able to withstand these deformations.

To increase the electron conductivity, the current collector's lead area and number per unit electrode area may be augmented [86], increasing the weight fraction of LIC-inactive materials and, hence, decreasing the specific energy.

Using electrodes with thicker active layers increases the active material's share of the cell weight and volume. Table 18 presents example electrode parameters for two lithium-ion cells with high specific energies. To form electric conductivity channels in thick electrodes, it may become necessary to use conductive additives with a well-developed structure (carbon fibre, graphite), ensuring the formation of long electric conductivity channels [87,88].

Table 18. Parameters of the electrode and separator of LICs with high specific energies (based on data from [2]).

Active Materials / Electrode Parameters	LIC, Lam. Foil, Spec. Energy ≈580 Wh/L Uoper 4.35 ÷ 2.75 V			LIC 18650 Spec. Energy ≈720 Wh/L Uoper 4.2 ÷ 2.65 V		
	Cathode	**Anode**	**Separ.**	**Cathode**	**Anode**	**Separ.**
	LCO/ NCM	**Gr/SiO$_x$**	–	**NCA or NCM**	**Gr/SiO$_x$**	–
Thickness, μm	122	162	12	160	186	12
Length, cm	102.5	96.2	213.6	61.6	69.7	76.5
Width, cm	8.2	8.4	8.5	5.9	6.0	6.1
Coating density, mg/cm^2	42.79	25.24	–	52.51	29.61	–
Density (g/cm^3)	4.00	1.64	–	3.67	1.70	–
Foil thickness, μm	15	8	–	17	12	–

Source: Authors' compilation based on data from references cited in the caption.

The denser the electrode-active layer, the larger the active material's share in the lithium-ion cell and, hence, its specific energy. A high active layer density can be achieved by increasing the electrode calendaring pressure. Nevertheless, calendaring should be performed so as to prevent the degradation of the active material particles. Otherwise, it will negatively affect the capacity and cycle life. Furthermore, the density of the electrode-active layer may also be elevated by increasing the weight and volume share of the active materials and by using active materials with high tap density. The tap density, in turn, is defined by the true density of the materials, particle densities (the absence of pores in secondary particles of the cathode slurry or graphite), shape, size, and size distribution.

The density of the applied active layer may be increased by using a mixture of active materials (artificial graphite/natural graphite [6] or lithiated cobalt oxide/lithiated mixed oxide (NCM)) [6]. The electrode density may also be increased by reducing the share of the binder (using binders with high molecular weights and functional groups) and conductive additive. Suppose that conductive additives with well-developed structures are used (specific surface area, small particle size, aggregate structure of agglomerates). Then, due to the large contact area (number of connections between the conductive additive and active material) and lengths of the conducting paths (particles with electric contacts), a lower weight fraction is needed to ensure the necessary electric conductivity. In this case, it is essential to consider that using conductive additives with well-developed structures (↑S_{BET}, ↑OAN) leads to a considerable increase in the paste viscosity, which can require an increase in the solvent (n-methylpyrrolidone) share.

The permissible electrode density is limited by the necessity of wetting the active material particles with electrolyte. The active material particles that are not wetted with electrolyte are excluded from the electrochemical reaction and do not contribute to the total LIC capacity. In the case of active carbon cathode materials, lithium deposition may occur in the electrolyte-depleted regions (agglomerates of particles non-dispersed in preparing the suspension) [89].

The LIC's specific energy may also be increased by using high-energy active materials. There are several current trends. The lithium-ion cells for portable electronic devices fabricated with modified lithiated cobalt oxide remain stable at high potentials. The structure stability allows a more significant number of lithium ions to be extracted per unit volume (weight) of the material, thus obtaining a greater charging/discharging capacity. The higher potential of the cathode material also provides a higher voltage (the cathode–anode potential difference). The stability is achieved by adding dopants (Mn, Mg, Ti, Al) [11], applying an artificial solid-electrolyte film [6], and/or introducing functional additives that become oxidised during charging with the formation of a film that prevents electrolyte oxidation and cathode-active material recovery. It is also necessary to take into account the extent of the stability of the electrolyte components [90], binder [91], and conductive additives [87] at high potentials.

In fabricating cylindrical and large-size prismatic LICs, layered nickel-rich cathode materials are used. The increment in the nickel share promotes an increase in the material density (as compared with NCM 523 => NCA [92]) and also in the specific discharge capacity. Along with this, there are the following restrictions:

- Nickel ions accelerate electrolyte dissociation, and, therefore, special functional additives for the surface passivation are needed [93];
- The increase in the nickel fraction leads to a reduction in the cathode's stability. The stability may be improved by introducing stabilising dopants and using materials of the core–shell structure (including those with the composition gradient).

Carbon anode materials differ in crystallinity. Soft carbon and hard carbon are less crystalline than artificial graphite (artificial-graphite-based materials also differ in crystallinity from each other) and natural graphite. Reversible capacity depends on the number of lithium ions intercalating/deintercalating in/from the graphite structure. The increase in the graphite share in the carbon anode material increases the reversible specific capacity. The average graphite–lithium potential is lower than soft carbon and hard carbon potentials. Hence, the difference between cathode and anode potentials is higher for graphite, and the cell's specific energy with a graphite-based anode is higher. The graphite's true and tap densities are higher than those of soft carbon and hard carbon. Hence, its weight fraction may be more significant at the same active layer volume and, therefore, the specific energy is higher.

The introduction of silicon–carbon (SiO_X/C, Si/C) and tin-cobalt (SnCo/C) composites into the graphite-based negative-electrode-active layer increases the anode capacity and specific capacity. However, in the charged state, the potential of these substances is higher than that of graphite, which makes the effect of the specific energy growth somewhat weaker due to a decrease in the average cell voltage. When using these composites, it is necessary to consider that their volume variations in charging/discharging are more significant than in the case of graphite. To assess the effect of the introduction of silicon into the electrode, the capacity, normalised to the volume of the lithiated active layer of the negative electrode, should be considered [94]. Mechanical deformations are undergone not only by the active layer, but by the electrochemical block. In this regard, it following points are necessary:

- The structure of the electrochemical unit should be resistant to possible deformations in the process of LIC charging/discharging;
- The binder should retain the active layer particles during cycling;
- The solid-electrolyte film should remain nondegraded when the volume of active material particles changes.

The problem may be solved by using special binders introduced into the electrolyte functional additive fluorine ethylene carbonate (and other additives ensuring the creation of an elastic solid-electrolyte film). However, it is necessary to keep in mind that side effects may accompany the improvement of the anode performance by introducing functional additives. For instance, the presence of fluorine ethylene carbonate in the electrolyte may enhance gas generation at elevated temperatures [94].

3.4. *Advanced Designs of Lithium-Ion Cells with High Specific Energy*

Besides major leading manufacturers, many startups devote themselves to developing high-specific-energy LICs; the characteristics of some cells will be considered below (Table 19).

Employees at Envia have developed LICs with lithium- and manganese-enriched cathodes ($xLi_2MnO_3 \cdot yNCM$, $x + y = 1$, HCMR^{TM} or LMR). This material's advantages are higher capacity (after charging to above ≈4.6 V, the reversible capacity can exceed 250 mAh/g), increased safety, and lower cost due to the large manganese fraction. The drawbacks are the lower average discharge voltage (compared with LCO and NCM materials), voltage decrease during cycling, and low electrode density due to low tap density of the active material [95].

Table 19. Advanced designs of high-specific-energy LICs and rechargeable lithium metal cells.

Manufacturer	Brand	Active Materials	U, V	C, Ah	m, g	E_{nom}		Cycles	Ref.
						Wh/L	Wh/kg		
LICs with advanced active materials									
Amprius	–	LCO/Si	3.55	6.2	–	930	355	–	[96]
Amprius	–	LCO/Si	3.55	–	–	950	370	280	[96]
Amprius	–	LCO/Si	3.55	3.8	–	1000	400	220 ÷ 300	[96,97]
Amprius	ANW4.0-455056	LCO/Si	3.55	4	34	1150	424	250	[98]
Amprius	ANW3.6-405056	LCO/Si	3.55	3.6 3.4	31.8	1000 870	415 360	–	[98]
Amprius	ANW2.6-405056	LCO/Si	3.55	2.8 2.65	27.6	915 815	365 325	–	[98]
Amprius	–	622/Si	–	46	450.7	925	359	–	[98]
Amprius	–	NCM811/Si	–	60	–	1200	450	–	[97]
Enevate	EN9072120	NCM/Si	–	17	–	800	300	>500	[32]
Envia	ENV 234426-XP	LMR (HCMR XP)/Gr	3.69	26.5	407	438	234	1000	
Farasis	–	H-Ni NCM/Si/Gr	3.67	35	482	≈600	265		[99]
Zenlabs	Range EV	NCM/Gr, SiO_x	3.46	51	592	640	305	1000	[100–102]
Zenlabs	Glide drone	LCO (CRC)/Gr, SiO_x	3.63	12.7	129	810	350	>150	[103,104]
Cells with lithium anode									
Embatt	3.0	LNMO/Li	–	–	–	800	–	–	[105]
Ganfeng	5889C05	–	3.75	10	–	581	350	300	[106]
Hydro Quebec	Gen2 30Ah	LFP/Li	3.37	32	455	280	225	–	[107]
SEEO	L13-001-001	LFP/Li	3.42	11	175	–	220	–	[108]
Sion Power	Licerion	HE-NMC/Li	3.85	0.4	–	1000	500	>450	[32,109]
Sion Power	Licerion	HE-NMC/Li	–	26 21	–	700 1000	500 500	>500	[32,109,110]
Cells with lithium anode									
Sion Power	Licerion HP	–	–	–	–	700	400	>320	[109]
Sion power	Licerion HE	–	–	–	–	1400	650	–	[109]
Solid Energy	Hermes	H-Ni NMC/Li	3.8	3.4	29	1157	450	>130	[111]
Ganfeng	7289C05	–	3.65	12		545	235	1000	[112]

Table 19. *Cont.*

Manufacturer	Brand	Active Materials	U, V	C, Ah	m, g	E_{nom}		Cycles	Ref.
						Wh/L	Wh/kg		
Lithium-ion cells with solid electrolyte									
Ilika	Goliath A6	NMC/oxide electrolyte/Si	3.6	30	270	1390	400	–	[113–115]
Prologium	4360A5AAMA	NMC/oxide electrolyte/C, SiO_x	3.75	1.95	60	257	123	–	
Prologium	36D3L8AAJA	LCO/oxide electrolyte/C, SiO_x	3.85	8.9	240	312	140	–	
Prologium	32D3L8ABKA	NMC/oxide electrolyte/C, SiO_x	3.65	11	220	420	182	–	
Prologium	59D3L8ABXA	NMC/oxide electrolyte/C, SiO_x	3.58	24	392	494	219	–	

Source: Authors' compilation based on data from references cited in the table.

It was also shown that the composition and structure of cathode-active layers enriched with lithium and manganese affect the dependences of impedance per electrode unit area (area-specific impedance [116]) on the charge rate and, hence, on the LIC's usable energy [117,118]. Based on the performed studies [95,117–122], Envia Co. has developed an LIC (ENV23426-XP, where XP is the designation of material with a low content of phase Li_2MnO_3, C = 200 ÷ 220 mAh/g, 4.6 ÷ 2.0 V, C/10) with a positive electrode comprising active material HCMR^{TM} and with a specific energy of 234 Wh/kg and 438 Wh/L (4.3 ÷ 2.5 V).

Considering the test results [99,116] for several high-specific-energy cathode (nickel-rich NCM, NCA, Cam7, and others) and anode (3M [117–121], OneD material [122,123], Shin-Etsu [14,124–127], and XG science [128]) materials, the company Farasis developed LICs with a high specific energy of 265 Wh/kg, (≈600 Wh/L [99], Figure 22).

Lithium-ion cells with high specific energies (305 Wh/kg, 640 Wh/L and higher) were developed by Zenlabs (ex. Envia) using the active cathode material NCM with a high nickel mole fraction and an anode based on SiO_x (>50%) [100–102,129] (Figure 22).

The company Enevate creates LICs using a positive electrode based on lithiated mixed nickel cobalt manganese oxide and a negative electrode comprising silicon produced according to a licenced manufacturing procedure [130]. At the first stage, the active layer contains silicon (70%), and carbon material provides conductivity. The obtained "self-standing" silicon-containing active layer is transferred on copper foil [130,131]. The capacity of the used silicon can reach 3000 mAh/g, but only about half of the capacity is used in operation for the sake of cycle life extension. The recommended operating voltage ranges are 4.2 ÷ 3.3 V and 4.2 ÷ 2.75 V [132]. In the extended operating voltage range and at the nominal charge/discharge current, the specific energy reaches 300 Wh/kg (750 ÷ 800 Wh/L). Such LICs may also be charged with relatively high currents at room and reduced (–20°C) temperatures [131].

The fabrication of the cathode using lithiated cobalt oxide (CRC [103,133]) that remains stable at high potentials, and the use of SiO_x-based anodes, enabled Envia (Zenlabs) to develop LICs with a specific energy of 350 Wh/kg (>810 Wh/L) [104] (Figure 22).

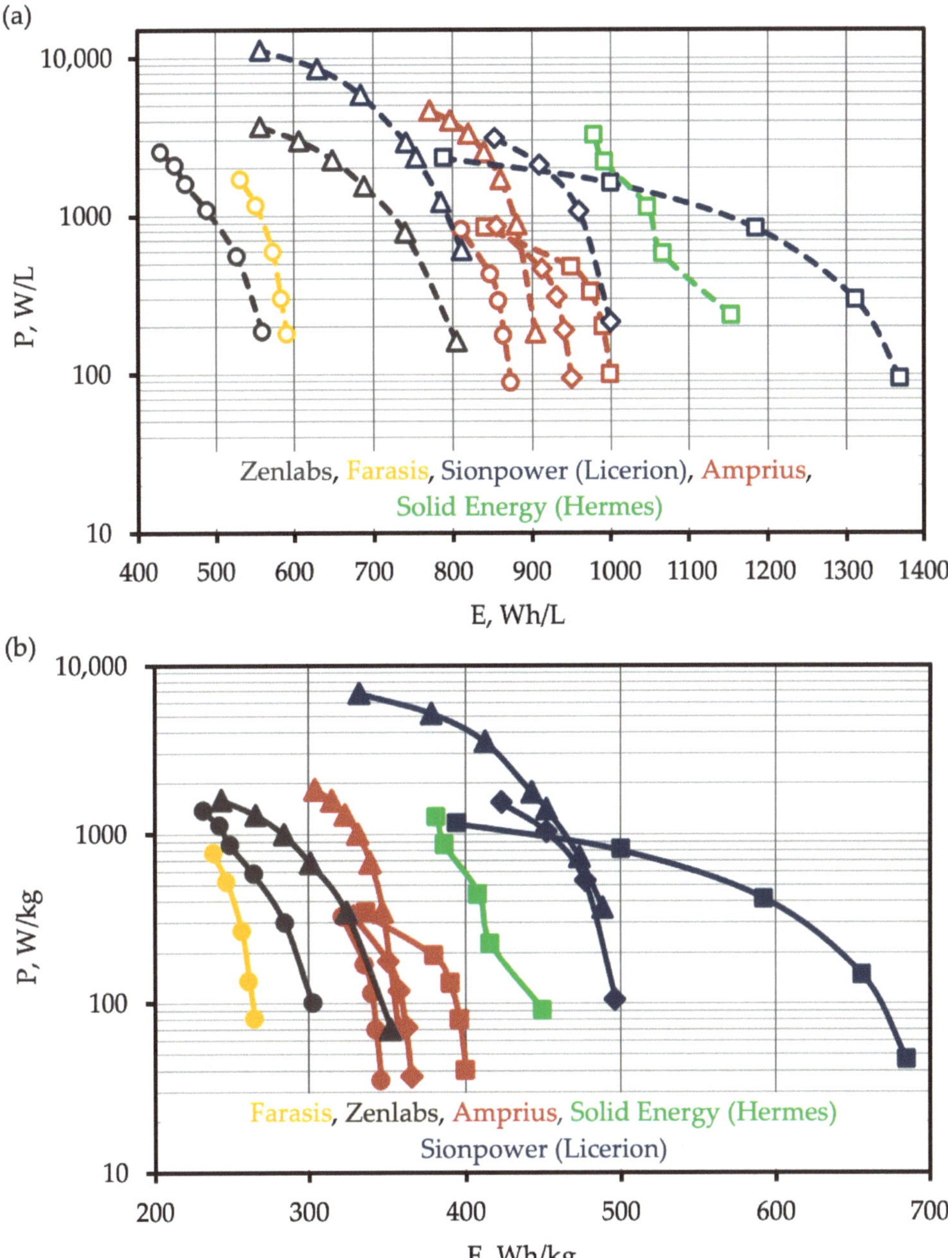

Figure 22. Energy (Wh/L, Wh/kg) and power (W/L, W/kg) of advanced LICs produced by Amprius [96], Farasis [99], Sionpower [109], Solid Energy Systems [111], and Zenlabs [100,104] (initial data were taken from references given in square brackets). Energy and power related to volume (**a**) and weight (**b**) of the cells, correspondingly. Source: Figure by authors.

3.5. Rechargeable Cells with Lithium Anodes

One of the trends to increase the specific energy is to use a lithium film in place of the negative-electrode-active layer. Lithium has a high specific capacity of 3860 mAh/g (2068 mAh/cm^3) [134], which allows the share of active materials in the cell composition to be increased because it is necessary to balance the cathode and anode capacities. The low potential ensures high voltage (significant potential difference). Contrary to LICs with carbon and lithium titanate anodes whose charge/discharge gives rise to lithium-ion intercalation/deintercalation into the material structure, in the case of rechargeable lithium metal cells, lithium plating/stripping takes place. Under certain conditions, this process may be highly reversible (high Coulombic efficiency), which is confirmed by literature data on the cathode cycle life obtained in testing the material relative to lithium foil.

In the literature, several techniques for applying the lithium active layer are described:

- Lithium is not applied on a copper current collector, but is formed in the first cycle of charge. Lithium ions leave the cathode structure, passing through the electrolyte and separator and then plating on the copper current collector. In discharge, lithium ions move in the opposite direction. Cells of this type are designated as lithium-free or anode-free [135–141]. The negative electrode capacity per square centimetre can reach 3 mAh/cm^2 [136] or more [142].
- Lithium is applied onto copper foil. The application may be performed by printing with ink containing encapsulated lithium [143,144] or by evaporating on the current collector [145,146].
- Lithium foil is used simultaneously as an active layer and a current collector placed between two cathode-active layers. Sufficiently thin lithium foil (about 50 μm thick) may be obtained by calendaring through rolls [147,148].

The lithium areal capacity may exceed 2 mAh/cm^2 [149], 3.2 mAh/cm^2 [150], 5 mAh/cm^2 [151], or 9.8 mAh/cm^2 (50 μm) [152]. The most commonly mentioned thicknesses of the lithium active layer range from 20 to 25 μm [94,151,153] (the lithium anode—50 μm, 25 μm of lithium on each cathode side [149]). The lithium anodes of large thickness may restrict the cycle life because of the intense growth of mossy lithium and dendrites [94].

The lithium anode cells may be fabricated using the following:

- A polymer separator (used in fabricating LICs, including ceramic-coated ones) and liquid electrolyte [94,152,154].
- A solid-state electrolyte separator, either polymer-based [148,149,155,156] or based on inorganic compounds [150,157–159].

In the literature, there are also descriptions of the use of hybrid electrolytes [160].

In the cyclic charge/discharge of cells with a lithium anode and liquid electrolyte, the lithium atoms' distribution over the film surface becomes non-uniform because of concentration gradients. The main factors are the nonuniformity of the cell compression, SEI film composition and thickness, ions flowing through the separator, and a reduction in the salt concentration near the lithium electrode. A coating of a lower density and larger volume (the so-called mossy lithium) is formed, whose thickness (volume) increases during cycling (an SEI film emerges on the lithium deposited during charge). The mossy lithium structure is formed by lithium particles surrounded by the SEI film [94,161] or dendrites [162] with a high density of distribution over the lithium film. As a result, a portion of the lithium loses electrical contact with the anode and becomes excluded from the electrochemical reaction, forming so-called dead lithium [161].

The primary efforts of researchers and designers aim to ensure high Coulomb efficiency (the plating of a homogeneous lithium film and its small growth in charging/discharging). The mossy lithium density, structure, and composition may be varied by fitting the salt concentration [162] (the localised concentration of salt [138] or its anion [163] or a combination of salts [139]) and the solvent composition [138] (the replacement of carbonates with esters [163] (1,2-dimethoxyethane, [tris(2,2,2-trifluoroethyl)orthoformate]), and by introducing functional additives into the electrolyte [94,161] by using lithium protecting layers [164, 165], modifying the anode design [166], and plating on the current collector composites alloying with lithium (e.g., Ag/C [141]), etc.

In charging/discharging, the lithium anode composition and structure depend on the operating mode, temperature, charge/discharge current density [162], pressure, and compression uniformity [152], etc. If the critical current value is exceeded (this depends on both the external factors and the cell design), lithium plates may develop—not in the form of a mossy coating, but in the form of freestanding dendrites [162] that may cause an internal short circuit after reaching a certain height. Therefore, for the sake of operational safety, it is necessary to use appropriate design solutions when fabricating the cell and to control its state in the process of operation (the battery management system [165]).

During cycling, the anode thickness increase is accompanied by a growth in mechanical stresses, cell swelling, and pressure upon the outer retaining plates [167], and depends on the electrochemical block design and used materials. For instance, the results provided in paper [167] for a less advantageous sample (a liquid electrolyte cell in a pouch case) showed that the thickness increased by more than 110% after 10 charge/discharge cycles. The thickness of the most advantageous samples grew by about 30% after 300 ÷ 400 cycles. The capacity retention increased in cycling with decreasing cell thickness increments. The operating lifetime and specific energy also depend on the electrolyte amount (for prototypes below 3 ÷ 4 g (1 M solution)) per cell Ah [94,152,168] and the anode–cathode capacity ratio [152,168].

Research and development institutions such as the Pacific Northwest National Laboratory [167,169] have devised rechargeable cells with lithium anodes and liquid electrolyte within Project Battery 500 [154,168], as have companies including Sion Power (Licerion [109,164]) and Solid Energy [111,140,170]. Depending on the design and manufacturing procedure, the specific energy may range from 300 to 700 Wh/kg (Figure 22).

The use of polymer electrolyte separators provides the following advantages [171]:

- The absence (or reduction) of the risk of electrolyte and solvent leaks;
- Low-cost materials and manufacturing processes;
- Adaptation to volume variations.

On the other hand, low mechanical strength, resistance at the electrode/separator interface, low ionic conductivity, a narrow electrochemical stability window, and a high operating temperature [171,172] hold back the commercial application of polymer electrolytes in lithium anode rechargeable cells.

To overcome the barriers mentioned above, various polymers and salts are suggested for use [148,172]. Among the significant number of proposed variants, it is possible to distinguish polymer electrolytes based on polyethene oxide because they ensure good solubility of the electrolyte salts and the high mobility of lithium ions due to the flexibility of polymer chains [173]. Lithium ions are coordinated with ester oxygen bridges on the polymer chain segments, similar to the solvation with organic carbonates in the liquid electrolytes of lithium-ion cells. The lithium ions move due to desolvation/solvation, which implies that lithium ions jump along the polymer chain or between polymer chains [155,173].

A relatively low ionic conductivity at room and lower temperatures makes it necessary to heat the battery (or battery pack). For instance, the regular operation of commercially produced batteries with polyethene-oxide-based electrolyte requires maintaining the temperature at 60 ÷ 80 °C [155,174]. Since there is a risk of polymer degradation by the products of salt $LiPF_6$ decomposition (especially at elevated temperatures), it is necessary to use other salts for cells of this type, e.g., LiTFSI [148,175]. This leads to a requirement for limiting the upper potential relative to lithium since, at potentials above 3.8 V, salt LiTFSI and its decomposition products can destroy the aluminium current collector [176,177]. Therefore, low-voltage cathode materials should be used, e.g., $Li_xV_3O_8$ [178] and LFP/C [174,179].

Rechargeable lithium anode cells are produced and developed by the following companies: Bollore [174,180] (polyethene oxide), CATL [174], HydroQuebec [148], SEEO [181]). The nominal specific energy of rechargeable batteries with polymer electrolyte and lithium anodes is approximately equal to 220 ÷ 225 Wh/kg (>280 Wh/L, LFP cathode, Hydro Quebec [107], Seeo [108]). More detailed information on rechargeable lithium metal cells may be found elsewhere [148,149,155].

One of the main advantages of using inorganic electrolytes is the safety improvement (the reduction in the combustible material content in the electrolyte) and the broadening of the operating temperature range (especially the increase in its upper limit). Inorganic electrolytes may be crystalline (perovskites, garnets, materials with structures of the NASICON and thio-LISICON types) or amorphous (LiPON, sulphide glasses). Among the commercial products or substances mentioned in manufacturing advanced solid electrolytes of solid-state cells, there are oxides (Li_2O-Al_2O_3-SiO_2-P_2O_5-TiO_2-GeO_2, Li_2O-Al_2O_3-SiO_2-P_2O_5-TiO_2 [182–185], $Li_7La_3Zr_2O_{12}$ (LLZO) [186–189], $Li_{7-2x+y}Mg_xLa_{3-y}Sr_yZr_2O_{12}$, $(0.1 \leq x \leq 0.3, 0 \leq y \leq 0.5)$ (LLZO-SM) [190], $Li_{6.4}La_3Zr_{1.4}Ta_{0.6}O_{12}$ (LLZTO) [191]), sulphur-containing compounds (Li_3PS_4 [192], $Li_7P_3S_{11}$ [192], $Li_{9.54}Si_{1.74}P_{1.44}S_{11.7}Cl_{0.3}$ [193], LGPS: $Li_{10}GeP_2S_{12}$ [192,193] $Li_{10.35}Ge_{1.35}P_{1.65}S_{12}$ [194], Li_6PS_5Cl [192,195]), phosphates ($Li_{1+x}Al_xGe_{2-x}(PO_4)_3$ [196]—LAGP, $Li_{1.3}Al_{0.3}Ti_{1.7}P_3O_{12}$ [191,196], $Li_{1.5}Al_{0.5}Ti_{1.5}(PO_4)_3$ [105,197]—LATP), compounds based on $LiBH_4$ ($LiBH_4 \times LiI$, $LiBH_4 \times LiNH_2$) [198,199], and others [200–203].

The types of solid electrolytes used by different companies in developing the cells are shown in references [159,174]. Lithium-ion conductivity is one of the key characteristics of solid electrolytes. Sulphur-containing solid electrolytes have high room-temperature ionic conductivity ($10^{-3} \div 10^{-2}$ S/cm, record values 25 mS/cm [193]). To reach the specified high conductivity in fabricating layers containing solid electrolyte, high-temperature ($280 \div 300$ °C) pressing may be required. As a result, the manufacturing procedure for the electrode block and the cell becomes more complicated.

In fabricating LICs with solid electrolytes, the positive-electrode-active layer is first applied on both sides of the current collector. The active layer contains a cathode, binder, conductive additive, and electrolyte. After that, hot pressing is performed (SolidPower informs that it is possible to achieve the sufficiently high ionic conductivity of sulphur-containing electrolytes without heating at pressing—$2 \div 9$ mS/cm [204]). Then, a solid electrolyte layer is applied (on both sides), and high-temperature pressing is repeated. After that, a buffer layer (on both sides) and lithium are applied. Finally, to increase the capacity, the obtained multilayer structures are placed onto one another, thus being united in an electrode block [174].

Sulphur-containing electrolytes can interact with water molecules. Thus, in manufacturing, they should be handled in a dry atmosphere. Modification can increase their resistance to water vapours [174]. In addition, a narrow electrochemical stability window is characteristic of many electrolytes based on sulphur compounds [205]. In operation, sulphur-based electrolyte becomes oxidised on the cathode slurry surface and the surface of conductive additives existing in the positive-electrode-active layer [206,207]. As a possible solution, the creation of a solid electrolyte buffer layer on the cathode material surface may be considered [174]. Another variant is that, due to a low process rate, materials formed on the solid electrolyte particle surfaces (or specially applied coatings) can inhibit the electrolyte reduction and oxidation [208]. For instance,

Mitsui Kinzoku presents test results demonstrating the stability of a solid electrolyte based on inorganic sulphur compounds up to 10 V [209].

On the contrary, solid electrolytes based on oxides and phosphates interact with water more weakly, but have ionic conductivity that is lower by order of magnitude. Therefore, their fabrication requires heating at relatively high temperatures. In addition, they have an exceptionally high theoretical specific density (LLZO—about 5.4 g/cm^3 [210], LATP—2.94 g/cm^3 [211]; for comparison, the Li-argyrodite density is 1.64 g/cm^3 [212]), which may reduce specific energy and specific power of cells due to high density (weight) of applied for fabrication solid-electrolyte.

As shown in experiments with Li-argyrodite, the use of solid inorganic electrolytes is accompanied by the formation of cavities in the lithium/solid electrolyte. Some of these cavities are filled with lithium in subsequent charge, and the remainder passes into the lithium bulk. As the charging/discharging continues, the cavities may move towards the anode/solid electrolyte, which finally results in the exfoliation of the anode and solid electrolyte regions and, thus, in a large lithium plating current, the growth of dendrites, and internal short circuits [213].

Because of the high lithium plasticity, one of the ways to reduce the content of cavities may be the application of external pressure. An increase in external pressure may also ensure an increase in the permissible current density [214]. In addition, the performance of cells with the lithium anode and solid electrolyte may be ameliorated by improving the solid electrolyte wettability with lithium [215–217]. The creation of an intermediate layer (polymer layer [205], boron nitride [197] (LATP protection against interaction with lithium), etc.) is achieved using three-layer electrolyte separators (two outer layers are porous, the inner one has lower porosity [216,218]).

To increase the ionic conductivity (power, cycle life) at the interface between the solid electrolyte and positive-electrode-active layer, the latter may be modified as follows:

- Particles of an ionic conductor similar to that used in the solid electrolyte may be introduced [105,219];
- The cathode particles may be covered with solid electrolyte [220];
- The active layer and solid electrolyte may be impregnated with liquid electrolyte [170,174].

A review of some solutions for creating the cathode-active layer/solid electrolyte interfaces may be found elsewhere [207]. More detailed information on the scientific achievements in the development and application of solid electrolytes may be found in presentations [170,221,222] and reviews [160,200,223,224].

The nominal specific energy of lithium anode cells being developed by Ganfeng [106], LG [225], Lionano SE [226], SDI [225], Sion Power [32,109,110], SolidEnergy Systems [32,111,170], and Solid Power [180,227] ranges from 320 to 700 Wh/kg (700 ÷ 1400 Wh/L). The specific characteristics of the battery modules and packs assembled from these cells will probably decrease more than conventional lithium-ion batteries because it is necessary to apply external pressure to provide the operating lifetime. In addition, the energy density values are sometimes given by only taking into account the volume occupied by the electrode block, ignoring the sealing unit and terminals.

Besides the high nominal specific energy, some of the innovative species of cells being produced by Amprius (anode—silicon structures), Zenlabs (Envia) (graphite with SiO_x), Sion Power (anode—Li), Solid Energy (anode—Li), and others have high specific power (Figure 22) comparable with or sometimes exceeding the values characteristic of state-of-the-art LICs. It seems that among the barriers hindering their entry into the market are their high cost (small-batch manufacturing), unsatisfactory performance at low temperatures, short cycle life, and others. To reach the cycle life of 200 charge/discharge cycles or more, some lithium anode cells should be clamped between plates to create external pressure [152,154].

For conventional LICs, the cycle life given in the specifications is typically a conditional value limited by the number of charge/discharge cycles before the capacity decreases to 80% of the initial one. If the permissible capacity is reduced to 70%, the number of charge/discharge cycles essentially increases. The discharge capacity of lithium anode cells may drastically decrease after reaching 80% [164]. A similar situation is observed in the case of the degradation of the lithium-ion cell anode performance. Considering the long cycle life of cathode materials (expressed in cycles) and the character of variations in the capacity retention depending on the charge/discharge cycle number, it is possible to conclude that the lithium anode cell cycle life is restricted by the performance of lithium (negative electrode).

In some applications, specific energy and power may be of greater importance than cycle life. As the main fields of application, lithium anode battery manufacturers mention special equipment such as different kinds of unmanned aerial vehicles: atmospheric satellites, quadcopters, etc.

3.6. Solid-State Lithium-Ion Cells with Inorganic Electrolyte and Non-Lithium Anode

In the literature, it is possible to find descriptions of solid cells (bulk-type) free of a lithium film and with a capacity of 1 Ah or more. Among the companies developing solid LICs of this type, the following may be distinguished: CATL [174], Hitachi Zosen [199, 228–230], Ilika (Goliath, NCM/Silicon) [231,232], Libtec [233–237], Prologium, Samsung [238], Toyota [174,239], and others [234,240,241]. The often-mentioned solid electrolytes used in manufacturing solid-state LICs (LIBs) are Li-argyrodite [209] (Libtec [233]), $75Li_2S \times 25P_2S_5$ (Toyota) [242], $80Li_2S{*}20P_2S_5$ (Samsung) [238], $Li_{10}GeP_2S_{12}$ (Toyota) [242–244], and oxide-based inorganic electrolytes (Prologium, Ilika [231]). At present, batteries of this type are not in high demand, probably due to their high cost [245].

Descriptions of the characteristics of cells of this type can rarely be found. For instance, in 2021, Hitachi Zosen presented an all-solid-state lithium-ion cell with a sulphide electrolyte (Mitsui Kinzoku) and an energy density of about 91 Wh/L [230]. Prologium produces lithium-ion cells with a solid oxide electrolyte (with a solid electrolyte content of more than 90% [159]) of different types. Assumably, LCO, NCM, graphite, and silicon-containing substances are used as active materials. The considered cells operate at low and high pressures and in a wide range of temperatures. Based on the weights and dimensions (Table 19) and the discharge curves given in the specifications, dependencies were plotted. The curves interrelate specific power and specific energy and demonstrate the discharge temperature effect on the specific energy (energy density). The calculation shows that the room-temperature maximal power of the LICs under consideration ranges from 175 to 975 W/kg (400 ÷ 2195 W/L, Figure 22). The nominal specific energy of Prologium cells varies from 125 to 215 Wh/kg (263 ÷ 484 Wh/L, Figure 23a,b). As the temperature decreases, the specific energy and power during discharge with a fixed current also diminish (Figure 23c,d).

The application of a solid electrolyte (separator) stable at high potentials allows the LIC voltage increment. Embatt (BMW [246]) and Prologium are designing LICs, a part of whose electrodes may be connected in series, and the average discharge voltage may be increased twice or more (a bipolar battery) [247,248]. The use of bipolar batteries allows the management system to be simplified, reducing the internal resistance and increasing the energy density of the module. In the current project, Embatt 3.0 [105] uses lithium nickel manganese spinel ($LiNi_{0.5}Mn_{1.5}O_4$) [249–251] as the active cathode material and a lithium-based anode. The planned energy density may reach 800 Wh/L.

3.7. Conclusions

This chapter considered the characteristics of high-specific-energy LICs in prismatic and cylindrical cases. The specific energy of LICs used to supply power to modern electronic devices may reach 260 ÷ 300 Wh/kg and 700 ÷ 760 Wh/L. The chapter also presented the results of the influence of discharge conditions, current, and temperature on the LICs' functional characteristics (capacity, average discharge voltage, etc.). In addition, designing and manufacturing practical methods for increasing the specific energy were demonstrated. In fabricating prismatic LICs, the modified (high-voltage and/or higher density) lithiated cobalt oxide is used in manufacturing the cathode to increase the energy density. As an active layer of the negative electrode, graphites of high capacity (and density), graphite-based mixtures, and silicon-containing particles are used. In fabricating cylindrical cells (18650, 21700), nickel-rich layered oxide (NCM, NCA) is used as the cathode-active material. The negative-electrode-active layer contains graphite and silicon-containing particles.

The specific energy of advanced LIC prototypes may reach 300 ÷ 700 Wh/kg and 720 ÷ 1400 Wh/L. As cathode-active materials, lithiated cobalt oxide and nickel-enriched layered oxides (NCM, NCA) are used. As the anode-active layer, silicon fibres, mixtures of silicon-containing materials, graphite, and lithium film are used.

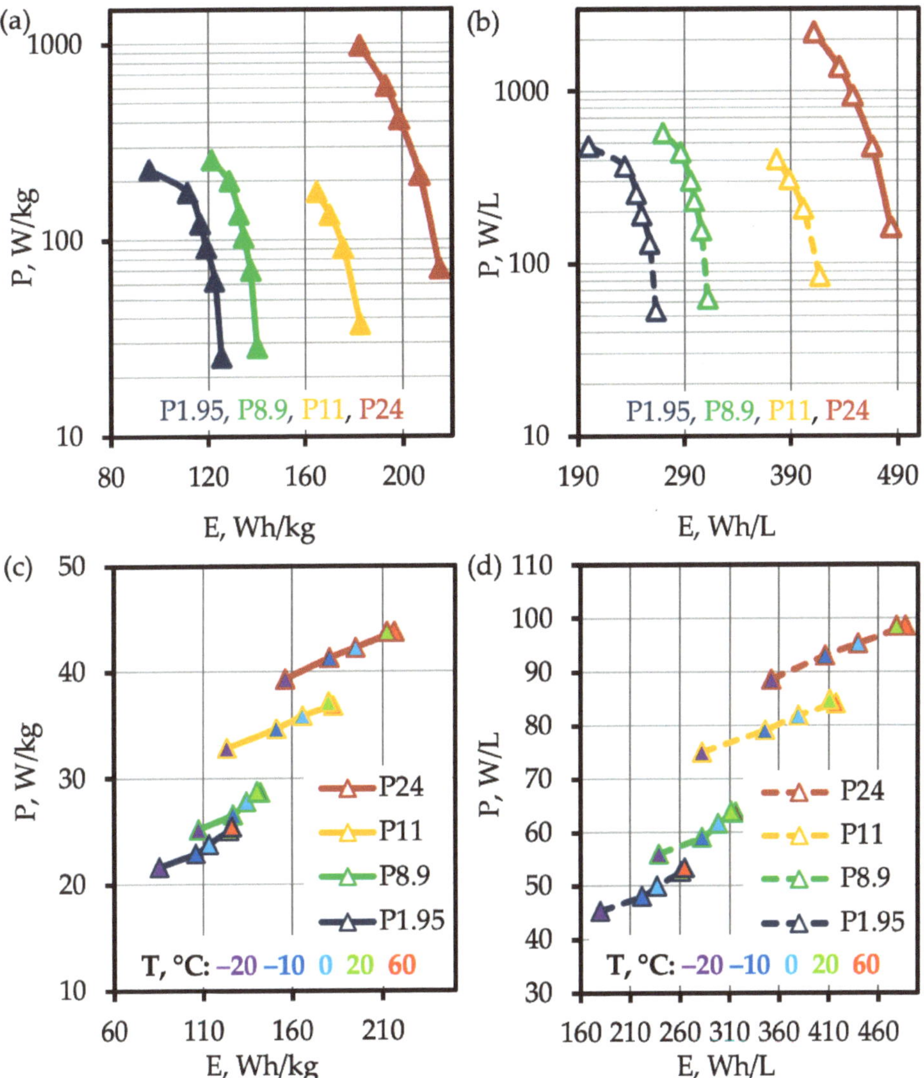

Figure 23. Variations in power and energy (**a**,**b**) and the effect of temperature (**c**,**d**) on the power and energy at a fixed discharge current of lithium-ion cells P1.95, P8.9, P11, and P24 (initial data were taken from references given in square brackets). Power and energy related to volume (**a**,**c**) and weight (**b**,**d**) of the cells, correspondingly. Source: Figure by authors.

References

1. Sony Energy Storage System Using Olivine Type Battery. Available online: https://na.eventscloud.com/file_uploads/89b02d8a4305f5ffe09d0c27691441af_O-302YasudaMasayuki.pdf (accessed on 14 November 2024).
2. Lu, H.-L. Commercial Technology & Product Development Trends of Cathode & Anode Materials for LIB in 2015. In Proceedings of the Second International Forum on Cathode & Anode Materials for Advanced Batteries, Shenzhen, China, 20–22 May 2015; p. 55.
3. de Leon, S. Lithium Rechargeable Pouch Cells the Hidden Secret. Available online: https://www.sdle.co.il/wp-content/uploads/2018/12/1-Lithium-Rechargeable-Pouch-Cells-The-hidden-Secret-ver2.pdf (accessed on 24 May 2019).
4. Hybrid Gel Technology (Sony). Available online: https://www.facebook.com/1599927223611707/photos/rpp.1599927223611707/1599953643609065/?type=3&theater (accessed on 26 July 2018).
5. Takeshita, H. Latest LIB Market Trends (The Automotive and Power Storage Markets Move Into Full-scale Growth. In Proceedings of the 10th Int'l Rechargeable Battery Expo Battery Japan 2019, Tokyo, Japan, 28 February–1 March 2019; pp. 1–6.
6. IEK, I. 鑑往知來: 由全球鈕電池黨與市場發展動向 看中國黨轉型與升級之路 (Learn from the past and learn from the future: Looking at the transformation and upgrading of the Chinese Communist Party from the global button cell industry and market development trends). In Proceedings of the 2012 中國化學與物理電源行業協會理事會議 (China Chemical and Physical Power Supply Industry Association Board of Directors Meeting), Beijing, China, 14–15 December 2012; p. 126.
7. Galaxy S21 Teardown: Best of the New Galaxies. Available online: https://youtu.be/ElJ9S-aZCG8?t=149 (accessed on 13 May 2021).
8. Whitson, G. iPhone 11 Teardown. Available online: https://ru.ifixit.com/News/33016/iphone-11-teardown (accessed on 2 January 2021).
9. Samsung Galaxy S20 Battery Replacement. Available online: https://guide-images.cdn.ifixit.com/igi/N44clOS4pOgkmXRj.huge (accessed on 13 May 2021).
10. Clap, K.K. iPhone XS Has an Upgrade Apple Didn't Mention. Available online: https://ifixit.org/blog/11463/iphone-xs-and-max-teardown/ (accessed on 24 May 2019).
11. Lu, M. Future Trends and Key Issues in the Global Lithium-ion Batteries Market and Related Technologies. In Proceedings of the 13th China International Battery Fair (CIBF 2018), Shenzhen, China, 22–24 May 2018; p. 30.
12. Lee, H. The future of new energy vehicles and automotive batteries (Samsung). In Proceedings of the 13th China International Battery Fair (CIBF2018), Shenzhen, China, 22–24 May 2018; p. 19.
13. Nagai, R.; Kita, F.; Yamada, M.; Katayama, H. Development of Highly Reliable High-capacity Batteries for Mobile Devices and Small- to Medium-sized Batteries for Industrial Applications. *Hitachi Rev.* **2011**, *60*, 28–32.
14. Park, J.; Park, S.S.; Won, Y.S. In situ XRD study of the structural changes of graphite anodes mixed with SiOx during lithium insertion and extraction in lithium-ion batteries. *Electrochim. Acta* **2013**, *107*, 467–472. [CrossRef]

15. EB-BS901ABY (Ningde Amperex Technology Limited). Available online: https://www.safetykorea.kr/release/certDetail?certNum=YU10427-21026A&certUid=5206706 (accessed on 18 February 2022).
16. Samsung G975 galaxy S10+ 4100 mAh (GH82-18827A). Available online: https://telefon-service.ru/part/gh82-18827a/ (accessed on 17 December 2020).
17. SAMSUNG Original Battery EB-BG973ABE EB-BG973ABU For Samsung GALAXY S10 Galaxy S10 X S10X SM-G973F SM-G9730 G9730 G973 3400mAh. Available online: https://www.aliexpress.com/item/33004882016.html (accessed on 2 January 2021).
18. Quandt, R. Samsung Galaxy S9 3000mAh battery EB-BG960. Available online: https://twitter.com/rquandt/status/959050499843608576 (accessed on 9 April 2018).
19. Teardown of S8 Galaxy. Available online: https://d3nevzfk7ii3be.cloudfront.net/igi/kYOneqatYDwUxPya.huge (accessed on 9 April 2018).
20. [In-Depth Look] What's Inside the Galaxy S7 and S7 Edge. Available online: https://news.samsung.com/global/in-depth-look-whats-inside-the-galaxy-s7-and-s7-edge (accessed on 9 April 2018).
21. Galaxy S6 SM-G920/G920A/G920P/G920R4/G920T/G920F/G920V Battery Replacement—Grade R. Available online: https://www.etradesupply.com/samsung-galaxy-s6-sm-g920-g920a-g920p-g920r4-g920t-g920f-g920v-battery-grade-r.html (accessed on 18 February 2022).
22. Battery (Galaxy S5). Available online: https://www.ifixit.com/products/galaxy-s5-replacement-battery (accessed on 22 May 2024).
23. Samsung B600BC. Available online: https://parts-plus.ru/product/akb-samsung-b600bc-galaxy-s4-i9500-3-8v-gray-2600mah-9-88wh?gclid=Cj0KCQiApL2QBhC8ARIsAGMm-KH7w1uDUnozE18oo-8T7umNPgHIEovX-ZFOQKmaH56r3i0bx2OgG7UaAmRYEALw_wcB (accessed on 18 February 2022).
24. Samsung Galaxy S3 Replacement Battery (2100 mAh) for AT&T, Sprint & T-Mobile Models (Discontinued by Manufacturer). Available online: https://www.amazon.com/Samsung-Replacement-T-Mobile-Discontinued-Manufacturer/dp/B0089VO7OM/ref=sr_1_3?keywords=samsung+galaxy+s3+battery&qid=1645175821&sr=8-3 (accessed on 18 February 2022).
25. EB-F1A2GBU Battery for Samsung Galaxy S2, I9100-1650mAh / 3.7V / 6.11Wh / Li-Ion. Available online: https://www.impextrom.com/en/eb-f1a2gbu-battery-for-samsung-galaxy-s2-i9100-1650mah-3-7v-6-11wh-li-ion-p1000041134 (accessed on 14 November 2024).
26. Samsung Original Battery for Galaxy S I9000. Available online: https://www.amazon.co.uk/Samsung-Original-Battery-Galaxy-I9000-Black/dp/B003UWZ6M2 (accessed on 14 June 2023).
27. *IIT LIB-related Study Program 10–11 (August 2010)*. 2010, p. 23. Available online: http://www.docin.com/p-117656255.html (accessed on 22 May 2024).
28. *LIB 材料市場速報 (15Q1)*; B3 Corporation: São Paulo, Brazil, 2015; p. 30. Available online: https://www.docin.com/p-1282371064.html?docfrom=rrela (accessed on 22 May 2024).
29. Graphic for iPad. Available online: https://apps.apple.com/ru/app/graphic-for-ipad/id363317633 (accessed on 10 January 2022).

30. GetData Graph Digitizer. Available online: http://getdata-graph-digitizer.com/index.php (accessed on 10 January 2022).
31. First iPhone 11 Pro Max Teardown Confirms 4000 mAh Battery. Available online: https://www.gsmarena.com/first_iphone_11_pro_max_teardown_confirms_4000mah_battery-news-39242.php (accessed on 2 January 2021).
32. De Leon, S. Beyond Li-Ion High Energy & Power Cells Market Review 2018 (>300Wh/kg). Available online: https://www.sdle.co.il/wp-content/uploads/2018/12/27-Beyond-Li-Ion-battery-High-Energy-and-Power-Cells-Market2018-for-conferences.pdf (accessed on 14 June 2023).
33. IT & New Application Battery. Available online: https://www.lgchem.com/global/small-battery/it-device-battery/product-detail-PDEA0001 (accessed on 24 May 2019).
34. de Leon, S. XFC—Extra Fast Charging Li-Ion Battery Market Review 2017. Available online: https://www.sdle.co.il/wp-content/uploads/2018/12/22-Fast-Charging-Li-Ion-Batteries-ver-4.pdf (accessed on 2 June 2019).
35. What Can We Learn from the Samsung Galaxy Note7 Battery Safety Event. Available online: https://www.sdle.co.il/wp-content/uploads/2018/12/18-What-can-we-learn-from-the-Samsung-Galaxy-Note-7-battery-safety-event-ver-5.pdf (accessed on 14 November 2024).
36. Applications Drones & Robots (6860C5 specification). Available online: https://www.atlbattery.com/en/details.html?product=uav (accessed on 12 November 2021).
37. Investigation of Causes of Deterioration & Failure of Batteries and Capacitors. Available online: https://www.jfe-tec.co.jp/en/battery/analysis/degradation/degradation.html (accessed on 24 May 2019).
38. Zhou, J. Progress of CE and xEV Li-ion Batteries at Lishen. In Proceedings of the 33rd International Battery Seminar & Exhibit 2016, Fort Lauderdale, FL, USA, 21–24 March 2016; p. 38.
39. Wang, J. 电池粘结剂的技术动向及对电池性能的影响 (Technical trends of battery binders and their impact on battery performance) (Zeon). In Proceedings of the 2nd HEV Market & Advanced Battery Technology development Seminar, Hangzhou, China, 13–14 October 2011; p. 35.
40. リチウムイオン電池応用・実用化先端技術開発事業」 事後評価）分科会 資料 (Lithium-Ion Battery Application/Practical Advanced Technology Development Project Ex-Post Evaluation) Subcommittee Materials) 7-1. Available online: https://www.nedo.go.jp/content/100873514.pdf (accessed on 13 February 2019).
41. Hummel, P.; Bush, T.; Gong, P.; Yasui, K.; Lee, T.; Radlinger, J.; Langan, C.; Lesne, D.; Takahashi, K.; Jung, E.; et al. *UBS Q-Series Tearing Down the Heart of an Electric Car: Can Batteries Provide an Edge, and Who Wins?* Report; UBS: New York, NY, USA, 2018.
42. Baldwin, R. Inside the Factory Building GM's Game-Changing Bolt EV. Available online: https://www.engadget.com/2016/12/09/inside-the-factory-building-gm-s-game-changing-bolt-ev/?guccounter=1# (accessed on 24 July 2019).
43. Nisewanger, J. Jaguar and Chevy Have LG in Common. Available online: https://electricrevs.com/2018/03/09/jaguar-and-chevy-have-lg-in-common/ (accessed on 24 July 2019).

44. GM Explains 57 vs 60 kWh Bolt EV Battery Ratings. Available online: https://electricrevs.com/2018/03/29/gm-explains-57-vs-60-kwh-bolt-ev-battery-ratings/ (accessed on 14 November 2024).
45. CATL Achieves 304 Wh/kg in New Battery Cells. Available online: https://pushevs.com/2019/03/30/catl-achieves-304-wh-kg-in-new-battery-cells/ (accessed on 24 May 2019).
46. Cheng, P. Research Progress of Advanced Polymer Lithium-Ion Polymer Battery. In Proceedings of the 33rd Annual International Battery Seminar and Exhibit, Fort Lauderdale, FL, USA, 21–24 March 2016; p. 40.
47. Wolfenstine, J.; Allen, J.L.; Read, J.; Foster, D. *Chemistry and Structure of Sony's Nexelion Li-ion Electrode Materials (ARL-TN-0257)*; USA Army Research Laboratory: Adelphi, MD, USA, 2006; p. 12.
48. Sony Nexelion 14430W1. Available online: https://image2.slideserve.com/4175542/sony-nexelion-n.jpg (accessed on 25 May 2019).
49. Kubota, T. Solvent-Dependent Solid Electrolyte Interphases on Nongraphite Electrodes. In Proceedings of the International Conference on the Frontier of Advanced Batteries (CIBF2014), Shenzhen, China, 20–22 June 2014; p. 31.
50. Foster, D.; Wolfestine, J.; Read, J.; Allen, J.L. *Performance of Sony's Alloy Based Li-Ion Battery (ARL-TN-0319)*; USA Army Research Laboratory: Adelphi, MD, USA, 2008; p. 14.
51. Nishi, Y. Past, Present and Future of Lithium-ion Betteries: Can New Technologies Open up New Horizons. In *Lithium-Ion Batteries: Advances and Applications*; Pistoia, G., Ed.; Elsevier: Amsterdam, The Netherlands, 2014; pp. 21–40.
52. Yang, M.-H. Li-ion Battery: Development Status and Perspective in Electric Vehicles. In Proceedings of the A3PS Conference 2010 Vehicle Integration and System Optimization, Vienna, Austria, 18–19 November 2010; p. 35.
53. LG 18650 MJ1. Available online: https://lygte-info.dk/review/batteries2012/LG%2018650%20MJ1%203500mAh%20(Green)%20UK.html (accessed on 6 May 2018).
54. ソニー、ノートPC市場向けに「スズ系アモルファス負極」を採用した3.5Ahの高容量リチウムイオン二次電池"Nexelion(ネクセリオン)"を開発 (Sony Develops "Nexelion," a 3.5 Ah High-Capacity Lithium-Ion Secondary Battery That Uses a "Tin-Based Amorphous Negative Electrode" for the Notebook PC Market). Available online: https://www.sony.co.jp/SonyInfo/News/Press/201107/11-078/ (accessed on 20 July 2018).
55. PRODUCT SPECIFICATION Rechargeable Lithium Ion Battery Model : ICR18650 E1 12.0 Wh. Available online: https://www.dnkpower.com/wp-content/uploads/2022/07/lg-icr18650-dame1-datasheet.pdf (accessed on 14 November 2024).
56. Available online: http://queenbattery.com.cn/our-products/196-lg-mj1-inr18650mj1-18650-3500mah-10a-37v-li-ion-battery-cell.html (accessed on 6 May 2018).
57. LG INR 21700 M50. Available online: https://www.batterydesign.net/lg-21700-m50/ (accessed on 14 November 2024).
58. Wilson, J. Tesla Model S + 125kWh Battery Delivers 420 Mile Range • Range Anxiety Is Abated. Available online: https://fau4u2.wordpress.com/2017/01/29/%F0%9F%94%98-tesla-model-s-125kwh-battery-delivers-420-mile-range-%E2%80%A2-range-anxiety-is-abated-%F0%9F%94%98/ (accessed on 4 May 2018).

59. Waldmann, T.; Scurtu, R.G.; Richter, K.; Wohlfahrt-Mehrens, M. 18650 vs. 21700 Li-ion cells—A direct comparison of electrochemical, thermal, and geometrical properties. *J. Power Sources* **2020**, *472*, 9. [CrossRef]
60. Quinn, J.B.; Waldmann, T.; Richter, K.; Kasper, M.; Wohlfahrt-Mehrens, M. Energy Density of Cylindrical Li-Ion Cells: A Comparison of Commercial 18650 to the 21700 Cells. *J. Electrochem. Soc.* **2018**, *165*, A3284–A3291. [CrossRef]
61. LG 18650 E1 3200 mAh (Green). Available online: https://lygte-info.dk/review/batteries2012/LG%2018650%20E1%203200mAh%20(Green)%20UK.html (accessed on 10 May 2018).
62. Sanyo/Panasonic NCR18650GA 3500 mAh (Red). Available online: https://lygte-info.dk/review/batteries2012/Sanyo%20NCR18650GA%203500mAh%20%28Red%29%20UK.html (accessed on 10 May 2018).
63. Nguyen, T.T.D.; Abada, S.; Lecocq, A.; Bernard, J.; Petit, M.; Marlair, G.; Grugeon, S.; Laruelle, S. Understanding the Thermal Runaway of Ni-Rich Lithium-Ion Batteries. *World Electr. Veh. J.* **2019**, *10*, 79. [CrossRef]
64. Samsung INR18650-35E 3500 mAh (Pink). Available online: http://lygte-info.dk/review/batteries2012/Samsung%20INR18650-35E%203500mAh%20%28Pink%29%20UK.html (accessed on 30 August 2019).
65. Sony VC7. Available online: http://queenbattery.com.cn/index.php?controller=attachment&id_attachment=83 (accessed on 4 May 2018).
66. Technical Report of INR18650-35E. Available online: https://www.akkuparts24.de/mediafiles/Datenblaetter/Samsung/Samsung%20INR18650-35E.pdf (accessed on 4 May 2018).
67. Specifications for NCR18650GA. Available online: http://www.nkon.nl/sk/k/ncr18650ga.pdf (accessed on 4 May 2018).
68. Dahn, J.; Li, H.; Zhang, N.; Liu, A.; Cormier, M. An Unavoidable Challenge for Ni-Rich Positive Electrode Materials for Li-Ion Batteries. In Proceedings of the 37th International Battery Seminar&Exhibit, Vitual, 28–30 July 2020; p. 27.
69. Li, H.; Cormier, M.; Zhang, N.; Inglis, J.; Li, J.; Dahn, J.R. Is Cobalt Needed in Ni-Rich Positive Electrode Materials for Lithium Ion Batteries? *J. Electrochem. Soc.* **2019**, *166*, A429–A439. [CrossRef]
70. Lu, Y.; Zhang, Y.; Zhang, Q.; Cheng, F.; Chen, J. Recent advances in Ni-rich layered oxide particle materials for lithium-ion batteries. *Particuology* **2020**, *53*, 1–11. [CrossRef]
71. Maehliss, J. Lithium-Ionen Batterietechnologie. Available online: https://www.hs-rm.de/fileadmin/persons/khofmann/Gastvortraege/Vortragsfolien/20160603-Maehliss-Lithium-Ionen-Batterietechnologie.pdf (accessed on 14 May 2018).
72. Anderson, N.; Tram, M.; Darcy, E. 18650 Cell Bottom Vent: Preliminary Evaluation into its Merits for Preventing Side Wall Rupture. In Proceedings of the S&T Meeting, San Diego, CA, USA, 7 December 2016; p. 53.
73. 【高工锂电•总工札记】LG和三星圆柱21700电芯和去钴化 ([High Engineer Lithium Battery·Chief Engineer's Notes] LG and Samsung Cylindrical 21700 Batteries and Reducing Cobalt Consumption.). Available online: http://www.sohu.com/a/255303019_740349 (accessed on 13 February 2019).
74. CELL—BMZ 21700 50EL. Available online: https://bmz-group.com/images/PDF-Downloads/Zelldaten/BMZ-21700-50EL_36629.pdf (accessed on 31 May 2019).

75. Available online: https://keeppower.com.ua/download/2015-09/141224-inr18650mj1-ti-lgc.pdf (accessed on 4 May 2018).
76. Battery Safety 101: Anatomy—PTC vs. PCB vs. CID. Available online: https://batterybro.com/blogs/18650-wholesale-battery-reviews/18306003-battery-safety-101-anatomy-ptc-vs-pcb-vs-cid (accessed on 14 May 2018).
77. Morgan Advanced Materials. Thermal Ceramics. Available online: http://admin.morganadvancedmaterials.com/media/5458/battery-system-solutions-for-electric-vehicle.pdf (accessed on 26 May 2019).
78. Clark, R.; Thomas, A. Addressing the Safety Issues Related to Air Transportation of Lithium-Ion Batteries with Effective Engineered Thermal Management Solutions (Morgan Advanced Materials). In Proceedings of the 33rd Annual International Battery Seminar & Exhibit, Fort Lauderdale, FL, USA, 21–24 March 2016.
79. Cadenza Innovation, Inc. Final Scientific/Technical Report Novel Low Cost and Safe Lithium-ion Electric Vehicle Battery (DE-AR0000392). Available online: https://cadenzainnovation.com/wp-content/uploads/2018/05/2018-03-28-Cadenza-ARPA-e-DE-AR0000392-Final-Report-webversion-002.pdf (accessed on 29 November 2019).
80. XING Mobility™ Immersion Cooled Modular Battery Pack System. Available online: https://xing.hellojcc.tw/XING-Mobility-Battery-Brochure-Download-en.pdf (accessed on 14 November 2024).
81. Performance, Sustainability, Safety. (3MTM NovecTM Engineered Fluids). Available online: https://www.acota.co.uk/wp-content/uploads/2018/11/Novec-Cleaning-White-Paper-2018.pdf (accessed on 26 May 2019).
82. High Performance Materials for Batteries (Solvay). Available online: https://www.solvay.com/sites/g/files/srpend221/files/2018-10/High-Performance-Materials-for-Batteries_EN-v1.7_0_0.pdf (accessed on 6 June 2019).
83. Topics. GSユアサによる大型リチウムイオン電池 の開発の歩み(その1) -世界初の角形ケースと車載用に有利な 縦巻き電極体の発想 - (Progress in the Development of Large Lithium-Lon Batteries by GS Yuasa (Part 1)—The World's First Square Case and the Idea of a Vertically Wound Electrode Body That Is Advantageous for Automotive Use -). *GS Yuasa Tech. Rep.* **2016**, *1*, 37–50.
84. Morin, B. 3rd Generation Separators: Using Thermally Stable Separators to Turn the Aluminum Current Collector into a Fuse (Dream Weaver). In Proceedings of the 34th Annual International Battery Seminar and Exhibit 2017, Fort Lauderdale, FL, USA, 20–23 March 2017; pp. 27, 135.
85. Koshtyal, Y.M.; Rumyantsev, A.M.; Zhdanov, V.V. High power lithium-ion cells—State of art and challenges of design. In Proceedings of the XV International Conference Topical Problems of Energy Conversion in Lithium Electrochemical Systems, Saint-Petersburg, Russia, 17–20 September 2018; pp. 49–53.
86. Zhang, N. Progress of Lishen High Power Li Ion Battery for HEV Application. In Proceedings of the International Seminar on HEV Market and Advanced Battery Technology Development, Beijing, China, 18 April 2014; p. 46.

87. Lei, H.; Blizanac, B.; Atanassova, P.; DuPaskier, A.; Oljaca, M. Carbon materials to Advance Li-ion and Lead-acid Battery Performance. In Proceedings of the CIBF 2014, International Conference on the Frontier of Advanced Batteries, Shenzhen, China, 20–22 June 2014; p. 29.
88. 锂电池用高性能导电材料 “DENKA BLACK Li” 高能量密度化技术 (High-Performance Conductive Material for Lithium Batteries “DENKA BLACK Li” High Energy Density Technology). Available online: http://www.achatestrade.com/filedownload/541261 (accessed on 14 October 2018).
89. LiBの性能を向上する高粘度CMC 第一工業製薬 社報 No.574 拓人2015秋 (High Viscosity CMC That Improves LiB Performance Daiichi Kogyo Seiyaku Company Newsletter No.574 Takuto 2015 Autumn). Available online: https://www.dks-web.co.jp/catalog_pdf/574_1.pdf (accessed on 29 October 2018).
90. Lingli, K. Advance of R&D of high voltage electrolyte system (Lishen). In Proceedings of the 2nd International Forum on Electrolyte and Separator for Advanced Batteries, Shenzhen, China, 11–13 November 2015; p. 27.
91. Ugawa, S.; Masuda, K.; Kajiwara, I. リチウムイオン電池用水系バインダーWater-Based Binder for LIB Application. *JSR Tech. Rev.* **2014**, *121*, 10–15.
92. Fetcenko, M. 适用于小型移动、固定以及交通运输的金属氢化物镍电池技术进展 (Nickel Metal Hydride Batteries for Portable, Stationary and Transportation Application). In Proceedings of the 12th China International Battery Fair (CIBF2016), Shenzhen, China, 24 May 2016; p. 37.
93. Chintawar, P. Automotive Grade NCM Cathodes for Lithium-Ion Batteries. In Proceedings of the Advanced Automotive Battery Conference (AABC2011), Pasadena, CA, USA, 24–28 January 2011; p. 24.
94. Strand, D. Engineering Lithium Metal Surface to Enable Long-Term Cycling with Carbonate-Based Electrolytes. In Proceedings of the Advanced Automotive Battery Conference (AABC 2020, USA), Virtual, 3–5 November 2020; p. 30.
95. Hernandez, P.; Li, I.; Vankatachalam, S.; Kumar, S.; Sinkula, M.; Lopez, H. High Energy Lithium Batteries for Electric Vehicles (Es247). In Proceedings of the DOE Annual Merit Review, Washington, DC, USA, 5–9 June 2017.
96. Stefan, I.C. High Energy Density and Specific Energy Silicon Anode-Based Batteries (13-2). In Proceedings of the 48th Power Sources Conference, Denver, CO, USA, 11–14 June 2018; pp. 223–226.
97. Stefan, I.C. High-Performance Li-Ion Cells with Silicon Nanowire Anode. In Proceedings of the 37th International Battery Seminar&Exhibit, Vitual, 29 July 2020; p. 33.
98. Stefan, I.C. High Energy Density and Specific Energy Batteries with Silicon Nanowire Anodes (Amprius). In Proceedings of the NASA Aerospace Battery Workshop, Huntsville, AL, USA, 27–29 November 2018; p. 29.
99. Thakur, M. Development of HighPerformance Lithium-Ion Cell Technology for Electric Vehicle Applications (bat355). In Proceedings of the DOE Vehicle Technologies Office Annual Merit Review, Arlington, VA, USA, 8–12 June 2019; p. 24.

100. High Energy EV Pouch Cells (Range EV Cells) Zenlabs. Available online: https://static1.squarespace.com/static/59dbcd906f4ca35190c9aeb4/t/5cac3272 23d25a000188813b/1554788979381/EPS+Range.pdf (accessed on 22 November 2019).
101. Dong, Y.; Zhang, W.; Sharma, S.; Vankatachalam, S.; Kumar, S.; Sinkula, M.; Lopez, H. High Energy Lithium Batteries for Electric Vehicles (bat247). In Proceedings of the 2018 DOE Vehicle Technologies Office Annual Merit Review, Arlington, VA, USA, 18–21 June 2018; p. 17.
102. Zhang, W.; Dong, Y.; Vankatachalam, S.; Kumar, S.; Sinkula, M.; Lopez, H. High Energy Lithium Batteries for Electric Vehicles (bat247). In Proceedings of the 2019 DOE Vehicle Technologies Office Annual Merit Review (AMR), Arlington, VA, USA, 12 June 2019; p. 19.
103. Sharma, S.; Karthikeyan, D.K.K.; Bowling, C.A.; Li, B.; Hernandez Gallegos, P.A.; Venkatachalam, S.; Lopez, H.A.; Kumar, S. Positive Electrode Active Materials with Composite Coatings for High Energy Density Secondary Batteries and Corresponding Processes. U.S. Patent 10193135, 29 January 2016.
104. High Energy Drone Cells (Glide Drone Cells) Zenlabs. Available online: https://static1.squarespace.com/static/59dbcd906f4ca35190c9aeb4/t/5cabe184 0d92979d597e6f43/1554768262216/DPS+Glide.pdf (accessed on 22 November 2019).
105. Wolter, M.; Nikolowski, K.; Wätzig, K.; Schilm, J.; Partsch, U. Application of Ceramic Technologies in All Solid State Batteries. Available online: https://www.embatt.de/filea dmin/downloads/160928_GBD_MW_04.pdf (accessed on 29 November 2019).
106. Solid-State Lithium Metal Cell (Ganfeng Lithium). Available online: http://www.ganf englithium.com/pro3_detail_en/id/182.html (accessed on 21 March 2021).
107. Zaghib, K.; Mauger, A.; Julien, C.; Armand, M.; Goodenough, J. Solid State Lithium Batteries: Past, Present and Future. In Proceedings of the 19th International Meeting on Lithium Batteries (IMLB2018), Kyoto, Japan, 17–22 June 2018.
108. Fehrenbacher, K. Why Bosch Is Buying Silicon Valley Battery Startup Seeo. Available online: https://fortune.com/2015/08/27/bosch-buys-battery-startup-seeo/ (accessed on 29 November 2019).
109. Mikhaylik, Y.; Kovalev, I.; Scordilis-Kelley, C.; Liao, L.; Laramie, M.; Schoop, U.; Kelley, T. 650 Wh/kg, 1400 Wh/L Rechargeable Batteries for New Era of Electrified Mobility (Sion Power). In Proceedings of the 2018 NASA Aerospace Battery Workshop, Huntsville, AL, USA, 27–29 November 2018; p. 20.
110. Licerion. Available online: https://sionpower.com/products/ (accessed on 2 December 2019).
111. HermesTM High Energy Rechargeable Metal Cells for Space. Available online: https://ru.scribd.com/document/421754567/Hermes-spec-shit (accessed on 14 November 2024).
112. Solid State Lithium Ion Battery (Ganfeng Lithium). Available online: http://www.ganf englithium.com/pro3_detail_en/id/181.html (accessed on 21 March 2021).
113. Ilika (IKA)—Capital Markets Day December 2019: Solid-State Battery Technology. Available online: https://www.youtube.com/watch?v=u9_6uFBhxrc (accessed on 15 August 2020).

114. Ilika Solid State Battery Technology Program. Available online: https://assets.website-files.com/5d96e3bc5ff1326aba5a2d2f/5def8119b67e21f8c7b72149_BESS%202019%2C%20John%20Tinson.pdf (accessed on 17 December 2020).
115. Ilika's Goliath Solid State Battery Technology for Electric Vehicles & Cordless Domestic Appliances. Available online: https://www.youtube.com/watch?v=1sxJd7-NTLs&feature=emb_logo (accessed on 17 December 2020).
116. Kepler, K. Development of High-Performance Lithium-Ion Cell Technology for Electric Vehicle Applications (bat355). In Proceedings of the 2018 DOE Vehicle Technologies Office Annual Merit Review, Arlington, VA, USA, 18–21 June 2018; p. 25.
117. Jiang, J.; Obrovac, M.; Lamanna, W.; Gardner, J.; Dahn, J. Advanced Lithium-ion Battery Material Developments in 3M. In Proceedings of the 4th Southern China Li-ion Battery Top Forum (CLTF2009), Shenzhen, China, 23–24 May 2009; p. 22.
118. Christensen, L.; Obrovac, M.N. Electrode Compositions Based on an Amorphous Alloy Having a High Silicon Content. U.S. Patent 7,972,727 B2, 8 June 2011.
119. Christensen, L.; Le, D.B.; Singh, J.; Obrovac, M.N. 3M Alloy Anode Materials. In Proceedings of the 27th International Battery Seminar & Exhibit, Ft. Lauderdale, FL, USA, 15–18 March 2010; p. 27.
120. Krause, L.J.; Chevrier, V.L.; Le, D.B.; Liu, L.; Jensen, L.; Singh, J.; Eberman, K.W. Implementation and Failure Mechanisms of Alloy Materials in Commercially Relevant Electrodes. In Proceedings of the 30th International Battery Seminar & Exhibit, Fort Lauderdale, FL, USA, 11–14 March 2013; p. 33.
121. Kaiser, J.; Chevrier, V.; Figgemeier, E.; Eberman, K. Commercialization of Silicon Anodes for Electric Vehicle Applications. In Proceedings of the 7th Advanced Automotive Battery Conference Europe (AABC), Mainz, Germany, 31 January 2017; p. 12.
122. SINANODE® Materials: Key Concepts. Available online: https://onedsinanode.com/wp-content/uploads/2021/05/SiNANOdeR-materials-Key-Concepts-OneD-Material.pdf (accessed on 14 November 2024).
123. SINANODE® Materials: Evaluation And Cell Design. Available online: https://onedsinanode.com/wp-content/uploads/2021/05/SiNANOdeR-Materials-Evaluation-and-Cell-Design-OneD-Material.pdf (accessed on 14 November 2024).
124. Yuan, Q.F.; Zhao, F.G.; Zhao, Y.M.; Liang, Z.Y.; Yan, D.L. Evaluation and performance improvement of Si/SiOx/C based composite as anode material for lithium-ion batteries. *Electrochim. Acta* **2014**, *115*, 16–21. [CrossRef]
125. Yoshida, S.; Okubo, T.; Masuo, Y.; Oba, Y.; Shibata, D.; Haruta, M.; Doi, T.; Inaba, M. High Rate Charge and Discharge Characteristics of Graphite/SiOx Composite Electrodes. *Electrochemistry* **2017**, *85*, 403–408. [CrossRef]
126. Yuan, Q.F.; Zhao, F.G.; Zhao, Y.M.; Liang, Z.Y.; Yan, D.L. Reason analysis for Graphite-Si/SiOx/C composite anode cycle fading and cycle improvement with PI binder. *J. Solid State Electrochem.* **2014**, *18*, 2167–2174. [CrossRef]
127. Nakanishi, T. Recent Development of Silicon-Based Anodes at Shin-Etsu. In Proceedings of the AABC Asia 2014, Kyoto, Japan, 20 May 2014; p. 20.

128. XG SiG™ Energy Storage Materials Silicon-Graphene Li-Ion Battery Anode (XGScience). Available online: https://xgsciences.com/wp-content/uploads/2018/02/XG-SiG-Data-Sheet.pdf (accessed on 29 November 2019).
129. Silicon Anodes & Prelithiation for Fast Charge Batteries | Sun, Cushing & Kumar | StorageX Symposium. Available online: https://www.youtube.com/watch?v=S5P6SUW4pnc (accessed on 21 January 2021).
130. Park, B.; Choilan, S.W.; Browne, I.; Schank, W. Electrodes, Electrochemical Cells, and Methods of Forming Electrodes and Electrochemical Cells. U.S. Patent 9,397,338B2, 19 July 2016.
131. Park, B.; Bonhomme, F. Ultrafast Charge Silicon-dominant Composite Anode and Cell for EV Applications. In Proceedings of the Advanced Lithium Batteries for Automobile Applications (ABAA- 10), Chicago, IL, USA, 22–25 October 2017; p. 1.
132. eBoost™ Feature. Available online: http://www.enevate.com/technology/eboost-feature/ (accessed on 6 December 2019).
133. Lopez, H.; Bowling, C.; Hernandez, P.; Karthikeyan, D.; Kumar, S.; Li, B.; Sharmaa, S.; Venkatachalama, S. Novel Cobalt-Rich Composite (CRC) Cathode and Siox Anodes for High-Energy Li-Ion Batteries. In Proceedings of the ECS Meeting Abstracts, San Diego, CA, USA, 29 May–2 June 2016; p. MA2016–01 85. [CrossRef]
134. Xin, F.X.; Whittingham, M.S. Challenges and Development of Tin-Based Anode with High Volumetric Capacity for Li-Ion Batteries. *Electrochem. Energy Rev.* **2020**, *3*, 643–655. [CrossRef]
135. QuantumScape: Solid-State Battery Showcase. Available online: https://www.youtube.com/watch?v=dGnPSkXKb0I (accessed on 5 February 2021).
136. Wang, M.J.; Carmona, E.; Gupta, A.; Albertus, P.; Sakamoto, J. Enabling "lithium-free" manufacturing of pure lithium metal solid-state batteries through in situ plating. *Nat. Commun.* **2020**, *11*, 9. [CrossRef]
137. Xie, Z.K.; Wu, Z.J.; An, X.W.; Yue, X.Y.; Wang, J.J.; Abudula, A.; Guan, G.Q. Anode-free rechargeable lithium metal batteries: Progress and prospects. *Energy Storage Mater.* **2020**, *32*, 386–401. [CrossRef]
138. Ren, X.D.; Zou, L.F.; Cao, X.; Engelhard, M.H.; Liu, W.; Burton, S.D.; Lee, H.; Niu, C.J.; Matthews, B.E.; Zhu, Z.H.; et al. Enabling High-Voltage Lithium-Metal Batteries under Practical Conditions. *Joule* **2019**, *3*, 1662–1676. [CrossRef]
139. Weber, R.; Genovese, M.; Louli, A.J.; Hames, S.; Martin, C.; Hill, I.G.; Dahn, J.R. Long cycle life and dendrite-free lithium morphology in anode-free lithium pouch cells enabled by a dual-salt liquid electrolyte. *Nat. Energy* **2019**, *4*, 683–689. [CrossRef]
140. Hu, Q. Blind Men and the Elephant—The Need for SolidEnergy's Integrated Approach in a Post Li-Ion Era. Available online: https://media.nature.com/original/nature-cms/uploads/ckeditor/attachments/3339/2016-09-08_Focus_on_Post_Lithium_-_Solid_Energy_web.pdf?platform=hootsuite (accessed on 6 February 2021).
141. Lee, Y.G.; Fujiki, S.; Jung, C.; Suzuki, N.; Yashiro, N.; Omoda, R.; Ko, D.S.; Shiratsuchi, T.; Sugimoto, T.; Ryu, S.; et al. High-energy long-cycling all-solid-state lithium metal batteries enabled by silver-carbon composite anodes. *Nat. Energy* **2020**, *5*, 299–308. [CrossRef]

142. Holmes, T. A Discussion of QuantumScape's Battery Technology Performance Results. Available online: https://www.quantumscape.com/blog/a-discussion-of-quantumscapes-battery-technology-performance-results/ (accessed on 20 February 2021).
143. Printable Lithium. Enabling Technology for Sustainable and Disruptive Innovation in Lithium Batteries (Livent). Available online: https://vexpo-data.eventxtra.com/92iw4l3r1lsuqmud6uib8v7ib71z?response-content-disposition=inline%3B%20filename%3D%22Livent%20SLMP_Printable%20Lithium_English.pdf%22%3B%20filename%2A%3DUTF-8%27%27Livent%2520SLMP_Printable%2520Lithium_English.pdf&response-content-type=application%2Fpdf&X-Amz-Algorithm=AWS4-HMAC-SHA256&X-Amz-Credential=AKIAQM43SQZCYXSQ4S4Q%2F20210218%2Fap-southeast-1%2Fs3%2Faws4_request&X-Amz-Date=20210218T060844Z&X-Amz-Expires=604800&X-Amz-SignedHeaders=host&X-Amz-Signature=488b2826c0c7d32e085a614045be01c4869061130573501c5cdd894e070aac65 (accessed on 18 February 2021).
144. Yakovleva, M. Printable Lithium Technology for LIB and SSB Applications. In Proceedings of the 37th International Battery Seminar&Exhibit, Vitual, 28–30 July 2020.
145. Ulvac Lithium Evaporation System. In Proceedings of the 37th International Battery Seminar&Exhibit, Vitual, 28–30 July 2020; p. 1.
146. Sasaki, S. Introduction of Thin Film Battery/Lithium Metal Anode Films Developed Using ULVAC Vacuum Deposition Equipment. In Proceedings of the 38th Annual International Battery Seminar & Exhibition, Virtual, 11 March 2021.
147. Month 12 Meeting in San Sebastian (ES). Available online: https://www.h2020-image.com/m12-meeting (accessed on 23 February 2021).
148. Garcia, I.; Armand, M.; Shanmukaraj, D. Li Metal Polymer Batteries. In *Solid Electrolytes for Advanced Applications Garnets and Competitors*; Murugan, R., Weppner, W., Eds.; Springer Nature: Cham, Switzerland, 2019; pp. 347–373. [CrossRef]
149. Armand, M. Polymer electrolyte as separator in solid-state Li batteries. In Proceedings of the Second International Forum on Electrolyte & Separator for Advanced Batteries, Shenzhen, China, 11–13 November 2015; p. 24.
150. QuantumScape, Next-Generation Solid-State Batteries. Available online: https://www.quantumscape.com/wp-content/uploads/2021/10/Data-Launch-Updated-Post-Presentation-20210107-2.pdf (accessed on 14 November 2024).
151. Glossman, T. Quest for Energy Density—Aspects of Li-Metal Anodes. In Proceedings of the 37th International Battery Seminar&Exhibit, Vitual, 28–30 July 2020; p. 15.
152. Niu, C.J.; Lee, H.; Chen, S.R.; Li, Q.Y.; Du, J.; Xu, W.; Zhang, J.G.; Whittingham, M.S.; Xiao, J.; Liu, J. High-energy lithium metal pouch cells with limited anode swelling and long stable cycles. *Nat. Energy* **2019**, *4*, 551–559. [CrossRef]
153. Bernard, P. Saft's Advanced & Beyond Lithium-Ion Technologies for Mobility Applications. In Proceedings of the Advanced Automotive Battery Conference (AABC 2021 Europe), Virtual, 20 January 2021; p. 22.
154. Liu, J. The B500 Battery Approach for Future High Energy Batteries. In Proceedings of the 10th Int'l Rechargeable Battery Expo (Battery Japan), Tokyo, Japan, 15–17 November 2023; pp. 1–18.

155. Armand, M. Polymer electrolytes, the safe and green option. In Proceedings of the Advanced Batteries for xEV/ESS Conference (ABEC), Yichun, China, 11–14 November 2013; p. 25.
156. Arya, A.; Sharma, A.L. A glimpse on all-solid-state Li-ion battery (ASSLIB) performance based on novel solid polymer electrolytes: A topical review. *J. Mater. Sci.* **2020**, *55*, 6242–6304. [CrossRef]
157. Garret, J. Solid Electrolytes—The Key to All Solid-State Batteries. Available online: https://www.youtube.com/watch?v=bHay2K3432k&feature=emb_rel_pause (accessed on 18 February 2021).
158. Solid Power Company Update: December 2020. Available online: https://5if66.r.a.d.sendibm1.com/mk/mr/90yqLoXClaJm0ga0gNAlfvClii0nHBoI-SnvB92Ai_aSMp_Eg0p-FWkFua23EzpLkUb0gxAyaLVd_XP_75bhe5Z89D-EP7goPu9kK7pOgeKp (accessed on 18 February 2021).
159. Hsu, L.; Chang, S. Better understanding of solid state battery and highlights of ProLogium. In Proceedings of the 38th Annual International Battery Seminar & Exhibition, Virtual, 12–15 March 2021; p. 15.
160. Liang, J.N.; Luo, J.; Sun, Q.; Yang, X.F.; Li, R.Y.; Sun, X.L. Recent progress on solid-state hybrid electrolytes for solid-state lithium batteries. *Energy Storage Mater.* **2019**, *21*, 308–334. [CrossRef]
161. Fang, C.C.; Li, J.X.; Zhang, M.H.; Zhang, Y.H.; Yang, F.; Lee, J.Z.; Lee, M.H.; Alvarado, J.; Schroeder, M.A.; Yang, Y.Y.C.; et al. Quantifying inactive lithium in lithium metal batteries. *Nature* **2019**, *572*, 511. [CrossRef] [PubMed]
162. Bai, P.; Li, J.; Brushett, F.R.; Bazant, M.Z. Transition of lithium growth mechanisms in liquid electrolytes. *Energy Environ. Sci.* **2016**, *9*, 3221–3229. [CrossRef]
163. Cao, X.; Ren, X.D.; Zou, L.F.; Engelhard, M.H.; Huang, W.; Wang, H.S.; Matthews, B.E.; Lee, H.; Niu, C.J.; Arey, B.W.; et al. Monolithic solid-electrolyte interphases formed in fluorinated orthoformate-based electrolytes minimize Li depletion and pulverization. *Nat. Energy* **2019**, *4*, 796–805. [CrossRef]
164. Fetcenko, M.A. Metallic Lithium Metal Anode for Ultra-High Energy Batteries for xEV and Aerospace Applications. In Proceedings of the 37th International Battery Seminar&Exhibit, Vitual, 28–30 July 2020; p. 24.
165. Hu, Q. 400 Wh/Kg Is Here, a Practical Approach to Solid-State Lithium Metal Cells. In Proceedings of the Advanced Automotive Battery Conference (AABC 2020, USA), Virtual, 3–5 November 2020; p. 17.
166. Liu, B.; Zhang, J.G.; Xu, W. Advancing Lithium Metal Batteries. *Joule* **2018**, *2*, 833–845. [CrossRef]
167. Xiao, J. High Energy Rechargeable Lithium-Metal Cells: Fabrication and Integration (Bat369). In Proceedings of the 2020 DOE Vehicle Technologies Program Annual Merit Review, Virual, 4 June 2020; p. 23.
168. Liu, J.; Bao, Z.N.; Cui, Y.; Dufek, E.J.; Goodenough, J.B.; Khalifah, P.; Li, Q.Y.; Liaw, B.Y.; Liu, P.; Manthiram, A.; et al. Pathways for practical high-energy long-cycling lithium metal batteries. *Nat. Energy* **2019**, *4*, 180–186. [CrossRef]

169. Zhang, J.G.; Xu, W.; Henderson, W.A. *Lithium Metal Anodes and Rechargeable Lithium Metal Batteries*; Springer International Publishing: Cham, Switzeland, 2016. [CrossRef]
170. How SolidEnergy Is Transforming the Future of Transportation and Connectivity. Available online: http://s3-service-broker-live-19ea8b98-4d41-4cb4-be4c-d68f4963b7dd.s3.amazonaws.com/uploads/ckeditor/attachments/6903/WEB_PDF_2017-10-00_-_Batteries_Materials_Collection-_SolidEnergy.pdf (accessed on 14 November 2024).
171. Barchasz, C.; Tarnopolskyi, V.; Picard, L.; Bloch, D.; Patoux, S.; Perraud, S. Progress and challenges Generation 4 (Liten CEA tech). In Proceedings of the Battery Workshop Bruxelles, Bruxelles, Belgique, 11–12 January 2018; p. 26.
172. Long, L.Z.; Wang, S.J.; Xiao, M.; Meng, Y.Z. Polymer electrolytes for lithium polymer batteries. *J. Mater. Chem. A* **2016**, *4*, 10038–10069. [CrossRef]17
173. Xue, Z.G.; He, D.; Xie, X.L. Poly(ethylene oxide)-based electrolytes for lithium-ion batteries. *J. Mater. Chem. A* **2015**, *3*, 19218–19253. [CrossRef]
174. Liang, C. Challenges and Progresses of Solid-State Li Metal Batteries (CATL). In Proceedings of the CIBF 2016 the International Conference on the Frontier of Advanced Batteries, CIBF2016, Shenzhen, China, 24-26 May 2016; p. 21.
175. Lascaud, S.; Perrier, M.; Vallee, A.; Besner, S.; Prudhomme, J.; Armand, M. Phase-DIAGRAMS and Conductivity Behavior of Poly(ethylene Oxide) Molten-Salt Rubbery Electrolytes. *Macromolecules* **1994**, *27*, 7469–7477. [CrossRef]
176. Younesi, R.; Veith, G.M.; Johansson, P.; Edstrom, K.; Vegge, T. Lithium salts for advanced lithium batteries: Li-metal, Li-O_2, and Li-S. *Energy Environ. Sci.* **2015**, *8*, 1905–1922. [CrossRef]
177. Di Censo, D.; Exnar, I.; Graetzel, M. Non-corrosive electrolyte compositions containing perfluoroalkylsulfonyl imides for high power Li-ion batteries. *Electrochem. Commun.* **2005**, *7*, 1000–1006. [CrossRef]
178. Lithium-metaL-PoLymer Battery (EASE). Available online: https://ease-storage.eu/wp-content/uploads/2016/07/EASE_TD_Electrochemical_LMP.pdf (accessed on 24 February 2021).
179. Hovington, P.; Lagace, M.; Guerfi, A.; Bouchard, P.; Manger, A.; Julien, C.M.; Armand, M.; Zaghib, K. New Lithium Metal Polymer Solid State Battery for an Ultrahigh Energy: Nano C-$LiFePO_4$ versus Nano $Li_{1.2}V_3O_8$. *Nano Lett.* **2015**, *15*, 2671–2678. [CrossRef]
180. Campbell, D. Advances and remaining challenges in electrolytes for solid state batteries (Solid power). In Proceedings of the AABC Europe, Mainz, Germany, 31 January 2017; p. 26.
181. Bullis, K. A Prototype Battery Could Double the Range of Electric Cars. Available online: https://www.technologyreview.com/s/533541/a-prototype-battery-could-double-the-range-of-electric-cars/ (accessed on 29 November 2019).
182. Lithium-Ion Conducting Glass-Ceramics (LICGC™). Available online: https://www.oharacorp.com/lic-gc.html (accessed on 29 November 2019).
183. Nakajima, K.; Katoh, T.; Inda, Y.; Hoffman, B. Lithium-Ion Conductive Glass Ceramics: Properties and Application in Lithium Metal Batteries. In Proceedings of the Symposium on Energy Storage Beyond Lithium Ion, Oak Ridge National Laboratory, Oak Ridge, TN, USA, 7–8 October 2010; p. 28.

184. 日本 小原 OHARA LICGC AG-01. 小原锂金属电池电解质 (Lithium Metal Battery Electrolyte). Available online: http://www.achatestrade.com/ (accessed on 16 December 2020).
185. Iriyama, Y.; Yada, C.; Abe, T.; Ogumi, Z.; Kikuchi, K. A new kind of all-solid-state thin-film-type lithium-ion battery developed by applying a DC high voltage. *Electrochem. Commun.* **2006**, *8*, 1287–1291. [CrossRef]
186. Awaka, J.; Kijima, N.; Hayakawa, H.; Akimoto, J. Synthesis and structure analysis of tetragonal $Li_7La_3Zr_2O_{12}$ with the garnet-related type structure. *J. Solid State Chem.* **2009**, *182*, 2046–2052. [CrossRef]
187. Kotobuki, M.; Munakata, H.; Kanamura, K.; Sato, Y.; Yoshida, T. Compatibility of $Li_7La_3Zr_2O_{12}$ Solid Electrolyte to All-Solid-State Battery Using Li Metal Anode. *J. Electrochem. Soc.* **2010**, *157*, A1076–A1079. [CrossRef]
188. Ohta, S.; Kobayashi, T.; Asaoka, T. High lithium ionic conductivity in the garnet-type oxide $Li_{7-X}La_3(Zr_{2-X},Nb_X)O_{12}$ (X = 0–2). *J. Power Sources* **2011**, *196*, 3342–3345. [CrossRef]
189. Murugan, R.; Thangadurai, V.; Weppner, W. Fast lithium ion conduction in garnet-type $Li_7La_3Zr_2O_{12}$. *Angew. Chem.-Int. Ed.* **2007**, *46*, 7778–7781. [CrossRef]
190. Seo, J.H.; Nakaya, H.; Takeuchi, Y.; Fan, Z.M.; Hikosaka, H.; Rajagopalan, R.; Gomez, E.D.; Iwasaki, M.; Randall, C.A. Broad temperature dependence, high conductivity, and structure-property relations of cold sintering of LLZO-based composite electrolytes. *J. Eur. Ceram. Soc.* **2020**, *40*, 6241–6248. [CrossRef]
191. Oxide Electrolyte Powder (Ganfeng Lithium). Available online: http://www.ganfenglithium.com/pro3_detail_en/id/177.html (accessed on 21 March 2021).
192. Sulfide Electrolyte Powder (Ganfeng Lithium). Available online: http://www.ganfenglithium.com/pro3_detail_en/id/176.html (accessed on 21 March 2021).
193. Kato, Y.; Hori, S.; Saito, T.; Suzuki, K.; Hirayama, M.; Mitsui, A.; Yonemura, M.; Iba, H.; Kanno, R. High-power all-solid-state batteries using sulfide superionic conductors. *Nat. Energy* **2016**, *1*, 7. [CrossRef]
194. Kim, H.; Hikima, K.; Watanabe, K.; Matsui, N.; Suzuki, K.; Obokata, S.; Muto, H.; Matsuda, A.; Kanno, R.; Hirayama, M. Mechanical Properties of Li10.35Ge1.35P1.65S12 with Different Particle Sizes. *Mater. Trans.* **2024**, *65*, 861–866. [CrossRef]
195. High Performance Solid Electrolyte for Next-Generation Lithium Battery Developed. Available online: https://www.mitsui-kinzoku.com/Portals/0/resource/uploads/topics_161124e.pdf (accessed on 14 November 2024).
196. Compact Solid Electrolytes Sheet (Ganfeng Lithium). Available online: http://www.ganfenglithium.com/pro3_detail_en/id/180.html (accessed on 21 March 2021).
197. Cheng, Q.; Li, A.J.; Li, N.; Li, S.; Zangiabadi, A.; Li, T.D.; Huang, W.L.; Li, A.C.; Jin, T.W.; Song, Q.Q.; et al. Stabilizing Solid Electrolyte-Anode Interface in Li-Metal Batteries by Boron Nitride-Based Nanocomposite Coating. *Joule* **2019**, *3*, 1510–1522. [CrossRef]
198. Development of Mass Production Technology for LiBH4-Based Solid Electrolytes. Available online: https://www.mgc.co.jp/eng/corporate/news/files/160120e.pdf (accessed on 28 March 2021).

199. World Smart Energy Week 2016 東京現場直擊系列報導一 (Tokyo Live Report Series 1). Available online: https://www.materialsnet.com.tw/DocView.aspx?id=24126 (accessed on 7 November 2018).
200. Nair, J.R.; Imholt, L.; Brunklaus, G.; Winter, M. Lithium Metal Polymer Electrolyte Batteries: Opportunities and Challenges. *Electrochem. Soc. Interface* **2019**, *28*, 55–61. [CrossRef]
201. 各種固体電解質の開発に性能比較用導電メンブレン検体TREKION LiMe-CP (Conductive Membrane Sample for Performance Comparison in the Development of Various Solid Electrolytes TREKION LiMe-CP). Available online: https://www.piotrek-il.co.jp/productsindex3053.html (accessed on 14 November 2024).
202. Zhao, Q.; Stalin, S.; Zhao, C.Z.; Archer, L.A. Designing solid-state electrolytes for safe, energy-dense batteries. *Nat. Rev. Mater.* **2020**, *5*, 229–252. [CrossRef]
203. Ding, Z.; Li, J.; Li, J.; An, C. Review—Interfaces: Key Issue to Be Solved for All Solid-State Lithium Battery Technologies. *J. Electrochem. Soc.* **2020**, *167*, 070541. [CrossRef]
204. Buettner-Garrett, J. Manufacturing Multi-Layer, Multi-Ah All Solid-State Lithium Metal Cells Using Lithium-Ion Industry Standard Processes and Equipment. In Proceedings of the 38th Annual International Battery Seminar & Exhibition, Virtual, 10 March 2021.
205. Yang, C.P.; Fu, K.; Zhang, Y.; Hitz, E.; Hu, L.B. Protected Lithium-Metal Anodes in Batteries: From Liquid to Solid. *Adv. Mater.* **2017**, *29*, 28. [CrossRef]
206. Walther, F.; Koerver, R.; Fuchs, T.; Ohno, S.; Sann, J.; Rohnke, M.; Zeier, W.G.; Janek, J. Visualization of the Interfacial Decomposition of Composite Cathodes in Argyrodite-Based All-Solid-State Batteries Using Time-of-Flight Secondary-Ion Mass Spectrometry. *Chem. Mater.* **2019**, *31*, 3745–3755. [CrossRef]
207. Janek, J. Cathode Materials for Solid State Batteries and Their Requirements. In Proceedings of the Advanced Automotive Battery Conference (AABC 2021 Europe), Virtual, 19 January 2021; p. 19.
208. Zhu, Y.Z.; He, X.F.; Mo, Y.F. Origin of Outstanding Stability in the Lithium Solid Electrolyte Materials: Insights from Thermodynamic Analyses Based on First-Principles Calculations. *ACS Appl. Mater. Interfaces* **2015**, *7*, 23685–23693. [CrossRef] [PubMed]
209. Washida, D.; Omura, J.; Ito, T.; Mistsumoto, T.; Takahashi, T.; Ide, H.; Miyashita, N.; Yasuda, K. High voltage charge/discharge performance of All solid-state battery using lithium Argyrodite solid electrolyte. In Proceedings of the Advanced Automotive Battery Conference Europe, Mainz, Germany, 29 January–1 February 2018.
210. Yu, S.; Schmidt, R.D.; Garcia-Mendez, R.; Herbert, E.; Dudney, N.J.; Wolfenstine, J.B.; Sakamoto, J.; Siegel, D.J. Elastic Properties of the Solid Electrolyte $Li_7La_3Zr_2O_{12}$ (LLZO). *Chem. Mater.* **2016**, *28*, 197–206. [CrossRef]
211. Waetzig, K.; Rost, A.; Heubner, C.; Coeler, M.; Nikolowski, K.; Wolter, M.; Schilm, J. Synthesis and sintering of $Li_{1.3}Al_{0.3}Ti_{1.7}(PO_4)_3$ (LATP) electrolyte for ceramics with improved Li^+ conductivity. *J. Alloys Compd.* **2020**, *818*, 153237. [CrossRef]
212. Zhou, L.D.; Park, K.H.; Sun, X.Q.; Lalere, F.; Adermann, T.; Hartmann, P.; Nazar, L.F. Solvent-Engineered Design of Argyrodite Li_6PS_5X (X = Cl, Br, I) Solid Electrolytes with High Ionic Conductivity. *ACS Energy Lett.* **2019**, *4*, 265–270. [CrossRef]

213. Ulissi, U.; Bruce, P. Solid-State Batteries for xEV Applications—An Academic and Industrial Perspective. In Proceedings of the Advanced Automotive Battery Conference (AABC 2021 Europe), Virtual, 20 January 2021.
214. Sakamoto, J. Transitioning Solid-State Batteries from Lab to Market. In Proceedings of the Advanced Automotive Battery Conference (AABC 2020, USA), Virtual, 3 November 2020; p. 23.
215. Sharafi, A.; Kazyak, E.; Davis, A.L.; Yu, S.H.; Thompson, T.; Siegel, D.J.; Dasgupta, N.P.; Sakamoto, J. Surface Chemistry Mechanism of Ultra-Low Interfacial Resistance in the Solid-State Electrolyte $Li_7La_3Zr_2O_{12}$. *Chem. Mater.* **2017**, *29*, 7961–7968. [CrossRef]
216. Wachsman, E.D. Beyond Dendrites: Cycling Li-Metal at High Current Density. In Proceedings of the Advanced Automotive Battery Conference (AABC 2020, USA), Virtual, 3 November 2020; p. 13.
217. Han, X.G.; Gong, Y.H.; Fu, K.; He, X.F.; Hitz, G.T.; Dai, J.Q.; Pearse, A.; Liu, B.Y.; Wang, H.; Rublo, G.; et al. Negating interfacial impedance in garnet-based solid-state Li metal batteries. *Nat. Mater.* **2017**, *16*, 572. [CrossRef] [PubMed]
218. Hitz, G.T.; McOwen, D.W.; Zhang, L.; Ma, Z.H.; Fu, Z.Z.; Wen, Y.; Gong, Y.H.; Dai, J.Q.; Hamann, T.R.; Hu, L.B.; et al. High-rate lithium cycling in a scalable trilayer Li-garnet-electrolyte architecture. *Mater. Today* **2019**, *22*, 50–57. [CrossRef]
219. Turner, G. Dearborn's Ford Partners with Colorado's Solid Power to Develop Solid-State Batteries for Electric Vehicles. Available online: https://www.dbusiness.com/tech-mobility-news/dearborns-ford-partners-with-colorados-solid-power-to-develop-solid-state-batteries-for-electric-vehicles/ (accessed on 29 November 2019).
220. Gurung, A.; Pokharel, J.; Baniya, A.; Pathak, R.; Chen, K.; Lamsal, B.S.; Ghimire, N.; Zhang, W.H.; Zhou, Y.; Qiao, Q.Q. A review on strategies addressing interface incompatibilities in inorganic all-solid-state lithium batteries. *Sustain. Energ. Fuels* **2019**, *3*, 3279–3309. [CrossRef]
221. Winter, M. Opportunities and Challenges for Electrochemical Energy Storage. In Proceedings of the 10th International AVL Exhaust Gas and Particulate Emissions Forum, Ludwigsburg, Germany, 20–21 February 2018; p. 31.
222. Placke, T.; Kloepsch, R.; Duhnen, S.; Winter, M. Lithium ion, lithium metal, and alternative rechargeable battery technologies: The odyssey for high energy density. *J. Solid State Electrochem.* **2017**, *21*, 1939–1964. [CrossRef]
223. Manthiram, A.; Yu, X.W.; Wang, S.F. Lithium battery chemistries enabled by solid-state electrolytes. *Nat. Rev. Mater.* **2017**, *2*, 16. [CrossRef]
224. Janek, J.; Zeier, W.G. A solid future for battery development. *Nat. Energy* **2016**, *1*, 4. [CrossRef]
225. 趙章恩 (Zhao Zhangen) 次世代はリチウム硫黄電池か全固体電池か、EV火災で安全性に脚光(The Next Generation Will Be Lithium-Sulfur Batteries or All-Solid-State Batteries, and the EV Fire Puts Safety in the Spotlight). Available online: https://xtech.nikkei.com/atcl/nxt/column/18/01231/00018/ (accessed on 26 March 2021).
226. Yu, A. Transformational Solid State Battery Technology. In Proceedings of the Advanced Automotive Battery Conference (AABC 2021 Europe), Virtual, 19 January 2021; p. 17.

227. Solid Power Media Gallery. Available online: https://solidpowerbattery.com/media.twig (accessed on 29 November 2019).
228. Sunayama, S. Development of All-solid-state Battery for Commercialization (Hitachi Zosen). In Proceedings of the 10th Battery Japan Technical Conference (Battery Japan 2019), Tokyo, Janpan, 27 February 2019; pp. 19–31.
229. Hitz As-Lib (Battery Japan 2021). Available online: https://vexpo-assets.eventxcdn.com/7p48yoz3haz8qtnn6vhkmtm8slom (accessed on 26 February 2021).
230. 日立造船が1Ahの全固体LIBを開発　宇宙利用に (Hitachi Zosen Develops 1 Ah All-Solid-State LIB for Space Use). Available online: https://xtech.nikkei.com/atcl/nxt/news/18/09822/ (accessed on 9 March 2021).
231. Hayden, B. Solid-State Batteries: Composite Materials Formulations & Manufacturing Processes. In Proceedings of the Advanced Automotive Battery Conference (AABC 2020, USA), Virtual, 3 November 2020; p. 17.
232. Pasero, D. Can solid state batteries be a good fit for aerospace applications? In Proceedings of the 2020 NASA Aerospace Battery Workshop, Virtual, 18 November 2020; p. 20.
233. 全固体電池の開発はどこまで進んだ？...新機能材料展 (How Far Has the Development of All-Solid-State Batteries Progressed? ...New Functional Materials Exhibition) 16 February 2018. Available online: https://response.jp/article/2018/02/16/306141.html (accessed on 13 March 2021).
234. 「先進・革新蓄電池材料評価技術開発(第2期)」 中間評価分科会 (2018年度~2020年度 3年間) プロジェクトの概要 (公開). ("Development of Advanced and Innovative Storage Battery Material Evaluation Technology (2nd Phase)" Interim Evaluation Subcommittee (3 years from FY2018 to FY2020) Project Overview (Public)). Available online: https://www.nedo.go.jp/content/100927227.pdf (accessed on 14 March 2021).
235. 「先進・革新蓄電池材料評価技術開発(第 2 期)」 (中間評価)分科会 資料 ("Advanced and Innovative Storage Battery Material Evaluation Technology Development (2nd Period)" (Interim Evaluation) Subcommittee Materials) 7-1 (NEDO). Available online: https://www.nedo.go.jp/content/100927228.pdf (accessed on 14 March 2021).
236. 先進・革新蓄電池材料評価技術開発」 前倒し事後評価 (2013年度~2017年度 5年間) プロジェクトの概要 (公開) ("Advanced and Innovative Storage Battery Material Evaluation Technology Development" Advance Ex-Post Evaluation (5 Years from FY2013 to FY2017) Project Overview (Released)) (NEDO). Available online: https://www.nedo.go.jp/content/100867609.pdf (accessed on 26 May 2024).
237. 先進・革新蓄電池材料評価技術開発」 (前倒し事後評価)分科会 資料 ("Advanced and Innovative Storage Battery Material Evaluation Technology Development" (Advanced Ex-Post Evaluation) Subcommittee Materials) 7-1 (NEDO). Available online: https://www.nedo.go.jp/content/100867610.pdf (accessed on 14 March 2021).
238. Ito, S.; Fujiki, S.; Yamada, T.; Aihara, Y.; Park, Y.; Kim, T.Y.; Baek, S.W.; Lee, J.M.; Doo, S.; Machida, N. A rocking chair type all-solid-state lithium ion battery adopting Li2O-ZrO_2 coated $LiNi_{0.8}Co_{0.15}Al_{0.05}O_2$ and a sulfide based electrolyte. *J. Power Sources* **2014**, *248*, 943–950. [CrossRef]

239. Fabrication of an All-Solid-State Thin-film Lithium-Ion Battery Prototype Using a Room Temperature Process. Available online: https://www.aist.go.jp/aist_e/list/latest_research/2010/20101224/20101224.html (accessed on 27 February 2021).
240. Roadmaps for Developing Solid-State Batteries. Available online: https://twitter.com/zenkotaidenchi/status/1364453980466675714/photo/1 (accessed on 14 March 2021).
241. 哲生, 野. (Tetsuo, No.) 中国NIO発表の衝撃の車載向け固体電池、製造はあのメーカーか (Shocking Automotive Solid-State Battery Announced by China's NIO, Is That the Manufacturer?). Available online: https://xtech.nikkei.com/atcl/nxt/column/18/00001/05229/ (accessed on 15 March 2021).
242. 【トヨタ 環境技術発表】出力密度を5倍に高めた全固体電池の電解質を開発 3枚目の写真（全4枚） (Toyota Environmental Technology Announcement: Development of Electrolyte for All-Solid-State Batteries with 5 Times Higher Output Density 3rd Photo (4 Photos in Total)). Available online: https://response.jp/article/img/2012/09/25/181894/478562.html (accessed on 13 March 2021).
243. 【トヨタ 環境技術発表】出力密度を5倍に高めた全固体電池の電解質を開発 1枚目の写真（全4枚） ([Toyota Environmental Technology Announcement] Developed Electrolyte for All-Solid-State Batteries with 5 Times Higher Output Density 1st Photo (4 Photos in Total)). Available online: https://response.jp/article/img/2012/09/25/181894/478564.html (accessed on 13 March 2021).
244. 【トヨタ 環境技術発表】出力密度を5倍に高めた全固体電池の電解質を開発 2枚目の写真（全4枚） ([Toyota Environmental Technology Announcement] Development of Electrolyte for All-Solid-State Batteries with 5 Times Higher Output Density 2nd Photo (4 Photos in Total)). Available online: https://response.jp/article/img/2012/09/25/181894/478565.html (accessed on 13 March 2021).
245. Secondary Battery System (Vol.8): Cost Evaluation and Technological Challenges of an All-Solid-State Lithium-Ion Battery. Available online: https://www.jst.go.jp/lcs/pdf/fy2019-pp-12.pdf (accessed on 8 March 2021).
246. Tsiouvaras, N.; Schmidt, J.P. Lithium-Ion Cell. U.S. Patent US2017/027916A1, 9 June 2017.
247. Takami, N.; Yoshima, K.; Harada, Y. 12 V-Class Bipolar Lithium-Ion Batteries Using $Li_4Ti_5O_{12}$ Anode for Low-Voltage System Applications. *J. Electrochem. Soc.* **2017**, *164*, A6254–A6259. [CrossRef]
248. Yoshima, K.; Harada, Y.; Takami, N. Thin hybrid electrolyte based on garnet-type lithium-ion conductor $Li_7La_3Zr_2O_{12}$ for 12 V-class bipolar batteries. *J. Power Sources* **2016**, *302*, 283–290. [CrossRef]
249. Zhong, Q.M.; Bonakdarpour, A.; Zhang, M.J.; Gao, Y.; Dahn, J.R. Synthesis and electrochemistry of $LiNi_xMn_{2-x}O_4$. *J. Electrochem. Soc.* **1997**, *144*, 205–213. [CrossRef]
250. Manthiram, A.; Chemelewski, K.; Lee, E.S. A perspective on the high-voltage $LiMn_{1.5}Ni_{0.5}O_4$ spinel cathode for lithium-ion batteries. *Energy Environ. Sci.* **2014**, *7*, 1339–1350. [CrossRef]
251. Wise, R. BASF Advanced Battery Materials for xEV and High-Performance Consumer Products. In Proceedings of the International Conference on the Frontier of Advanced Batteries, CIBF 2014, Shenzhen, China, 21 June 2014; p. 19.

4 Advanced High-Power Lithium-Ion Cells for Electric Hand Tools

High-power lithium-ion cells in the 18650 case (and currently in the 21700 case) are used to manufacture lithium-ion battery packs for the power supply of electric hand tools, vacuum cleaners, and electric vehicles (push scooters, bicycles, etc.). There is also some evidence of attempts to use the cells to manufacture batteries for hybrid transport vehicles [1]. A battery cell must suit the application, and not all devices are equipped with superior performance cells. For example, according to presentation [2] in the row pertaining to vacuum cleaners—Stick vac, Upright, Handvac, Robot—the requirement for a maximum permissible discharge current of cylindrical cells decreases.

Table 20 presents the characteristics of some high-power LICs fabricated in the 18650 case, whose nominal energy varies from 3.1 to 11.2 Wh and depends on the applied materials and manufacturing technologies. According to the specifications, the maximum continuous discharge current of high-power 18650 LICs may reach 30 A. However, some LICs can be discharged with a current of up to 100 A for a short time (1 s) [3].

Figure 24 shows the discharge curves under the 1C discharge current. The minimum energy is typical for the 18650 LICs with lithium titanate anodes (cathodes are possibly made of lithium manganese oxide spinel—LMO). The A123 (18650M1A) cells made of lithium iron phosphate (LFP/C, active cathode material) and graphite (active anode material) have slightly higher energy. The IMR18650e lithium-ion cells with positive electrodes based on lithium manganese oxide spinel (with graphite as the active anode material) feature is an average discharge voltage of 3.8 V, the maximum among those reviewed in this section. The energy of the high-power 18650 LICs lies within the range of 5.3 to 11.2 Wh depending on the nickel mole fraction in active cathode materials—the mixed lithium nickel cobalt manganese (NCM) and lithium nickel cobalt aluminium (NCA) oxides. To improve functionality, the use of mixtures of active cathode materials is permitted, for example, LMO and NCM in IBR18650BC cells. Nowadays, most companies seldom use lithium cobalt oxide (LCO) to manufacture high-power 18650 LICs.

Table 20. Characteristics of high-power 18650 lithium-ion cells with various active materials.

Manufacturer	Huahui	A123	LG	E-One Moli	E-One Moli	Sony	LG	E-One Moli	Samsung	LG	Sony (Murata)
Brand	HTC1865	18650 M1A	18650 HB2	IMR 18650e	IBR 18650BC	US18650 VTC3	18650 HD2	IHR 18650C	INR 18650-25R	18650HG2	US18650
Designation	HTC	M1A	HB2	IMR	IBR	VTC3	HD2	IHR	25R	HG2	VTC6
Year	–	2009	2011	2007	2012	2011	2012	2013	2013	2014	2015
Capacity at 0.2C, Ah	1.3	1.1	1.5	1.4	1.5	1.6	2.0	2.05	2.5 (2.56)	3.0 (2.998)	3.125
Active Materials	/LTO	LFP/Gr	NMC/Gr	LMO/Gr	NMC, LMO/Gr	M.O./Gr	NMC/Gr	NMC/Gr	NCA/Gr	H-NMC/Gr + SiO	–
Mass, g	38.5	39.3	48 (44.7)	44 (42)	45	(45)	<48 (45.2)	45 (44.5)	<45 (43.8)	<48 (45)	(46.6)
Overall dimensions ø, H, mm	18.8/65.9	18.4/65.15	18.4/65.2	18.4/65.15	18.4/65.2	18.35/65.1	18.5/65.2	18.6/65.2	18.4/65	18.5/65.2	18.5/65.2
AC resistance (1 kHz), mΩ	–	15	15 (10 ÷ 12)	–	–	12	<20	13	<18 (13.2)	<17 (15)	(13)
DC resistance (10 A ÷ 1 A), mΩ	30	–	30 (24 ÷ 26)	32	–	–	<30 (22 ÷ 24)	20	<30 (22.15)	<30 (25)	–
Nominal voltage, V	2.4	3.2	3.65	3.8	3.6	3.7	3.65	3.6	3.6 (3.64)	3.6	3.6
Voltage range, V	0.5 ÷ 2.85	2.0 ÷ 3.6	2.5 ÷ 4.2	2.5 ÷ 4.2	2.0 ÷ 4.2	2.5 ÷ 4.2	2.5 ÷ 4.2	2.0 ÷ 4.2	–	2.0 ÷ 4.2	2.0 ÷ 4.2
Max. discharge current, A	13	30	30 (45)	20	30	30	25	20	20	20	20
Energy (nom.), Wh	3.12	3.62	5.3	5.32		5.92	7.61	7.4	9.29	11	11.2
Specific energy (nom.), Wh/kg	81	92	118	126	129	130	168	160	212	240	240
Ref.	[4–6]	[7–9]	[10,11]	[12,13]	[14,15]	[16]	[17–19]	[14,20]	[3,21]	[22,23]	[24,25]

Source: Authors' compilation based on data from references cited in the table.

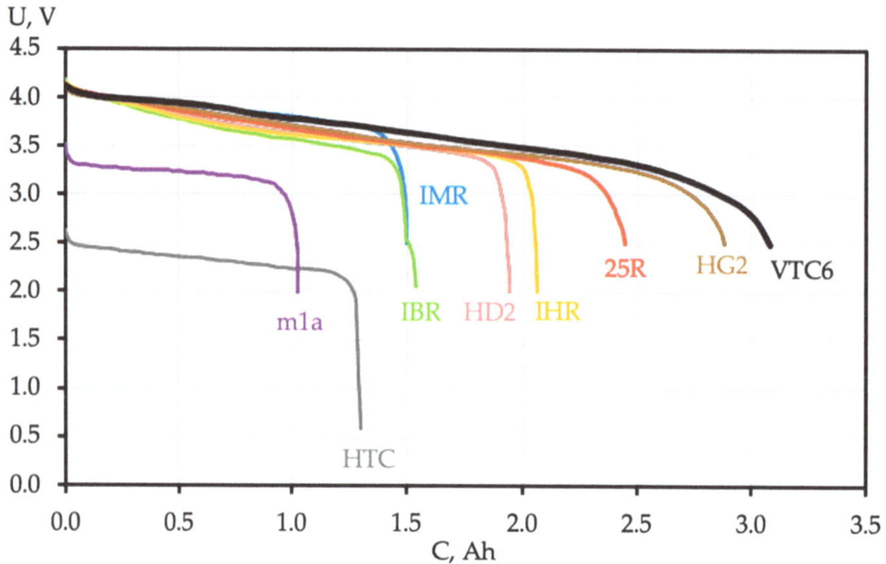

Figure 24. Discharge curves of 18650 HTC [5,6], M1A [7,9], IBR [15], IMR [13], HD2 [17], IHR [14,20], 25R [3], HG2 [22], and VTC6 [24] LICs (initial data were taken from references given in square brackets). Source: Figure by authors.

4.1. Discharge Current Effect on LIC Functional Characteristics

A brief technical description of LICs usually specifies the energy (or specific energy) measured at nominal current discharge (low current charge under constant current, and further charge with a dropping current under constant voltage). A high current discharge (per one LIC) is required to operate the lithium-ion battery packs for portable electric tools.

It is necessary to keep in mind that an increase in the discharge current leads to a drop in the LIC energy (Table 21) and therefore operation time (Table 22). The capacity drops due to the diffusion constraints of the mass transfer of lithium ions under an increased discharge current. On the other hand, since the lithium-ion cell has a relatively high internal resistance, a current rise results in a simultaneous increase in the internal cell temperature, monitored by the LIC outer wall's temperature (Table 23). An increased internal temperature of the cell reduces the diffusion constraints and contributes to the mass transfer of lithium ions. Thus, there are two processes observed that have opposite effects on the capacity and internal resistance. Since both processes can have a different effect, the capacity under an increased discharge current may remain constant, decrease, or increase. On the other hand, when the discharge current is high, the voltage (at the beginning of discharge) may drop sharply due to increased internal resistance, and a local minimum can be observed on the discharge curve (Figure 25).

Table 21. Variation in energy, capacity, and average discharge voltage depending on discharge current of high-power 18650 cells.

	$I_{discharge}$	1 C	5 A	10 A	15 A	20 A	25 A	30 A	35 A	40 A	45 A	Ref.
	Brand	Wh	% of the value at 1C discharge current									
Energy	HTC	2.99	87.8	79.1	67.7	61.9						[5]
	M1A	3.27	96.6	91.6		82.3		69.4				[8,9]
	HB2	5.26	94.1	92.8	92.6	92.6	92.4	91.3	89.3	87.4	84.5	[10]
	IMR	5.74	95.2	90.4	86.9	82.2	89.7	84.2				[12]
	IBR	5.61	97.3	96.8	94.5	93.5		87.2				[15]
	VTC3	5.65		93.2		89.8		85.1				[16]
	HD2	7.09	96.2	94.8	94.4	94.2	91.7	89.2				[17]
	IHR	7.52	101	101	100	97.6						[14,20]
	25R	8.80	98.1	98.1	97.5	94.7	90.9					[3]
	HG2	10.25	98.0	95.5	92.6	89.2	85.0	79.8				[22]
	VTC6	11.03		94.7		89.8						[24]
	Brand	Ah	% of value at 1C discharge current									
Discharge capacity	HTC	1.30	95.8	93.7	91.8	90.0						[5]
	M1A	1.02	101	101		99.4		92.0				[8,9]
	HB2	1.44	96.9	98.6	101	103	105	106	107	107	107	[10]
	IMR	1.50	99.7	98.5	97.5	95.5	98.9	96.6				[12]
	IBR	1.54	99.4	101	102	102		101				[15]
	VTC3	1.53		99.2		101		100				[16]
	HD2	1.94	98.6	100	103	105	105	104				[17]
	IHR	2.06	103	106	108	108						[14,20]
	25R	2.44	99.8	103	105	104	103					[3]
	HG2	2.88	100	101	101	99.6	97.2	93.8				[22]
	VTC6	3.08		99.0		98.3						[24]
	Brand	V	% of the value at 1C discharge current									
Average discharge voltage	HTC	2.31	92.2	84.8	74.9	70.1						[5]
	M1A	3.23	94.7	89.8		82.4		74.9				[8,9]
	HB2	3.67	97.0	93.7	91.6	89.4	87.7	85.6	83.7	81.5	79.0	[10]
	IMR	3.85	95.8	92.5	89.4	86.5	91.4	88.1				[12]
	IBR	3.66	97.8	95.6	92.9	91.3		86.1				[15]
	VTC3	3.70		94.1		89.2		85.1				[16]
	HD2	3.67	97.5	94.3	91.8	89.6	87.7	85.6				[17]
	IHR	3.66	97.8	95.4	92.6	89.9						[14,20]
	25R	3.62	98.1	95.3	92.5	90.6	88.1					[3]
	HG2	3.58	97.8	94.1	91.6	89.4	87.4	84.9				[22]
	VTC6	3.59		95.5		91.1						[24]

Source: Authors' compilation based on mathematical processing of plots from references cited in the table.

It seems that the maximum value of the discharge current specified in the technical description is limited by allowed wall temperature at the end of discharge. The test results indicate that LICs can be discharged at currents exceeding those specified in the technical description, but this leads to a significant increase in the LIC wall temperature (Table 23). Since safety is a must at high current discharge, sometimes a temperature cutoff is set. The cell discharge is stopped once the proper wall temperature is reached (for example, 70 °C [8]). Then, the cell, cooled down to lower temperatures (for example, 50 °C [8]), is discharged again if it is capable of energy release. The wall temperature of HB2 and VTC3 cells with a capacity of approximately 1.5 Ah is below 70 °C under a discharge current of 20 A, while the case wall temperature of HD2, 25R, and HG2 cells with higher specific energy is above 70 °C, and it may be dangerous to operate the cells at a discharge current of 20 A or more.

Table 22. Operation time (in minutes) of M1A [7], HB2 [11], IMR [13], HD2 [17], 25R [3], and VTC6 [25] cells under 5 ÷ 30 A discharge current.

Brand	U, V	5 A	10 A	15 A	20 A	25 A	30 A
M1A	2.0 ÷ 3.5	11.0	5.5	3.7	2.8		1.8
HB2	2.8 ÷ 4.2	16.8	8.7	5.7	4.2		2.7
IMR	2.0 ÷ 4.2			5.9	4.3		
HD2	2.8 ÷ 4.2	23.2	12.1	7.9	5.7		3.2
25R	2.5 ÷ 4.2	29.3	15.1	10.3	7.7	6.0	
VTC6	2.8 ÷ 4.2	35	17.2	11.2	7.4		

Note: The capacity of M1A and VTC6 cells under 1C current was equal to 0.93 Ah and 3.0 Ah, respectively. Source: Authors' compilation based on mathematical processing of plots from references cited in the caption of the table.

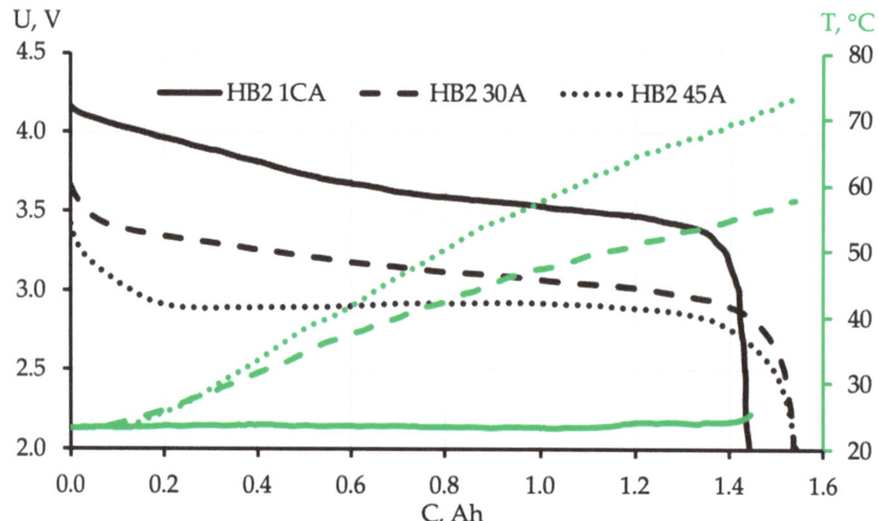

Figure 25. Discharge current effect on discharge curve shape and HB2 LIC wall temperature increase (based on data from [10]). Source: Figure by authors.

A high temperature of the electrochemical block increases the risk of in-service emergencies and reduces the cycle life. In addition, there may be a decomposition of salt ($LiPF_6$) at high temperatures and the partial dissolution of SEI film [26,27]. The repeated formation (overgrowth on the accessible areas) of the SEI film reduces the lithium salt concentration in the electrolyte and increases resistance. When developing lithium-ion battery packs, keep in mind that the closely spaced LICs isolated in the typical case will be warmed up quickly and to higher temperatures because the heat dissipation is lower. In this regard, a tighter limitation of discharge current or a more advanced heat dissipation system may be required.

Table 23. Temperature variation measured on the walls of HB2 [10], IMR [13], VTC3 [16,28], HD2 [19], IHR [14], 25R [3], HG2 [29], and VTC6 [24] cells at the end of discharge, depending on discharge current.

Brand	U, V	1 C	5 A	10 A	15 A	20 A	25 A	30 A	35 A	40 A	45 A
HB2	2.0 ÷ 4.2	25.0	28.8	34.4	41.0	46.6	52.4	57.4	62.7	67.0	74.0
IMR	2.5 ÷ 4.2				60.0	71.0					
VTC3	2.5 ÷ 4.2			50.0		68.0		84.8	87.0	87.0	
HD2	2.8 ÷ 4.2			52.0	67.0	76.0	91.0	96.0			
IHR	2.0 ÷ 4.2	29.9	41.3	58.6	75.0	91.0					
25R	2.5 ÷ 4.2	30.9	38.2	53.9	69.0	83.7	97.8				
HG2	2.8 ÷ 4.2			51.0	66.0	78.0	90.0				
VTC6	2.5 ÷ 4.2	37.9		55.0		89.1					

Source: Authors' compilation based on mathematical processing of plots from references cited in the caption of the table.

Since the average discharge voltage has decreased, the relationship between the discharge current and LIC average power is not linear (though it is close to it). The relationships of specific power and energy for the LICs under consideration are shown in Figure 26. The HTC (LMO/LTO) and M1A (LFP/Gr) LICs have the minimum specific energy at maximum power.

Despite the ability to be discharged at high currents (30C) and a lower weight, due to a lower average discharge voltage, the specific power (W/L, W/kg) of the M1A LIC is comparable or less than the specific power of the 18650 LIC with other active cathode materials (LMO, NCM, NCA, or probably LCO). The maximum specific power of the 18650 LICs in question is within the range of 3.5 ÷ 5.5 kW/L (1.3 ÷ 2.1 kW/kg). The exception is the HB2 LIC with a maximum specific power of 7.5 kW/L (2.7 kW/kg) under a specific energy of 257 Wh/L (93 Wh/kg). The maximum specific energy of 565 Wh/L (205 Wh/kg) at a specific power of 3.7 kW/L (1.36 kW/kg) is typical for VTC6 LICs.

The LICs in case 21700 have specific power and energy close to the values typical for the LICs in case 18650. When comparing the relationships, it can be noted that an increase in the case volume has a positive effect on the values of the energy density and power density.

An increase in the discharge current leads to a drop in LIC service life (Table 24). At low-current (10A) discharge, the LICs with a capacity below 2.2 Ah, except for IMR LICs (and HTC—data not available), have a relatively high cycle life of over 600 charge/discharge cycles. The IMR LICs (with $LiMn_2O_4$ cathodes) are rarely used, probably due to their short cycle life. The cycle life is reduced dramatically for 18650 LICs with high specific energy, and for HG2 and VTC6 LICs is 370 and 130 cycles, respectively, with a discharge current of 10A (about 3.3 C).

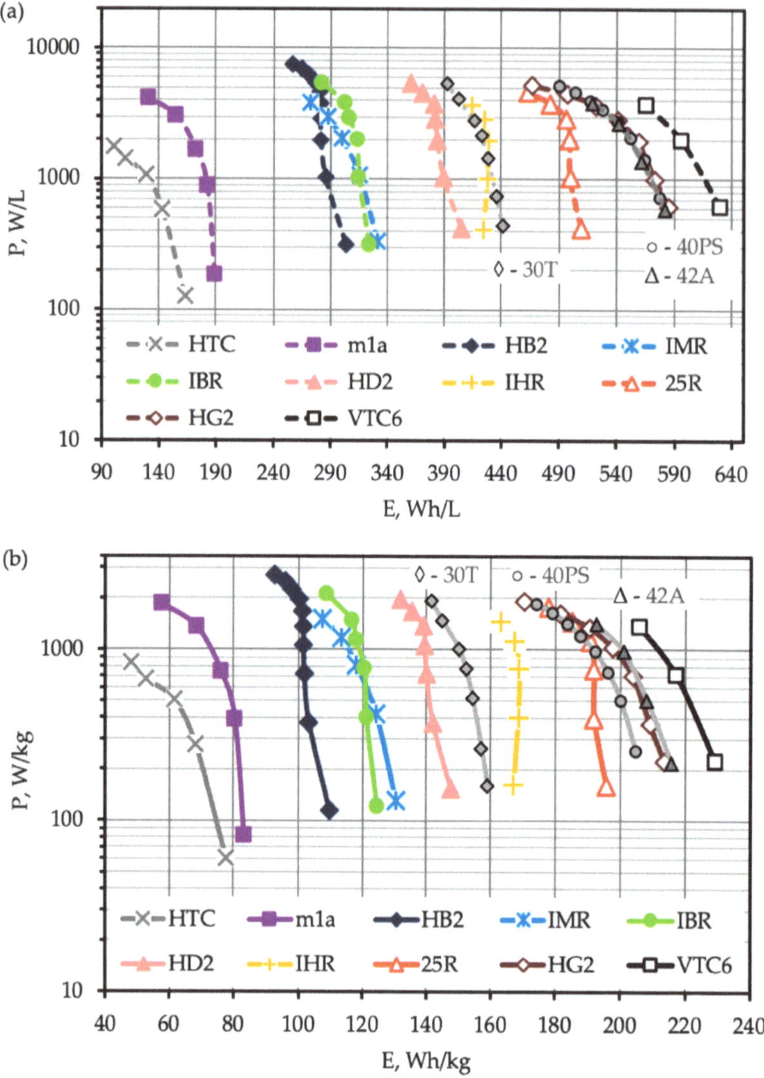

Figure 26. Energy (Wh/L, Wh/kg) vs. power (W/L, W/kg) of LICs in 18650 cylindrical cases (HTC [5], M1A [8,9], HB2 [10], IMR [12,13], IBR [15], HD2 [17], IHR [14,20], 25R [3], HG2 [22], VTC6 [24]) and 21700 (30T [30,31], 40PS [32], 42A [33]). (**a**) energy density and power density are expressed in Wh/L and W/L, correspondingly. (**b**) Specific energy and specific power are expressed in Wh/kg and W/kg, correspondingly. Source: Figure by authors.

Table 24. Discharge current effect on cycle life of M1A [9], HB2 [10], IMR [13], IBR [15], VTC3 [16], HD2, IHR [14,34], 25R [3], HG2 [22], and VTC6 [24] cells.

Brand	U, V	Number of Cycles (CL(A))					Capacity of the First Cycle				
		10 A	**15 A**	**20 A**	**25 A**	**30 A**	**10 A**	**15 A**	**20 A**	**25 A**	**30 A**
M1A	2.0 ÷ 3.5	*625*		*140*		63	1.03		0.99		0.93
HB2	2.0 ÷ 4.2	*1000*		*940*		*364*	1.44		1.51		1.55
IMR	2.5 ÷ 4.2	*200*					1.47				
IBR	2.0 ÷ 4.2			*790*		*665*			1.57		1.56
VTC3	2.5 ÷ 4.2	*865*				227	1.52				1.43
HD2	2.8 ÷ 4.2	*680*		*404*	280		1.98		2.07	2.03	
IHR	2.0 ÷ 4.2	*878*	*550*	290			2.18	2.23	2.24		
25R	2.5 ÷ 4.2			135		130			2.53		2.32
HG2	2.5 ÷ 4.2	367	294	173			3.01	3.06	3.02		
VTC6	2.5 ÷ 4.2	132					3.05				

Note: The *italicised* values are those obtained by Forecast (Excel) function-based extrapolation of the curves given in the references. The regular font denotes the values obtained by interpolation. The M1A cells were discharged by a current of 22 A (not 20 A). When discharging a 25R cell, the discharge was stopped upon reaching 70 °C and was resumed after cooling down to 50 °C. CL(A) is the number of charge/discharge cycles of the cell sufficient to reduce the capacity by 20% of the initial value. Source: Authors' compilation based on mathematical processing of plots from references cited in the caption of the table.

Let us consider a change in the discharge curve shape (Figure 27) during cyclic charge (low current)/discharge (high current) using the example of the A123 M1A LIC [9]. The capacity and average discharge voltage drop in proportion to an increase in the completed charge/discharge cycles. At high current discharge, the detected variation in the discharge curve shape becomes more prominent. A decrease in the average discharge voltage, capacity, and energy can be expressed as a percentage of the respective values typical for the first cycle (Figure 28).

The variations become more pronounced in the "LIC average discharge voltage (average power), capacity, and energy" series. It is generally more correct to determine the LIC cycle life through energy retention, but a change in capacity and energy may be similar at low discharge currents. In such a case, the cycle life (number of charge/discharge cycles) can be determined using a capacity retention curve depending on the number of charge/discharge cycles.

An increased charge current has a negative effect on cycle life and operational safety. The capacity of IHR LICs does not drop for 500 cycles (discharge current of 10 A), with the charge current varying in the range of 1 C ÷ 3 C (2 ÷ 6 A) [14,20]. If the charge current is increased up to 4 C (8 A) after 200 charge/discharge cycles, one can observe a capacity drop that further increases proportionally to the number of cycles. The capacity is equal to 80% of the initial value after approximately 400 cycles. At high charge currents, lithium plating on the anode is possible, followed by dendrite formation.

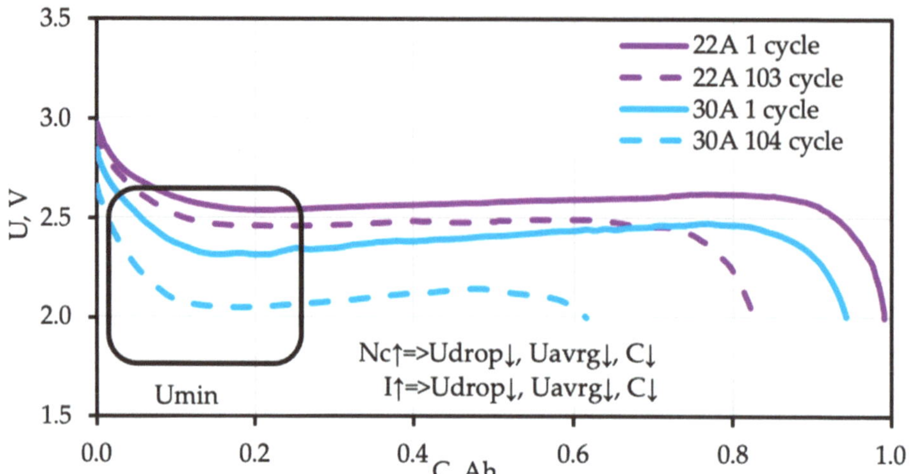

Figure 27. Effect of current and number of charge (low current)/discharge (high current) cycles on the shape of 18650 M1A LIC discharge curves (the initial data for the plots were taken from reference [9]). Source: Figure by authors.

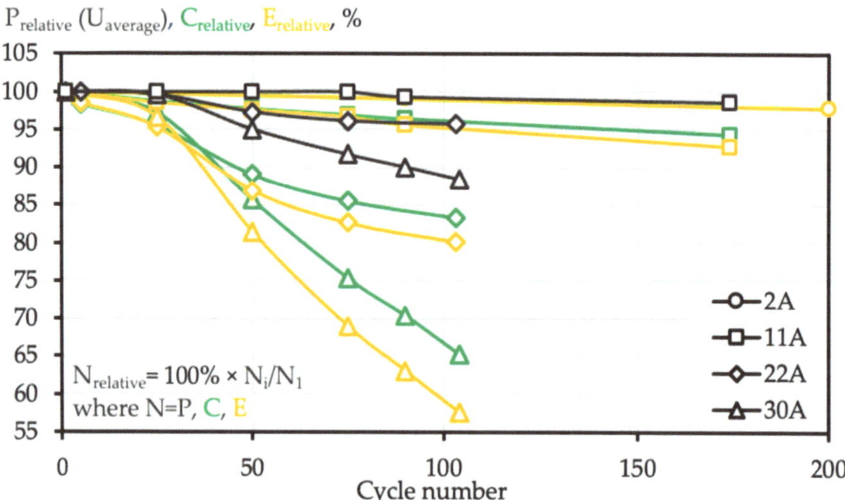

Figure 28. Variation in average discharge voltage (average power), capacity, and energy depending on the number of charge/discharge cycles and discharge current of M1A (the initial data for plots were taken from reference [9]). Source: Figure by authors.

4.2. Discharge Temperature Effect on LIC Functional Characteristics

The processed discharge curves given in the LIC technical descriptions were referenced to determine an average discharge voltage and capacity (Table 25) for the 18650 LICs reviewed in this section. According to the data given in Table 25, the capacity and average discharge voltage become lower with decreasing temperature if the discharge current is fixed. Conversely, when the temperature is set, the value of the total discharge capacity can decrease or increase when the discharge current rises, but the average discharge voltage drops almost in all cases.

Table 25. Temperature effect on discharge capacity and average discharge voltage.

		C, Ah, % of the Value						U, V, % of the Value					
	T, °C	0.5 C	1 C	5 A	10 A	15 A	20 A	0.5 C	1 C	5 A	10 A	15 A	20 A
HTC [5]	23	1.32						2.30					
	−20	91.4						91.3					
	−30	81.9						63.5					
IMR [12,13]	45		101						101				
	23		1.50						3.85				
	0		100						96.6				
	−20		98.7	99.5	98.9	96.2	98.2		90.9	88.3	86.2	82.9	75.6
IBR [15]	45					102						94.3	
	23		1.54			101			3.66			92.9	
	0					97.1						89.6	
	−10					96.5						88.3	
	−20					94.4						85.8	
	−30					94.3						84.2	
IHR [14,20]	45		107	107	109	109	109		100	98.6	95.9	93.4	90.7
	23		2.06	103	106	108	108		3.66	97.8	95.4	92.6	89.9
	0		92.3	94.4	99.1	103	107		97.3	95.1	92.6	90.2	87.4
	−20		84.0	88.5	94.8	101	105		91.8	90.2	88.8	86.9	84.4
	−30		80.9	86.8	94.0	100	97.5		88.3	87.7	86.6	84.4	82.5
	−40		70.8	68.7	69.7	75.1	83.4		86.9	82.2	80.1	78.7	79.2
25R [3]	23		2.44		103				3.62		95.6		
	0				94.4						89.2		
	−10				98.3						84.8		
	−20				98.0						81.5		
HG2 [22]	23		2.88		101				3.58		94.1		
	0				101						89.4		
	−10				100						86.6		
	−20				99.2						83.8		
VTC6 [24]	60				98.6						97.2		
	45				98.7						97.2		
	23		3.08		98.7				3.59		95.8		
	0				98.1						93.0		
	−10				97.7						91.6		
	−20				97.1						89.7		

Note: Cells were charged at low current at room temperature. Absolute C and V values are underlined. Source: Authors' compilation based on mathematical processing of plots from references cited in the caption of the table.

The curves of LIC energy vs. power (Figure 29) were plotted based on technical specifications to compare performance at low temperatures. The specific energy and power drop when the temperature decreases while the discharge current is fixed. When the discharge current is 10A within the temperature range of −20 °C ÷ 0 °C, the specific energy of IMR, IHR, 25R, HG2, and VTC6 cells becomes higher. The specific energy reduces with increasing discharge current (average power) at temperatures above 0 °C. In the case of discharge at temperatures below 0 °C, the specific energy can both decrease and increase. The specific energy increases due to a rise in the discharge capacity caused by the internal heating of the LIC, as internal resistance increases at low temperatures. The average specific power of the 1.5 ÷ 2 Ah LICs manufactured in the 18650 case can exceed 1 kW/kg at −30 °C and −20 °C under a specific energy of 95 ÷ 145 Wh/kg. The maximum average specific power was recorded for IHR LICs as follows: 640 W/kg (1640 W/L) when discharging (or charging at room temperature) at a temperature of −40 °C under a specific energy of 110 Wh/kg (280 Wh/L).

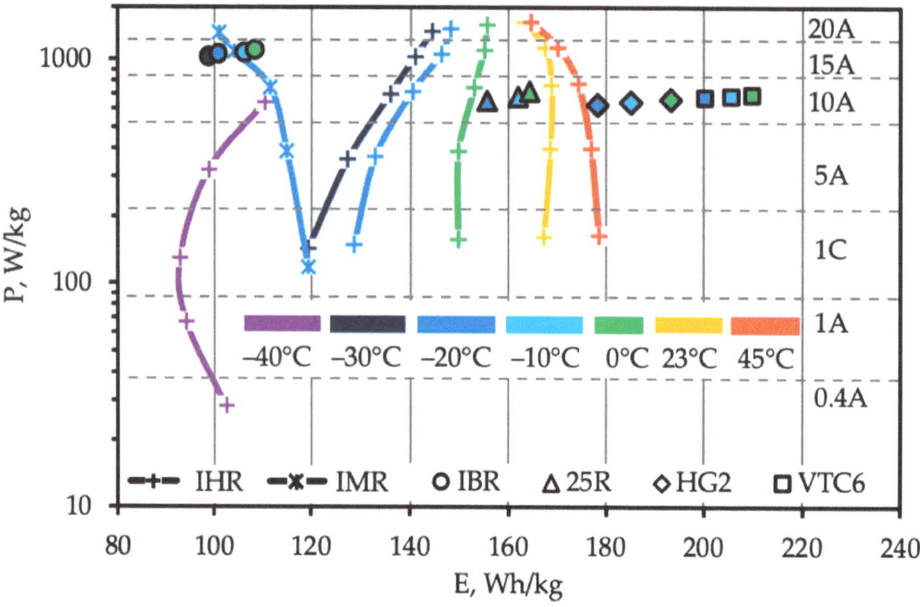

Figure 29. Discharge temperature effect on specific energy and power of IHR [14], IMR [12], IBR [15], 25R [3], HG2 [22], and VTC6 [24] LICs (the initial data for the plots were taken from references in square brackets). Source: Figure by authors.

A minimum value of instant power should be considered when developing batteries. In the case of discharging at low temperatures, a minimum can be observed on the discharge curve at the beginning of discharge. The minimum value reduces simultaneously with the increase in the discharge current and drop in temperature (Figure 30). If the voltage drops below the preset value (the voltage interval monitored by the battery management system, the minimum voltage value, the power required to operate the powered product), this may result in the battery or powered device switching off. For example, in the case of discharge of an IRH cell at –40 °C and 10A, a minimum voltage reaches 1.98 V at the beginning of discharge. If the battery management system is preset to a voltage interval of 4.2 ÷ 2.0 V, a continuous battery discharging with 10A current/LIC (IHR) will be impossible. Forced short-term attempts to discharge the battery can increase the LIC's internal temperature, thereby lowering the internal resistance and reducing the voltage drop during discharge. High-power cells may be discharged by low currents at lower temperatures. A keynote [35] presents the discharge curves of M42A (21700, Molicel) cells at −50 °C (0.84 A, 0.2 C current) and –60 °C (0.1 A, 0.024 C current). At the beginning of discharge, the LIC voltage drops down to 2.47 and 2.28 V, respectively. The output capacity (specific energy) is 1.78 Ah (65 Wh/kg) and 1.13 Ah (39 Wh/kg), respectively.

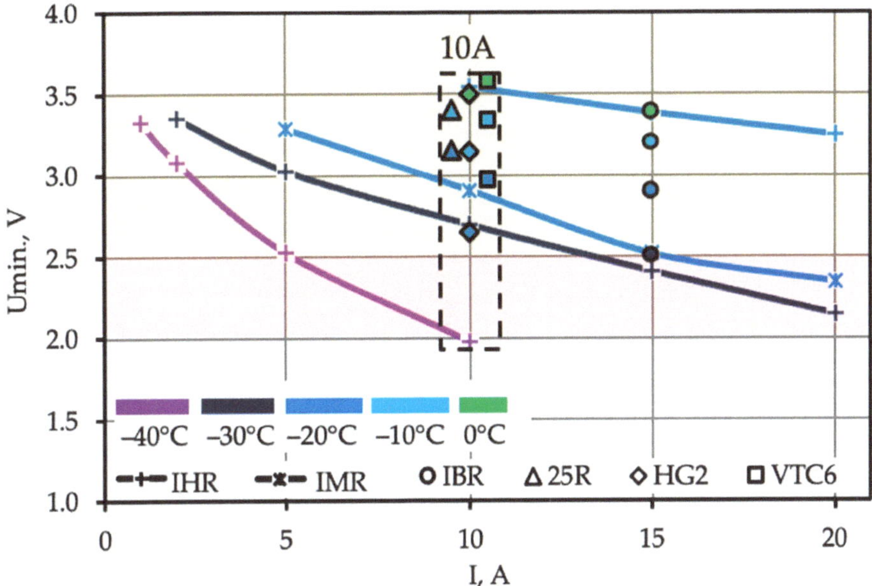

Figure 30. Effect of temperature and discharge current on the voltage minimum observed at the beginning of IHR [14], IMR [12], IBR [15], 25R [3], HG2 [22], and VTC6 [24] LIC discharge (the initial data for the plots were taken from references in square brackets). Source: Figure by authors.

It is usually recommended to charge 18650 LICs at a temperature above 0 °C to reduce the risk of lithium plating on the anode and of dendrite formation (the higher the discharge current and the lower the temperature, the higher the probability of dendrite formation). The upper charge temperature limit is necessary as when charged with relatively high currents, the LIC heats up, leading to an increase in the LIC temperature beyond the range in which safe operation is possible.

4.3. Conclusions

This chapter described the characteristics of several types of high-power LICs in 18650 and 21700 cases differing in terms of the specific energy and active materials applied in their manufacture. The effect of discharge current on the specific energy (Wh/L, Wh/kg), specific power (W/kg, W/L), average discharge voltage, capacity, wall temperature, discharge time, and cycle life were described. In addition, the effect of temperature on LICs' functional characteristics was demonstrated for cases of discharging with various currents.

References

1. de Leon, S. A Game Changer PHEV Battery. Available online: https://www.sdle.co.il/wp-content/uploads/2018/12/8-A-Game-Changer-PHEV-Battery-Presentation-for-conferences-ver-4_2.pdf (accessed on 6 June 2019).
2. Lee, C.-Y. Connecting the Dots between Li-Ion Battery Technology, Supply Chain, and Portable Home Appliances. In Proceedings of the 37th Annual International Battery Seminar & Exhibition, Virtual, 29–30 July 2020; p. 14.
3. Introduction of INR18650-25R Oct. 2013 Energy Business Division. Available online: https://www.powerstream.com/p/INR18650-25R-datasheet.pdf (accessed on 14 November 2024).
4. Products from HuaHui New Energy Co., Ltd. Available online: https://www.globalsources.com/si/AS/HuaHui-New/6008851808132/pdtl/Lithium-Ion-Battery-2.4V-8mAh/1168568392.htm (accessed on 12 July 2019).
5. HTC-钛酸锂系列超级电容式锂电池 (HTC-Lithium Titanate Series Supercapacitor Lithium Battery). Available online: https://www.huahuienergy.com/?post_type=products&page_id=17326(accessed on 14 November 2024).
6. Lithium-Titanate Battery HTC18650 2.4 V 1300 mAh. Available online: https://www.alibaba.com/product-detail/HTC18650-2-4V-1300mAh-Lithium-titanate_60728895871.html (accessed on 14 November 2024).
7. A123 18650 1100mAh (Yellow). Available online: https://lygte-info.dk/review/batteries2012/A123%2018650%201100mAh%20(Yellow)%20UK.html (accessed on 14 November 2024).
8. High Power Lithium Ion APR18650M1A. Available online: http://www.batteryspace.com/prod-specs/6612.pdf (accessed on 6 June 2016).
9. Jeevarajan, J. Performance and Safety Evaluation of High-rate 18650 Lithium Iron Phosphate Cells. In Proceedings of the NASA Battery Workshop, 19 November 2009; p. 36.
10. Technical Information of LG 18650HB2 (1.5Ah). Available online: https://telit.co.rs/PDFDokumenti/Li-Ion/LG18650HB2.pdf (accessed on 2 June 2016).
11. LG 18650 HB2 1500mAh (Green). Available online: https://lygte-info.dk/review/batteries2012/LG%2018650%20HB2%201500mAh%20(Green)%20UK.html (accessed on 14 November 2024).
12. Product Data Sheet MODEL: IMR-18650E (Molicel). Available online: http://www.molicel.com/ca/pdf/IMR18650E.pdf (accessed on 3 June 2019).
13. Product Datasheet Model IMR18650E (Molicel). Available online: http://forum.drc.su/files/IMR18650E.pdf (accessed on 3 June 2019).
14. MOLICEL® IHR18650C 2Ah Power Cell (2013-01-30). Available online: https://www.master-instruments.com.au/file/64178/1/Molicel-IHR18650C.pdf (accessed on 3 June 2019).
15. Product Data Sheet Model IBR18650BC. Available online: http://www.molicel.com/ca/pdf/Certificates/High%20Power%20Cells/IBR18650BC/DM_IBR18650BC.pdf (accessed on 3 June 2019).

16. Lithium Ion Rechargeable Battery Technical Information US18650VTC3. Available online: http://keeppower.com.ua/download/file01_US18650VTC3.pdf (accessed on 14 June 2023).
17. LG 18650 HD2 2000mAh (Magenta). Available online: https://lygte-info.dk/review/batteries2012/LG%2018650%20HD2%202000mAh%20(Magenta)%20UK.html (accessed on 14 November 2024).
18. PRODUCT SPECIFICATION Rechargeable Lithium Ion Battery Model: 18650HD2 2000mAh. Available online: http://keeppower.com.ua/download/2015-04/datasheet-LG-18650HD2.pdf (accessed on 14 June 2023).
19. Mooch. LG HD2 25A 2000mAh 18650 Constant-Current Tests.bmp. Available online: https://www.dropbox.com/s/d5rbbaqclv7t821/LG%20HD2%2025A%202000mAh%2018650%20Constant-Current%20Tests.bmp?dl=0 (accessed on 8 June 2016).
20. PRODUCT DATA SHEET MODEL IHR-18650C. Available online: https://www.molicel.com/wp-content/uploads/DM_IHR18650C-V4-80073.pdf (accessed on 14 November 2024).
21. Introduction of INR18650-25R Oct. 2013 Energy Business Division. Available online: https://www.powerstream.com/p/INR18650-25R-datasheet.pdf (accessed on 12 July 2019).
22. Technical Information of LG 18650HG2 (3.0 Ah). Available online: http://www.nkon.nl/sk/k/hg2.pdf (accessed on 24 May 2016).
23. Nguyen, T.T.D.; Abada, S.; Lecocq, A.; Bernard, J.; Petit, M.; Marlair, G.; Grugeon, S.; Laruelle, S. Understanding the Thermal Runaway of Ni-Rich Lithium-Ion Batteries. *World Electr. Veh. J.* **2019**, *10*, 79. [CrossRef]
24. Lithium Ion Rechargeable Battery Technical Information, Revision 0.2, 30 June 2015. Available online: https://www.kupifonar.kz/upload/manuals/batteries/sony-us18650vtc6-techinfo.pdf (accessed on 13 February 2019).
25. Sony US18650VTC6 3000mAh (Green). Available online: https://lygte-info.dk/review/batteries2012/Sony%20US18650VTC6%203000mAh%20(Green)%20UK.html (accessed on 14 November 2024).
26. Hammami, A.; Raymond, N.; Armand, M. Runaway risk of forming toxic compounds. *Nature* **2003**, *424*, 635–636. [CrossRef] [PubMed]
27. Campion, C.L.; Li, W.T.; Euler, W.B.; Lucht, B.L.; Ravdel, B.; DiCarlo, J.F.; Gitzendanner, R.; Abraham, K.M. Suppression of toxic compounds produced in the decomposition of lithium-ion battery electrolytes. *Electrochem. Solid State Lett.* **2004**, *7*, A194–A197. [CrossRef]
28. Sony VTC3 1500 mAh 18650 Retest Results...a Hard Hitting 28A 1600mAh Battery. Available online: https://www.e-cigarette-forum.com/threads/sony-vtc3-1500mah-18650-retest-results-a-hard-hitting-28a-1600mah-battery.733199/ (accessed on 14 November 2024).
29. Bench Retest Results: LG HG2 20 A 3000 mAh 18650 … a Fantastic 20 A Battery! Available online:. Available online: https://www.e-cigarette-forum.com/threads/bench-retest-results-lg-hg2-20a-3000mah-18650%E2%80%A6a-fantastic-20a-battery.846005/ (accessed on 12 July 2019).

30. Lu, H.-l. The Development Status & Trends of Main Materials for the WW Power LIB. In Proceedings of the 2016′ 第五届中国电池市场年会, 暨第一届动力电池应用国际峰会, 第二届中国电池行业智能制造研讨会 (The 5th China Battery Market Annual Conference, the 1st International Power Battery Application Summit, and the 2nd China Battery Industry Intelligent Manufacturing Seminar) (CBEA), Beijing China, 14–15 November 2016; p. 37.
31. Samsung INR21700-30T 3000 mAh (Gray). Available online: https://lygte-info.dk/review/batteries2012/Samsung%20INR21700-30T%203000mAh%20(Gray)%20UK.html (accessed on 14 November 2024).
32. Cell—BMZ 21700 40PS1. Available online: https://bmz-group.com/images/PDF-Downloads/Zelldaten/BMZ-21700-40PS1_33644.pdf (accessed on 15 July 2019).
33. Product Data Sheet Model INR-21700-P42A. Available online: https://www.imrbatteries.com/content/molicel_p42a.pdf (accessed on 18 December 2020).
34. BU-208: Cycling Performance. Available online: https://batteryuniversity.com/learn/article/battery_performance_as_a_function_of_cycling (accessed on 3 June 2019).
35. Craig, P. Lithium-ion cells for Medical and Mission Critical Applications. In Proceedings of the 1st Medical Battery Conference, Dusseldorf, Germany, 17 November 2017; p. 15.

5 Lithium-Ion Cells and Batteries for Hybrid Electric Passenger Vehicles

One of the ways to relieve the pressure on the environment and reduce carbon dioxide emissions is to develop vehicles that include the ability to move on electric traction. Many countries have announced programmes to co-finance the development of electric transport. Due to popularisation, lowering the cost of ownership, and the development of infrastructure, the demand for transport involving electric driving has increased. For ease of classification, three types of electric vehicles are mentioned. Hybrid electric vehicles (HEVs) have a small, high-power battery that assists the internal combustion engine. HEVs do not need charging infrastructure and have a simpler design. Their primary function is to economise fuel and decrease gas emissions from the internal combustion engine. Plug-in electric vehicles (PHEVs) consume less fuel and emit less gas while moving. Battery electric vehicles (BEVs) do not consume fuel but require access to installed charging infrastructure. However, there is a need for establishing charging infrastructure. Various parameters influence the choice of type of vehicle. Each of these types of cars is in demand. In the USA, EU countries, and especially Japan, the demand for hybrid electric vehicles (HEVs) prevails. In contrast, in China, the share of battery electric vehicles and plug-in hybrid electric vehicles is significantly higher (Figure 31). In the following sections, we will consider the characteristics of the lithium-ion cells (LICs) and lithium-ion batteries (LIBs) used in various types of cars.

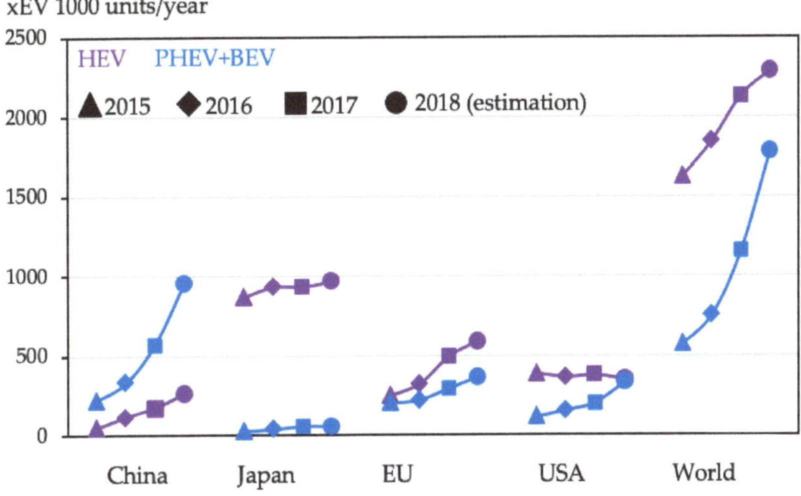

Figure 31. Evolution of demand for HEVs, PHEVs, and BEVs in the USA, European Union, Japan, and China in 2015–2018 (the initial data for the plot were taken from [1]). Source: Figure by authors.

5.1. *Low-Voltage Lithium-Ion Batteries*

The creation of energy storage based on lithium-ion cells with high energy and power characteristics made it possible to significantly expand the range of vehicles that use electric energy for driving. Such vehicles include electric bicycles, gyro scooters, hybrid and electric vehicles, buses, and trolleybuses. In recent years, there has been a noticeable growth in global demand for lithium-ion battery packs (LIB packs) for passenger cars [1–4].

Lithium-ion batteries can be divided into low-voltage (<60 V [5], Table 26) and high-voltage batteries. Low-voltage batteries are used as a replacement for lead–acid batteries (LABs), as well as in hybrids with parallel circuits (micro (12 V)—Suzuki Wagon R [6], and mild (48 V)—Audi A8 [7], Volkswagen Golf 8 [8], etc.) [9]).

The 12 V LIB packs used as replacements for LABs are mainly manufactured using LFP/Gr LICs (except for Lishen; the battery characteristics are summarised in Table 26). The battery capacity for this application exceeds 18 Ah (NCM/LTO) and 30 Ah (LFP/Gr) to provide the required current (power) to start the engine. In comparison with LABs, LIB packs have a longer cycle life (calendar life), higher (LFP/Gr) or similar (LTO) specific energy, and are capable of cold cranking (-18 °C, 10 s, voltage drop not lower than 7.5 V, maximum current: A123 60 Ah Gen3—900 A [10], GS Yuasa 69 Ah—780 A [11]). However, they are more expensive, less safe in abnormal situations, and harder to recycle [12]. In this regard, there is currently little demand for LIB packs as a replacement for 12 V LABs. However, they are used in premium car brands such as Mercedes (Mercedes AMG—A123) [13], BMW (M3—GS Yuasa) [13], and Hyundai (Ioniq HEV—LG Chem) [14].

In micro and mild hybrids (Figure 32), LABs work with LIB packs with 12 V and 48 V voltages, respectively. More information on the design and function of batteries in these types of vehicles can be found in [15,16]. Depending on the configuration, 12 V LIB packs in a micro hybrid can provide the following functions [5,12,17,18]: start–stop, advanced start–stop, crank to idle speed, coasting/sailing, regenerative braking (up to $\approx$3 kW), alternator assist, and active alternator management. The use of 48 V LIB packs in the mild hybrid allows [5,12,17] the regenerative braking power to be increased (up to $\approx$13 kW), broadens the options for start–stop systems (engine shutdown for a long time, engine shutdown at high travel speeds), and provides acceleration (direct boost, torque smoothing).

It follows from Table 26 that the 12 V LIB packs used in micro hybrids are mainly made with NCM/LTO LICs (exception NCM/C, Audi), which seems to be due to the significantly longer service life (number of charge/discharge cycles) at higher charge/discharge currents of NCM/LTO batteries.

Table 26. Characteristics of low-voltage LIB packs (12 V and 48 V) used in vehicle manufacturing.

	Manufacturer	Configuration	U_{nom}	C_{nom}	L × W × H	m	E		P_{in}	P_{out}	DCIR	Active Materials of LICs	Ref.
			V	Ah	mm	kg	Wh/L	Wh/kg	kW		mOhm		
12 V	A123	2P4S	13.2	40	160 × 175 × 190	6.5	99	81	–	–	–	LFP/Gr	[19]
	A123	3P4S	13.2	60	278 × 175 × 190	12	86	66	–	–	–	LFP/Gr	[20–22]
	Audi	1P4S	13.6	11	110 × 175 × 190	5	37	27	5	–	–	NCM/C	[23]
	Denso	1P5S	12	10	305 × 200 × 90	–	22	–	4	2.7	–	LMO/LTO	[24]
	Denso	1P5S	12	3	200 × 178 × 70	2.5	14.5	14.4	–	–	–	LMO/LTO	[18,25]
	GS Yuasa	1P4S	13.2	69	353 × 175 × 190	13.6	78	67	–	–	–	LFP/Gr	[11]
	JC	1P5S	12	10	240 × 189 × 91	3.8	29	32	3.5	3.5	–	NCM/LTO	[5,12]
	LG	3P4S	13.2	60	278 × 175 × 190	–	86	–	–	–	1.5	LFP/Gr	[26]
	LG	2P4S	12.8	30	–	–	–	–	–	1.8		LFP/Gr	[14]
	LG	1P6S	13.8	10	110 × 175 × 190	–	38	–	–	–	0.73	NCM/LTO	[26]
	Lishen	1P5S	12.5	18	–	–	–	36	–	–	–	/LTO	[27]
	Lishen	2P5S	12.5	36	–	14	–	<32	–	–	–	/LTO	[27]
48 V	A123	1P14S	46	8	304 × 180 × 96	8	70	46	16	15	–	LFP/Gr	[10,28–30]
	Bosch	1P12S	44.4	8	309 × 175 × 90	6	72	59	14	11	–	NCM/Gr	[31,36]
	Hitachi	1P12S	44.4	5.5	280 × 175 × 90	5	55	49	13	10	–	NCM/Gr	[32]
	Hitachi	1P12S	44.4	8	300 × 175 × 90	8	75	44	15	12	–	NCM/Gr	[33,34]
	JC	–	48	10	354 × 175 × 100	7	77	69	14	13	–	NCM/Gr	[5,12]
	LG	1P20S	48	4.5	343 × 143 × 143	–	31	–	–	–	1.5	NCM/LTO	[26]
	LG	1P13S	48	9.5	394 × 175 × 110	–	60	–	–	–	1.8	NCM/Gr	[26]
	Toshiba	1P20S	48	5	–	–	–	–	–	–	1.8	NCM/LTO	[35]
	Toshiba	1P20S	48	20	407 × 193 × 155	12	–	77	–	–	–		[35]

Note: P—parallel connection, S—serial connection. P_{in}, P_{out}, DCIR—power, direct current internal resistance. The underlined values are determined at the degree of charge—50%, at the end of 10 s charge/discharge, at room temperature. Source: Authors' compilation based on data from references cited in the table.

In addition, 48 V LIB packs can be manufactured using different types of LICs. The choice of LIC type depends on the specific requirements of the battery. The results of the optimisation of the type and size of LICs to reduce CO_2 emissions per kilometre when using 48 V LIB packs (mild hybrid) are given in presentation [36].

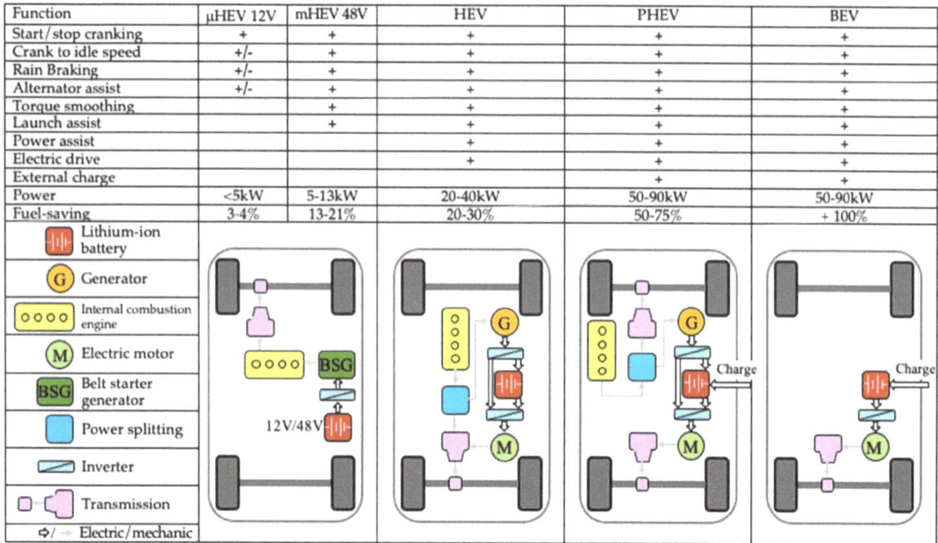

Function	µHEV 12V	mHEV 48V	HEV	PHEV	BEV
Start/stop cranking	+	+	+	+	+
Crank to idle speed	+/-	+	+	+	+
Rain Braking	+/-	+	+	+	+
Alternator assist	+/-	+	+	+	+
Torque smoothing		+	+	+	+
Launch assist		+	+	+	+
Power assist			+	+	+
Electric drive			+	+	+
External charge				+	+
Power	<5kW	5-13kW	20-40kW	50-90kW	50-90kW
Fuel-saving	3-4%	13-21%	20-30%	50-75%	+ 100%

Figure 32. Structure schemes of micro hybrid (12 V), mild hybrid (48 V, P0), full hybrid (series–parallel), plug-in hybrid, and battery electric vehicle. Design based on data presented elsewhere [17,37,38]. Source: Figure by authors.

The lower limit of the temperature interval of high-power LICs' performance is lower than that of high-energy ones, and it varies within $-40 \div -30$ °C. Nevertheless, when developing the batteries, it is necessary to consider that the energy and power of LICs deteriorate with decreasing temperature, and charging, if possible, must be conducted at lower currents.

5.2. High-Voltage Lithium-Ion Batteries

In the case of the series–parallel circuit (Figure 32), the internal combustion engine (ICE) and the electro drive (ED) are connected through a planetary gear (full hybrid, HEV). The design of this type of electric vehicle includes a generator that charges the battery when the internal combustion engine is running. Batteries for this vehicle type can also be charged and discharged with high currents and are classified as high voltage because, depending on design, their voltage varies in the $115 \div 300$ V range.

In plug-in hybrid vehicles (PHEVs, Figure 32), the battery has more significant energy (and specific energy) and can provide the vehicle movement due to the ED operation for relatively long distances ($18 \div 85$ km, Figure 33, Audi A3 e-tron [39], BYD

F3DM [40], Chevrolet Volt G1 [41], Chevrolet Volt G2 [42], Ford C-max Energi [43], Ford Fusion Energi [44,45], Hyundai Sonata [46], Kia Optima [47], Toyota Prius PHV [48]).

A battery electric vehicle (BEV) design does not include an internal combustion engine, but has a battery with high energy capacity, charged from the external network, and provides movement over longer distances (61 ÷ 485 km, Figure 34, Audi R8 e-tron [49,50], BMW I3 [51,52], BYD e6 [53,54], Chevrolet Spark [55], Chevrolet Bolt [56], Coda [57], Fiat 500e [58], Honda Fit [59,60], Kia Soul [61], Kia Ray ev [62,63], Mazda Demio [64], Mercedes B250e [65], Mercedes Smart [66,71], Mitsubishi I-Miev G [67], Mitsubishi I-Miev-M [68], Nissan Leaf [69], Nissan Leaf [70], Peugeot C-zero [72], Renault Twizy 45 [73], Renault Zoe [72,74], Renault Zoe [75], Tesla Model S 85 [76,77], Tesla Model S P100D [77], Toyota Rav 4 [78], Toyota Scion IQ [79], VW e-Golf [80], VW e-up [81,82]) compared to PHEVs.

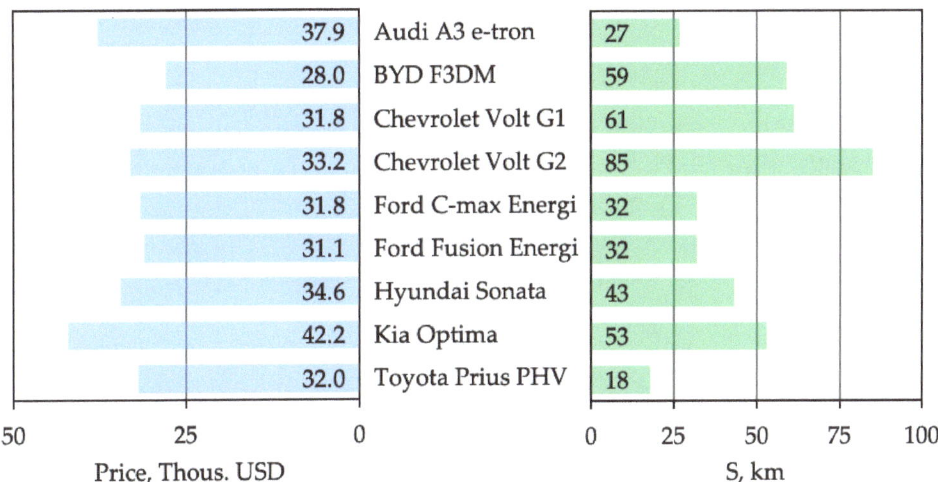

Figure 33. Price of a hybrid vehicle (PHEV) and driving distance using a battery. (Note: Indicates an approximate minimum price per car, not including possible subsidy from the state; S is mainly determined by the EPA (Environmental Protection Agency) evaluation.) Source: Figure by authors.

A more detailed classification with a description of the main functions of batteries can be found in publications [12,17,18,49]. In addition, much literature is devoted to the design of hybrid and electric vehicles so that the following sources can be consulted for a more in-depth understanding of the design, operating principles, and technical solutions [37,50–54]. This review will focus on the characteristics of lithium-ion cells and battery packs used in the manufacture of passenger hybrid (HEVs, PHEVs) and electric vehicles (BEVs).

PHEVs combine the functions of HEVs and BEVs. Therefore, to obtain an overview of the structure, consider the design of the Chevrolet Volt I PHEV battery. The Chevrolet Volt I battery [55–57] includes 288 lithium-ion pouch cells (LG Chem) with a 15 Ah capacity and 3.8 V nominal voltage. Three LICs are connected in parallel to form a group of cells with a capacity of 45 Ah. The 96 cell groups are connected in series, providing a total nominal battery voltage of ≈360 V. Each LIC is in tight contact with a heat transfer plate with channels for the passage of liquid coolant. The LICs and the heat transfer plates are connected using plastic frames with channels for fluid passage. The liquid coolant is pumped by an electric pump and flows through the valve to the radiator or the liquid cooling system. The vehicle also features a coolant heater to increase the battery's internal temperature when ambient temperatures are low.

The battery includes nine modules placed in three segments and has a multilayer energy density [57,58]. The module monitoring systems (BMSs) of each segment monitor the temperature and voltage of each cell. Groups of cells with higher voltage (relative to others) can be discharged to equalise the voltage within the battery. The top-level monitoring system (BMS) monitors the charge level, temperature, battery current, and the insulation resistance between the high voltage conductors and the vehicle's neutral potential.

The battery is connected via a traction converter to the generator (and combustion engine) and the electric motor. The battery also provides internal mains power (12 V, DC) through connection to an auxiliary power module (DC/DC converter). More information on battery design can be found elsewhere [57].

The battery is charged during braking by energy recuperation or connecting the onboard charger to an external power cable. Since batteries for plug-in hybrid and electric vehicles can store large amounts of energy (15 ÷ 100 kWh), one essential feature that enhances usability is the charging speed. Some cars are designed to quickly change from a discharged battery to a pre-charged one (e.g., Renault Fluence). In most hybrid and electric vehicles, the batteries are non-removable. The battery charging time is determined by the parameters of the power supply, the degree of charge of the battery, the total energy of the battery, the cells used, the built-in charger, and the construction of the battery itself [59].

At a state of charge higher than ≈80%, the resistance grows, and it is reasonable to reduce the charging current, which increases the charging time. Therefore, it is advisable not to fully charge the battery when a fast charge is required, and in addition, the thermal management system must dissipate the heat from the batteries.

Suppose that a 10% to 80% charge requires 40 kWh to be transferred. Since the power supply at home is low and amounts to a few kilowatts (alternating current), charging the battery may take several hours (e.g., overnight charging).

Faster charging of batteries can be achieved (if technically possible) using dedicated charging stations of 11, 22, 43 kW (AC), 50 kW (400 V, CHAdeMO, CCS—Combo 1 (US), Combo 2 (De), GB/T (PRC), Tesla), 150 kW (400 V, CSS), 200 kW (600 V, CHAdeMO), or 350 kW (400 V/800 V, CSS).

ABB produces charging stations of different power parameters: 2÷25 kW (AC, charging time 4 ÷ 16 h), 20 ÷ 25 kW (DC, charging time 0.5 ÷ 2 h), 50 kW (DC, charging time 15 ÷ 45 min), and 150 ÷ 350 kW (DC, 5 ÷ 20 min) [60]. The charging channel voltage can reach 400 V and 800 V (920 V); the current strength may achieve 500 A [61,62]. Even more powerful chargers—pantographs (150 ÷ 600 kW)—are used to charge electric buses at bus stops. Considering that the station can provide charging for several vehicles (multiple vehicle connection points), the power supplied to the station can be quite large, which imposes restrictions on their location. Thus, the fast charging of LIB packs is limited not only by the battery design, but also by the network and the parameters of the charging stations (power supply).

One can find more information about charging stations in the presentations of ABB [62], CHAdeMO [61], and CharIN [63].

Due to the complexity of the design, the specific energy of lithium-ion batteries for transport applications is significantly lower than that of the cells and modules from which they are made (Figure 35).

To increase the battery's specific energy, the concept of "cell to pack" was proposed. According to this concept, the module, as a battery design element, is dismissed, and the cells are directly integrated into the battery. This technical solution reduces the number of components [64], decreasing the size of the battery and the weight of the entire vehicle. Several companies, such as BYD, CATL [65], and Svolt, [66], have announced the possibility of manufacturing batteries using this technology.

For example, BYD presented a battery [67] that includes long (about 90 cm), narrow (up to 11.8 cm), and thin (1.35 cm) lithium-ion cells (SK Innovations also developed long batteries in a laminated foil case with four current leads [68,69]) with no modules in its design. The longer length allows a single row of cells to fill the entire battery width. The small width allows for a shallow battery depth, allowing the battery to be placed in the bottom of the vehicle (cell to chassis, CTC). The small depth allows for better heat dissipation and, thus, a lower heat increase during operation. Nevertheless, the production of such long cells is a technological challenge. The design may include several connected electrode rolls to form a long-length accumulator. More information on prospective lithium-ion battery designs for passenger cars can be found in the presentation in [70].

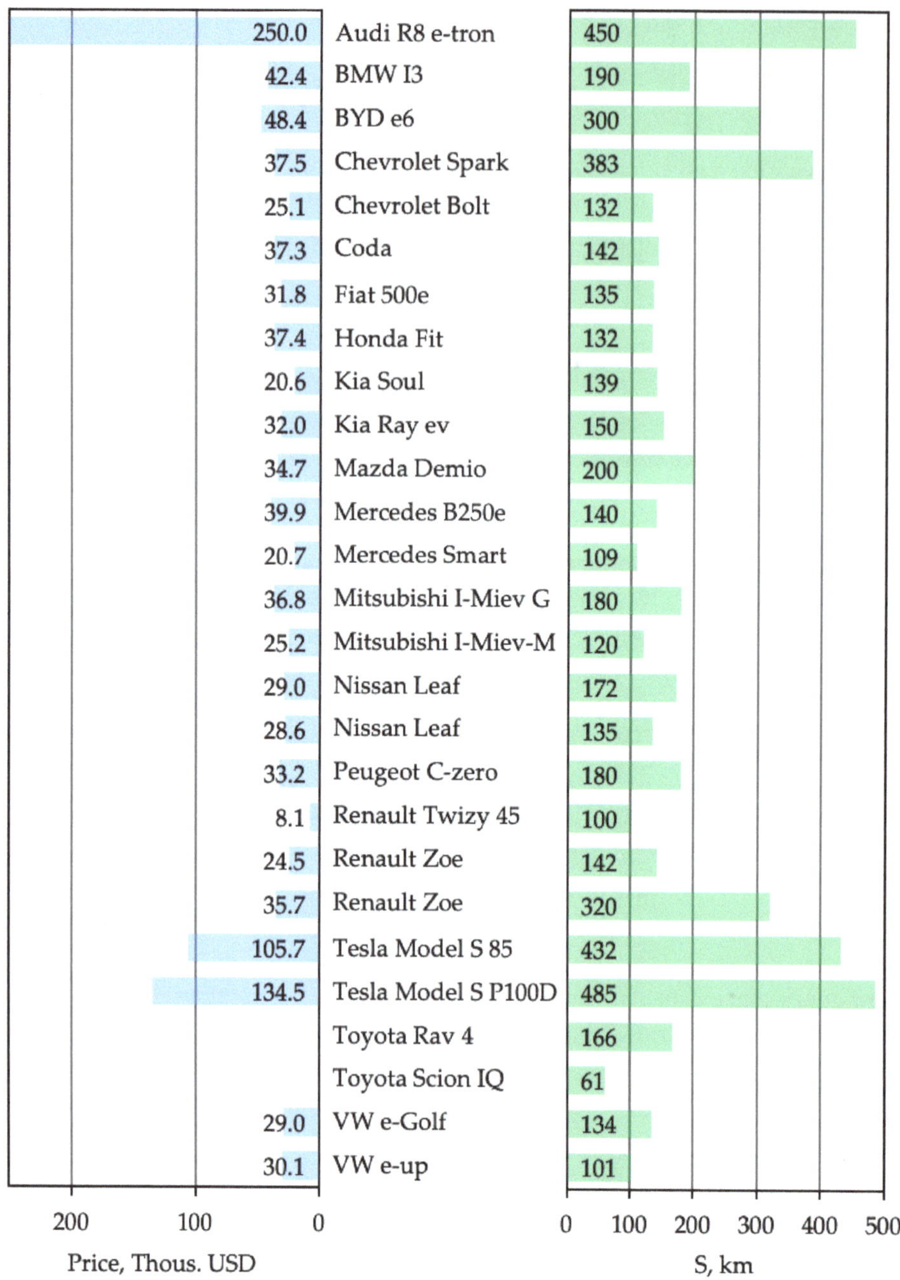

Figure 34. Electric vehicle price (BEV) and travel distance (S) when using a battery. (Note: Minimum price, per vehicle, not including possible government subsidy; for most cases, S is determined based on EPA data.). Source: Figure by authors.

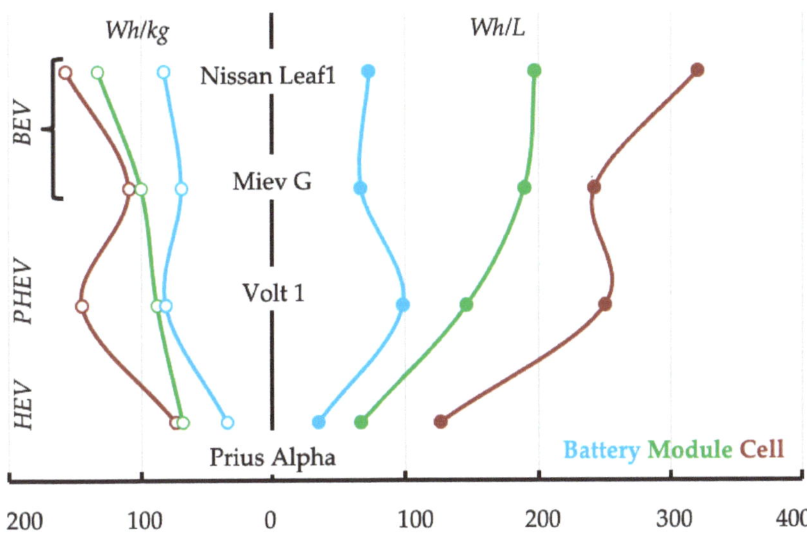

Figure 35. Specific energy (Wh/kg (left), Wh/L (right)) of lithium-ion cells, modules, and batteries of electric and hybrid vehicles according to reference [71]. Source: Figure by authors.

Batteries for HEVs, PHEVs, and BEVs differ in their capacity and the ratio of maximum specific power to rated specific energy (P/E, Figure 36). Hybrid vehicles are equipped with batteries with a power-to-energy ratio of 15 to 50 (more often around 40). In hybrid (PHEV) vehicles with the ability to drive on battery power for a more extended period, the P/E ratio varies from 7 to 15. For battery electric vehicles, the P/E ratio is less than 7.

In the HEV, PHEV, and BEV series, the weight and weight share of the battery increases and is 1 ÷ 3% (20 ÷ 60 kg), 4 ÷ 12% (80 ÷ 180 kg), and 17 ÷ 32% (200 ÷ 700 kg) [72], correspondingly.

The batteries used in the HEVs under consideration have a relatively low energy capacity of up to 1.4 kWh (Table 27). The nominal voltage of HEV batteries varies from 115 V to 300 V. High-power, low-capacity lithium-ion batteries are used for their manufacturing. The small capacity and, hence, the small size of LIBs ensures a more efficient heat dissipation.

The battery energy of the PHEVs in question ranges from 8.8 to 18 kWh. Some manufacturers indicate, along with the total energy of the battery, the value of energy that can be used when driving at the expense of electricity (indicated in parentheses in Table 28). The voltage rating of PHEV batteries is higher than that of HEV batteries and ranges from 207.2 V to 360 V. Batteries are mostly manufactured using medium-capacity LICs—the capacity ranges from 15 to 27.2 Ah. The exception is some BYD batteries, manufactured using lithium iron phosphate LICs with a relatively high capacity of 45 Ah.

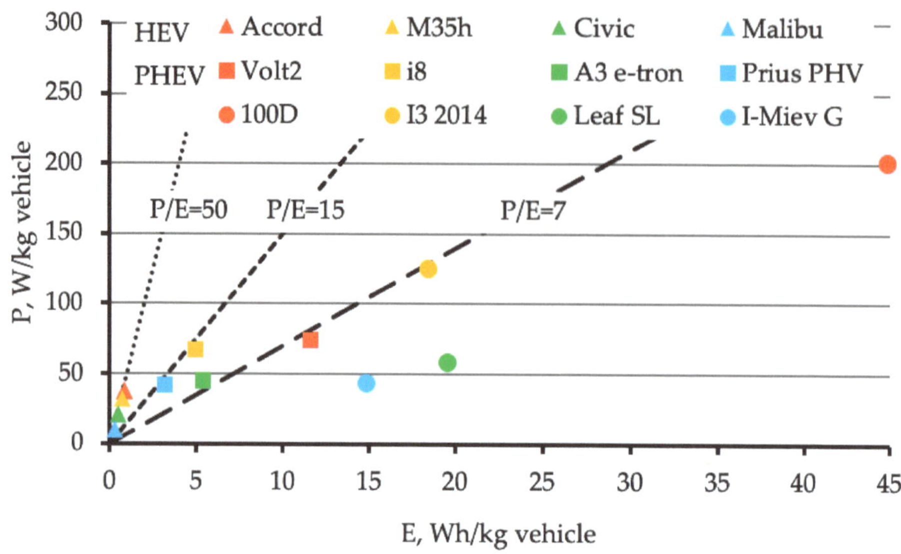

Figure 36. Specific (per kg of kerb weight) power and energy of HEVs (Accord, M35h [73], Civic [74,75], Malibu [76]), PHEVs (Volt2 [55,77], i8 [78], A3 e-tron [79,80], Prius PHV [48]), and BEVs (100D [81–83], i3 2014 [84], Leaf SL [85], I-Miev G [86]). Source: Figure by authors.

The Twizy and i-Miev small electric cars have relatively low energy batteries—6 ÷ 10 kWh (Table 29). On the other hand, the battery energy of premium electric cars can reach 100 kWh (Tesla S P100D). The battery nominal voltage ranges from 270 V to 400 V (except for miniature electric cars such as Twizy—58 V, Wuling Hong Guang MINI EV—96 V, etc., and the Porsche Taycan electric car—800 V). A higher battery voltage of 800 V compared to batteries with a lower voltage, when using a high-voltage charging station (capacity > 300 kW), allows faster charging at a lower charging current [87]. The cells' capacity to construct BEV batteries ranges from 3.2 Ah to 210 ÷ 225 Ah (Table 29).

5.3. *Lithium-Ion Batteries for Light Hybrid and Electric Vehicle Applications*

When using electric energy storage (LICs, LIB packs) for hybrid and electric transportation, many characteristics must be considered, some of which are shown in the radar chart (Figure 37) and will be discussed below (highlighted in bold and underlined). Tables 30 and 31 show the characteristics of the lithium-ion cells used in passenger cars. In the manufacture of all types of low-voltage batteries and batteries for hybrid (HEV) transport, LICs with relatively low specific energy and increased specific power are used, which can be discharged with high currents (in units of C). A higher power density can be achieved by using materials of different chemical compositions (cathode—LFP, LMO, NCM, LNO (NCA), anode—natural graphite, artificial graphite, hard carbon, soft carbon, lithium titanate).

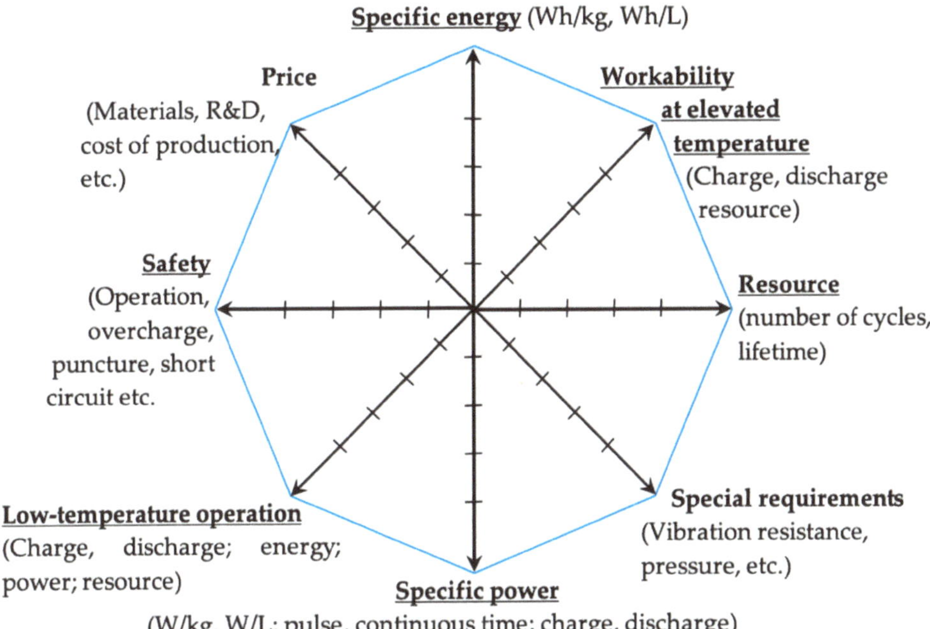

Figure 37. Radar chart—main characteristics of energy storage units for functional transport applications. Source: Figure by authors.

The characteristics of LICs depend not only on the chemical nature of the material, but also on the parameters of the structure and size of the particles, the electrode fabrication technology, etc. The A123 cells produced in 2011 had a specific power of about 2.4 kW/kg (Table 30). Due to changes in manufacturing technology and design [30], it was possible to increase the power of LICs to 5.2 kW/kg, slightly reducing the nominal specific energy from 133 to 126 Wh/kg. The power of LICs with LTO-based anodes, despite the possibility of discharge with high currents, is comparable to or less than that of LFP/Gr and NCM/Gr due to the essentially lower discharge voltage and comparatively higher mass and volume at the same nominal capacity.

Table 27. Characteristics of HEV lithium-ion batteries.

Brand	Model	Year	E, kWh	U, V	m, kg	V, L	LIB Producer	C, Ah	n, pcs.	Ref.
Chevrolet	Malibu Eco	2013	0.5	115.2	29.5	24.1	Hitachi	4.4	32	[76]
Chevrolet	Malibu	2016	1.5(0.45)	300	43.0	–	Hitachi	5.2	80	[88–90]
Ford	C-max	2013	1.4	281.2	34.5	47.7	Panasonic	5	76	[91]
GAC	Trumpchi	2012	$\approx$2.16	290	–	–	Lishen	7.5	90	[92]
Honda	Accord	2015	1.3	259.2	48.1	62	Blue Energy Corporation	5	72	[93–95]
Honda	Civic	2013	0.68	144	21.8	17.6	Blue Energy Corporation	4.7	40	[96]
Hyundai	Ioniq	2017	1.56	240	–	–	LG Chem	6.5	64	[14]
Kia	Optima	2016	1.4	270	–	–	LG Chem	5.3	75	[97]
Toyota	Prius alpha	2011	1	201.6	34	–	Primeearth EV Energy	5	56	[98,99]
Toyota	4 Generation	2016	0.75	207.2	24.5	30.5	Primeearth EV Energy	3.6	56	[18,99,100]
VW	Jetta SE	2014	1.1	220	36.3	28.7	Sanyo	5	60	[101]

Source: Authors' compilation based on data from references cited in the table.

Table 28. Characteristics of PHEV lithium-ion batteries.

Brand	Model	Year	E, kWh	U, V	m, kg	V, L	LIBs Producer	C, Ah	n, pcs.	Ref.
Audi	A3 e-tron	2014	8.8	355	125	–	Panasonic	25	96	[79]
Audi	A6 Phev	2017	14.1	382	140	–	–	37	104	[102]
BYD	F3DM	2009	16(13.5)	325	200	–	BYD	45	100	[103,104]
BYD	Qin DM I	–	16	330	220	240	BYD	50	100	[105]
BYD	Qin DM II	2013	10	500	99	108	BYD	20	152	[105]
BMW	I8	2014	7.1	355	98	-	SDI	20	96	[78]
Chevrolet	Volt G1	2011	17(11)	360	196	145	LG	15	288	[55]
Chevrolet	Volt G2	2016	18(14)	360	183	154	LG	25.5	192	[55,106]
Geely	博瑞 (Borui) GE	2018	11.3	306.6	116	–	–	37	84	[107]
Hyundai	Sonata	2016	9.8	360	–	–	LG Chem	27.2	–	[108]
Ford	C-max Energi	2013	7.6	310.8	123.4	117	Panasonic	26	84	[109]
Ford	Fusion Energi	2013	7.6	310.8	123.4	117	Panasonic	26	84	[110]
Kia	Optima	2016	9.8	360	–	–	LG Chem	–	–	[111]
Toyota	Prius PHV	2013	4.4	207.2	79.8	124	Primeearth EV Energy	21.5	56	[112]

Note: The E column shows the total energy of the battery, and the maximum energy that can be used from the battery when driving is given in parentheses. Source: Authors' compilation based on data from references cited in the table.

Table 29. Characteristics of BEV lithium-ion batteries.

Brand	Model	Year	E, kWh	U, V	m, kg	V, L	LIBs Producer	C, Ah	n, pcs.	Ref.
Audi	R8 e-tron	2015	90	385	595	–	Panasonic	3.2	7488	[113]
BMW	I3	2014	18.8	355.2	235	190.9	Samsung SDI	60	96	[84]
BMW	I3	2016	33(27.2)	360	204	–	Samsung SDI	94	96	[114]
BMW	I3	2018	42.2(37.9)	352	278	278	Samsung SDI	120	96	[115–117]
BYD	e6	2009	(57)	325	650	–	BYD	45	400	[103]
BYD	e6	2015	62	307	680	–	BYD	210	96	[118]
Chevrolet	Spark	2015	18.4	355.2	215	237.3	LG Chem	27	192	[119]
Chevrolet	Bolt	2017	60	360	436	306	LG Chem	59	288	[120,121]
Coda	Coda	2012	31	333	405	–	Lishen	15	624	[122]
Fiat	500e	2015	24	364	272	–	Samsung SDI	60–64	97	[123,124]
Ford	Focus	2013	23	318.2	302.6	268.3	LG Chem	15	430	[125]
Honda	Fit	2012	20	331	317	–	Toshiba	20	432	[126,127]
Hyundai	Kona	2019	64	356	452	390	LG Chem	59	294	[121,128]
Kia	e-Niro	2019	64	356	457	–	SK Innovation	59	294	[129]
Kia	Soul	2015	27	355.2	202.8	216.5	SK Innovation	75	96	[130]
Kia	Ray ev	2012	16.4	330	–	–	SK Innovation	–	–	[131]
Mazda	Demio	2012	20	346	–	–	Panasonic	3.2	6250	[132]
Mercedes	B250e	2015	28	300	289.9	286.7	Tesla	–	3696	[133]
Mercedes	Smart	2014	17.6	344.1	191	143.3	Li-Tec	52	93	[134]
Mitsubishi	I-Miev G	2012	16.3	325.6	164.7	192.2	LEJ	50	88	[135]
Mitsubishi	I-Miev-m	2012	10.5	270	–	–	Toshiba	20	234	[136]
Nissan	Leaf	2011	24(21)	360	295	350.6	AESC	33	192	[137–140]
Nissan	Leaf	2016	30	360	316	–	AESC	42	192	[141,142]
Nissan	Leaf	2018	40	350	300	–	Envision AESC	56	192	[143,144]
Nissan	Leaf e+	2019	62	350	–	–	Envision AESC	56	288	[143]

Table 29. *Cont.*

Brand	Model	Year	E, kWh	U, V	m, kg	V, L	LIBs Producer	C, Ah	n, pcs.	Ref.
Peugeot	C-zero	2011	14.5	300	–	–	LEJ	50	80	[145,146]
Porsche	Taycan	2020	93.4	800	–	–	LG Chem	64.6	396	[147,148]
Renault	Fluence Z.E.*	2013	(22)	360	280	–	–	–	–	[149]
Renault	Zoe	2012	23.3	360	280	–	LG Chem	36	192	[150]
Renault	Zoe ZE 40	2016	45	345.6	–	–	LG Chem	65.6	192	[151,152]
Renault	Twizy 45	2012	6.1	58	100	–	LG Chem	–	–	[153,154]
SGMC	Wuling Hong Guang MINI EV **	2020	9.6	96	88	–	Guoxuan	100	30	[155]
Tesla	Model S 85	2014	85	350	545	–	Panasonic	3.2	7104	[156,157]
Tesla	Model S P100D	2016	100	350	641	–	Panasonic/Tesla	3.4	8256	[81–83]
Tesla	Model 3	2017	75	350	478(457)	314	Panasonic/Tesla	4.6 ÷ 4.75	4416	[121,158,159]
Toyota	Rav 4	2014	(41.8)	386	380	–	Panasonic	3.2	4500	[160]
Toyota	Scion IQ	2012	12	277.5	166	–	–	21	150	[161,162]
VW	e-Golf	2015	24	320	313	229.4	Panasonic	25	264	[163–165]
VW	e-up	2013	18.7	374	230	–	LG Chem	–	204	[166]

Note: The E column shows the total energy of the battery, and the maximum energy that can be used from the battery when driving is shown in parentheses. * Renault Fluence—removable battery. ** The Wuling Hong Guang MINI EV comes with one of four different batteries produced by one of three manufacturers. Source: Authors' compilation based on data from references cited in the table.

High-power LICs have small capacity and dimensions to provide better commutation of discharge (charging) current and heat dissipation. LICs' capacity for hybrid transport batteries (for micro HEVs, mild HEVs, HEVs) is less than 11 Ah.

Typically, the use of high-power cells involves short charge/discharge intervals, so manufacturers usually specify the power (and/or DCIR resistance) values in their LIC specifications, not at continuous discharge, but determined by an HPPC (hybrid pulsed power characterisation) test (e.g., IEC 62660-1). In other words, the power is determined based on the maximum allowed current discharge-out (charge-in) of LICs (with a given state of charge, more often 50%) observed at the end of the time (more often 10 s) of discharge (charge) carried out at a selected temperature (more often 20 ÷ 25 °C). When the temperature increases (in the allowed temperature range) and the discharge time of the tested LICs decreases, the determined power value increases (Figure 38). The specific power values obtained from the HPPC test may differ from the power values for continuous discharge (or charge).

Figure 39 shows the discharge curves of two HEV cells with NMC/C (BEC) and NMC/Gr (PEVE) chemistries. The cells were taken from decommissioned vehicles. Thus, new ones might show better performance. The PEVE cells (current generation, estimated year of vehicle issue—2019) showed higher nominal capacity—4.1 Ah than communicated in the paper (3.6 Ah) [99] and the specific power at 60 C (3.68 kW/kg) was close to the value given in the paper—3.92 kW/kg. The BEC cells (older, previous generation; estimated year of vehicle issue—2016) had smaller nominal energy and rated energy, but exhibit close values of specific power at 60 C—3.59 kW/kg. To understand the performance limits, we attempted to discharge BEC cells at even higher C-rates. The current elevation to 70 C led to an increment of specific power to 4.11 kW/kg and a slight decrease in the specific energy from 48 (60 C) to 47 Wh/kg. The subsequent increase in the current up to 100 C resulted in the elevation of specific power up to 5.75 kW/kg and a significant decrease in energy down to 27 Wh/kg.

Table 30. Characteristics of Li-ion cells used in transport: 12 V and 48 V batteries.

Manufacturer	Active Materials	Case	U, V	C, Ah	m, g	Th/d × W × L mm	E_{nom} Wh/L	E_{nom} Wh/kg	P_{max} kW/L	P_{max} kW/kg	DCIR mOhm	T °C	Ref.
12 V													
A123	LFP/Gr	Lp	3.3	20	496	7.25 × 227 × 160	251	133	<u>4.6</u>	<u>2.4</u>	–	−30 ÷ 55	[167]
A123	LFP/Gr	Lp	3.3	20	520	14.2 × 150 × 144	208	126	<u>9.6</u>	<u>5.2</u>	–	−30 ÷ 55	[168]
GS Yuasa	LFP/Gr	Pr	3.3	69	2200	55 × 116 × 171	207	104	–	–	–	–	[169]
Lishen	LMO/LTO	Cy	2.5	18	780	60 × 144 × –	105	55	<u>3.3</u>	<u>1.7</u>	–	−40 ÷ 60	[27]
Micro HEV 12 V													
Toshiba	LMO/LTO	Pr	2.4	10	510	22 × 106 × 116	89	47	<u>6.7</u>	<u>3.5</u>	0.65	−30 ÷ 55	[170,171]
Toshiba	LMO/LTO	Pr	2.4	2.9	150	14 × 97 × 63	80	46	<u>4.9</u>	<u>2.8</u>	2.6	−30 ÷ 55	[25,172]
–	NCM/C	Pr	3.4	11	600	22.2 × 86.5 × 174	112	62	–	–	–	−30 ÷ 75	[23]
Mild HEV 48 V													
A123	LFP/Gr	Lp	3.3	8	330	4.8 × 160 × 227	151	80	<u>11</u>	<u>5.8</u>	1.4	−30 ÷ 55	[173]
CATL	NCM/Gr	Pr	3.65	10	246	17.5 × 85 × 120	201	146	<u>6.2</u>	<u>4.5</u>	–	–	[174]
Lishen	NCM/Gr	Pr	3.65	10	384	13 × 97 × 148	188	95	<u>7.9</u>	<u>4.0</u>	–	−40 ÷ 75	[27]
Toshiba	NCM/LTO	Pr	2.3	5	165	13.5 × 63 × 97	140	70	<u>7.8</u>	<u>3.9</u>	1.8	−30 ÷ 55	[35,175]
Toshiba	NCM/LTO	Pr	2.3	20	545	22 × 106 × 116	170	84	<u>7.8</u>	<u>3.9</u>	0.6	−30 ÷ 55	[35,175]

Note: (Lp)—prismatic accumulator in laminated foil casing; (Pr)—prismatic case made of metal; Cy—cylindrical; U—voltage; C—capacity; m—weight; Th/d—thickness (Lp, Pr)/diameter (Cy); W—width; L—length; E_{nom}—specific nominal energy; P_{max}—maximal power. <u>Underlined values of power</u> obtained according HPPC test: degree of charge—50%, discharge time—10 s, room temperature; DCIR—direct current internal resistance; T—temperature. Source: Authors' compilation based on data from references cited in the table.

Table 31. HEV, PHEV, and BEV lithium-ion battery cells used in passenger cars.

Manufacturer	Active Materials	Case	U, V	C, Ah	m, g	Th/d × W × L mm	E_{nom} Wh/L	E_{nom} Wh/kg	P_{max} kW/L	P_{max} kW/kg	Cycle Life cycles	T, dchrg, (chrg) °C	Ref.
						HEV							
AESC	–	Lp	3.7	4.3	206	3.3 × 136 × 251	139	76	6.6	3.6	–	−20 ÷ 55	
BEC	NCM/C	Pr	3.6	5.0	280	15.7 × 78.5 × 112.2	135	67	4.6	2.3	–	−30 ÷ 55	[176,177]
BEC	NCM/C	Pr	3.6	5.0	229	12.5 × 85 × 120	141	79	8.5	4.7	–	−30 ÷ 55	[94,178, 179]
CATL	NCM/Gr	Pr	3.6	6.9	285	12.5 × 85 × 120	195	88	9.7	4.4	–	−30 ÷ 55	[180]
Hitachi	NCM/Gr, C	Pr	3.7	5.2	240	12.0 × 80.5 × 120	166	80	10.3	5.0	–	−30 ÷ 55	[90,181]
Hitachi	LMO/C	Pr	3.7	5.3	250	12.5 × 85 × 120	149	76	6.7	3.4	–	−30 ÷ 55	[182]
Hitachi	LMO/C	Cy	3.7	4.4	260	40 × 90 × –	137	61	6.8	3	–	–	[181,183]
Hitachi	LMO/C	Cy	3.7	5.5	300	–	–	68	–	2.6	–	–	[183]
Lishen	NCM/Gr	Pr	3.7	5.5	233	12.5 × 85 × 120	155	85	5.8	3.2	–	–	[92]
Lishen	LFP/Gr	Pr	3.2	7.5	387	27 × 70 × 112	113	62	2.5	1.3	>4500	–	[92]
PEVE	LNO/NGC	Pr	3.6	5.0	245	14.1 × 91.8 × 111	125	73	5.0	2.9	–	–	[99]
PEVE	NCM/NGC	Pr	3.6	3.6	204	13.3 × 63.3 × 137	115	65	6.9	3.9	–	–	[99]
SBL	NCM/NGC	Pr	3.6	5.2	220	12 × 80 × 120	169	90	7.7	4.1	262 kAh	−30 ÷	[184,185]

Table 31. *Cont.*

Manufacturer	Active Materials	Case	U, V	C, Ah	m, g	Th/d × W × L mm	E_{nom} Wh/L	E_{nom} Wh/kg	P_{max} kW/L	P_{max} kW/kg	Cycle Life cycles	T, dchrg, (chrg) °C	Ref.
						PHEV							
BYD	LFP/C	Pr	3.0	50	1700	28 × 375 × 100	150	95	1.4	0.8	4000	−25(0) ÷ 55	[103,104, 186]
EIG	NCM/Gr	Lp	3.65	20	428	7.2 × 129 × 217	370	174	4.6	2.3	1000	−30(0) ÷ (40)55	[187]
GS Yuasa	NCM, LFP/Gr	Pr	3.62	13	365	21 × 81 × 112	247	129	5.3	2.8	–	–	[188,189]
GS Yuasa	LFP/Gr	Pr	3.3	25	1030	27 × 100 × 171	173	78	0.6	1.3	–	−30 ÷ 45	[190]
GS Yuasa	NCM/Gr	Pr	3.7	25	1400	37 × 124 × 145	139	66	3.9	1.9	3000	−20 ÷ 45	[191]
Hitachi	–	Pr	3.7	28	720	26.5 × 91 × 148	282	140	4.6	2.3	–	–	[182]
LG	NCM, LMO/Gr, HC	Lp	3.7	15	383	5.5 × 165 × 226	270	150	–	–	–	–	[4,192–194]
LG	NCM, LMO/Gr	Lp	3.7	26	583	8.3 × 170 × 230	*295*	*165*	–	–	–	–	[4,194, 195]
Lishen	NCM, LMO/Gr	Pr	3.7	25	730	26.5 × 97 × 148	264	129	–	–	1500	−20 ÷ 55	[196]
Microvast	NCM/Gr (activated)	Lp	3.5	9	290	6.5 × 130 × 210	185	113	–	–	10,000	−20(−5) ÷ (45)60	[197–200]
Microvast	NCM/Gr	Lp	3.6	15	300	7.2 × 126 × 220	280	170	*2.1*	*1.3*	–	−20(−5) ÷ 45	[201,202]
SBL	NCM, LMO/NG, AG	Pr	3.7	20.5	640	21 × 85 × 173	250	116	5.2	2.4	4000	−30 ÷ 60	[185]
SBL	NCM, LMO/NG, AG	Pr	3.7	24.5	728	26.5 × 91 × 148	257	126	4.7	2.3	4000	−30 ÷ 60	[184,185]
Panasonic	NCM 4.35 V/Gr	Pr	3.7	20	360	15 × 65 × 270	454	200	5.4	3.3	<1500	–	[158,203]

Table 31. *Cont.*

Manufacturer	Active Materials	Case	U, V	C, Ah	m, g	Th/d × W × L mm	E_{nom} Wh/L	E_{nom} Wh/kg	P_{max} kW/L	P_{max} kW/kg	Cycle Life cycles	T, dchrg, (chrg) °C	Ref.
						BEV							
AESC33	LNO, LMO/Gr	Lp	3.8	33	799	7.1 × 216 × 290	281	156	1.1	0.6	–	–	[140]
AESC56	NCM/Gr	Lp	3.65	56	914	7.9 × 216 × 261	460	224	–	–	–	–	[144]
Boston P	NCM, NCA/Gr	2Cy	3.65	5.3	93.5	18.5 × 37 × 65	490	207	2.3	1	–	−40(−20) ÷ (60)70	[204,205]
BYD	LFP/NG	Pr	3.2	225	6100	58 × 146 × 410	207	118	3.5÷ 4.3	2÷ 2.5	–	−20(0) ÷ (50)55	[105,180, 206,207]
BYD	LFP/	Pr	3.2	202	3.92	13.5 × 118 × 905	448	165	–	–	–	–	[64,70, 208]
Cadenza	NCM/Gr	24Cy	3.7	83	2000	53 × 122 × 175	271	154	1.3	0.75	–	−20(−10) ÷ (45)60	[209]
CATL	NCM/Gr	Pr	3.7	50	990	27 × 97 × 148	477	187	–	–	–	−20(0) ÷ (45)55	[210,211]
CATL	NCM/Gr	Pr	3.7	70	1250	38 × 91 × 149	517	213	–	–	–	–	[211,212]
DFD	NCM/Gr	Pr	3.73	55	800	11.5 × 102 × 313	559	256	–	–	2000	−20(−10) ÷ (55)	[213]
Guoxuan	LFP/	Pr	3.2	105	2030	27 × 175 × 200	355	165	2.8	1.3	3000	–	[214]
Hitachi	H-NCM/Gr, Si-comp	Pr	3.6	40	700	–	–	206	–	2	–	–	[203]
LEJ	LMO/Gr	Pr	3.85	50	1700	43.8 × 113.5 × 171	220	110	*1.2*	*0.6*	>1000	–	[215,216]
LG	NCM/Gr	Lp	3.7	59	835	*13.5* × 100 × 338	470	260	–	–	–	–	[203,211, 217,218]
Microvast	NCM/Gr	Lp	3.7	43	690	*12* × 100 × 300	440	230	–	–	–	–	[219]
Panasonic	NCA/Gr,Si	Cy	3.7	3.3	48.5	18.5 × 65.3 × –	680	250	–	–	–	–	[220,221]
Panasonic	NCA/Gr,Si	Cy	3.7	4.85	69	21 × 70 × –	720	255	–	–	–	–	[158,194, 221–223]

Table 31. *Cont.*

Manufacturer	Active Materials	Case	U, V	C, Ah	m, g	Th/d × W × L	E_{nom}		P_{max}		Cycle Life	T, dchrg, (chrg)	Ref.
						mm	Wh/L	Wh/kg	kW/L	kW/kg	cycles	°C	
SBL	NCM, NCA, LMO/Gr	Pr	3.7	60	1923	45 × 125 × 173	243	123	<u>4.4</u>	<u>2.1</u>	5000	–	[184,185]
SDI	NCM/Gr	Pr	3.7	94	2010	45 × 125 × 173	360	174	1.1	0.5	4600	–	[224]
SDI	NCM/Gr	Pr	3.7	120	2000	45 × 125 × 173	454	220	–	–	–	–	
Toshiba	NCM/LTO	Pr	2.3	20	515	22 × 106 × 116	176	89	<u>4.1</u>	<u>2.1</u>	15,000	−30 ÷ 55	[172,225, 226]
Toshiba	NCM/$TiNb_2O_7$	Lp	2.25	49	800	14.5 × 111 × 194	350	138	<u>4</u>	<u>1.58</u>	>14,000	−30 ÷ 45	[227,228]
Zenlabs	NCM/SiO_x,Gr	Lp	3.46	51	592	10 × 102 × 320	640	305	2.0	0.85	1000	−30 ÷ 52	[229]

Note: *Italics show evaluative values*. Underlined values of power obtained according HPPC test: degree of charge—50%, discharge time—10 s, room temperature. Dchrg—discharge, chrg—charge. Perspective LICs with improved characteristics are highlighted **in bold**. Other designations are given in Note to Table 30. Source: Authors' compilation based on data from references cited in the table.

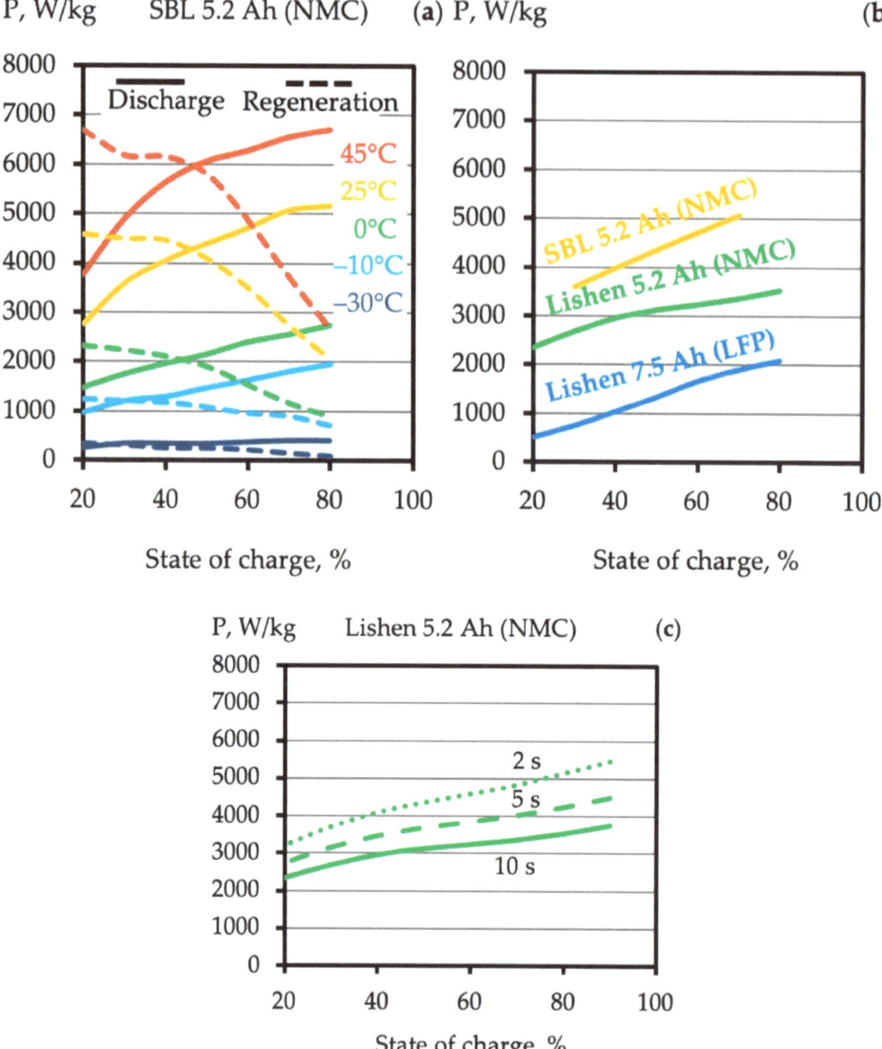

Figure 38. Effect of state of charge, temperature of SBL 5.2 Ah (NMC) cell (**a**) [185], state of charge of different (**b**) LICs [92,185] for HEVs and discharge time of Lishen 5.2 Ah (NMC) cell (**c**) [92] on the power determined by HPPC test [230]; Source: Figure by authors.

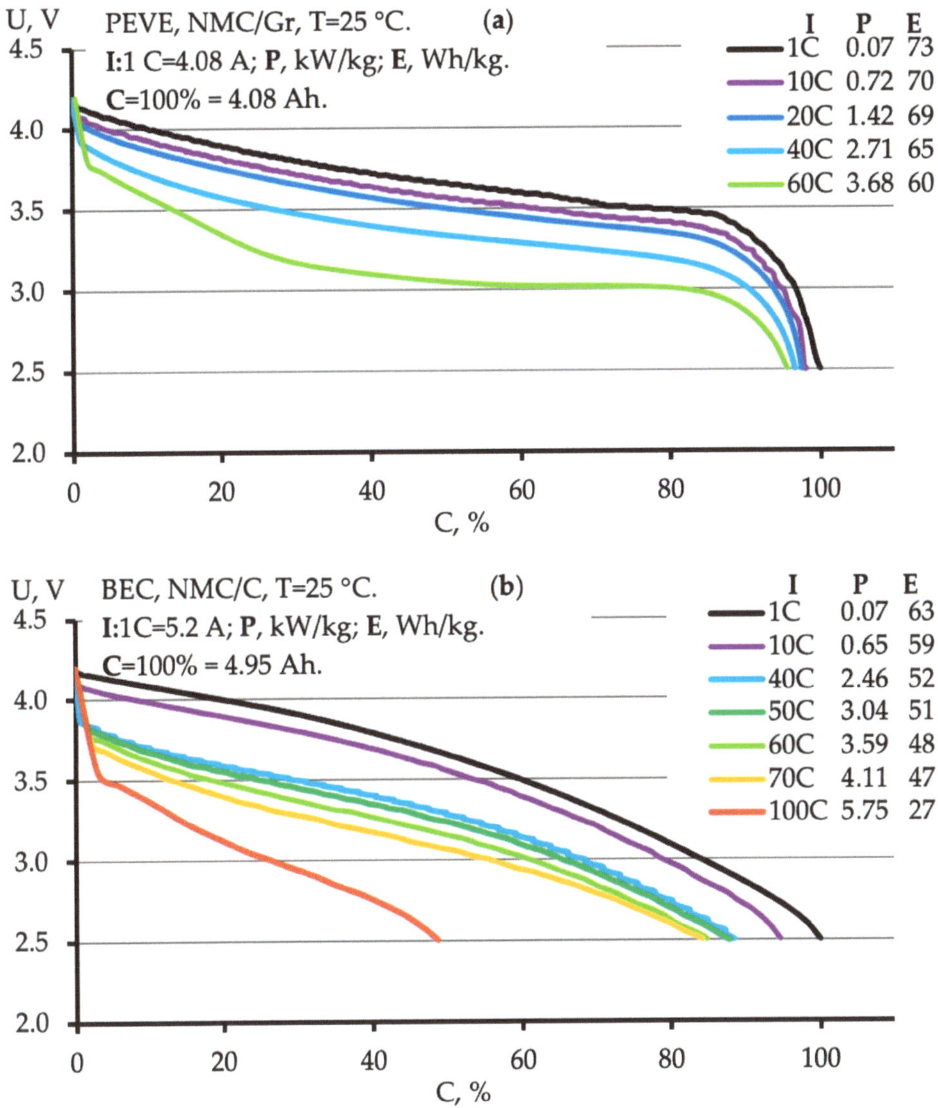

Figure 39. Discharge curves during laboratory tests of PEVE [90] (Toyota Prius) (**a**) and BEC [176,177] (Honda Civic) (**b**) cells. Source: Figure by authors.

In the manufacture of batteries for PHEVs, less powerful (value according to HPPC 50%, 10 s test does not exceed 3 kW/kg) LICs with higher specific energy (specific nominal energy—100 ÷ 170 Wh/kg) and a capacity varying in the range of 9 ÷ 50 Ah are used.

The specific energy of LICs used to manufacture batteries for electric vehicles (BEVs) tends to be even higher. Power values determined under short-term loads are rarely found in the descriptions of this type of LIC. The LICs' capacity ranges from

3.3 Ah to 225 Ah. Figure 40 shows a graph characterising the power values (W/kg) at short-term discharge, nominal specific energy (Wh/kg), and the capacity of LICs used to create batteries of different types of electric vehicles. The dashed black oval marks two points describing Panasonic (PHEV) and Hitachi (BEV) LICs developed under the NEDO programme with enhanced energy and power characteristics. An even higher specific energy (>300 Wh/kg, 640 Wh/L), but lower power, is characteristic of LICs developed by Zenlabs under the DOE programme (Table 31).

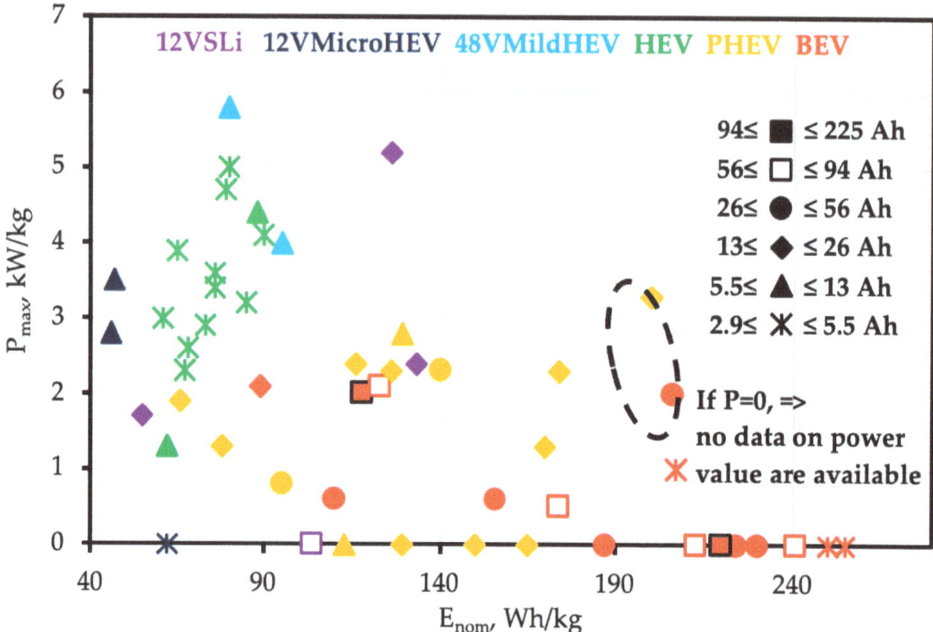

Figure 40. Specific power, rated energy, and capacity of LICs used in the manufacture of batteries for various types of passenger cars. Dashed black oval marks two points corresponding to LICs with enhanced energy and power characteristics. Source: Figure by authors.

The change in the shape of the charge/discharge curves and the temperature during the charge/discharge process are essential characteristics that are considered when designing a battery. The LICs charging curves presented in the specifications usually start at voltages significantly higher than the end-of-discharge voltage. The voltage at the LICs is determined by the potential difference between the surface of the cathode and anode particles. During the discharge (time-limited) in the particles of the active cathode and anode material, there is a gradient of lithium ions distribution across the particle cross-section. After the discharge, the external layers of the anode/cathode material are depleted/enriched with lithium ions. Thus, the difference between potentials of cathode and anode (voltage) is less than in equilibrium state. If charging is carried out some time after discharge, the concentration of lithium atoms in the cathode and anode material volume is equalised. On the surface of the active anode (cathode) material, the concentration of lithium increases (decreases) compared to the concentration observed immediately after the discharge. Thus, the potential at the surface of the active anode (cathode) material decreases (increases), and the potential difference grows.

Let us consider the effect of the charging current on the shape of discharge curves, capacity, and temperature changes using the example of Lishen's LICs used in PHEV batteries. Without increasing the capacity, the voltage increase is proportional to the internal resistance of the LICs and the charging current. In this regard, when the charging current is increased (Figure 41), a more considerable jump in voltage is observed in the charge curves.

At low currents (rates) of charge (0.3 C, Figure 41a), the temperature of LIBs decreases. Increasing the charging current from 1 C to 6 C decreases the charging capacity, increases the average charging voltage, and increases the temperature due to the heat released due to the internal resistance.

As the discharge current increases, the voltage across LICs decreases, the capacity usually decreases (but may remain unchanged or increase, depending on the design and manufacturing technology of the LICs). When LICs are discharged, their internal temperature always rises. The discharge (charge) current at which the temperature of the LIC wall at the end of discharge (charge) rises to the manufacturer's maximum allowable value (70 ÷ 75 °C, or lower, depending on the construction of the cell) is taken as the maximum allowable for this LIC (Figure 41b).

One can compare different types of LICs by using plots showing the relationship between specific energy and power (Figure 42). The values of specific energy and power were obtained by the mathematical processing of discharge curves (Figure 12) considering the weight and volume of the cells. The specific energy of LICs for hybrid transport ranges from 50 ÷ 110 Wh/kg (80 ÷ 180 Wh/L, power 1 ÷ 5 kW/kg (2 ÷ 8 kW/L) and higher. Batteries in plug-in hybrid vehicles are equipped with LICs with a specific energy of 50 ÷ 180 Wh/kg (80 ÷ 360 Wh/L) and a maximum power of 0.55 ÷ 0.9 kW/kg (1 ÷ 1.7 kW/L). Lithium-ion cells with a lower energy density rating are used in Mitsubishi i-Miev electric cars (M—90 Wh/kg (180 Wh/L), G—112 Wh/kg (225 Wh/L)). For comparison, Tesla electric cars use LICs with a specific rated energy (250 Wh/kg, 700 Wh/L) and specific power at constant discharge current in the range of 0.4 ÷ 0.8 kW/kg (0.8 ÷ 1.2 kW/L).

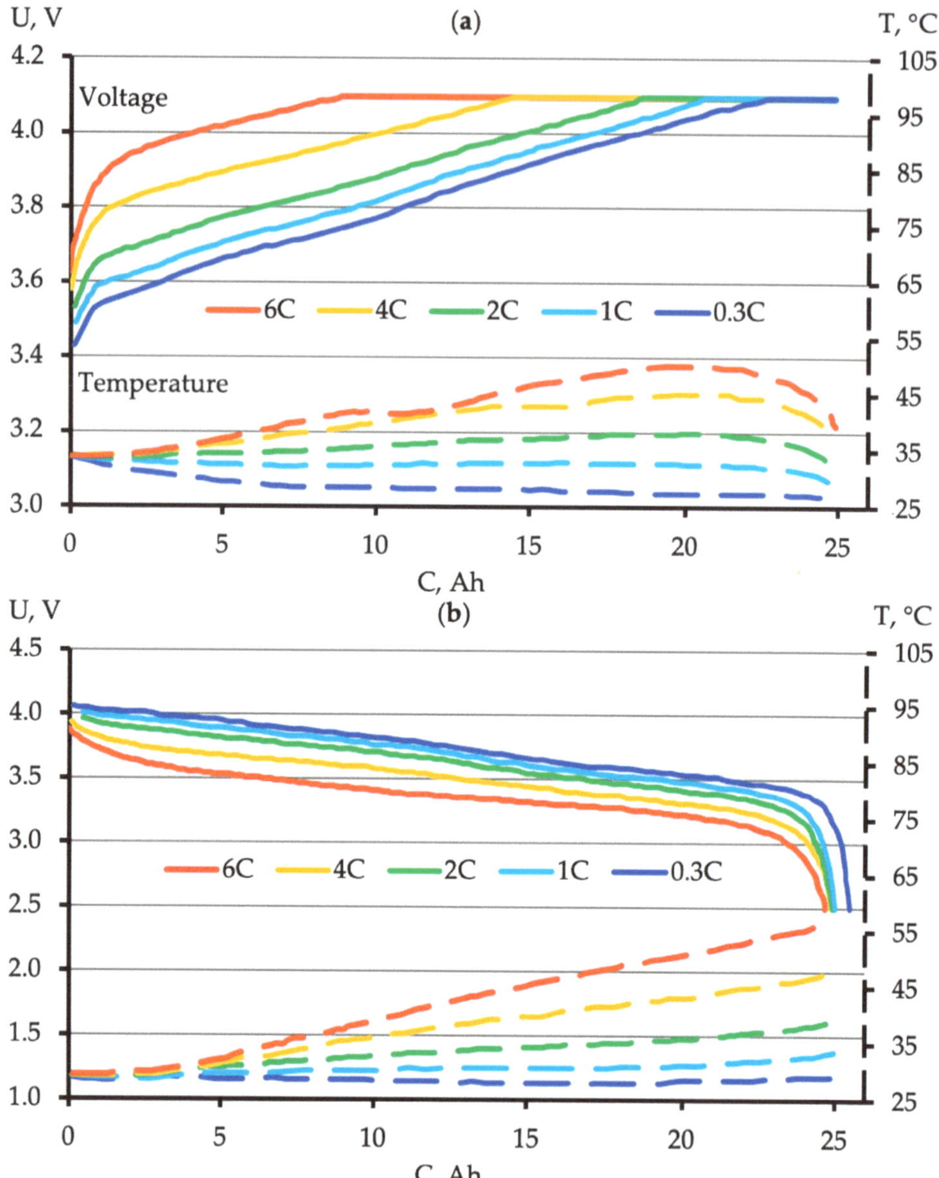

Figure 41. Effect of charge rate (**a**) and discharge rate (**b**) on the shape of the discharge curves and temperature of LICs (NCM:LMO, Lishen) in a PHEV battery (initial data for plots were taken from reference [231]). Source: Figure by authors.

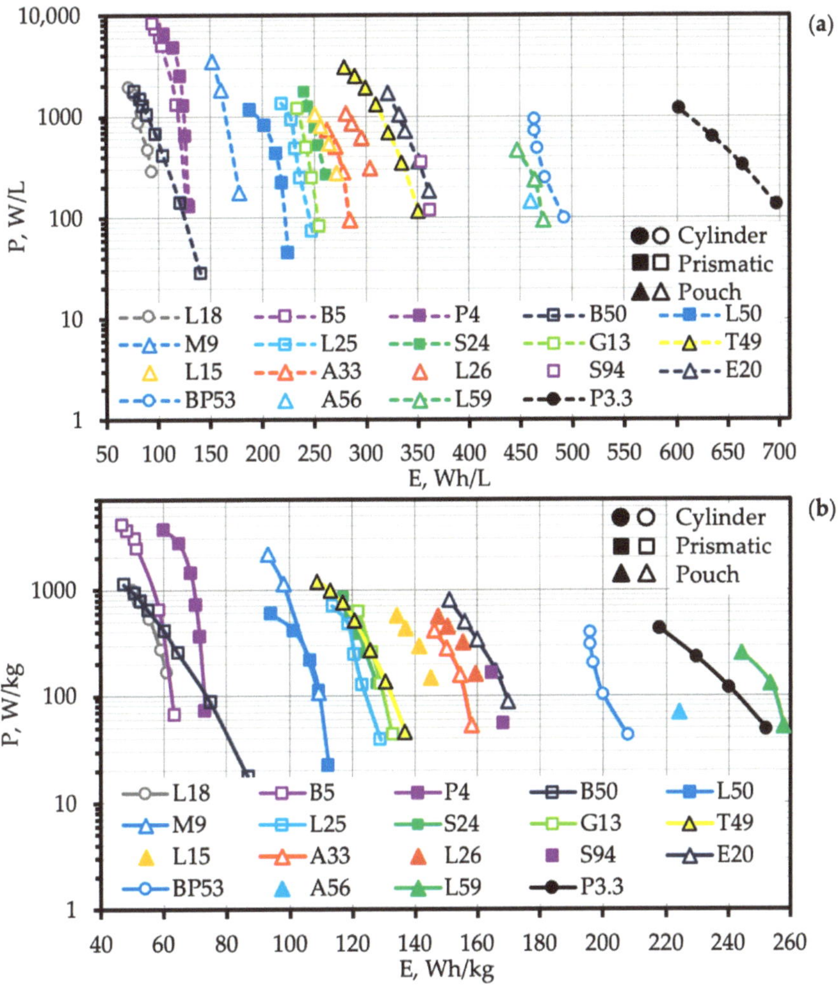

Figure 42. Energy (Wh/L, Wh/kg) vs. power (W/L, W/kg) of LICs used in hybrid (HEV, PHEV) and electric vehicles (BEV): L18—Lishen 18 Ah [27], B5—BEC 5 Ah [176,177], P4—PEVE 4 Ah [99], B50—BYD 50 Ah [103,104,186], L50—LEJ 50 Ah [215,216], M9—Microvast 9 Ah [197–200], L25—Lishen 25 Ah [196], S24—SBL 24Ah [184,185], G13—GS Yuasa 13Ah [188,189], L15—LG 15 Ah [4,192,193], A33—AESC 33 Ah [140], L26—LG 26 Ah [4,195], T49—Toshiba 49 Ah [227], S94—SDI 94 Ah [224], E20—EIG 20 Ah [187], BP53—Boston Power 5.3 Ah [205], A56—AESC 56 Ah [144], L59—LG 59 Ah [203,211,217,218], P3.3—Panasonic 3.3 Ah [220]. (**a**) energy density and power density are expressed in Wh/L and W/L, correspondingly. (**b**) Specific energy and power are expressed in Wh/kg and W/kg, correspondingly. Dots without lines indicate power and energy estimates. For both **a** and **b** figures the form of a marker on the plot reveals the type of cell case: circle, square, and triangle corresponds to cylindrical, prismatic and pouch cell cases. Source: Figures by authors.

5.3.1. Influence of Temperature on the Functional Characteristics of LICs

At low and high temperatures, the performance is reduced. For levelling the effect of ambient temperature, large batteries for PHEVs and BEVs can be equipped with a thermal control system, using a liquid heat transfer agent (Chevrolet Volt, Tesla, etc.). However, starter and hybrid batteries are not equipped with a fluid heat transfer agent thermal control system because of their small size.

The performance of cells for hybrid transport at low temperatures and short time intervals is characterised by the total or specific power (Figure 38a, [178]) obtained following the HPPC test or cold start (cold-cranking) test. In the VDA cold-cranking test, three short discharges for 5 s at a state of LIC charge of 50% [185] or 30% [232] is performed at −30 °C. The lower limit of cell discharge is 2.1 V. The power is determined based on the current measured at the end of the third cell (battery) discharge at a voltage of 2.1 V. LICs for hybrid vehicles (HEVs) are rated at 170 W (SBL, 5 Ah, 50% degree of charge, −30 °C) and 110 W (CATL, 6.9 Ah, 30% degree of charge, −25 °C) in the cold start test.

The value of the internal resistance during charging and discharging and at different states [185] of charge serves as an indicator of the operability of LICs in short time intervals. For example, when the LICs' temperature decreased from 25 °C to −10 °C, the DCIR resistance (50% degree of charge, 10 s) increased from 2.6 mOhm to 7.9 mOhm (discharge) and from 2.5 mOhm to 9.7 mOhm (charge). A further drop in temperature to −30 °C led to an increase in the internal resistance to 45.5 mOhm (discharge) and 51.7 mOhm (charge).

At low temperatures, continuous charging is possible, but not for all LICs, and manufacturers specify a maximum allowable current at a given temperature. Depending on the manufacturing technology and materials used, the extent of the change in the shape of the charge curve differs. In the case of high-energy LICs for electric vehicles compared to LICs for hybrid cars, there is a more significant increase in the average charging voltage and a decrease in the relative capacity during constant current charging to the cutoff voltage (Figure 43).

As the temperature diminishes, the permissible charging current decreases. For example, for LICs (NCM/Gr, 5 Ah), the maximum allowable charging currents are 35 A (7 C, −10 °C) and 7.5 A (1.5 C −30 °C). The current limitation for LICs with anodes based on carbon materials is caused by possible lithium deposition on the negative electrode, which reduces the cycle life and the occurrence of micro-short circuits. For LICs with a lithium titanate anode, this limitation is removed due to the high potential of the negative electrode relative to lithium. For example, Toshiba reports a 1 C cyclic charge/discharge capability at −20 °C, which reduces the capacity to 60% of the rated capacity [171].

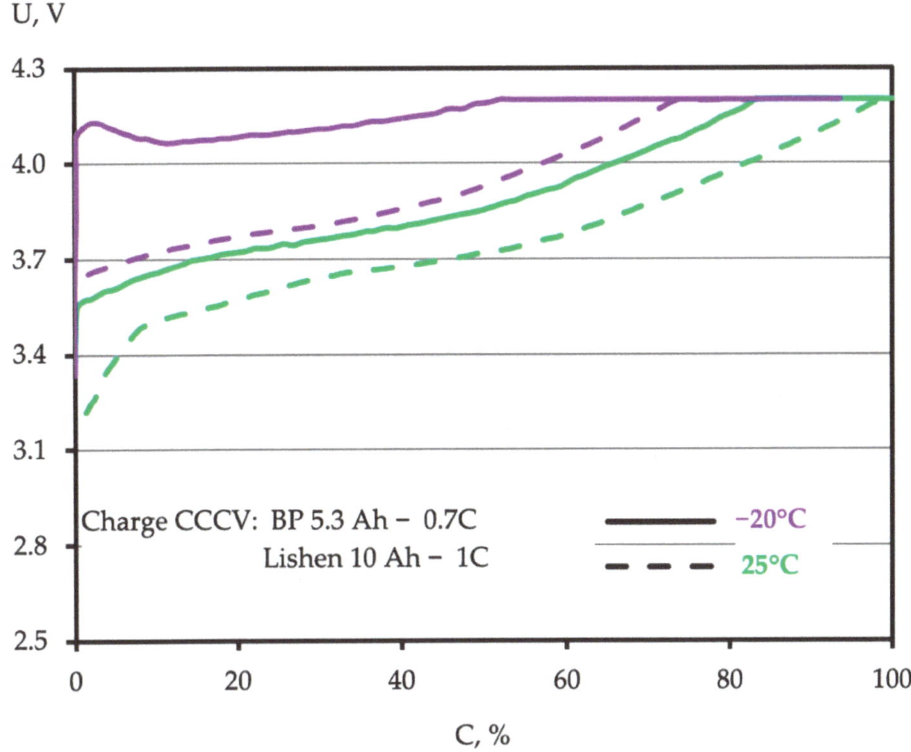

Figure 43. Charging curves of Boston Power Swing 5300 [205] and Lishen 10 Ah [27] lithium-ion cells obtained at −20 °C and 25 °C (CCCV—Constant current then constant voltage). Source: Figure by authors.

As the temperature goes down, the average discharge voltage decreases and the capacity of LICs during discharge diminishes. For high-power LICs (Figure 44a), the decline in power and energy output with decreasing discharge temperature is less intensive compared to the high-energy LICs of electric vehicle batteries (Figure 44b). If the internal resistance of LICs is relatively high, a minimum can be observed at the beginning of the discharge curve. The voltage increase at the subsequent discharge is related to the decrease in the resistance caused by the internal heating of the LICs.

LICs with lithium titanate anode-active material are also characterised by decrement of specific power and energy with decreasing temperature. Based on the discharge curves [27] presented in the presentation and the technical description, a diagram (Figure 45) showing the effect of discharge temperature on the specific capacity and energy of the Lishen 18 Ah battery (LTO/LMO) was plotted. For example, when there is a decrease in the discharge temperature from room temperature to −20 °C, the maximum specific power diminishes almost twofold (from 1200 to 630 W/kg), and the specific energy reduces by ≈15% from 47 to 40 Wh/kg.

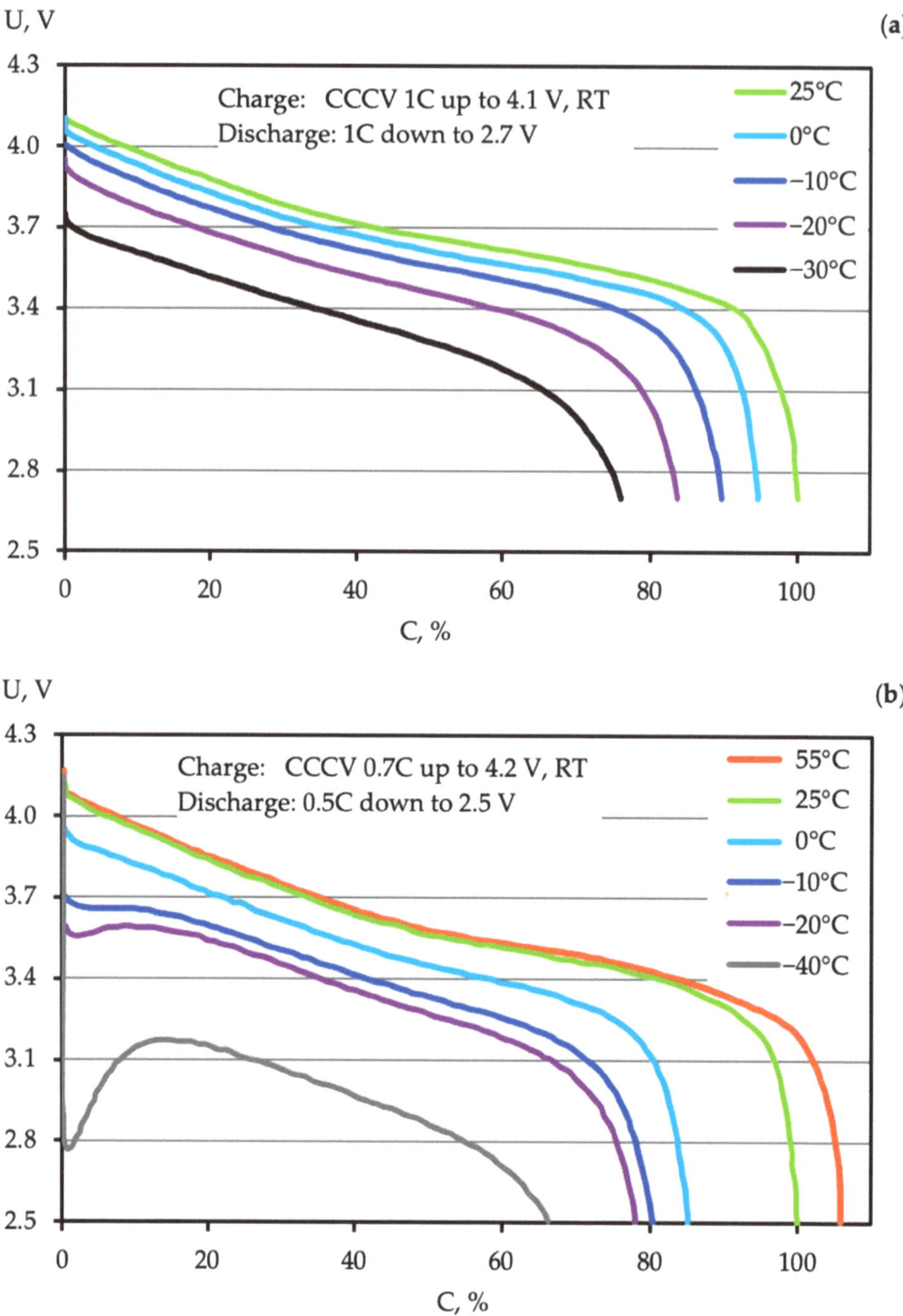

Figure 44. Influence of temperature on the shape of discharge curves for (**a**) CATL (HEV LIC) [180] and (**b**) Boston Power Swing 5300 (BEV LIC) [205] (the initial data were taken from references given in square brackets). Source: Figure by authors.

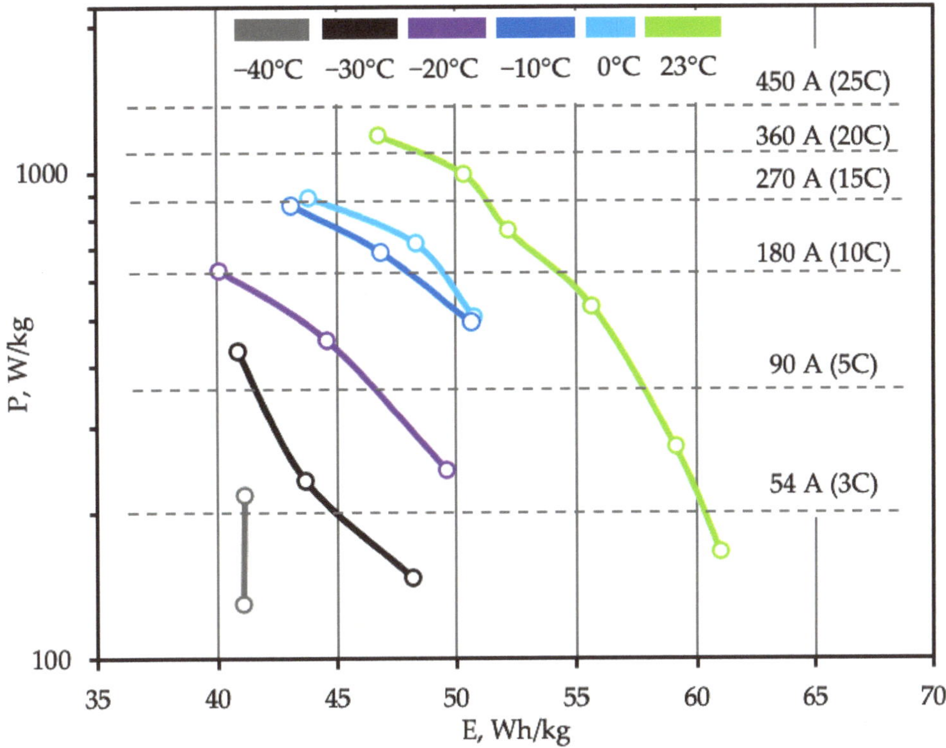

Figure 45. Influence of temperature and discharge current on specific energy and power of LTO Lishen 18 Ah LICs. Plots prepared based on data from [27]. Source: Figure by authors.

As the LIC temperature rises, the specific power and energy grow (Figures 38a and 44b). However, the degree of degradation of the electrochemical system components also increases, leading to the deterioration of the functional characteristics, even in the case of exposure at elevated temperatures only and without conducting charge and discharge (Figure 46). In the case of storage at 35 °C after 500 days, the capacity, power, and internal resistance of LICs undergo insignificant changes. Increasing the storage temperature up to 60 °C leads to an intensive decrease in the functional characteristics of the LICs over time. After 500 days at 60 °C, the capacity and power are reduced by ≈20% and 25%; the internal resistance augments by 32% compared to the initial values. More information about the effect of elevated temperature on the retention of HEV LIC performance during testing (VDA cycling, exposure, effect of state of charge) can be found in the presentation in [185].

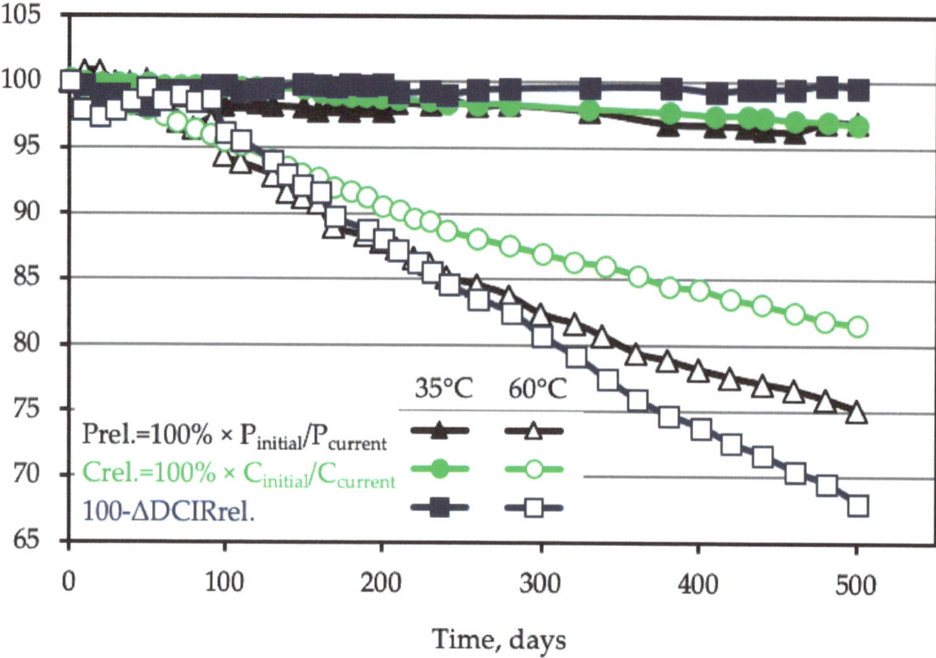

Figure 46. Effect of storage times at 35 and 60 °C on power, capacity, and internal resistance of LICs for hybrid vehicles. Plots prepared based on data from [185]. Source: Figure by authors.

5.3.2. Influence of Operating Conditions on the Life Cycle of LICs

The choice of operating conditions affects not only the energy and power characteristics, but also the life (the number of charge/discharge cycles that the battery or cell can withstand before its capacity, energy, or power reaches 80 (70)% of the nominal—end of life (EOL)). The choice of 70% instead of 80% of the nominal capacity as a criterion for EOL can lead to almost double the number of charge/discharge cycles from the initial state (beginning of life—BOL) [233]). LICs' operation resource depends primarily on the materials used in the manufacture, the functional additives (quality of the solid electrolyte film) in the electrolyte, the proportion of electrolyte in the LIC mass (more/longer) [234,235], and the manufacturing technology (and control of its compliance). During fabrication, to increase the service life, it is necessary to reduce the moisture ingress into the LICs (reducing the probability of side reactions leading to the formation of hydrofluoric acid). A reduced service life can be caused by the non-uniform application of active layers of positive and negative electrodes, leading to the local excess of cathode material (relatively anode capacity) compared to the value specified in the manufacturing technology [236]. An excess of positive electrode charge capacity leads to a drop in the performance of the anode due to lithium deposition on its surface. The operating life is often estimated based on the capacity drop caused by many possible side reactions that take place during the operation of LICs. For example, a gradual decrease in capacity was found after 500 charge/discharge cycles at room and elevated temperatures [235]. The batteries were disassembled and the capacity of positive and negative electrode samples in the CR2016 cell relative to lithium was determined. It was found that the cathode capacity was almost unchanged (tests at room and elevated temperature). The capacity of the negative electrode decreased noticeably, but still significantly exceeded the capacity of the positive electrode.

Consequently, the gradual decrease in capacity can be caused by the loss of lithium ions active in the electrochemical reaction [237]. The loss of active lithium can be caused by the process of solid electrolyte film formation involving lithium ions, the deposition of lithium metal on the anode, and the exclusion of active material from the electrochemical reaction (loss of electrical contact, destruction of structure) [238]. Lithium metal (as an additional lithium source) may be introduced into the battery to reduce the effect of this process on lifetime [239]. A decrease in capacity can also be caused by side reactions unrelated to the loss of electrochemically active lithium, in particular, forming part of the solid electrolyte film without involving lithium ions in its structure, or transition into solution and subsequent deposition of transition metal ions, etc. [238]. The list of adverse reactions and their intensity depends on the operating conditions of the LICs.

One can estimate the effect of discharge conditions on capacity retention from the graphs shown in Figure 47. According to data from Renault, increasing the battery charging current up to 2 C (43 kW, fast-charging battery ZE) has little effect on capacity reduction during cycling (Figure 47a). Battery life decreases in the range of operating temperatures of 25 °C, 35 °C, 10 °C, and 0 °C (Figure 47b). The number of charge/discharge cycles can be increased by reducing the operating voltage interval (Figure 47c) and the depth of discharge (Figure 47d, $LiFePO_4$). In addition, the service life can be affected by the range of states of charge in which the batteries are cycled. For instance, when Kokam cells (16 Ah) are discharged/recharged at 80%, the service life is increased in the following intervals of cell capacity: 20 ÷ 100% < 0 ÷ 80% < 10 ÷ 90% [240].

Employees of GS Yuasa have found [241] that when cycling in the range of 20 ÷ 80% is performed, the capacity reduction is more intense than when tripling the number of cycles in the 40 ÷ 60% range. Although the amount of transferred Ah is the same, with a broader range, there is a more significant change in the volume of graphite during charge/discharge, which leads to the more intense destruction and growth of the solid electrolyte film and an increase in internal resistance.

One of the advantages of lithium titanate cells is their long cycle life. If we consider two batteries with the same capacity (10 Ah) and different average discharge voltages (2.4 V and 3.7 V), at an equal number of charge/discharge cycles, the amount of energy transferred/discharged will be higher for the LIC with a higher voltage, 24 Wh, and 37

Wh, respectively. Suppose we equate the amount of energy transferred per cycle (24 Wh, for LICs with an average discharge voltage of 3.7 V—65% energy) and select the voltage ranges in which the most negligible degradation of materials occurs. In that case, the number of LIC cycles (NCM/Gr) may exceed the number of cycles of an LTO-based LIC with an anode. Thus, it is reasonable to compare the service life of LICs made using different active anode and cathode materials not by the number of cycles, but by the amount of transmitted energy in Wh.

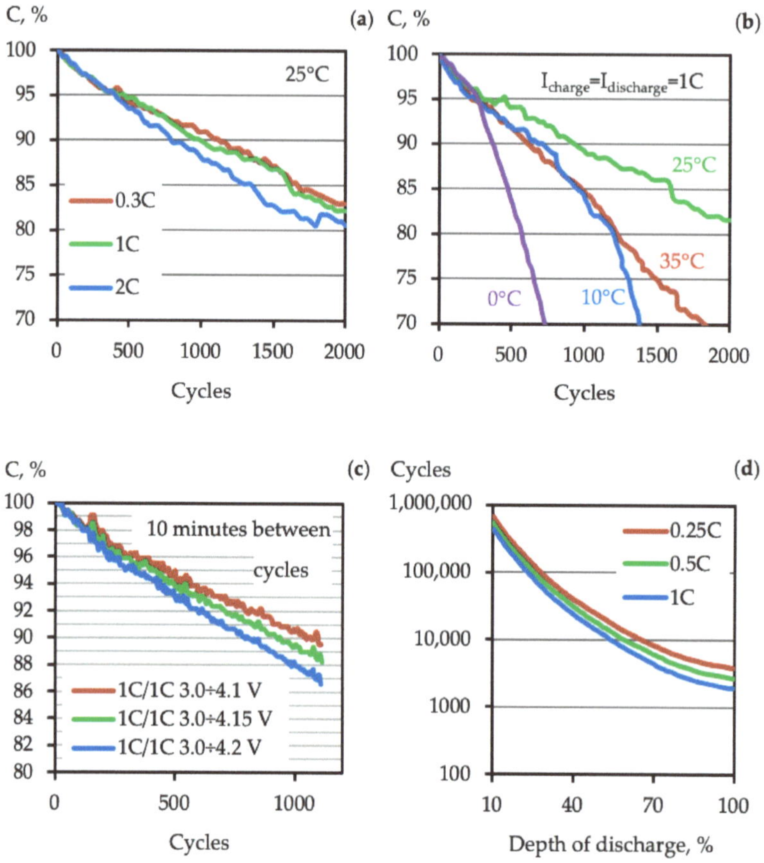

Figure 47. Influence of charging current (**a**) [242], temperature (**b**) [242], width of operating voltage range (**c**) [231], and depth of discharge (**d**) [243] on service life, expressed in the number of charge/discharge cycles (the initial data for the plots were taken from references given in square brackets). Source: Figure by authors.

Considering the increased cost per Wh of LICs with LTO-based anodes and the greater number of cells required to make a battery with a given voltage, it may not be cost-effective to use these batteries. However, in applications requiring operation at low temperatures, the use of LTO batteries may be appropriate due to the lower allowable charge temperature.

The battery life is also determined by the change in capacity during cyclic testing. In a performance study of a battery developed by Saft (95 kW capacity, 41 Ah capacity) and designed for hybrid vehicles (PHEVs), it was found that the power to a value of 80% of the original value is reduced in a smaller number of charge/discharge cycles compared to the capacity (Figure 48). It has also been shown that as the degree of discharge decreases, the lifetime determined based on peak power values increases. The observed decline in battery power is due to increased internal resistance during operation (Figure 49).

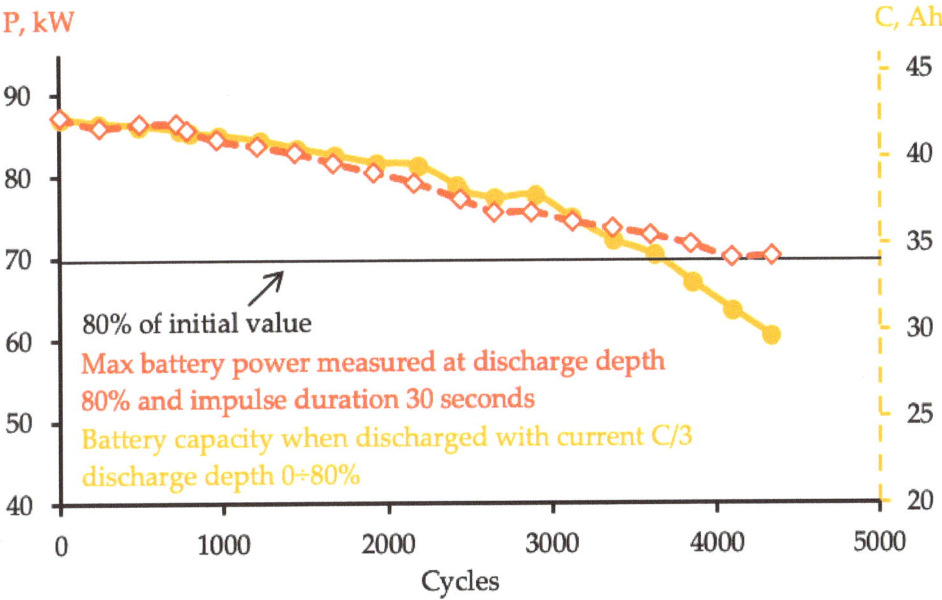

Figure 48. Comparison of change in capacity and a maximum power of lithium-ion battery for PHEVs during cycling (the initial data for the plot were taken from [244–246]). Source: Figure by authors.

In Table 32 and the explanatory text below, an attempt is made to describe the mutual influence of LIC parameters and LIB operation parameters on the lifetime (in a simpler way in the form of blocks and links, the effect of parameters on the lifetime is presented elsewhere [247]).

The cell lifetime (nC, τ) depends on the design, the battery control and management system (and operating algorithms), the materials and technologies used to manufacture the LICs (C), and the modes of operation.

A greater remaining battery life can be achieved by the following:

- Optimal specific energy (E) of LICs;
- Low (relative to LICs' capability and LIB pack design) charge/discharge current (I) (slow charge (τ.1), low intensity and time (τ.2) of vehicle acceleration);
- Smaller number of charge/discharge cycles (nC, nChg, nDchg);

- Optimal specific power (P) of LICs;
- Lower resistances (internal LICs (R.1) and conductive battery components (R.2), load resistance) and their increase during operation;
- Optimal interval of the state of charge (SoC, depth of discharge (DoD)) and voltage (U);
- Low self-discharge (SD);
- Optimum temperature (T) of LICs (temperature range) during operation;
- Shorter elapsed calendar time (τ) and optimum not working period (τ.3) of the cell in the average interval of degrees of charge.

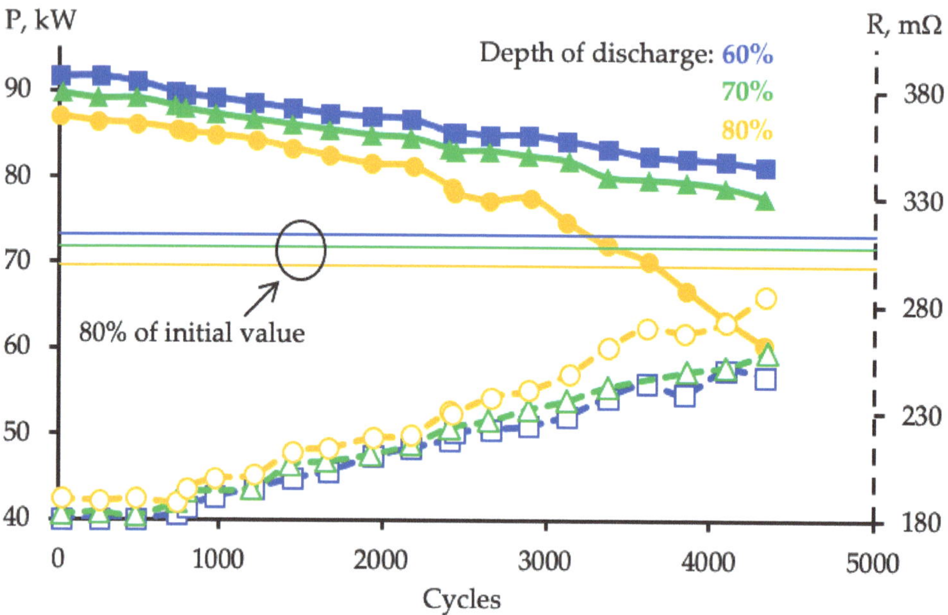

Figure 49. Change in maximum power and internal resistance of the battery during cycling as a function of the degree of discharge (the initial data for the plot were taken from [244,245]). Source: Figure by authors.

The design, materials (C), and manufacturing technology (MT) of LIBs affect the parameters described below, which, in turn, influence the structure and materials of LICs during operation.

The cell's specific energy (E) is higher when charge/discharge is carried out at low currents (I) when the cell has passed a smaller number of charge/discharge cycles (nC) and the cell operated less time (τ). For cells with the same design and chemistry, lower resistance (internal resistance LICs R.1 and load resistance R.2) allows lower voltage drop (ΔU) and, thus, higher specific energy. A smaller discharge time (τ.2) at a fixed discharge current strength (lower discharge depth, DoD, higher state of charge, SoC, and voltage, U) provides a higher remaining specific energy of the cell. The lower the resistance, the lower the voltage drop (ΔU). At low temperatures, the ionic resistance increases, and there is an increase in the internal resistance of the LICs (R.1). The specific energy augments with increasing temperature, but may decrease near the permissible upper limit of the temperature range. In case of prolonged operation at elevated temperatures, the probability of side reactions increases, leading to material degradation (lower capacity), solid electrolyte film redeposition (increased resistance), and a reduction in specific energy.

Batteries with the same electrochemical system and higher energy density (E) may have a higher internal resistance (R.1). Therefore, they are characterised by lower allowable charge/discharge currents (I), lower power (P), and less self-discharge (SD). On the other hand, increased internal resistance leads to higher self-heating and longer charge and discharge times due to current (I) and LIC temperature (T) limitations.

The battery resistance (R) is determined by the internal resistance of the LIB cells (R.1), the resistance of the current-carrying battery components (R.2), and the method of their connection (ultrasonic, laser welding). The resistance of the battery's current lead components depends on the design and manufacturing technology (quality of the welds) of the battery. The internal resistance of LICs increases as the temperature (T) decreases and the holding time at the elevated temperature (T) increases. The growth of the internal resistance R.1 is observed at small and large states of charge ($SoC(U) < A$, $SoC(U) > B$), with the increase in the specific energy (E) of the LICs (due to the use of denser electrodes, smaller volume of electrolyte, conductive additives, etc.). Furthermore, due to the course of side processes with the increasing number of charge/discharge cycles (nC) and operation time (τ, including idle time τ.3), the internal resistance (R.1) increases.

Table 32. Mutual influence of operation parameters and characteristics on the lifetime of LICs.

Life Time↑	Parameter	(A) Impact on Parameter	(B) Effect of the Parameter
↔	C—Design, LIC materials	Mutual influence of LIC design and the following parameters E, I, nC, P, R.1, SD, T.1, T.2, T.3, τ, τ.1, τ.2, τ.3, U	
↔	MT—LIC manufacturing technology		E, I, nC, P, R.1, SD, T.1, T.2, T.3, τ, τ.1, τ.2, U
↔	E—Specific energy↑	C, MT, *I*, *nC*, *P*, *R.1*, *τ*, *τ.2* **SoC**, **T(↔)**, **τ.1**, **U**	*I*, *P*, *SD* **R.1***, **T.1**, **τ.1**, **τ.2**.
↓	I—Amperage↑	Battery operating modes, C, MT, SoC, T **nC**, **R**, **τ**, **τ.1**, **τ.2**	U, E, C, *nC*, SoC, *τ.1*, *τ.2*. **P**, **T.1**
↓	nC—Number of charge/discharge cycles↑	Battery type (HEV, PHEV, BEV), etc., battery energy, operating modes. C, MT, E, P, SoC, U, T, *I*, *ΔSoC*, *τ*	U, C, *E*, *I*, *P*, *τ.1*, *τ.2* **R.1**, **T.1**
↔	nChg—Charge	Product operating modes	
↔	nDchg—Discharge	Product operating modes	
↔	P—Specific power↑	C, MT, SoC (U), T, *nC*, *R.1*, *T.1*, *τ*, *τ.1*, *τ.2* **I**	*E*, *τ.1*, *τ.2* **SD**
↓	R—Resistance↑	R.1, R.2	R.1 at battery level
↔	R.1—Internal resistance↑	C, MT, *SoC(U)<A*, *T* **E**, **nC**, **SoC(U)>B**, **T (at exposure)**, **τ**, **τ.3**	SoC, SD, τ.1, τ.2, U *E*, *I*, *nC*, *P*, *τ*, **T.1**

Table 32. *Cont.*

Life Time↑	Parameter	(A) Impact on Parameter	(B) Effect of the Parameter
↓	R.2—Resistance of current-carrying parts↑	Battery design, operating modes	**R**
↔	SoC—State of charge LIC↑	Battery operating modes, nChg, nDchg, I, R.1, SD, τ.1, τ.2, τ.3	C, nC, P (U), R.1, *I*, *τ.1* E, SD, τ.2, U
↓	SD—Self-discharge↑	C, MT, *R.1* **nC**, **SoC**, **T**, **τ**, **τ.3**	nChg, *SoC (U)*, *τ*
T↔	LIB temperature (T)	T.2 **T.1**, **T.3**	C, E, I, nC, P, R.1, SoC, τ, τ.1, τ.2 **SD**, **U**
↓	T.1—Self-Heating↑	C, MT, *P*, **E**, **I**, **nC**, **R.1**, **τ**, **τ.1** (at low currents *T.1*↓), **τ.2**	See T
↔	T.2—Thermal management	C, MT, battery design	See T
↔	T.3—Ambient temperature		See T
↓	Calendar time (τ)		U, *C*, *E*, *I*, *P*, *SoC* **R.1**, **SD**
↔	τ.1—Charge duration↑	C, MT, *I*, *nC*, *P*, *T*, *SoC*, *τ* **E**, **R.1**	C, nC, R.1, *I*, *P (charging)* **E**, **P (U)**, **SoC**, **T.1** (at low currents *T.1*)
↔	τ.2—Charge duration↑	C, MT, R.1, *I*, *nC*, *P*, *T*, *τ* **E**, **SoC**	C, nC, *E*, *P*, *I*, *SoC* **T.1**, **P** (charging)

Table 32. *Cont.*

Life Time↑	Parameter	(A) Impact on Parameter	(B) Effect of the Parameter
↔	τ.3—Idle time↑	Battery operating modes	C **SD**
↔	Voltage (U)↑	C, MT, I, R.1, T, *SD*, *τ*, *τ.2* **τ.1**, **SoC**	C, I, nC, P, R.1, *τ.1*, **E**, **SD**, **SoC**, **τ.2**

Note: *—correlation, increase (**bold green**) (↑), *decrease* (*italicized blue*) (↓), ↔ increase or decrease. Source: Table by authors.

In turn, an increase in internal resistance (R.1) affects the change in the degree of charge. With the selected charge parameters, to ensure a high state of charge (SoC), the charging current (I) is reduced, and the charging time τ.1 is increased. The growth of the internal resistance (R.1) reduces the following parameters:

- Specific energy (stored and given out, which leads to a reduction in the charge time (τ.1) and discharge time (τ.2));
- Allowable charge/discharge current (I);
- Charge/discharge power (P);
- Calendar operation time (τ); a number of cycles (nC);
- Discharge voltage (U).

On the other hand, a higher internal resistance (R.1) increases the charge voltage (U) and self-heating (T.1) of the LICs and battery.

At the beginning of operation (low number of charge/discharge cycles—nC, short-term use—τ, and low internal resistance—R.1), LICs are capable of charging/discharging with currents (I) of higher strength than after long-term operation. When the charge/discharge time (τ.1/τ.2) is reduced, the allowable charge/discharge current (I) increases. As the LIC temperature (T) grows (not exceeding the specified acceptable temperature at a charge), the ionic conductivity increases, and the maximum allowable charge/discharge current (I) augments. However, the maximum charging current (I, hence, power P) is limited by the heat dissipation at the internal resistance (R.1), the deposition probability during the charging of lithium dendrites (graphite-based anodes), and the achievement of the allowable operating temperature of the LIC. The maximum discharge current (I and power P) is generally limited by the internal resistance (R.1) and the upper limit of the safe operating temperature range resulting from self-heating (T.1).

With an increase in the charging current (I), the charging time decreases (τ.1), and at full charge, the stored energy (E) diminishes since lithium diffuses into smaller anode depths and the given LIC voltage is reached with fewer lithium ions transferred from the cathode to the anode. Increasing the discharge current (I) decreases the specific energy (E), diminishes the value of the state of charge (SoC), and reduces the discharge time (τ.2). The specific power (P) during charge/discharge increases in proportion to the charge/discharge's current intensity (I) of the charge/discharge. As the charge/discharge current increases, the average charge/discharge voltage (U) increases/decreases and the temperature (T.1) of the LIC increases.

The expected number of charge/discharge cycles (nC) is influenced by LIC design (C) and manufacturing technology (MT). In the case of optimisation with an increase in specific energy (E), the number of cycles nC can decrease due to the optimisation of design (C) and manufacturing technology (MT) for the increase in the specific energy (a growth in internal resistance, a decline in electrolyte fraction). The optimisation of design (C) and manufacturing technology (MT) to achieve maximum power can also have a negative impact on the lifetime (nC, τ) due to the intensification of side reactions due to high-porosity electrodes, the developed material surface, etc.

The number of charge/discharge cycles (nC) can be increased by operating under the optimum conditions of temperature (T; close to room temperature, the battery design may include a thermal management system), lower charge/discharge currents (I), lower change in the state of charge (SoC, may also be affected by the values of the range of change in the state of charge during cycling [231], Figure 38c), a shorter residence time (τ) at high and low degrees of charge (SoC, U; the battery monitoring and control system can limit the voltage range during cycling), and a shorter calendar time (τ) of operation, as this also contributes to the ageing of LICs.

An increase in the number of conducted charge/discharge cycles (nC) leads to the decay of energy (E; at the beginning of cycling, there may be an increase in energy, apparently due to an increase in LIC [248] temperature) and the permitted charge/discharge current (I), power (P), charge/discharge duration (τ.1/τ.2), and average discharge voltage (U). The internal resistance (R.1), average charge voltage (U), and self-heating (T.1) increase during cycling.

The specific power (P) of the LIC on charge/discharge is higher in the case of a low state of charge (SoC, lower limit of voltage (U) interval)/high degree of charge (SoC, high voltage (U) values). For the effect of temperature on the specific power (P) of charge/discharge, see the impact of temperature (T) on the maximum current (I) of charge/discharge. At the beginning of operation (a small number of charge/discharge cycles nC, calendar time τ), the power (P) at charge/discharge is higher than after the operation time interval. As the charge/discharge duration (τ.1/τ.2) decreases, the charge/discharge current (I) can be increased, and the voltage rise/drop (U) is smaller; hence, the power (P) during the charge/discharge of LICs also increases. Improved (close to room temperature or higher) heat dissipation (through thermal management T.2 or LICs C design) (reduced self-heating (T.1) of LICs) allows charging/discharging with a higher current (I), thereby increasing power (P). At lower temperatures (T, T.3), poor heat dissipation contributes to the more significant heating of the LIC and, therefore, a greater increase in the internal temperature of the LICs (T) and, therefore, lower internal resistance and greater power (P).

The design of high-power LICs implies the use of electrodes with a thin layer of active mass of increased porosity; in this connection, the share of active materials in terms of the mass and volume of LIC decreases, and, consequently, the specific energy (E) and time (τ.1/τ.2) of the operation (charge/discharge) decrease. Due to lower internal resistance (because of the more developed surface of the active materials and layers), the self-discharge (SD) in high-power LICs is higher.

The state of charge (SoC) increases with an increasing/decreasing number of cycles and duration (τ.1/τ.2) of incomplete charge/discharge (nChg/nDchg, for example, observed with HEV batteries). At small states of charge (SoC), the capacity growth is proportional to the charging current (I) and the charging time (τ.1). Due to the increase in the internal resistance, a decrease in the charging current (I) and a prolongation of charging time (τ.2) are necessary to achieve a higher state of charge (SoC). The lower current (I) and discharge time (τ.2) allow less reduction of the LICs' (SoC) charge. Due to an increase in the internal resistance (R.1), the charge/discharge capacity (specific energy—E) decreases, which reduces the degree of charge/discharge with direct current. The state of charge (SoC) is higher with a shorter idle time (τ.3) and lower self-discharge (SD).

The range of variation in the state of charge (SoC) affects the number of charge/discharge cycles (nC) and the calendar time (τ) of the LICs' operation. The intensity of side processes at a low and high state of charge increases, and maintaining the LIC in these states can decrease the service life (nC, τ). The power (P) in the case of a discharge at low states of charge (SoC) is lower than at high degrees due to the lower voltage (U) and the higher internal resistance R.1. The power (P) (current, I) at charge is more elevated at a low state of charge (SoC) than at large SoCs due to the lower internal resistance (R.1). The resistance (R.1) in the region of the state of charge close to full discharge and full charge is higher than in the average range of the state of charge. The higher the state of charge (SoC), the shorter the time to achieve full charge (τ.1), but the charge at a high state of charge (SoC) is slower than at the minimum and medium state of charge for the LIC. The voltage (U), specific energy (E), and discharge time (τ.2) increase with increasing state of charge (SoC).

Self-discharge (SD) at the beginning of the operation is higher for LICs with lower internal resistance (R.1). Self-discharge increases with an increasing state of charge (SoC), LIC temperature (T), number of charge/discharge cycles (nC), operation time (τ), and idle time (τ.3) of the LICs.

At the beginning of the operation, the self-discharge (SD) will be less in LICs with higher internal resistance in close design. However, at the end of life, the self-discharge (SD) may be higher due to the lower total energy of the LICs.

As the self-discharge value (SD) (and the idle time—τ.3) increases, the degree of charge (SoC) (voltage—U) decreases. Therefore, the loss of energy (capacity) can lead to the need to charge the LIC more frequently (nChg) and, consequently, to a reduction in the operating time (τ).

The temperature of LIBs (T) depends on the ambient temperature (T.3), on self-heating (T.1) or self-cooling (cells can be cooled at low charging currents), and on the effect of the thermal management system (T.2).

The operation of an LIC at temperatures close to the allowed operating temperature limits or at temperatures outside the operating temperature range may lead to the destruction of the materials and construction of the LIC and, consequently, the reduction in its operation life (number of charge/discharge cycles (nC), time (τ) of operation). As the LIC temperature rises from the lower limit of the operating interval to the upper limit, there is an increase in the following:

- Specific energy (E);
- Charge/discharge current (I);
- Average discharge voltage (U);
- Power (P);
- Self-discharge (SD) and discharge time (τ.2, at a fixed discharge current (I)).

At the same time, the decline in internal resistance (R.1), average charge voltage (U), and charge time (τ.2, at a fixed charge current (I) to a given state of charge (SoC)) is observed.

When approaching and exceeding the upper limit of the operating temperature range, the specific energy (E), allowable charge/discharge current (I), and power (P) may decrease due to potentially exceeding the permissible temperature due to the heating (T.1) of the LIC. As a result of holding the LIC at elevated temperatures due to the possible re-precipitation of the solid electrolyte film (growth of thickness, irregularity), the internal resistance (R.1) may increase. As the temperature of LICs increases (T), the yielded capacity may be higher, which is reflected in the state of charge (SoC) of the LICs relative to the state of charge (SoC) at a lower temperature.

The temperature of LIBs due to self-heating (T.1) augments with rise in the following parameters:

- Internal resistance (R.1);
- Charge/discharge current (I);
- Charge/discharge time (τ.1/τ.2) at a fixed current (I) of charge/discharge.

With the deterioration in the state of health (SOH) with the growth of conducted charge/discharge cycles (nC) and operation time (τ), due to an increase in internal resistance, a reduction in power (P) and increase in the self-heating (T.1) of LICs is observed.

With increasing operating time (τ), the destruction of LICs' structure (C) and a decrease in energy (E), permissible charge/discharge current (I), average discharge voltage (U), power (P), and state of charge (SoC) occur. Moreover, a rise in internal resistance (R.1), self-discharge (SD), and average charging voltage (U) is observed.

The duration of charge (τ.1) at fixed conditions is longer at the beginning of operation (nC, τ) since the LIC has more energy (E). This duration increases with a decrease in the charging current (I), power supply (P), temperature (T), and state of charge (SoC) of LICs and an increase in internal resistance (R.1). If the internal resistance (R.1) is high, the charging time can be shortened by reaching the set cutoff voltage before the LIC's capacity is reached, but in this case, the stored energy will also be shorter.

Prolonged charging (τ.1) with a low current (I) can increase the number of charge/discharge cycles (nC) and the operating time (τ), whereas reducing the charge time by increasing the current (I) reduces the state of health and life of the LICs (nC, τ). Under constant charge (float charge) conditions, the internal resistance (R.1) of LICs can rise due to the destruction of cathode materials and the growth of solid electrolyte film. As the charging time interval (τ.1) lengthens, the allowable charging current (I) and power (P) decrease. During prolonged charging with small currents (I), the value of stored energy (E) increases and tends towards the maximum for the LICs under consideration. The state of charge (SoC), voltage (U), and power (P) increase with the increasing time of charge (τ.1) at fixed values of current (I). The temperature (T, T.1) of LIBs can decrease when charging with low currents (or descending current). Due to the presence of internal resistance (R.1) in LIBs at a fixed current value (I, above the minimum), the temperature (T) of the LIBs rises.

The duration of discharge (τ.2) is longer at the beginning of LICs operation (nC, τ) because the cell has more energy (E). It augments with the decrease in the discharge current (I), power (P), temperature (T) (unless the cutoff voltage (U) is reached earlier in case of relatively high discharge current (I) and internal resistance (R.1)), and higher LIC's state of charge (SoC). If the internal resistance (R.1) is high, the discharge time can be shortened by reaching the set cutoff voltage before the LIC's capacity is discharged.

Prolonged discharge (τ.2) with a low current (I) can increase the number of charge/discharge cycles (nC) and operating time (τ), while reducing the discharge time by increasing the current (I) reduces state of health and the life of the LICs (nC, τ). With the lengthening of the discharge time interval (τ.2), the strength of the allowable constant current (I) discharge charge and power (P) diminishes. At long discharges with low currents (I), the value of energy output (E) increases and tends towards the maximum for the LICs under consideration. With the increase in the discharge time (τ.2) at fixed values of current (I) and state of charge (SoC), the voltage (U) decreases. Due to internal resistance (R.1), the temperature (T, T.1) of the LICs rises as the discharge time increases.

The average charging/discharging voltage (U) grows/drops with increasing internal resistance (R1), decreasing the LIC's temperature (T). The LIC's voltage is higher with lower self-discharge (SD), idle time (τ), and discharge duration (τ.2) and a higher charge time (τ.1) and state of charge (SoC).

In LICs (with the same electrochemical system), a higher voltage (U) indicates a higher state of charge (SoC), the possibility of self-discharge (SD), duration of discharge (τ.2), specific energy (E), and power (P), and the possibility to discharge with a higher current (I). On the other hand, the charging current (I), the charging power (P), and the time to full charge (τ.1) decrease because the internal resistance (R.1) in the areas near the limits is higher than in the middle of the operating voltage interval (U). To increase the number of cycles (nC) and the operating time (τ), the charge/discharge must be carried out in the optimal voltage range (U, SoC). However, the specific energy is reduced compared to the energy stored and given during the full charge/discharge of LICs.

Almost all LIC components, including the housing and current leads, are subject to ageing during operation [249,250]. A description of the influence of the main factors (high and low temperatures, overcharge, deep discharge, etc.) on the electrochemical system of LICs is provided in Table 33.

5.3.3. Electrochemical LIC Unit

The active layer of positive LIC electrodes contains a conductive additive, a binder, and active cathode material. The lithium-ion cells used for various HEVs, PHEVs, and BEVs are listed in Table 34. According to the presented information, the mass fraction of conductive additive in the active layer composition of the positive electrode is higher than 8 wt.%; the binder is about 3 wt.%. Active cathode materials are selected from the LMO, NCM, LNO, and NCA list. Each of the materials mentioned is found in the manufacture of LICs used in HEV, PHEV and BEV batteries. The materials may have the same chemical composition, but differ in porosity, tap density, and other parameters.

The active layer of the negative electrode typically includes an active anode material and a binder. Exceptions such as those outlined below are possible:

- The introduction of conductive graphite into graphite-based electrodes (increases the density and mass fraction of the material into which lithium intercalates).
- The active layer of cells with lithium titanate used as the anode material may include significant amounts of conductive additives (up to 15 wt.% of carbon black to increase electronic conductivity).

The mass fraction of the binder in negative electrodes varies from 3 ÷ 5 wt.%. Both graphite and carbon materials with a structure different from graphite—hard carbon, soft carbon—are used as the active anode material for HEV LICs. The negative electrodes of the less powerful PHEV and BEV LICs mainly contain different types of graphite (artificial, natural with conductive coating).

Table 33. Li-ion battery ageing mechanisms (based on data presented elsewhere [249–256]).

Cause	Effect	Result	Weakening	Reinforcing
		Anode		
Li^+ intercalation/deintercalation, temperature fluctuation, vibration	Changes within the graphite structure	CD, PR	Temperature control	LT, HT, vibration
Change in surface layer structure due to intercalation of solvents into graphite structure	Loss of active material (graphite delamination), loss of lithium	CD	Electrolyte additives, SEI film stabilisers, carbon coatings	OChg
SEI film growth	Increase in resistance, overpotential, a decrease in the available active material surface, loss of lithium	CD, PR	Introduction of additives into the electrolyte that stabilises the SEI layer	HR chg/dchg, HT, HSoC
SEI film degradation	Resistance growth	PR	Introduction of additives into the electrolyte that stabilises the SEI layer	HT
Loss of contact with the active material due to volume changes during cycling	Loss of active material	CD	External pressure	HR chg/dchg, LDD
Binder dissolution	Loss of lithium, loss of mechanical strength	CD	Correct binder selection	HSoC, HT
Changes in porosity due to volume changes during intercalation/de-intercalation	Increased resistance	PR	External pressure	HR chg/dchg, HSoC

Table 33. *Cont.*

Cause	Effect	Result	Weakening	Reinforcing
Deposition of lithium metal and subsequent electrolyte decomposition	Lithium loss, electrolyte loss, increase in resistance	CD, PR, thermal runaway due to dendrite formation	Narrowing the operating voltage range, the anode area is larger than the cathode area The use of hard carbon, soft carbon, $Li_4Ti_5O_{12}$ Ceramic coating at the anode/separator interface and thick separator can reduce the effect of dendrites	HR chg/dchg, LT not balanced cell, defects at jointing, contamination on the anode surface
Gassing when using $Li_4Ti_5O_{12}$ as an anode	Ti^{4+} ions can accelerate electrolyte decomposition	swelling of cells due to gassing, decomposition of electrolyte solvents	Stabilising additives in electrolyte, carbon coating of anode material	–
Corrosion of the Cu current collector	Overvoltage, uneven current and potential distribution, loss of contact	RI, overvoltage growth, PR, CD (active material exclusion), dendrite growth (micro-short circuits)	–	LDD
		Cathode		
Li^+ intercalation/deintercalation *	Distortion of structure	CD, PR	Combining materials	Materials without stabilising impurities

Table 33. *Cont.*

Cause	Effect	Result	Weakening	Reinforcing
Li^+ intercalation/deintercalation *, electrode calendering	Particles cracking in the electrode-active layer	CD, RI	Lower calendering pressure, increased particle strength (for NCM, increased crystallite size, etc. [203])	Lower particle strength
Phase transitions	Distortion of crystal structure Subsequent mechanical stresses	CD	Control of the charge cutoff voltage	OChg
Dissolution and transition of metals into the electrolyte	Re-precipitation of new phases, loss of active material, formation of surface layer	CD, RI	Temperature control	HT
Binder dissolution	Exclusion of active material, reduced mechanical stability	CD	Selection of a better binder	HSoC, HT
Corrosion of Al current collector in the presence of $LiPF_6$	Overvoltage, uneven current distribution, potential, loss of contact	RI, PR, CD (active material exclusion) Reinforces other ageing mechanisms,	Current collector machining	ODchg, LSoC
Morphological changes in electrodes	Change in surface porosity	PR	–	–

Table 33. *Cont.*

Cause	Effect	Result	Weakening	Reinforcing
		Separator		
Separator seal	Changing the porous structure of the separator	–	–	–
		Electrolyte		
Electrolyte decomposition (cathode—oxidation; anode—reduction)	Emission of gases, surface layer formation	RI	Use of alternative conductive salts and solvents	Protic impurities, HT, storage at high potentials
Decomposition of electrolyte with the formation of SEI film	Lithium loss, increased resistance	CD, PR	Additives stabilising SEI layer	HT, HSoC
Hydrolysis of salt by interaction with H_2O	Interaction of LICs components with HF	Reduced capacity for work	Cell operation atmosphere control	Increased humidity

Note: SEI—solid electrolyte interface; solid film—electrolyte; LDD—large depth of discharge; HSoC—high state of charge; HR chg/dchg—high charge/discharge rate; HT—high temperature; LSoC—low state of charge; LT—low temperature; OChg—overcharge; ODchg—over-discharge; CD—decrease in capacity; PR—power reduction; RI—resistance increase. * Information about the changes taking place in the structure of cathode materials of various types can be found in [255]. Source: Table by authors.

When comparing the compositions of positive electrodes of LICs for transport and for portable equipment, it can be seen that the latter have a significantly higher proportion of the active material (about 96 wt.% in the active layer). The percentages of active anode material and binder in the negative electrodes of transport and portable LICs are close. However, the negative electrodes of LICs for portable equipment may include more artificial graphites (or graphite-SiO_x composites) with higher capacity.

The capacity of LICs is formed by the mass and specific capacity of the active cathode material. The mass of the active anode material is chosen so that the capacity of the negative electrode is 20 ÷ 25% (or more) higher than the capacity of the positive electrode. Because the discharged capacity of the cathode material is much smaller than the capacity of the anode material, the former's mass fraction exceeds the latter's mass fraction (Figure 50).

For high-power LICs, the mass fraction of electrolyte in the weight of active anode material, active cathode material, and electrolyte is about 30% (Figure 50). For LICs with an anode based on lithium titanate, the mass fraction of electrolyte reaches 40%. In the less powerful cells, the mass fraction of electrolyte varies from 20 to 30 wt.%. For high-specific-energy LICs, the electrolyte fraction is reduced to 14 ÷ 15 wt.% and the anode and cathode material fractions are increased.

Electrolytes for the LICs of hybrid and electric vehicles, due to requirements for power and operability at low temperatures, differ in composition from the LIC electrolytes used in portable equipment [257]. The application of lower-viscosity electrolytes with an increased content of dimethyl carbonate, ethyl methyl carbonate, and esters (e.g., ethyl acetate, methyl acetate etc.) and the addition of lithium bis(fluorosulfonyl)imide (LiFSI) and lithium difluorophosphate ($LiPO_2F_2$) can provide a higher discharge current (and improve the specific power). To enhance the stability of electrolytes at elevated voltages (and increase the average discharge voltage), additives from the series 3,3,3-trifluoropropylmethyl sulfone (3,3,3-trifluoropropylmethyl sulfone), tris (trimethylsilyl) phosphite (TMSB (Tris (trimethylsilyl) phosphite)), salt difluoro (bisoxalate) lithium phosphate (lithium difluoro (bisoxalato) phosphate), etc., can be applied.

The lowest share (Figure 51) of active anode and cathode materials in the battery mass is typical for the high-power LICs used to manufacture batteries for HEVs. On the other hand, the mass fraction of active LIC materials for PHEVs and BEVs is higher and varies in the range of 37 ÷ 52%. The highest proportion of active materials is characteristic of LICs used for portable power equipment.

In addition to the higher proportion of LIC active materials, the use of positive and negative electrodes with a higher density and thickness of active layers is characteristic for portable electronics (Table 35). Furthermore, a thinner separator is applied in LICs for mobile applications to ensure a higher specific energy density.

Table 34. Comparison of LIC compositions for transport applications and portable equipment.

Type of LICs	Positive Electrode					Negative Electrode				CM + El + AM	Ref.
	CA %	Binder %	CM %	Cathode	Ratio % of Capacity	Binder %	AM %	AM	Ratio % of Capacity	g	
HEV1 (L)	9.9	3.0	87.1	LMO/LNO	80/20	3	97	HC	100	20.2	
HEV2 (L)	8.0	3.0	89.1	LMO/NCM	70/30	3	97	SC	100	19.9	
HEV3 (P)	8.1	3.0	88.9	NCM 111	100	3	97	Gr	100	19.5	
HEV4 (P)	10.1	3.0	86.9	NCM 111	100	5.1	94.9	MAGX	100	18.8	[258]
HEV5 (P)	10.0	3.0	87.0	NCM	100	3	97	HC	100	22.6	
HEV6 (P)	10.1	4.9	85.0	LNO	100	5.1	94.9	Gr	100	18.8	
PHEV1 (L)	8.0	3.0	89.0	LMO/NCM	70/30	3	97	HC/Gr	10//90	56.3	
PHEV2 (L)	6.0	4.0	90.0	LMO/NCM	40/60	4	96	HC/Gr	15/85	57.7	[193]
PHEV3 (P)	10.0	3.0	87.0	NCM 523	100	5	95	SC/MAGX	50/50	74.5	[258]
BEV1(L)	10.0	3.0	87.0	LMO/LNO	80/20	3	97	NGC	100	120.4	
BEV2 (L)	5.0	6.0	89.0	LMO/LNO	70/30	5	95	Gr 3 types	100	120	[71]
BEV3 (P)	8.0	3.0	89.0	LMO/NCM	50/50	3	97	Gr	100	157.5	
BEV4 (P)	8.0	3.0	89.0	LMO/NCM	50/50	5	95	AG/NGC	50/50	184.7	[258]
BEV5 (P)	8.0	3.1	89.0	LMO	100	5	80[1]	LTO	100	133.3	
BEV6 (P)	4.2	3.9	91.9	LMO/NCA/NCM	53/21/26	4.4	95.6	Gr	100	244.4	[259,260]
3C1 (18650)	2.0	2.0	96.0	LCO	100	3	97	MAGE3	100	10.06	[193]
3C2 (L)	1.8	1.9	96.3	LCO	100	1.6	98.4	AG	100	17.5	[259]
3C3 (18650)	2.3	0.9	96.8	NCA	100	2.6	97.4	MAGE3	100	11.64	[261]

Note: [1]—the negative electrode contains 15 wt.% carbon; 3C—batteries for portable electronics; AG—artificial graphite; Gr—graphite; HC—hard carbon; LCO—$LiCoO_2$; LMO—$LiMn_2O_4$; LNO—$LiNiO_2$ or $LiNi_{0.8}Co_{0.15}Al_{0.05}O_2$; LTO—$Li_4Ti_5O_{12}$; MAGE3—artificial graphite; MAGX—artificial graphite; NCA—$LiNi_{0.8}Co_{0.15}Al_{0.05}O_2$; NCM—$Li_{1+x}Ni_aCo_bMn_cO_2$; NGC—natural graphite coated with carbon; SC—soft carbon; AM—anode-active material; CM—cathode-active material; CM+El+AM—the sum of the masses of the active cathode material, electrolyte and active anode material; CA—conductive additive; (Lp)—laminated foil case; (Pr)—prismatic body; 18650—a cylinder with a diameter of 18 mm and a height of 65 mm. Source: Authors' compilation based on data from references cited in the table.

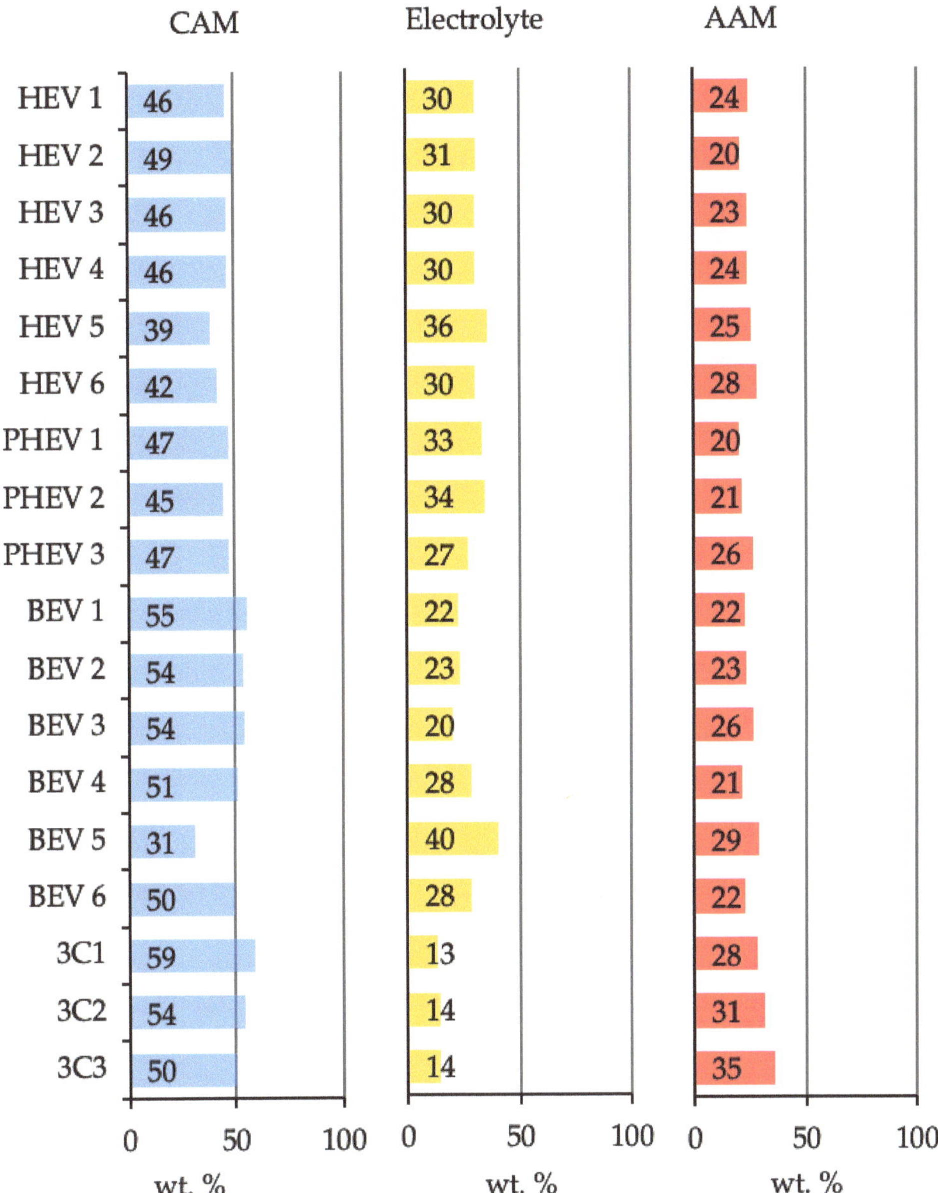

Figure 50. Distribution by mass of the cathode-active material (CAM), electrolyte, anode-active material (AAM), and sum of component weights is 100%. Histograms are based on data presented in references [71,193,257–259]). Source: Figure by authors.

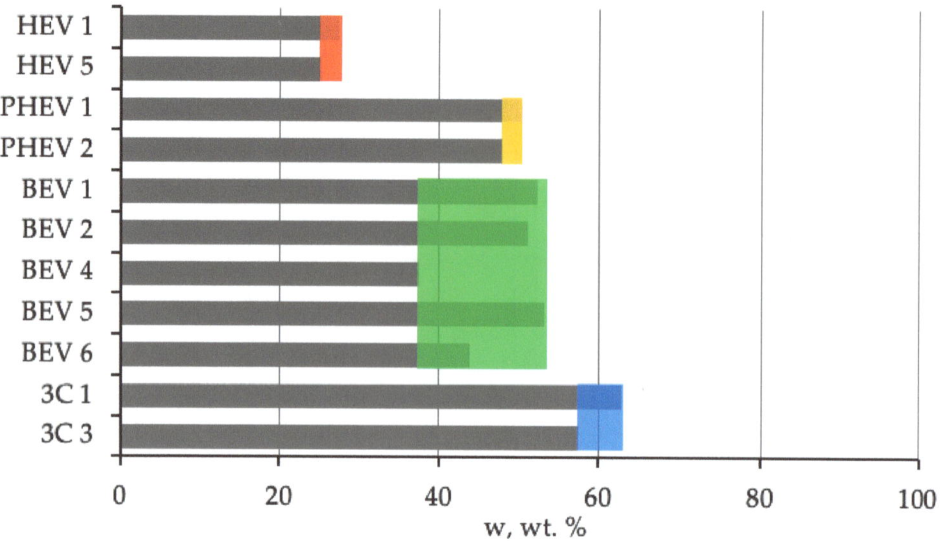

Figure 51. Proportion of active cathode and anode materials in the mass of batteries (the initial data were taken from references [71,193,257–259]). Colors mark range of proportions for HEV cells (red), PHEV cells (yellow), BEV cells (green), 3C (blue). Source: Figure by authors.

On the contrary to 3C cells, a lower density of positive- and negative-electrode active layers is used, as their higher porosity allows the higher power of BEV cells to be achieved, which is required when using LICs as a power source for electric motor vehicles. In turn, PHEV and, in particular, HEV cells have higher specific power and lower specific energy compared to BEV batteries. The positive- and negative-electrode-active layers of PHEV and HEV cells are even thinner and have higher porosity. Table 35 presents, for comparison, a description of a PHEV battery characterised by specific energy and power more closely resembling HEV batteries.

5.3.4. Technical and Design Parameters Affecting LICs' Power

The main parameters that determine the power density of a lithium-ion battery include the mass (M), the internal resistance (R), the voltage (U), and the heat dissipation (H). Ways to increase or decrease these parameters due to a LIC's design are given in Table 36.

Case. Peak current is limited by cutoff temperature. The heat dissipation depends, among other things, on the shape of the case. The external surface area of prismatic batteries is larger than that of cylindrical cells. Reducing the depth (diameter) of the case also increases the specific surface area and, hence, the heat dissipation. On the other hand, if the heat-dissipating plate only comes into contact with the cell base, the heat will be better dissipated from cells with a larger base area.

Table 35. Specific energy, specific power, and electrode and separator characteristics of LICs for electric vehicles and portable appliances.

Parameter		Unit	PHEV (Pr)	BEV 2 (Lp)	BEV 6 (Pr)	3C4 18650	3C5 (Lp)
			Cell characteristics				
Specific energy		Wh/kg	111	120 ÷ 160		**255**	–
Energy density		Wh/L	223	230 ÷ 280		**740**	**640**
Specific power		kW/kg	–	2.4 ÷ 2.8		≈0.77	–
Power density		kW/L	–	4.4 ÷ 4.9		≈2.2	–
U		V	4.1 ÷ 2.5	4.2 ÷ 2.8	4.1 ÷ 3.0	**4.2 ÷ 2.65**	**4.35 ÷ 2.75**
			Positive electrode				
Material			NCM	LMO/NCA	LMO/NCM/NCA	**NCA**	**LCO/NCM**
Specific capacity		mAh/g	≈150	103.7	108.7	**≈180**	**≈180**
Application density (both sides)		mg/cm^2	24.6	48.5	32.6	52.51	42.79
Electrode thickness		μm	115	197	150	160	122
Foil		μm	15	20	13	17	15
Active Layer	w	μm	100	177	137	143	107
	ρ	g/cm^3	–	2.74	2.38	**3.67**	**4**
	P	%	52	29.8	31	–	–
			Negative electrode				
Material			NGC	Gr	Gr	**Gr + SiO$_x$**	**Gr + SiO$_x$**
Application density		mg/cm^2	13	18.7	14.5	**29.61**	**25.24**
Electrode thickness		μm	110	149	129	186	162
Foil		μm	10	11	9	12	8
Active layer	w	μm	50	138	115	**174**	**154**
	ρ	g/cm^3	–	1.36	1.2	**1.7**	**1.64**
	P	%	52	37.7	33	–	–

Table 35. *Cont.*

Parameter	Unit	**PHEV (Pr)**	**BEV 2 (Lp)**	**BEV 6 (Pr)**	**3C4 18650**	**3C5 (Lp)**
			Separator			
Thickness	μm	32	20÷45		**12**	**12**
Reference		[262]	[260]		[259]	

Note: Gr—graphite; Gr + SiO_x—composite of graphite and silicon oxide (<5%); LMO—$LiMn_2O_4$; Lp—Al-laminated foil case; NCA—$LiNi_aCo_bAl_cO_2$ (a + b + c = 1); NCM—$LiNi_aCo_bMn_dO_2$ (a + b + d = 1); NGC—natural graphite core; Pr—prismatic aluminium case; P—porosity; U—voltage; w—thickness; ρ—density. Parameters contributing to higher **energy (values in bold)** and power (underlined values). The PHEV battery chosen for comparison is closer to an HEV than a BEV LIC. LIC HEVs are characterised by even lower deposition density and electrode thickness (5.0 Ah 1st generation (Prius Alpha) cathode Al foil ≈ 15 μm, electrode ≈ 75 μm (LNO), 3.6 Ah Prius 2nd generation (Prius gen 4), cathode: Al foil—15 μm, electrode—65 μm (hollow secondary particles ≈ 4 μm NCM [99])). "–" means no data available. Source: Authors' compilation based on data from references cited in the table.

Table 36. LIC parameters affecting power.

	Method	Description	M	R	U	H
Case	Case shape (case depth, diameter)	**Heat dissipation**	*			⇧
	Enclosure material	**Heat dissipation**: plastic < laminated foil < steel < aluminium *Case weight*: steel > aluminium > plastic > laminated foil	*			⇧
	Case wall thickness	*Case strength*	⇩			⇧
	Location and cross-sectional area of the current lead (electronic conductivity)	For prismatic batteries with roll construction, the weld seam of the current collector and the current collector is perpendicular to the length of the case. For LICs in laminated foil cases, the current leads are located on opposite sides of the case and occupy a significant portion of the case sides and are of sufficient thickness (area). For cylindrical batteries, a connection between the current collector and external current lead may be effectuated through the disc plate (EAS batteries, Tesla 4680)	⇧	⇩		⇧
	Welding quality of the current collector and current lead			⇩		
	Use of two pressed-out laminated foil containers (compared to one)	*Less bending of current leads means* better **quality of welded seam current collector—lead**		⇩		

Table 36. *Cont.*

	Method	Description	M	R	U	H
Case	*Share of unused space (several electrode rolls instead of one in a deep prismatic housing)*	**increasing complexity of manufacturing technology**	⇩	⇩		
	Not using additional structural parts that provide increased safety (puncture protection plates, tapes insulating part of the electrodes, etc.)	*Safety in case of possible emergency triggers (puncture, short circuit, crushing, etc.)*	⇩			
Separator	*Separator thickness*	**Difficulty of fabrication** due to reduced *separator strength*, *safety* in case of possible abnormal triggers	⇩			
	Gas permeability of the separator (Gurley)	**Challenging to fabricate** due to reduced *separator strength*		⇩		
	Separator porosity	**Difficult to fabricate** due to reduced *separator strength*	⇩	⇩		
	The functional coating on the separator to increase adhesion between separator and electrodes (PVDF, PVDF-HFP)	**Price**, *internal resistance* reduction	⇧	⇩		
	The functional coating on the separator to increase the wettability of the separator with electrolyte	**Price**, ceramic coating may reduce *ionic conductivity*	⇧	⇩ ⇧		

Table 36. *Cont.*

	Method	Description	M	R	U	H
Electrolyte	**Electrolyte fraction** in the battery volume (mass)	**Wettability of electrodes, flammability**, *reduction in specific energy*	⇧	⇩		
	Addition of functional additives and use of low-viscosity solvents	**Wettability of electrodes**, **price**, need to ensure sealed case or subsequent polymerisation of electrolyte		⇩		
	Electrolyte ionic conductivity (variation in solvent ratio, an increase in $LiPF_6$ concentration, the introduction of LiFSI, $LiPO_2F_2$ salts, etc.)	**Cost**, passivation of aluminium foil when $LiPF_6$ concentration decreases due to the introduction of other lithium-containing salts		⇩		
	Functional additives in electrolyte that reduce electrode impedance (for cathode—BP, for anode—FEC, PS), etc.	**Cost**, provide SEI film growth with improved lithium-ion conductivity. Allow higher discharge and charge currents		⇩		
	Functional additives in the electrolyte for passivation of the cathode surface	**LICs voltage increase**, **price**			⇧	
Electrodes	**Thickness of current collectors**	Selection of optimum current collector thickness. Less thickness, less mass, but may reduce electronic conductivity	⇧	⇩		
	Degree of current collector purity	**Electronic conductivity** of current collectors (Cu, Al) grows with an increase in purity. Cu foils can contain additives that increase strength but reduce electrical conductivity		⇩		

Table 36. *Cont.*

	Method	Description	M	R	U	H
Electrodes	**Number and/or area (width) of leads from current collectors**	As the surface area of the current collectors increases, in addition to the **electronic conductivity**, the **mass** also increases, an optimum ratio is required	⇧	⇩		
	Conductive layer between the current collector and active mass layer	Increased electrical **contact area** between the active layer and the current collector, increase in the **mass**	⇧	⇩		
	Decrease in *the active mass thickness on the current collector*	*Path length of ions and electrons decreases*, but the share of active materials in the mass and volume of the cell decreases => lower *specific energy*	⇩	⇩		
	Density of the applied active mass (**porosity**).	Set by the composition of the active mass, the slurry deposition process parameters, and the *degree of calendaring* and it changes during drying, electrolyte filling, and formation. The lower the *degree of compaction*, the better the wettability by the electrolyte, but the lower the *electronic conductivity* since part of the active mass of the electrode is not connected to the network of conductive additive particles	⇩	⇩ ⇧		
	Gradient of porosity of active electrode layer	Density increases closer to the current collector (special machines for applying the active mass, double slot die)	*	⇩		

Table 36. *Cont.*

	Method	Description	M	R	U	H
Electrodes	Uniform or defined distribution of conductive additives in the depth of the active layer	**Dispersing process** becomes more complex (increasing **time** and requiring careful selection of parameters for preparation and transfer of the active mass), and slurry deposition (viscosity, speed)		⇩		
	Application of conductive graphite and/or carbon fibres	For relatively thick active layers, there is an increase in **electronic conductivity along the depth of the active mass of the electrode**		⇩		
	Mass and volume fraction of conductive additives	Conductive additives affect not only electrical and thermal conductivity, but also the porosity of the active electrode layer	*	⇩		⇧
	Mass and volume fraction of binder (and thickener)	Binder is generally dielectric, less *binder (thickener)*, higher **electronic conductivity**	⇩	⇩		
Active materials	*Bulk density of active materials*	*Density of the applied active mass* (**porosity**): see **Electrodes**	*	⇩		
	Particle (secondary) **porosity**	**Large contact area of the active material with the electrolyte**, but may increase **the internal resistance** due to the solid electrolyte film formation	*	⇩ ⇧		

Table 36. *Cont.*

	Method	Description	M	R	U	H
Active materials	*Particle size (secondary and primary)*	Shorter *diffusion path length of lithium ions*, *specific energy*, **resistance contribution of the solid electrolyte film**		⇩ ⇧		
	Specific surface	**Large contact area of the active material with the electrolyte**, but may increase the internal **resistance** due to the formation of a solid electrolyte film		⇩ ⇧		
	Predominant orientation of crystallites	Increase in **the fraction of the area of planes through which lithium is introduced/removed to/from the active material structure**		⇩		
	Narrow particle size distribution	Electrode with higher **porosity**	⇩	⇩		
	Presence of carbon coating on the particles	For cathode materials—an increase in **electronic conductivity**. For anode materials—an increase in **ionic conductivity** by the introduction of lithium ions (charge)		⇩		

Table 36. *Cont.*

	Method	Description	M	R	U	H
	Degree of crystallinity of conductive additives	Provide improved **electrical** and **thermal conductivity**		⇩		⇧
	Branching of the structure of conductive additives (↑SBET,↑OAN)	A smaller *mass of conductive additives* is required to provide the required electronic conductivity of the active layer. The branched structure of the conductive additives allows more **electrolyte accumulation** and increases the active material's **electrolyte wettability**	⇩	⇩		
Other active layer materials		The quantity is limited by the viscosity of the slurry to be applied, which is acceptable for the selected application parameters				
	↑Particle size of carbon blacks (mass fraction↑)	Larger particle size allows increase in **mass fraction of conductive additive** in slurry composition until a given paste viscosity is reached; consequently, **electronic conductivity** of active layer can be increased	⇧	⇩		
	Use of a binder with a high molecular weight (strength required for peeling, at equal content)	The higher the molecular weight, the smaller the quantity required to make the electrode. But as the molecular weight increases, the electronic conductivity of the active layer decreases	⇩	⇩ ⇧		

Table 36. *Cont.*

	Method	Description	M	R	U	H
Other active layer materials	**Binder swelling**	When pouring the electrolyte, the binder can swell, and when the distance between the particles increases, the *electronic conductivity* can decrease. On the other hand, **ionic conductivity** may increase due to electrolyte accumulation by the binder		↓ ↑		
Positive-electrode -active layer materials	Introducing dopants and coating to extend the stress operating range of active materials	Dopants and coatings for NCM and LCO. An increase in mole fraction of Mn in materials of $LiMn_aFe_bPO_4$ composition can increase voltage, but decreases the diffusion coefficient and electronic conductivity		↑	↑	
	Inclusion of materials capable of providing power in the active layer composition, e.g., $LiMn_2O_4$			↓	↑	
(+) electrode-active layer	Application of conductive additives and binders **stable at high potentials**	Binder and conductive additives may oxidise at high voltages, decreasing electrode-active layer performance			↑	

Table 36. *Cont.*

	Method	Description	M	R	U	H
	Active anode materials with low average charge/discharge potential relative to lithium	Risk of lithium deposition at high charge currents or reduced temperatures			↑	
	Graphites with a smaller C004/C110 plane ratio	A high proportion of the **surface of crystallographic planes to/from which lithium intercalates/deintercalates**		↓		
	Active anode materials with relatively high average charge/discharge potential relative to lithium (LTO, HC, SC)	Energy and power are diminished by reducing the potential difference; it may be necessary to use a non-aqueous binder. The **charging current** can be increased	↑	↓	↓	
Materials of the active layer of the negative electrode	Non-graphitisable (HC) carbon materials, graphitisable (SC) carbon materials, carbon-coated graphites or their mixtures with graphite	It may be necessary to use non-aqueous solvents to dilute anode materials, including HC and SC Increase in **electrolyte wettability of negative electrode surface**	↑	↓	↓	
	Use of aqueous binders	For graphite-based electrodes using aqueous binder (SBR) and thickener, the **electronic conductivity** is higher than that using PVDF (at the same mass fraction or peel strength)	↓	↓		
	Use of aqueous binders with a high glass transition temperature	**Porosity of the active layer** is higher due to the more rigid structure of the active electrode layer		↓		

Note: **Increase** (**bold**, green), *decrease* (*italicized*, blue), ↓ ↑—increase power, ↓ ↑—decrease power, *—affects, M—mass, R—resistance, U—voltage, H—heat dissipation, OAN—oil absorption number, S_{BET}—specific surface area. Source: Table by authors.

Shallow depth is typical for lithium-ion batteries in cases made of laminated foil (the heating of the LIC case is not uniform during discharge; the highest temperature is in the area proximate to the positive current lead [263]). The increase in temperature due to the excellent heat removal from LICs in cases made of laminated foil is less than for cylindrical cases [264] (in the central part of the cylinder, the temperature is higher than in the areas where electrodes are located close to the walls of the case). On the other hand, for LICs in a laminated foil case, the viscosity of the electrolyte must be increased, and the quantity should be reduced to prevent its escape through the current lead sealing unit; thus, the allowable cell temperature must be lower. Furthermore, gel-like electrolytes have less ion conductivity.

Heat dissipation also depends on the thermal conductivity of the material and the thickness of the case walls. The highest thermal conductivity is typical for aluminium (>200 W/m K), steel ($\approx$50 W/m K), laminated foil (thickness 88 $\div$ 153 μm, aluminium 40 μm, the rest consisting of polymers), and plastic. The thickness of the case grows with the capacity of the cell. For example, the wall thickness of 18650 cases made of steel is 0.15 $\div$ 0.2 mm [265], the wall of prismatic case 36135235 is made of 0.8 mm aluminum, and of small cases made of laminated foil is 0.088 mm for small cells (0.067 mm sometimes is mentioned) and 0.153 mm for large batteries used in electric vehicles.

Increasing the specific power is also possible at the expense of reducing the mass of each of the LIC components (in the absence of a decrease in functional parameters) and case elements, in particular. For example, making a case from a material of lower density and thickness, not including in the structure of the case components, increases safety in emergencies (for example, additional plates for strengthening the case).

The design of prismatic [266] (steel or aluminium) elliptical and cylindrical housings provides [267] the possibility of [268] installing current leads with a large cross-sectional area, which makes it possible to reduce the electrical resistance of LICs. Furthermore, in LICs with a metal casing, the volume fraction of the electrolyte can be increased compared to LICs in laminated foil casing, which allows the wettability of electrodes to be increased and the internal resistance to be decreased [234].

A quality weld of sufficient area is required to reduce electrical resistance and increase heat dissipation. In aluminium prismatic batteries for hybrid and electric vehicles, the electrode roll (or rolls—increasing the proportion of the case volume used) is placed in the case so that the axis around which the coil is twisted is parallel to the long side of the cell [266]. The positive and negative current collectors (parts of the aluminium and copper foil free of active layer) are located on different bases of the roll. On the base side, the positive current collector is welded to a positive lead, and the negative current collector is welded to a negative lead [266]. By spacing out the welds on opposite sides of the battery, the contact area between the current collector and the current lead increases. For cylindrical cells, the current collectors are led out from opposite bases of the housing cylinder. To increase the contact area, discs and an appropriate current lead are welded on each side [268,269].

Current collectors in an LIC pouch cell can be positioned on one or different sides of the case. The location of the current collector/current lead joint on different sides of the cell allows the width (area) of the current collector to be increased (to increase the ratio of the weld width of the current collector to the width of the electrode) and provides more uniform current withdrawal from the area of the electrodes [263]. To reduce mechanical stresses and increase the thickness of the electrode unit, the laminated foil LIC body can be assembled from two pressed-out forms.

Separator: Since the thickness of the active layer is small and the area of the electrodes is higher, respectively, the area and mass fraction of the separator in the design of high-power LICs is higher than that of LICs with high specific energy [270]. The use of a thinner separator with greater porosity reduces the weight of the cell. On the other hand, the manufacturing process becomes more complicated due to the reduced strength of the separator. Separators with high gas permeability (depending on porosity, pore tortuosity, and thickness) are also better wetted by the electrolyte. The application of functional coatings (PVDF, ceramic particles) on separators allows the wettability of the separator with electrolyte to be increased and ensures closer contact between the electrodes and the separator, thereby lowering the internal resistance [186].

Electrolyte: The increase in the electrolyte mass fraction in an electrochemical block of LICs and the introduction of low-viscosity solvents and functional additives into the electrolyte composition allows the wettability of the separator and active electrode layer with electrolyte to be increased, thus decreasing the internal resistance of the LICs. Furthermore, increasing the salt concentration ($LiPF_6$), introducing lithium salts with higher solubility, and including functional additives that provide more effective lithium conductivity of the SEI film also reduce the internal resistance of the LICs. In addition, functional additives for the passivation of the cathode surface allow cutoff charging (average discharge) voltage to be increased and the power of the LICs to be elevated. More information on electrolytes and functional additives can be found in Materials for Lithium-Ion Cells (Section Electrolytes) and references [271–274]).

Electrodes: One can achieve a decrease in the electronic resistance of the block of electrodes by increasing the following:

- The purity (material conductivity) and thickness of the current collector;
- The ratio of the output width (the total width of the leads, if there are several) from the current collector to the current output to the width (active layer);
- The uniform distribution of the leads (and their number [275]) along the length of the electrode [92].

The electronic resistance between the active layer and the current collector can be reduced by increasing the contact area between the active layer and the current collector (intermediate carbon conductive layer, scratching), as well as by removing (scratching, formation of aluminium carbide [276] nanowhiskers breaking through the oxide film) the oxide layer (aluminium current collector).

With the decrement in the active layer thickness, the path length of ions and electrons also decreases, leading to an increment in the allowed charge/discharge rate. The improved wettability of the active material particles with the electrolyte can be achieved due to the high porosity of the active electrode layer. Calendering augments the active layer compaction degree, i.e., the number of contacts between the conductive additive and the active material, increasing the electronic conductivity [277]. Calendering also decreases the porosity (and increases the tortuosity of the pores [278]) of the electrode-active layer, thereby decreasing electrode wetting by the electrolyte and reducing ionic conductivity. Thus, the optimum pressure and compaction degree are necessary for providing the required characteristics of the cell. For example, it was found [235] that when using positive electrodes (LFP/C) with the same application density of 11 mg/cm^2 and different bulk density (1.7 g/cm^3, 1.9 g/cm^3, 2.1 g/cm^3), thickness (82 ÷ 83 microns, 75 ÷ 76 microns, 69 ÷ 70 microns), and porosity (38.8%, 31.6%, 28.1%), the electronic conductivity was 0.04 S/cm, 0.061 S/cm, and 0.072 S/cm, respectively. As the active layer density increased, the ohmic polarisation decreased, but it differed insignificantly even at discharge currents of 10 C. The diffusion polarisation at currents up to 5 C was almost the same for electrodes with different active layer densities. With further elevation in the current strength, the diffuse polarisation increased with the increase in the density of the active layer of the electrode. Moreover, the difference in the overpotential values when determining the diffusion polarisation significantly exceeded the difference in overvoltages observed when determining the ohmic polarisation. In this regard, a balance between the degree of compaction (electronic conductivity), porosity (ionic conductivity), pore tortuosity, and electrode thickness is necessary.

The active layer may have a density gradient along the depth (denser then closer to the current collector), which may also be accompanied by a change in the concentration of binder and conductive additives. The outer part of the active electrode layer is less dense and has a larger contact area with the electrolyte. This active layer structure allows the peak charge current to be increased.

The conductive additives also influence the electronic conductivity and the structure of the active layer. Their mass fraction and distribution in the active layer volume should ensure effective charge transfer from the surface of the active material to the current collector (and back). If electrodes with a thick active layer are used, conductive graphite or carbon fibres may need to be used. Due to their larger size, they can provide more efficient conductivity through the depth of the active layer.

The binder (and thickener) is a dielectric and, therefore, reduces the electronic conductivity of the electrode [279]. To reduce the internal resistance and increase the power of the LIC, the mass and volume fraction of the binder in the active layer should be minimised.

Active materials: Greater porosity of the active layer of electrodes can be achieved by using materials with lower tap density, narrow particle size distribution, smaller particle size, and higher porosity of particles.

Manufacturers claim that it is possible to regulate the arrangement of crystallites in the composition of active material particles. For example, if the atomic planes between which the lithium ions move are predominantly directed toward the particle's outer surface, the surface area through which the lithium ions move increases. Some grades of cathode materials with a layered structure and spherical shape (NCM [280]) and artificial graphites of the MAG series (Hitachi) are among the materials with directed crystallite orientation [281].

In addition to the orientation of crystallites in the composition of secondary particles of cathode materials, the small size of primary particles [282], as well as the presence of pores (porous structure [283] or cavities [99,284]) in secondary particles, has a positive effect on the charge/discharge rate.

A high specific surface area (small particle size, porous structure) provides an increased active material/electrolyte contact area [285]. However, materials with a higher specific surface area require more binder (dielectric) and may increase resistance more intensively during operation due to the thickening of the solid electrolyte film.

Carbon-conductive coatings on graphites help to reduce the intensity of structure destruction during the intercalation of electrolyte components and increase the allowable charging current, both at room and low temperatures. Carbon coatings on lithium titanate and cathode (for example, for LFP/C and NMC manufactured by Daejung) materials allows their electronic conductivity to be increased, therefore reducing the required concentration of conductive additives in the active layer and reducing the internal resistance of LICs.

Conductive additives: The use of conductive additives with an increased degree of crystallinity improves the active layer's thermal and electrical conductivity [286]. The branched structure of conductive additives allows more electrolyte to accumulate in the pores of the active electrode layer and provides more effective charge removal (supply) from (to) the active material particle(s) [287]. The viscosity of the electrode mass rises with the increasing concentration of conductive additives, branching their structure and reducing their particle size. The use of carbon with large particles and a small branching structure enhances the concentration of the conductive additives in the slurry and electrode-active layer, consequently growing the electronic conductivity [288]. Thus, ensuring sufficient electronic conductivity is possible by using large amounts of carbon blacks in the active layer composition with a small, branched structure and large particle size and by using smaller amounts of carbon blacks with a branched structure and smaller particle size.

Binder: CMC/SBR binder has a smaller contact area with the active material particles than PVDF and reduces electrode conductivity to a lesser extent [289]. When manufacturing a negative electrode with a PVDF binder, it may be necessary to introduce conductive additives into the active electrode layer composition. When wetting the active electrode layer, the binder can swell, leading to increased distances and a decreased electrical contact area between the particles of the conductive additives and the active material (or only the active material). Butadiene styrene rubber with higher glass transition temperature has higher stiffness, which provides lower density (higher porosity and ionic conductivity) of the applied, rolled-up active layer [290].

A binder with a higher molecular weight reduces electronic conductivity and provides improved adhesion of the active layer to the current collector. In this regard, the mass fraction of high-molecular-weight binders in the active layer composition is lower.

Positive electrode materials: Coating, the introduction of dopants into the structure of layered cathode materials (lithiated cobalt oxide, lithiated nickel cobalt manganese oxide), and the substitution of iron for manganese in cathode materials with olivine structures ($LiFePO_4$, $LiMn_{0.66}Fe_{0.34}PO_4$) allows elevation of the average discharge voltage, and, hence, the power, to be increased. However, it should be taken into account that the electronic conductivity and diffusion coefficient of $LiMn_{0.66}Fe_{0.34}PO_4$ are lower than those of $LiFePO_4$. Therefore, when increasing the charging voltage, the degree of stability of the conductive additives at high potentials must also be taken into account [291].

To increase the power of the positive electrode, a cathode material, lithium manganese spinel ($LiMn_2O_4$), which has a larger diffusion coefficient and higher average discharge potential than cathode materials with a layered structure, can be added to the active layer composition. It should be considered that $LiMn_2O_4$ has lower capacity and electronic conductivity than cathode materials with a layered structure [292].

Negative electrode materials: When lithium ions are intercalated into the graphite structure, the potential approaches the lithium potential, increasing the potential difference (voltage, and, hence, the power). On the other hand, the proximity of the anode potential to the lithium potential increases the risk of the deposition of lithium films in case of charging with high currents or when charging at low temperatures [196,293,294]. This problem can be partially solved when using graphites with a lower ratio of planes C004/C110 [295], graphites with a carbon coating, and electrolytes with functional additives that provide improved conductivity of the solid electrolyte film, or when using anode materials with a high average charge/discharge material [294]. Carbon materials (HC, SC) with high potential relative to lithium, due to the set of functional groups on the surface, are better wetted by the electrolyte [296]. However, when making electrodes based on these materials, it may be necessary to use non-aqueous binder solutions—PVDF/n-methylpyrrolidone. The active layer with a PVDF binder has lower electronic conductivity at equal mass concentrations than the active layer with SBR/CMC binder. An increase in the average charging/discharging potential of the anode also decreases the potential difference between the electrodes (decrease in voltage).

The usability of lithium-ion batteries depends on the charging rate. To reduce the charge time, LICs must be manufactured in such a way as to increase the permissible charge current. Charge currents may be limited by processes leading to reduced safety in the use of LICs. The charging current (or low-current charging at low temperature) is limited by the performance of the negative electrode. If the allowable charging current is exceeded, lithium dendrites may form on the surface of the anode material, which, as they grow, may pierce the separator, and provoke micro-short circuits and internal LIC short circuits. The following list describes some of the ways to increase the allowable charging current:

- Application of anode materials with the elevated specific surface (porous graphites, anode materials with small particle size, large particle surface) and/or small plane ratio of 004/110 [295]. This approach allows an increase in the area through which the intercalation of lithium ions into the structure of the anode material occurs. On the other hand, the paste preparation process becomes more complicated, and the internal resistance may increase. As a result, an SEI film with increased resistance may form on the surface (which requires the selection of functional electrolyte additives that provide an increased ionic conductivity of the solid electrolyte film).
- Growth of wettability of negative electrode with electrolyte: the high porosity of the active layer (the gradient of porosity of the active layer by depth) on the preservation of necessary electroconductivity, the small thickness of the active layer, the application of carbon coatings on graphite, the use of mixtures of graphite with SC and/or HC, the reduced viscosity of electrolyte solvents, and functional additives in the electrolyte.
- Increase the relative capacity of active anode material (not only total, but also local—there should not be areas where the cathode exceeds the capacity of the anode).
- The flow of lithium ions through the separator should not have areas with pronounced increases or decreases in lithium-ion throughput. There may be a local excess of lithium ions in some areas of the anode and, therefore, the formation of dendrites is possible.
- Carbon coatings and/or a solid electrolyte film applied to graphite (the introduction of functional additives into the electrolyte) should exhibit an increased conductivity towards lithium ions.
- Application of hard carbon and soft carbon. Due to the increased average charging potential relative to lithium (compared to graphite), the risk of lithium deposition on the anode surface is reduced (but the specific power and energy may decrease due to the smaller potential difference between the cathode and anode—LIC voltage).

- The use of lithium titanate as active anode material. Because of the high potential relative to lithium, the probability of lithium deposition when charged with high currents (or relative high currents at lower temperatures) is extremely small. The current per battery can be increased, but due to the lower voltage, the power density of the LIC on the charge may be lower. Suppose the battery is not intended to be charged with high currents at low temperatures. In that case, it may be advisable to make a battery of the same mass–size characteristics, but a higher capacity of cells with a carbon-based anode. The relative strength of the charging current on the LICs will decrease, and the probability of lithium dendrite formation may also decrease.
- Secondary porous electrode design and charge protocol selection [297].

5.3.5. Improving LIC Safety

Possible malfunctions in LICs can affect the safe use of the equipment even without a sudden temperature increase and thermal runaway. In this regard, the assessment [237] of the reliability and safety of cells and the batteries based on them is an important task.

LICs consist of a cathode (oxidant in the charged state), an anode (if made with carbon materials, a reducing agent), a separator (oxidising agent reductant), and an electrolyte (oxidising agent reductant). The presence of an oxidiser and reducing agent in a limited volume leads to the risk of their interaction with the release of large amounts of energy (depressurisation, fire, explosion). Under regular operation, the LIC design ensures the separation of the anode and cathode and, therefore, safe performance.

There are standards from various organisations (JIS (C8712, C8714, C8715), IEC (62660-2 (3), ISO (12405-1 (2), 12405-3), SAE (J2464, J2929), UL (1642, 2054, 2580, 9540, 9540A, and others) for testing battery, module, and module group safety. The factors of influence during testing are divided into mechanical (shock, falling from a height, penetration (piercing), immersion, crushing, rollover, vibration), electrical (external short circuit, internal short circuit (the introduction of a metal particle between electrodes during manufacture of LICs), overcharge, overdischarge), environmental (thermal stability, thermal shock and thermal cycling, overheating, low temperatures, exposure to fire), and chemical factors (exposure to chemicals (including seawater), flammability testing) [298]. The consequences of the impact of each of these factors are assessed on a point scale (safety level 0—no effect, 7—explosion [299]) and are reflected in the detailed technical specifications (at least for some impact factors).

Safety inspections can take place both at the manufacturer's plant and in special centres, for example, Sandia BATlab (USA) [300,301], Espec (Japan) [302], and TüV (Germany) [303]. The staff in these centres have relevant competencies, appropriate methodologies, software (including simulation software), and equipment located in specially prepared buildings and premises in which suitable fire and explosion hazard testing can be carried out with special fire extinguishing equipment (e.g., F-500 [304]).

Since safety checks are performed during the operation (or at certain states of charge) of LICs, LIB packs, modules, and module groups, the list of equipment includes charge–discharge battery testing systems (and other equipment allowing the determination of the operability of LICs and LIB packs), thermal imaging cameras, calorimeters, climatic chambers, pressure chambers, and specialised equipment.

To reduce the risk of emergency situations, precautions are taken at various levels: the level of the LIB pack in the car, the battery design (sealed and durable housing, thermal control system, multi-level battery control and management system (voltage, balance of state of charge, temperature), fuses, ensuring effective electrical insulation, the introduction of heat-resistant materials, heat-absorbing materials [305], the use of a low-combustible liquid in the thermal control system, etc.), and the design and manufacturing technology of LICs.

As a rule, the solutions used to improve the safety of LICs lead to a reduction in the specific power and energy characteristics. In this regard, LIC developers use a compromise solution that depends on the functional purpose of the LICs, operating conditions, risks, etc. Therefore, the methods and requirements described below are not comprehensive.

To increase the strength, the case can be made of steel (18650 [157]) or aluminium (3003, etc. [177,306]). In addition, an additional guard (nail safety device [234,307]) can be inserted into the housing to reduce the risk of puncture. When selecting the volume of the case and where to weld the current collectors, the increase/decrease in the electrode unit volume during charging/discharging must be taken into account. The list of body parts and design approaches providing enhanced LIC safety may include the following:

- A mechanical circuit breaker (current interruptive device [308,309]) and safety valve [308,309];
- A safety vent [234,307] (membrane);
- A design solution involving the local depressurisation of LICs in a laminated foil body [310];
- PTC (positive temperature coefficient) membrane (opening the circuit when the temperature rises) [308,311];
- An internal fuse [234] (including in LICs in a laminated foil case, Kokam [312,313] positive current terminal);
- A membrane that opens when overcharged [307,308] (due to increased gassing) [308];
- Reducing the thickness of the cell due to increasing its length (connecting several electrode blocks) [70,314];
- Manufacturing methods and structural elements that ensure LICs' tightness (for example, glass-sealed leads [315]).

If a laminated foil case is used, it is necessary to check the tightness of the weld sealing area (and, to increase the viscosity of the electrolyte, introduce functional additives preventing the processes leading to outgassing). In addition, the volume and shape of the housing must not cause the electrode unit current collectors to bend [316].

The electrode unit can be wrapped with a separator for additional insulation. To reduce the likelihood of a short circuit, part of the positive electrode (current collector) is covered [316,317] (Figure 52) with a special insulating tape (green tape, but can be of different colours; for example, 3M [318] offers polypropylene tapes with an acrylic adhesive). This tape is also used to secure the electrode assembly wrapped with a separator [319]. For the insulation of the aluminium current collector of the cathode, a coating (with a thickness of 18 microns) can be applied on the part of aluminium foil overlapped by the negative electrode and can include a binder and aluminium oxide particles [262]. On the other hand, a thinner coating containing ceramic particles (0.5 μm [259]) can reduce the oxidation of the separator at elevated potentials [320].

An aluminium foil with a PTC (positive temperature coefficient) film can be used to limit the flowing current, which has good electronic conductivity at the operating temperatures of the LIC and low conductivity when the operating temperature is exceeded (Figure 53). A temperature rise may be caused by a short circuit, punctured electrode block, overcharge, etc. The presence of the PTC film on the current collector allows the reversible regulation of the electrical resistance in the electrode block (Figure 54). Above a set temperature, the PTC film insulates (electrically) the cathode, thus ensuring that the battery stops being charged or discharged [321]. The breaking of contact between the active layer and the current collector may also be caused by the destruction of the intermediate layer (overcharge, temperature increase), applied (1÷10 μm) during electrode manufacturing [322,323]. The termination of charge/discharge at high temperatures can also be provided by functional coatings on cathode materials, e.g., the elevated-temperature-polymerising Stoba coating [324].

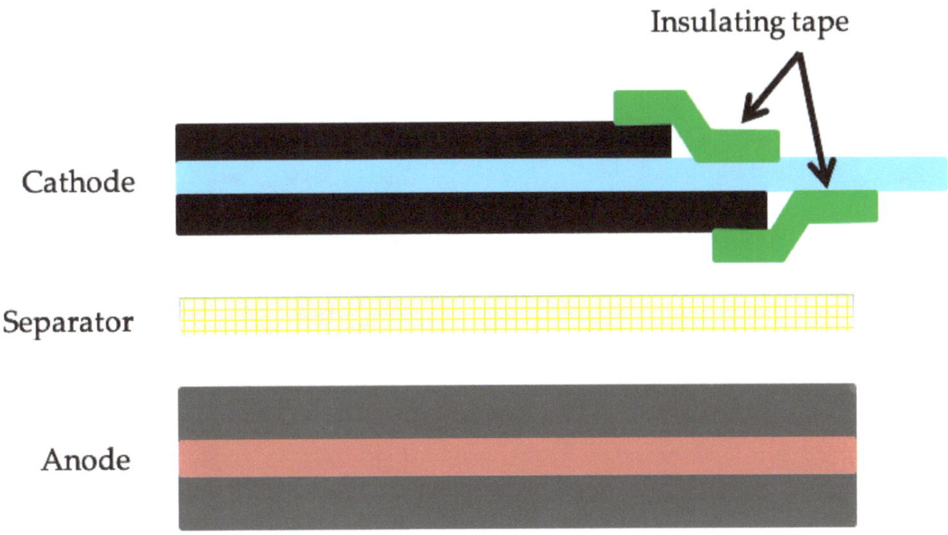

Figure 52. Additional insulation of the cathode (plotted on the basis of image presented in reference [317]). Source: Figure by authors.

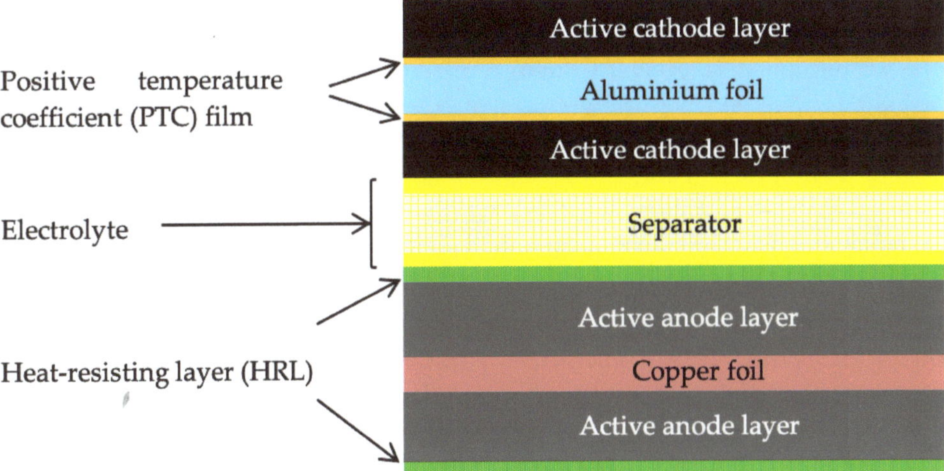

Figure 53. Schematic diagram of the Prius Alpha battery HEV electrochemical system (plotted on the basis of image presented in reference [71]). Source: Figure by authors.

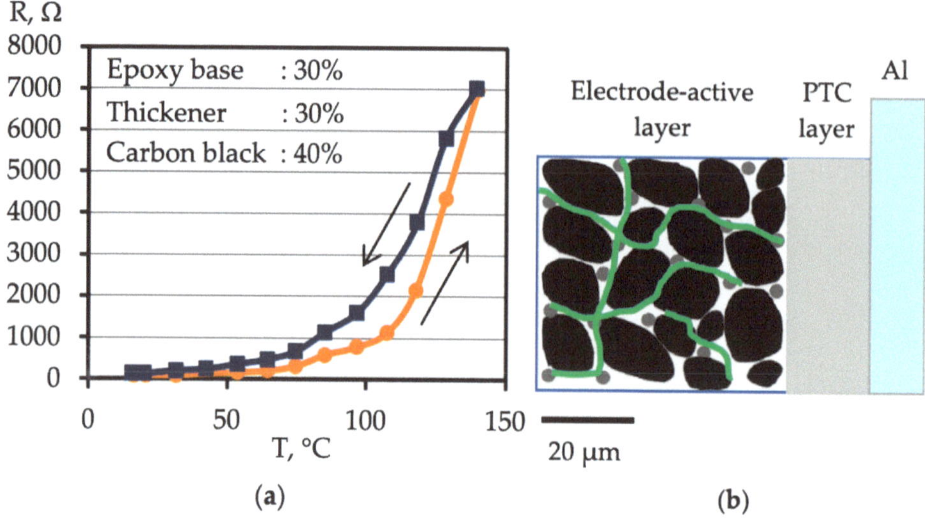

Figure 54. Change in conductivity (**a**) of the PTC layer in the composition of the positive electrode (**b**) (plotted on the basis of data provided in reference [325]). Source: Figure by authors.

When LICs are heated, three stages are distinguished that differ in the speeds of self-heating: initial stage ($\approx$0.2 °C/min), acceleration stage (increase in speed of heating up to 10 °C/min), and thermal runaway (>10 °C/min) [326]. The self-heating rate at the initial stage is low, and the heat generated can be dissipated by the battery's thermal management system. Starting from a particular temperature value (depending on the properties of the solid electrolyte film), the dissolution of the solid electrolyte film is observed, which stimulates the exothermic reactions of the electrolyte reduction on the active anode material. As a result, the partial oxidation of the electrolyte components is observed on the active cathode material. The reactions and their intensity depend on the nature of the active materials and the degree of cell charge. If the generated heat is not dissipated, then due to the increased intensity of reactions at the anode and cathode, an acceleration of self-heating (acceleration stage) is observed. There may be a depressurisation of the case at this stage, and smoke may escape. At the third stage (thermal runaway), the reactions on the active cathode or anode material proceed with even higher intensity, and the self-heating rate reaches 10 °C/min or more. There is very little chance of stopping thermal acceleration using external cooling at this self-heating rate.

When heating LICs with different cathode materials using reaction calorimetry (Accelerating Rate Calorimetry), it was found that the self-heating rate of LICs increases in the following series of active cathode materials: LMO < LFP < NCM < NCA < LCO [327,328]. Thermal runaway was practically not observed for LICs with LMO and LFP cathode materials. In the NCM < NCA < LCO series, the LIC temperature at which thermal runaway begins decreases. In the series of NCM cathode materials with different mole fractions of nickel of 0.83 (4.25 V), 0.7 (4.3 V), and 0.65 (4.35 V), the temperature of the beginning of the exothermic process in the cell increases. When conducting similar tests without electrolyte poured into the LIC, the temperature of the beginning of the exothermic reaction was approximately the same. In this regard, when evaluating the onset and magnitude of the thermal effect, it is advisable to test the cell rather than the aggregate of individual materials [329].

To reduce the probability of thermal runaway and its intensity, mixtures of cathode materials, such as NCM/LMFP, can be used to manufacture positive electrodes. The combustion reaction rate of propagation can be reduced by using lithium titanate, a non-flammable active anode material, or low-flammable additives in the electrolyte (see Section Electrolytes).

In case of the application of carbon-active material, a polymeric porous layer including aluminium oxide particles (HRL, heat-resistant layer) can be introduced between the negative electrode and the separator in a LIC design [71]. At elevated temperatures, the battery separator melts, and the aluminium oxide particles enter the separator structure and prevent contact between the anode and cathode. The ceramic coating on the SFL anode (SDI) has the same function [257].

The use of a three-layer separator (PP/PE/PP (Celgard [330]) or $PE_{HM}/PE_{LS}/PE_{HM}$ (Setela^{TM} Toray [331,332])), in which the melting point of the inner layer is lower than the melting temperature of the outer ones, can also reduce the likelihood of thermal runaway. At an uncontrolled increase in the LIC's temperature, polyethene melts ($\approx$120 °C) and closes the pores of the polypropylene layer (or polyethene with a higher melting temperature), stopping the LIC's operation (cutoff temperature) and reducing the possibility of further heating.

To increase safety, the coatings used contain layers of inorganic particles—CCS (SDI) [257], HRL (Panasonic) [333], SRS (LG Chem [334]) [257], boehmite coating [261] (Boemithe, Hitachi Chemical), etc.—which manufacturers apply in the process of cells manufacturing. Separators can have a thin ceramic coating on one side (anode or cathode) or both sides. The separator may also include a thick composite layer containing polymer fibres and ceramic particles [335].

- When a ceramic-coated separator is punctured, the expansion of the puncture area (which leads to overheating) is smaller than the expansion of the puncture area of a non-ceramic-coated separator [333]. The ceramic coating on the separator also promotes the following:
- Temperature melt uniformity and low thermal conductivity;
- Low shrinkage;
- Prevention of dendrite growth;
- Increase in wettability;
- Reduction of polarisation;
- Increment of tortuosity of the porous separator structure (more uniform lithium ions flow) [336].

BYD announced [186] the technology development for applying a high-temperature binder layer (HBL) to electrodes. The binder layer (probably PVDF-based) may withstand elevated temperatures. First, a solution containing polymer is applied to the electrodes, and then the solvent evaporates, and a porous binder layer is left on the surface (Figure 55).

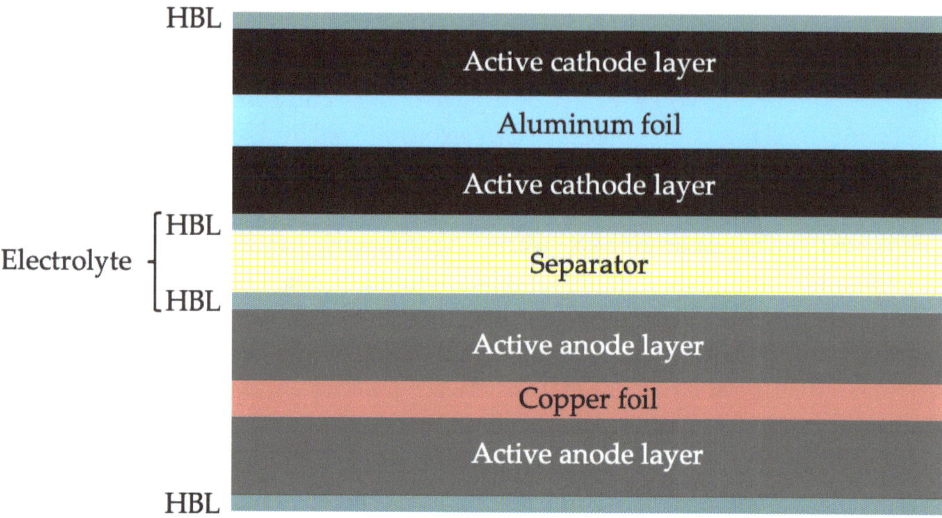

Figure 55. Schematic representation of LIC electrode block, including HBL (BYD) (plotted on the basis of an image presented in [186]). Source: Figure by authors.

The introduction of HBL into the electrode block provides the following:

- An increase in the number of protective layers between the electrodes.
- An improvement in the thermal stability of the electrode boundary.
- Tighter contact between the electrodes and separator (no free space, less chance of dendrite formation).

In the production of LICs, separators with a porous layer of PVDF [257] (Teijin, SKI, Asahi) can also be used, which can also include particles of inorganic materials (for example, aluminium oxide—SDI).

Incorporating polymers into the electrolyte (PVDF-HFP) increases the viscosity and reduces the probability of electrolyte leakage or evaporation from the LIC (in the laminated foil casing). The heat released during the combustion of electrolyte contributes significantly to the heat released during the combustion of LICs [337]. In case of electrolyte oxidation, dangerous reaction products are released [338]. In this regard, reducing the proportion of electrolyte in LICs may improve safety. On the other hand, a lack of electrolyte increases polarisation, reduces the service life, and may increase the risk of abnormal situations. Various functional phosphorus-containing additives are added to the electrolyte to reduce flammability. However, their introduction can reduce the ionic conductivity of the electrolyte. To protect against overcharging and high potentials, additives can be used to minimise gassing by forming a dense coating on the active materials. Conversely, additives can be used that degrade with the release of gas to increase internal pressure and open the circuit by moving or bursting the valve. Functional additives forming a uniform lithium-ion-conducting layer on the anode reduce the polarisation of the negative electrode at high charging currents, as well as charging at low temperatures, preventing the formation of lithium dendrites and, consequently, reducing the probability of an internal short circuit.

More detailed information on the processes taking place in the LIC under abnormal situations and measures reducing the risk can be found in references [256,337,339,340].

5.4. Conclusions

This section reviewed the parameters of lithium-ion cells and batteries used in hybrid and electric vehicles. First, the basic energy and power characteristics of lithium-ion cells used to manufacture batteries were shown. Then, differences in the composition, quantity of active materials, and attributes of electrodes leading to differences in energy and power characteristics of LICs for passenger electric vehicles and portable electronics applications were demonstrated. Next, the influence of LIC test conditions on the service life was considered. Finally, the main ways of increasing lithium-ion cells' specific power and safety were shown.

References

1. Takeshita, H. Latest LIB Market Trends (The Automotive and Power Storage Markets Move Into Full-scale Growth). In Proceedings of the 10th lnt'l Rechargeable Battery Expo Battery Japan 2019, Tokyo, Japan, 28 February 2019; pp. 1–6.
2. Alexander, M. Electric Vehicle Market Overview—What's Here, What's Coming, and What are Utilities Doing about It. In Proceedings of the SAE Hybrid and Electric Vehicle Symposium, San Diego, CA, USA, 20–22 February 2018; p. 25.
3. Tal, G. Exploring the global demand for pevs and the impact of policies and incentives on the market. In Proceedings of the SAE Hybrid and Electric Vehicle Symposium, San Diego, CA, USA, 20–22 February 2018; p. 30.
4. Lu, M. Future Trends and Key Issues in the Global Lithium-ion Batteries Market and Related Technologies. In Proceedings of the 13th China International Battery Fair (CIBF 2018), Shenzhen, China, 22–24 May 2018; p. 30.
5. Cheng, L. Electrification Strategies: Finding a Mass Market Solution. In Proceedings of the SAE 2016 New Energy Vehicle Forum, Shanghai, China, 22 September 2016; p. 16.
6. スズキ 6 代目「ワゴンRUTF8」詳細解説 ハイブリッドターボもある3 機種を展開 (Suzuki 6th Generation "Wagon R" Detailed Explanation Three Models Including Hybrid Turbo Are Available) (Suzuki). Available online: https://autoprove.net/suzuki/wagon-r/40300/ (accessed on 17 July 2019).

7. Looking Ahead to the New Audi A8: More Voltage for Enhanced Efficiency. Available online: https://www.dsf.my/2017/06/looking-ahead-to-the-new-audi-a8-more-voltage-for-enhanced-efficiency/ (accessed on 14 November 2024).
8. Volkswagen Bringing 48V MHEV to Eight-Generation Golf. Available online: https://www.greencarcongress.com/2019/05/201900518-golf48.html (accessed on 14 November 2024).
9. Green Car Congress (48 V). Available online: https://www.greencarcongress.com/48v/ (accessed on 17 July 2019).
10. Duren, A. 48 V Battery System Design for Mild Hybrid Applications. Available online: http://www.a123systems.com/wp-content/uploads/saehevs2016_anaheim_a123-48v-case-study_widescreen.pdf (accessed on 16 July 2019).
11. Konishi, D.; Ueda, J.; Kusunoki, T.; Ushijima, O.; Aoki, T.; Mizuta, Y. アイドリングストップ車用 12 V リチウムイオン電池の開発 (Development of 12 V Lithium-ion Battery for Start and Stop Vehicle System). *GS Yuasa Tech. Rep.* **2013**, *10*, 16–23.
12. Rosenkranz, C. Defining Future Li-Ion Stop-Start Battery Systems for Low-Voltage Hybrid Applications. In Proceedings of the 3rd World Mobility Summit, Munich, Germany, 17–19 October 2023; p. 16.
13. Review of ELV Exemption 5 (A123 Systems LLC, Fraunhofer ICT, LG Chem Ltd., and Samsung SDI Co. Ltd.). Available online: https://elv.exemptions.oeko.info/fileadmin/user_upload/Consultation_2014_1/Ex_5/20141217_A123_et.al.2015_ELV_exemption_5_review_-_Li-ion_stakeholders.pdf (accessed on 17 July 2019).
14. Gregeois, J. Ioniq HEV & Niro HEV presentation. In Proceedings of the Public Workshop on Lead-Acid Batteries and Alternatives, CalEPA Headquarters, Sacramento, CA, USA, 11 November 2016; p. 17.
15. Englisch, A.; Pfund, T.; Reitz, D.; Simon, E.; Kolb, F. Synthesis of various hybrid drive systems. In *Der Antrieb von morgen 2017*; Springer: Wiesbaden, Germany, 2017; pp. 61–78.
16. Cardoso, D.S.; Fael, P.O.; Espirito-Santo, A. A review of micro and mild hybrid systems. *Energy Rep.* **2020**, *6*, 385–390. [CrossRef]
17. Anderman, M. Assessing the Future of Hybrid and Electric Vehicles: The xEV Industry Insider Report. Available online: http://www.yano.co.jp/pdf/announce/Executive_Summary_Selections.pdf (accessed on 15 July 2019).
18. Lu, H.S. Overview on Technology & Market Progress of HEV and Advanced Batteries Outlook. In Proceedings of the 2nd HEV Market & Advanced Battery Technology development Seminar, Hangzhou, China, 13–14 October 2016; p. 29.
19. 12 V Starter Battery UltraPhosphate^{TM} Technology (A123, 40 Ah). Available online: http://www.a123systems.com/wp-content/uploads/12V-Starter-Battery-Flier_2017_40Ah.pdf (accessed on 13 July 2019).
20. Fehrenbacher, C. 12V Li-Ion Batteries—Ready for Mainstream Adoption. In Proceedings of the Advanced Automotive Battery Conference (AABC Europe 2017), Mainz, Germany, 1–2 February 2017; p. 18.
21. 12 V Starter Battery UltraPhosphate^{TM} Technology (A123, 60 Ah). Available online: http://www.a123systems.com/wp-content/uploads/12V-Starter-Battery-Flier_2016_Gen-3.pdf (accessed on 13 July 2019).

22. Kessen, J. Safety Requirements for Low Voltage Systems. Available online: http://www.a123systems.com/wp-content/uploads/A123-Low-Voltage-System-Safety_final.pdf (accessed on 17 July 2019).
23. Gehrmann, J.; Renner, D. Design of a 14V nominal dual battery system (Audi). In Proceedings of the 7th Advanced Automotive Battery Conference Europe (AABC), Mainz, Germany, 1–2 February 2017.
24. Kume, H. Denso Shows Off Li-ion Battery for Suzuki's Mild HEVs in India. Available online: https://tech.nikkeibp.co.jp/dm/atclen/news_en/15mk/031301964/?ST=print_en (accessed on 17 July 2019).
25. Lu, H.L. Recent Status & Future Outlook of World LTO Anode Material and Related LIBs (ITRI IEK). In Proceedings of the First LTO Battery Industrial Forum (LBIF2017), Wuxi, China, 6–7 July 2017; p. 27.
26. Gray, D. LG Chem's Current and Future Battery Technologies for xEVs. In Proceedings of the SAE 2018 Hybrid & Electric Vehicle Technologies Symposium, San Diego, CA, USA, 20–22 February 2018; p. 15.
27. Zhang, N. Lishen High Power Battery Technology for Start-stop. In Proceedings of the 2nd HEV Market & Advanced Battery Technology development Seminar, Hangzhou, China, 13–14 October 2016; p. 30.
28. 48 V Lithium-ion Battery—8 Ah. Available online: http://www.a123systems.com/wp-content/uploads/48V-Battery-Flier_2016.pdf (accessed on 16 July 2019).
29. Lee, S.D.; Cherry, J.; Safoutin, M.; McDonald, J.; Olechiw, M. *Modeling and Validation of 48 V Mild Hybrid Lithium-Ion Battery Pack*; SAE Technical Paper 2018-01-0433; SAE: Warrendale, PA, USA, 2018. [CrossRef]
30. Kessen, J. Technical Solutions and Value Analysis of Advanced Hybrid Batteries. In Proceedings of the 2nd HEV Market & Advanced Battery Technology Development Seminar, Hangzhou, China, 13–14 October 2016; p. 17.
31. Low-Voltage Hybrid Systems 48 V Battery (Bosch). Available online: https://www.bosch-mobility-solutions.com/en/products-and-services/passenger-cars-and-light-commercial-vehicles/powertrain-systems/electric-drive/48v-battery/ (accessed on 17 July 2019).
32. Hitachi Automotive Systems Has Developed a High-Output 48 V Lithium-Ion Battery Pack for Mild Hybrid Electric Vehicles. Available online: https://www.hitachi.com/New/cnews/month/2016/04/160420.html (accessed on 17 July 2019).
33. Development of 48 V Lithium-Ion Battery Pack with Improved Output Density and Energy Density Developed for Mild Hybrid Vehicles. Available online: http://www.hitachi.com/New/cnews/month/2017/05/170522.pdf (accessed on 17 July 2019).
34. Ogawa, K. Hitachi Automotive Showcases 48 V Li-ion Battery. Available online: https://tech.nikkeibp.co.jp/dm/atclen/news_en/15mk/061301392/ (accessed on 18 July 2019).
35. Ishiwa, K.; Yamamoto, D.; Sekino, M. Rising to the challenge: Rechargeable battery SCiB™ with LTO anode for LV-xEV applications. In Proceedings of the eMove360° Conference 2019 For Electric Mobility & Autonomous Driving Battery Conference, Munich, Germany, 15–16 October 2019; p. 32.

36. Grudmann, S. Bosch Standard Battery for 48 V BRS. In Proceedings of the 7th Advanced Automotive Battery Conference Europe (AABC), Mainz, Germany, 1–2 February 2017; p. 13.
37. Pistoia, G. *Lithium-Ion Batteries Advances and Applications*; Elsevier: Warsaw, Poland, 2014.
38. Kowalec, S. 12 V/48 V Hybrid Vehicle Technology. Available online: https://www.psma.com/sites/default/files/uploads/tech-forums-transportation-power-electronics/presentations/is115-12v-48v-hybrid-vehicle-technology.pdf (accessed on 5 June 2021).
39. Fuel economy 2016 Audi A3 e-tron ultra. Available online: http://www.fueleconomy.gov/feg/Find.do?action=sbs&id=37130 (accessed on 14 June 2024).
40. BYD F3DM. Available online: https://en.wikipedia.org/wiki/BYD_F3DM (accessed on 14 June 2024).
41. Chevrolet Volt. Available online: https://en.wikipedia.org/wiki/Chevrolet_Volt (accessed on 14 June 2024).
42. Chevrolet Volt (Second Generation). Available online: https://en.wikipedia.org/wiki/Chevrolet_Volt_(second_generation) (accessed on 14 June 2024).
43. 2016 Ford C-MAX Energi Plug-in Hybrid. Available online: https://www.fueleconomy.gov/feg/Find.do?action=sbs&id=36932 (accessed on 14 June 2024).
44. Ford Fusion Hybrid. Available online: https://en.wikipedia.org/wiki/Ford_Fusion_Hybrid (accessed on 14 June 2024).
45. 2014 Ford Fusion Energi Plug-in Hybrid. Available online: http://www.fueleconomy.gov/feg/Find.do?action=sbs&id=34089 (accessed on 14 June 2024).
46. Sonata Plug-in Hybrid. Available online: https://en.wikipedia.org/wiki/Hyundai_Sonata#Sonata_Plug-in_Hybrid (accessed on 14 June 2024).
47. Naylor, S. New Kia Optima PHEV plug-in hybrid 2016 review. Available online: http://www.autoexpress.co.uk/kia/optima/96813/new-kia-optima-phev-plug-in-hybrid-2016-review (accessed on 14 June 2024).
48. Toyota Prius Plug-in Hybrid. Available online: https://en.wikipedia.org/wiki/Toyota_Prius_Plug-in_Hybrid (accessed on 14 June 2024).
49. Understanding Micro, Mild, Full and Plug-In Hybrid Electric Vehicles. Available online: https://x-engineer.org/automotive-engineering/vehicle/hybrid/micro-mild-full-hybrid-electric-vehicle/ (accessed on 15 July 2019).
50. Emadi, A. *Advanced Electric Drive Vehicles*; CRC Press, Taylor & Francis Group: Boca Raton, FL, USA, 2015.
51. Jiang, J.; Zhang, C. *Fundamentals and Applications of Lithium-Ion Batteries in Electric Drive Vehicles*; John Wiley & Sons Singapore Pte. Ltd.: Pondicherry, India, 2015.
52. Pistoia, G. *Electric and Hybrid Vehicles Power Sources, Models, Sustainability, Infrastructure and the Market*; Elsevier: Devon, UK, 2010.
53. Scrosati, B.; Garche, J.; Tillmetz, W. *Advances in Battery Technologies for Electric Vehicles*; Woodhead Publishing (Elsevier): Amsterdam, The Netherlands, 2015; Volume 80.
54. Warner, J. *The Handbook of Lithium-Ion Battery Pack Design, Chemistry, Components, Types and Terminology*; Elsevier Science: Amsterdam, The Netherlands, 2015.

55. 2016 Chevrolet Volt Battery System. Available online: https://media.gm.com/content/dam/Media/microsites/product/Volt_2016/doc/VOLT_BATTERY.pdf (accessed on 14 March 2017).
56. GM Exec: Gen 3 Voltec Battery to Have Shortened Lifespan, Simpler Shape, and be Offered in Smaller Ranges. Available online: http://gm-volt.com/2010/03/26/gm-exec-gen-3-voltec-battery-to-have-shortened-lifespan-simpler-shape-and-be-offered-in-smaller-ranges/ (accessed on 9 February 2024).
57. Matthé, R.; Eberle, U. The Voltec System—Energy Storage and Electric Propulsion. In *Lithium-Ion Batteries Advances and Applications*; Pistoia, G., Ed.; Elsevier: Amsterdam, The Netherlands, 2014; pp. 151–176.
58. EVEngineering. Dissecting the Chevy Volt's Electric Drivetrain. Available online: https://www.greencarreports.com/news/1093708_whats-inside-chevrolet-volt-battery-pack-and-drivetrain-video-teardown-shows-all (accessed on 21 January 2021).
59. Nicholas, M.; Hall, D. Lessons Learned on Early Electric Vehicle Fast-Charging Deployments. Available online: https://theicct.org/sites/default/files/publications/ZEV_fast_charging_white_paper_final.pdf (accessed on 7 November 2019).
60. Hering, T. ABB Smarter Mobility Fast charging infrastructure to reduce CO_2 emissions. In Proceedings of the eMove360° Battery Conference 2019, Munich, Germany, 15–16 October 2019; p. 19.
61. Likar, U. CHAdeMO global footprint and advantages (Mitsubishi). In Proceedings of the World Mobility Summit and European EV Congress, Munich, Germany, 17–18 October 2017; p. 66.
62. Peeters, J. Fast charging just got faster Developments in High Power Charging (ABB). In Proceedings of the World Mobility Summit and European EV Congress, Munich, Germany, 17–18 October 2017; p. 23.
63. DC fast-charging—What is the right system? CCS—The one system approach of charging (CharIn). In Proceedings of the World Mobility Summit 2016, Munich, Germany, 18–20 October 2016; p. 19.
64. Lu (Mark), H.-L. Chinese xEV Market: Vehicle, Battery and Materials Impact. In Proceedings of the Advanced Automotive Battey Confernce (AABC Europe 2021), Virtual, 20 January 2021.
65. 355-390-590 Module and CTP Evolution Logic. Available online: https://daydaynews.cc/en/car/449159.html (accessed on 29 January 2021).
66. Lima, P. Simple Solution for Safer, Cheaper and More Energy-Dense Batteries. Available online: https://pushevs.com/2020/04/12/simple-solution-for-safer-cheaper-more-energy-dense-batteries/ (accessed on 29 January 2021).
67. Meng, X. The Next-Generation Battery Pack Design: From the BYD Blade Cell to Module-Free Battery Pack. Available online: https://medium.com/batterybits/the-next-generation-battery-pack-design-from-the-byd-blade-cell-to-module-free-battery-pack-2b507d4746d1 (accessed on 29 January 2021).

68. 趙章恩 (Zhao Zhang'en). 次世代はリチウム硫電池か全固電池か `EV 火災で安全性に脚光 (Next Generation Lithium-Sulfur or Solid-State Batteries? EV Fires Put Safety in the Spotlight). Available online: https://xtech.nikkei.com/atcl/nxt/column/18/01231/00018/ (accessed on 26 March 2021).
69. SK Innovation Shows Green Mobility Technologies at 2020 Green New Deal Expo. Available online: https://skinnonews.com/global/archives/2448 (accessed on 26 March 2021).
70. Lu (Mark), H.-L. Technical Development of the Chinese New Energy Vehicle Battery & Future Possibilities through 2025. In Proceedings of the 38th Annual International Battery Seminar & Exhibition, Virtual, 9–11 March 2021; p. 20.
71. Lu, M. Key Factors of the Power Battery Development in 2013: E-Motorcycle & EV. In Proceedings of the 2013 EV Market Trends Seminar (The 3rd International Taiwan Electric Vehicle Show), Taipei, Taiwan, 12 April 2013; p. 46.
72. Weiss, C. Electric Vehicle Battery Chemistry and Pack Architecture. In Proceedings of the Charles Hatchett Seminar High Energy and High Power Batteries for e-Mobility Opportunities for Niobium, London, UK, 8 July 2018; p. 31.
73. Gall, J. 2012 Infiniti M35h Hybrid. Available online: http://www.caranddriver.com/reviews/2012-infiniti-m35h-hybrid-road-test-review (accessed on 14 March 2017).
74. HEV Battery Testing Results 2013 Honda Civic Hybrid—VIN 1356. Available online: https://energy.gov/sites/prod/files/2015/02/f19/batteryCivic1356.pdf (accessed on 14 March 2017).
75. Stoklosa, A. 2013 Honda Civic Hybrid. Available online: http://www.caranddriver.com/reviews/2013-honda-civic-hybrid-test-review (accessed on 14 March 2017).
76. Battery Pack Laboratory Testing Results. 2013 Chevrolet Malibu Eco—VIN 3800. Available online: https://avt.inl.gov/sites/default/files/pdf/hev/batteryMalibu3800.pdf (accessed on 22 November 2016).
77. CHEVROLET VOLT—2016. Available online: http://media.chevrolet.com/media/us/en/chevrolet/vehicles/volt/2016.tab1.html (accessed on 14 March 2017).
78. Schoewel, F.; Hockgeiger, E. High Voltage Batteries of BMW Vehicles. In Proceedings of the AABC2014, Atlanta, GA, USA, 3–7 February 2014; p. 14.
79. Audi A3 e-tron—Detailed Specs. Available online: http://insideevs.com/audi-a3-e-tron-detailed-specs/ (accessed on 22 November 2016).
80. 2016 Audi A3 Sportback e-tron Plug-In Hybrid. Available online: http://www.caranddriver.com/audi/a3-sportback-e-tron (accessed on 14 March 2017).
81. Tesla's 100 kWh Battery Pack Is as Big as It Gets (for Now). Available online: http://www.teslarati.com/tesla-p100d-100kwh-battery-pack-size/ (accessed on 24 November 2016).
82. Tesla's Plan for Gigafactory Vehicle Battery Pack Rollout and Why It Matters. Available online: https://electrek.co/2016/07/27/tesla-gigafactory-vehicle-battery-pack-rollout/ (accessed on 24 November 2016).
83. Tesla's Battery Tech Explained: Part 3—The Pack (EV-Tech Explained). Available online: https://youtu.be/izUl28YtQbE (accessed on 20 July 2019).

84. Battery Pack Laboratory Testing Results. 2014 BMW i3 EV—VIN 5486. Available online: https://avt.inl.gov/sites/default/files/pdf/fsev/batteryi5486.pdf (accessed on 23 November 2016).
85. 2016 Nissan Leaf. Available online: http://www.nissanusa.com/electric-cars/leaf/ (accessed on 11 October 2016).
86. Mitsubishi i-MiEV. Available online: https://en.wikipedia.org/wiki/Mitsubishi_i-MiEV (accessed on 14 March 2017).
87. Ombach, G. Commercialization of a battery system for a premium car segment—From digital twin to product. In Proceedings of the 2nd Virtual Battery Exhibition, Virtual, 27 April–3 May 2021; p. 25.
88. Blanco, S. There's a Bit of the Chevy Volt Hidden in the Malibu Hybrid. Available online: https://www.autoblog.com/2016/04/29/chevy-volt-hidden-inside-malibu-hybrid/ (accessed on 15 July 2019).
89. Hitachi Automotive Systems Delivers 5000 W/kg High Output Power Density Prismatic Lithium-Ion Battery Cells for 2016 New Model GM Chevrolet Malibu Hybrid. Available online: http://www.hitachi.com/New/cnews/month/2015/05/150519a.pdf (accessed on 15 July 2019).
90. Saharan, V.; Nakai, K. *High Power Cell for Mild and Strong Hybrid Applications Including Chevrolet Malibu*; SAE Technical Paper 2017-01-1200; SAE: Warrendale, PA, USA, 2017. [CrossRef]
91. Battery Pack Laboratory Testing Results. 2013 Ford C-Max SE—VIN 2158. Available online: https://avt.inl.gov/sites/default/files/pdf/hev/batteryCMax2158.pdf (accessed on 22 November 2016).
92. Zhang, N. Progress of Lishen High Power Li Ion Battery for HEV Application. In Proceedings of the International Seminar on HEV Market and Advanced Battery Technology Development, Beijing, China, 22–24 April 2014; p. 46.
93. Battery Pack Laboratory Testing Results. 2015 Honda Accord—VIN 5774. Available online: https://avt.inl.gov/sites/default/files/pdf/hev/batteryAccord5774.pdf (accessed on 22 November 2016).
94. GS ユアサの電池をめぐる物語 (The story of GS Yuasa batteries) (Volume 5). Available online: https://www.gs-yuasa.com/jp/deepstory/vol5.html (accessed on 15 July 2019).
95. Nishida, Y.; Komoda, S.; Maruno, N. Development of Li-ion Battery Control Technology for HEV. *SAE Int. J. Alt. Power.* **2015**, *4*, 225–232. [CrossRef]
96. Battery Pack Laboratory Testing Results. 2013 Honda Civic—VIN 0594. Available online: https://avt.inl.gov/sites/default/files/pdf/hev/batteryCivic0594.pdf (accessed on 22 November 2016).
97. 2016 Kia Optima Hybrid Specifications. Available online: http://www.kiamedia.com/us/en/models/optima-hybrid/2016/specifications (accessed on 22 November 2016).
98. Hayashi, T. PEVE Supplies Li-ion Battery for Prius Alpha. Available online: http://techon.nikkeibp.co.jp/english/NEWS_EN/20110516/191786/ (accessed on 22 November 2016).
99. Nagai, H.; Morita, M.; Satoh, K. *Development of the Li-ion Battery Cell for Hybrid Vehicle*; SAE Technical Paper 2016-01-1207; SAE: Warrendale, PA, USA, 2016. [CrossRef]

100. 4th-Generation Toyota Prius Teardown (Part 1). Available online: https://www.marklines.com/en/report_all/rep1473_201602 (accessed on 15 July 2019).
101. Battery Pack Laboratory Testing Results. 2014 Volkswagen Jetta SE—VIN 0875. Available online: https://avt.inl.gov/sites/default/files/pdf/hev/batteryJetta0875.pdf (accessed on 22 November 2016).
102. Lehnert, S. Lithium-Ion Battery for Audi A6 PHEV. In Proceedings of the 7th Advanced Automotive Battery Conference Europe (AABC), Mainz, Germany, 1–2 February 2017; p. 15.
103. BYD Lithium Fe Battery for HEV/EV. Available online: https://girasole.co.za/product/byd-lithium-fe-block-12v-10-ah-b-bms/ (accessed on 14 November 2024).
104. Berman, B. New Wrinkle in Hybrid Cars. Available online: http://www.nytimes.com/2011/02/20/automobiles/20TECH.html?emc=eta1&_r=0 (accessed on 22 November 2016).
105. Xi, S. Recent Progress of BYD EV Batteries. Available online: https://wenku.baidu.com/view/44064e3beefdc8d376ee3234.html (accessed on 21 July 2019).
106. 2018 Chevrolet Volt 355.2V Li-Ion Battery—Deep Dive. Available online: https://www.youtube.com/watch?v=eWYtq0hxhQg (accessed on 22 July 2019).
107. 吉利新能源策略及博瑞GE 开发 (Geely's new energy strategy and Borui GE development). In Proceedings of the SAE 2018 新能源汽车国际论坛(第六届) New Energy Vehicle Forum, Shanghai, China, 11–12 September 2018; p. 38.
108. 2016 Hyundai Sonata PHEV Raises the Bar among Blended PHEVs. Available online: http://gm-volt.com/2015/05/22/2016-hyundai-sonata-phev-raises-the-bar-among-blended-phevs/ (accessed on 22 November 2016).
109. Battery Pack Laboratory Testing Results. 2013 Ford C-Max Energi—VIN 3817. Available online: https://avt.inl.gov/sites/default/files/pdf/phev/batteryCMax3817.pdf (accessed on 22 November 2016).
110. Battery Pack Laboratory Testing Results. 2013 Ford Fusion Energi SE—VIN 1518. Available online: https://avt.inl.gov/sites/default/files/pdf/phev/batteryFusion1518.pdf (accessed on 5 June 2024).
111. Kia Unveils Optima Plug-In Hybrid, The Next Step In It's 'Green' Car Revolution. Available online: https://press.kia.com/eu/products/optima%20phev/optima%20phev/ (accessed on 23 November 2016).
112. Battery Pack Laboratory Testing Results. 2013 Toyota Prius Plug-in—VIN 8663. Available online: https://avt.inl.gov/sites/default/files/pdf/phev/batteryPrius8663.pdf (accessed on 23 November 2016).
113. A Closer Look at Audi's New R8 e-Tron EV and Battery. Available online: https://www.greencarcongress.com/2015/06/20150612-r8etron.html (accessed on 14 November 2024).
114. Roper, L.D. BMW i3. Available online: http://www.roperld.com/Science/BMWi3.htm (accessed on 19 July 2019).
115. Samsung SDI Develops New Lithium-Ion Battery. Available online: http://english.etnews.com/20190517200002 (accessed on 15 July 2019).
116. Technical Specifications. BMW i3 (120 Ah). Available online: https://www.press.bmwgroup.com/global/article/attachment/T0284828EN/421721/Specifications_BMW_i3_(120_Ah)_BMW_i3s_(120_Ah).pdf (accessed on 14 November 2024).

117. Superior Battery Technology. Available online: https://www.torqeedo.com/en/technology-and-environment/battery-technology.html (accessed on 15 July 2019).
118. BYD e6 2015 model. Available online: https://iaspub.epa.gov/otaqpub/display_file.jsp?docid=33966&flag=1 (accessed on 19 September 2019).
119. Battery Pack Laboratory Testing Results. 2015 Chevrolet Spark—VIN 4878. Available online: https://avt.inl.gov/sites/default/files/pdf/fsev/batterySpark4878.pdf (accessed on 23 November 2016).
120. Kane, M. Deep Dive: Chevrolet Bolt Battery Pack, Motor and More. Available online: http://insideevs.com/deep-dive-chevrolet-bolt-battery-pack-motor-and-more/ (accessed on 23 November 2016).
121. Ramaka, V. Continued Glimpses into xEV Batteries on the Market—AVL Series Battery Benchmarking. In Proceedings of the eMove360° Battery Conference 2019, Munich, Germany, 15–16 October 2019; p. 22.
122. CODA 2012 Emergency Responder's Guide. Available online: http://135jik1bbhst1159ri1ax2pj.wpengine.netdna-cdn.com/wp-content/uploads/sites/20/2010/11/CODA_Emergency_Responders_Guide_ERG.pdf (accessed on 23 November 2016).
123. Fiat 500e Full Vehicle Specifications! Available online: http://www.fiat500usa.com/2013/04/fiat-500e-full-vehicle-specifications.html (accessed on 23 November 2016).
124. Anderman, M. *Assessing the Future of Hybrid and Electric Vehicles: The 2014 xEV Industry Insider Report*; Advanced Automotive Batteries (AAB): Oregon House, CA, USA, 2013.
125. 2013 Ford Focus Electric. Advanced Vehicle Testing—Baseline Testing Results. Available online: http://energy.gov/sites/prod/files/2015/02/f19/fact2013fordfocus.pdf (accessed on 23 November 2016).
126. Emergency Response Guide 2013 Fit EV. Available online: https://techinfo.honda.com/Rjanisis/pubs/web/ACI48132.pdf (accessed on 23 November 2016).
127. 2014 Honda Fit EV 1 Centennial to Denver, CO. Available online: http://www.kuni-honda.com/2014-honda-fit-ev-near-denver-colorado.htm (accessed on 23 November 2016).
128. Nisewanger, J. Exclusive: Details on Hyundai's New Battery Thermal Management Design. Available online: https://electricrevs.com/2018/12/20/exclusive-details-on-hyundais-new-battery-thermal-management-design/ (accessed on 26 July 2019).
129. Weintraub, S. The Electrek Review: Kia Niro EV—The New Normal... Electric Family Car. Available online: https://electrek.co/2019/05/07/kia-niro-ev-review/ (accessed on 26 July 2019).
130. Battery Pack Laboratory Testing Results. 2015 Kia Soul—VIN 1908. Available online: https://avt.inl.gov/sites/default/files/pdf/fsev/batterySoul1908.pdf (accessed on 23 November 2016).
131. Kia Ray EV. Available online: https://www.netcarshow.com/kia/2012-ray_ev/ (accessed on 23 November 2016).
132. Mazda to Lease 'Demio EV' Electric Vehicle in Japan from October. Available online: https://newsroom.mazda.com/en/publicity/release/2012/201207/120706a.html (accessed on 14 November 2024).

133. Battery Pack Laboratory Testing Results. 2015 Mercedes B-Class—VIN 4477. Available online: https://avt.inl.gov/sites/default/files/pdf/fsev/batteryBClass4477.pdf (accessed on 23 November 2016).
134. Battery Pack Laboratory Testing Results. 2014 Smart Electric Drive Coupe—VIN 2764. Available online: https://avt.inl.gov/sites/default/files/pdf/fsev/batteryElectric2764.pdf (accessed on 23 November 2016).
135. Battery Pack Laboratory Testing Results. 2012 Mitsubishi iMiev SE—VIN 4550. Available online: https://avt.inl.gov/sites/default/files/pdf/fsev/batteryiMiev4550.pdf (accessed on 23 November 2016).
136. Kariatsumari, K. Toshiba Displays i-MiEV M's Li-ion Battery at Trade Show. Available online: http://techon.nikkeibp.co.jp/english/NEWS_EN/20111103/200392/ (accessed on 23 November 2016).
137. Battery Pack Laboratory Testing Results. 2013 Nissan Leaf S—VIN 5045. Available online: https://avt.inl.gov/sites/default/files/pdf/fsev/batteryLeaf5045.pdf (accessed on 23 November 2016).
138. Nissan LEAF Battery Pack—Initial Analysis. Available online: http://www.slideshare.net/georgebetak/nissan-leaf-battery-pack-initial-analysis (accessed on 23 November 2016).
139. Inside The Battery of a Nissan Leaf. Available online: http://qnovo.com/inside-the-battery-of-a-nissan-leaf/ (accessed on 23 November 2016).
140. Ikezoe, M.; Hirata, N.; Amemiya, C.; Miyamoto, T.; Watanabe, Y.; Hirai, T.; Sasaki, T. *Development of High Capacity Lithium-Ion Battery for NISSAN LEAF*; SAE International 2012-01-0664; SAE: Warrendale, PA, USA, 2012. [CrossRef]
141. Taylor, J. Battery Boost for 2016 Nissan Leaf Increases Range by 25%. Available online: http://www.carmagazine.co.uk/car-news/industry-news/nissan/battery-boost-for-2016-nissan-leaf-increases-range-by-25-/ (accessed on 23 November 2016).
142. Tomioka, T. Nissan Improves Energy Density of Leaf EV's Battery Pack. Available online: https://tech.nikkeibp.co.jp/dm/atclen/news_en/15mk/100501625/ (accessed on 31 October 2019).
143. Kane, M. Here Is The Nissan LEAF e+ 62 kWh Battery: Video. Available online: https://insideevs.com/news/342009/here-is-the-nissan-leaf-e-62-kwh-battery-video/ (accessed on 31 October 2019).
144. Nissan LEAF 40-kWh Battery: Deep Dive. Available online: https://insideevs.com/nissan-leaf-40-kwh-battery-deep-dive/ (accessed on 13 February 2019).
145. 2015 Citroen C-Zero (model for Europe) Specifications & Performance Data Review. Available online: http://www.automobile-catalog.com/car/2015/2018870/citroen_c-zero.html (accessed on 23 November 2016).
146. Lima, P. GS Yuasa's Improved Cells: LEV50 vs. LEV50N. Available online: http://pushevs.com/2015/11/04/gs-yuasas-improved-cells-lev50-vs-lev50n/ (accessed on 23 November 2016).
147. Maluf, N. Available online: https://qnovo.com/132-why-does-the-porsche-taycan-use-800-v-battery-packs/ (accessed on 2 May 2021).

148. Wienkötter, M. The Battery: Sophisticated Thermal Management, 800-Volt System Voltage. Available online: https://newsroom.porsche.com/en/products/taycan/battery-18557.html (accessed on 2 May 2021).
149. Origuchi, M. Renault Electric Vehicles and Batteries. In Proceedings of the ABEC, Yichun, China, 13–15 November 2013; p. 23.
150. Vortrag von Masato Origuchi (Renault) auf der eCartec 2015 EV/HEV Batterys. Available online: https://youtu.be/N6sRWGjFx5w (accessed on 23 November 2016).
151. Renault ZOE ZE 40 Full Battery Specs. Available online: https://pushevs.com/2019/02/10/renault-zoe-ze-40-full-battery-specs/ (accessed on 24 May 2019).
152. Lampert, F. Renault Surprises with New 2017 ZOE w/ ~200 Miles of Range on 41 kWh Battery Available Next Month. Available online: https://electrek.co/2016/09/28/renault-surprises-with-new-2017-zoe-w-200-miles-of-range-on-41-kwh-battery-available-next-month/ (accessed on 11 October 2016).
153. Renault Twizy. Available online: https://en.wikipedia.org/wiki/Renault_Twizy (accessed on 11 October 2016).
154. Twizy, Z.E. Driver's Handbook. Available online: http://maben.homeip.net/static/auto/renault/reanault%20twizy%20users%20manual.pdf (accessed on 24 November 2016).
155. Lima, P. Wuling Hong Guang MINI EV Had a Strong First Full Sales Month (Update). Available online: https://pushevs.com/2020/08/26/wuling-hong-guang-mini-ev-had-a-strong-first-full-sales-month/ (accessed on 29 May 2021).
156. 2014 Tesla Model S 85 kWh. Advanced Vehicle Testing—Baseline Vehicle Testing Results. Available online: https://avt.inl.gov/sites/default/files/pdf/fsev/fact2014teslamodels.pdf (accessed on 24 November 2016).
157. de Leon, S. Tesla—Success Story or Hype. In Proceedings of the NASA Aerospace Battery Workshop, Huntsville, AL, USA, 17–19 November 2015; p. 22.
158. 「リチウムイオン電池応用・実用化先端技術開発事業」 事後評価 (2012 年度~2016 年度 5 年間) プロジェクトの概要(公開) (Lithium-ion Battery Application and Practical Use Advanced Technology Development Project" Post-Fact Evaluation (2012-2016 5-Year Project) Project Overview (Disclosed)). Available online: https://www.nedo.go.jp/content/100873513.pdf (accessed on 19 September 2019).
159. Bower, G. Tesla Model 3 2170 Energy Density Compared To Bolt, Model S P100D. Available online: https://insideevs.com/news/342679/tesla-model-3-2170-energy-density-compared-to-bolt-model-s-p100d/ (accessed on 19 July 2019).
160. Toyota RAV4, EV. Available online: https://en.wikipedia.org/wiki/Toyota_RAV4_EV (accessed on 11 October 2016).
161. iQ EV Electric Vehicle. Electric Vehicle Dismantling Manual. Available online: http://prius20.ru/instructions/dismantling/iqevdisman.pdf (accessed on 24 November 2016).
162. Toyota, iQ. Available online: https://en.wikipedia.org/wiki/Toyota_iQ (accessed on 24 November 2016).
163. Battery Pack Laboratory Testing Results. 2015 Volkswagen E-Golf—VIN 2012. Available online: https://avt.inl.gov/sites/default/files/pdf/fsev/batteryEGolf2012.pdf (accessed on 24 November 2016).

164. Volkswagen e-Golf: Batterie 35.8 kWh d'ici la fin de l'année. Available online: http://www.automobile-propre.com/breves/volkswagen-e-golf-batterie-35-kwh-fin-2016/ (accessed on 24 November 2016).
165. Volkswagen Launches the Battery-Electric e-Golf in Germany; "Das e-Auto". Available online: http://www.greencarcongress.com/2014/02/20140215-egolf.html (accessed on 24 November 2016).
166. Volkswagen E-Up! Air Cooled Battery, Featured In ViaVision Magazine. Available online: http://www.myelectriccarforums.com/volkswagen-e-up-air-cooled-battery-featured-in-viavision-magazine/ (accessed on 24 November 2016).
167. AMP20m1HD-A A123. Available online: http://liionbms.com/pdf/a123/AMP20M1HD-A.pdf (accessed on 28 February 2018).
168. A123 20Ah Lithium Ion UltraPhosphate Prismatic Pouch Cell. Available online: https://www.altertek.com/products/lithium-ion-pouch-cylindrical-cells/a123-li-ion-cells/a123-20ah-lithium-ion-ultraphosphate-pouch-cell/ (accessed on 14 November 2024).
169. Anderman, M. Future Batteries for 12V Automotive SLI+ Applications. Available online: https://dtsc.ca.gov/wp-content/uploads/sites/31/2018/10/Menahem-Anderman_Lead-acid-Batteries-Workshop_11-6-2017.pdf (accessed on 4 November 2019).
170. Saruwatari, H.; Yamamoto, D. 10 Ah-Class SCiB™ Lithium-Ion Battery for Idling Stop Systems and Micro Hybrid Vehicles. Technology Administration & Planning Office, Corporate Technology Planning Division, Toshiba Corporation, 1-1, Shibaura 1-chome, Minato-ku, Tokyo 105-8001, Japan. 東芝レビュ *(Toshiba Rev.)* **2016**, *71*, 44–47.
171. High-Power Type Cells. Available online: https://www.global.toshiba/ww/products-solutions/battery/scib/product/cell/high-power.html (accessed on 31 May 2024).
172. Rechargeable Lithium-Ion Battery SCiB™. Available online: https://www.global.toshiba/content/dam/toshiba/ww/products-solutions/battery/scib/pdf/ToshibaRechargeableBattery-en.pdf (accessed on 31 May 2024).
173. Durable Thin Prismatic System A123 3.2v 8 Ah lifepo4 Lithium-Ion Polymer Battery for Racing Car Forklift Electric Motorcycle. Available online: https://www.alibaba.com/product-detail/Long-Life-Thin-Prismatic-A123-System_1600316497208.html (accessed on 14 November 2024).
174. Xu, J.M. Progressive development for 48V ESS (CATL). In Proceedings of the 2nd HEV Market & Advanced Battery Technology Development Seminar, Hangzhou, China, 13–14 October 2016.
175. Watanabe, Y.; Naka, T.; Yamamoto, D. 高エネルギー密度と高入出力を両立させた高性能HV 向け リチウムイオン二次電池20 Ah ・5 Ah SCiB™ (20 Ah and 5 Ah SCiB™ Lithium-Ion Battery Cells Offering Both High Energy Density and High Input-Output Power Characteristics for High-Performance Hybrid Vehicles). Technology Administration & Planning Office, Corporate Technology Planning Division, Toshiba Corporation, 1-1, Shibaura 1-chome, Minato-ku, Tokyo 105-8001, Japan. *Toshiba Rev.* **2019**, *74*, 56–59.

176. Iguchi, T.; Ochiai, S.; Kozono, S.; Nitta, K.; Abe, Y.; Kohno, K. ハイブリッド自動車用リチウムイオン電池 EH5 (High Power and Long Life Lithium-ion Battery EH5 for HEVs). *GS Yuasa Tech. Rep.* **2014**, *11*, 24–30.
177. Oguri, K.; Maruno, N. *Development of Lithium-Ion-Battery System for Hybrid System*; SAE International SAE: Warrendale, PA, USA, 2011. [CrossRef]
178. Product Information. Available online: https://www.blue-energy.co.jp/jp/products/ (accessed on 14 November 2024).
179. Sogo, Y.; Araki, T.; Kashiwa, Y.; Matsuyoshi, T.; Kozono, S.; Ochiai, S.; Iguchi, T. HEV 用リチウムイオン電池 EHW5 の開発 (Development of Li-ion cell EHW5 for HEV). *GS Yuasa Tech. Rep.* **2019**, *16*, 33–39.
180. Lu, H.-L. Current Situation Regarding xEV-Batteries & Materials in the Chinese Market and Future Outlook (ITRI IEK). In Proceedings of the 7th Korea Advanced Battery Conference 2016, Coex, Seoul, Republic of Korea, 29–30 September 2016; p. 40.
181. Grewe, T.; Hubbard, G.; Cottrell, D. General Motors 3rd Generation Eassist Propulsion System. In Proceedings of the Hybrid and Electric Vehicle Technologies Symposium, San Diego, CA, USA, 20–22 February 2018; p. 30.
182. Namiki, F.; Maeshima, T.; Inoue, K.; Kawai, H.; Saibara, S.; Nanto, T. Lithium-ion Battery for HEVs, PHEVs, and EVs. *Hitachi Rev.* **2014**, *63*, 48–53.
183. Higashimoto, K.; Homma, H.; Uemura, Y.; Kawai, H.; Saibara, S.; Hironaka, K. Automotive Lithium-ion Batteries. *Hitachi Rev.* **2011**, *60*, 17–21.
184. 三星和博世SBL 动力电池介绍 SB Limotive Lithium ion Cells Introduction. In Proceedings of the Foton & SB Limotive Technical Conference, Beijing, China, 17 August 2011; p. 42.
185. Kim, K. 车用电池技术 (Automotive battery technology) SB LiMotive Automotive Battery Technology. In Proceedings of the 电动汽车科技创新国际论坛日程 (Agenda of the International Forum on Electric Vehicle Technology Innovation), Beijing, China, 12–13 July 2012; p. 34.
186. BYD Fe Battery for Electrical Energy Storage System and Automobile Applications. Available online: http://wenku.baidu.com/view/1161e6e3524de518964b7db8.html?from=search (accessed on 7 March 2017).
187. C020 ePLB C High Energy Product (EIG). Available online: http://www.ebaracus.com/sites/default/files/2012/12/EIG-ePLB-C020-Datasheet.pdf (accessed on 21 July 2019).
188. Sukino, K.; Matsui, H.; Inamasu, T.; Okuyama, R. High Performance 13 Ah-class Lithium-ion Cell with Mixed Positive Active Materials of $LiNi_xMn_yCo_zO_2$ ($x + y + z = 1$)/$LiFePO_4$ for PHEV Application. *GS Yuasa Tech. Rep.* **2011**, *8*, 16–21.
189. Nakajima, N.; Sukino, K.; Matsui, H.; Inamasu, T.; Okuyama, R. Development of Lithium-ion Battery Using $LiNi_xMn_yCo_zO_2$($x + y + z = 1$)-$LiFePO_4$ Composite Positive for Plug-in Hybrid Vehicle. *GS Yuasa Tech. Rep.* **2012**, *9*, 10–15.
190. Suzuki, I.; Mochizuki, T.; Nakamoto, T.; Uebo, Y.; Funabiki, A.; Nishiyama, K.; Sonoda, T. Improvement on Safety and High-rate Discharge Performances by Adoption of Carbon-loaded $LiFePO_4$ Positive Electrode for Large-sized Lithium-ion Cells. *GS Yuasa Tech. Rep.* **2009**, *6*, 20–24.

191. Iguchi, T.; Okamoto, K.; Kuratomi, J.; Ohkawa, K.; Kohno, K.; Izuchi, S. Development of Lithium-Ion Battery"EX25A"with New Positive Active Material of $LiCo_xMn_yNizO_2$ (x + y + z = 1). *GS Yuasa Tech. Rep.* **2004**, *1*, 25–31.
192. Darcy, E. Li-ion Pouch Cell Designs; Are They Ready for Space Applications? In Proceedings of the Large Li-Ion Battery Technology and Application Symposium (AABC12), Orlando, FL, USA, 8 February 2012; p. 20.
193. IEK, I. 鑑往知來: 由全球鈕電池黨與市場發展動向 看中國黨轉型與升級之路 (Learn from the past and learn from the future: Looking at the transformation and upgrading of the Chinese Communist Party from the global button battery industry and market development trends). In Proceedings of the 2012 中國化學與物理電源行業協會理事會議 (China Chemical and Physical Power Supply Industry Association Board of Directors Meeting), Beijing, China, 14–15 December 2012; p. 126.
194. 海外动力电池系列研究之一: 角力与共生---- 全球动力电池竞争格局 (Overseas Power Battery Series Research 1: Struggle and Symbiosis—Global Power Battery Competition Landscape) (Essence Securities). Available online: http://pdf.dfcfw.com/pdf/H3_AP201811291253422317_1.pdf (accessed on 21 January 2021).
195. Herrmann, M.; Matthe, R. Progress of battery systems at General Motors. In Proceedings of the World Mobility Summit, Munich, Germany, 18–20 October 2023; p. 24.
196. Gao, J. Progress of xEv LIB battery technology and its requirement on electrolyte development. In Proceedings of the 中国锂电池电解质、隔膜材料技术与 市场发展论坛 (China Lithium Battery Electrolyte, Diaphragm Material Technology and Market Development Forum), China, Guangzhou, 14–16 August 2013; p. 39.
197. Microvast Solutions for HEV and PHEV. Available online: https://wenku.baidu.com/view/df931c2d9b6648d7c1c74689.html (accessed on 19 May 2019).
198. Microvast Cell. Available online: http://www.microvast.com/index.php/solution/solution_cell (accessed on 20 March 2019).
199. Mattis, W.L. Nonflammable, Fast Charging and Superb Battery Cycle Life Battery Technology For Automotive Applications (Microvast). In Proceedings of the 18th International Meeting on Lithium Batteries (IMLB), Chicago, IL, USA, 19–24 June 2016; p. 39.
200. 65V 9Ah LpCO (Gen 2) Microvast. Available online: https://li-bat.ru/katalog/192-novie_litiyititanatnie_akkumulyatori_ili4ti5012i_kitay_imicrovasti_1520325636.html (accessed on 25 July 2019).
201. Microvast (gen3 MpCO). Available online: https://li-bat.ru/katalog/211-mnogokomponentnaya_litievaya_batareya_3i7vi.html (accessed on 24 July 2019).
202. MV07220126MPE-15Ah (Microvast MpCO). Available online: https://docviewer.yandex.ru/view/901289335/?page=2&*=a4uExDPfvTpSCr6rBwbuSqo8fLV7InVybCI6InlhLWRpc2stcHVibGljOi8vMXNQa0x0V29ST3RyVUFWWlI0N2tJbXh3SlhEdHhHVHBiYXI3YTRzVFZIcz0iLCJ0aXRsZSI6Ik1QUy5UUzIxNS0yMDE2IOS6p%2BWTgeaKgOacr%2BinhOagvOS5pu%2B8iE1WMDcyMjAxMjZNUEUtMTVBaO%2B8iS5wZGYiLCJub2lmcmFtZSI6ZmFsc2UsInVpZCI6IjkwMTI4OTMzNSIsInRzIjoxNTYzOTkzNDA3NTU4LCJ5dSI6IjMyNzc4MzE2MzE1MDM0OTA4NzkifQ%3D%3D(accessed on 24 July 2019).

203. リチウムイオン電池応用・実用化先端技術開発事業」 事後評価）分科会 資料 7-1 (Lithium-Ion Battery Application and Practical Use Advanced Technology Development Project" Post-Fact Evaluation) Subcommittee Materials 7-1.). Available online: https://www.nedo.go.jp/content/100873514.pdf (accessed on 13 February 2019).
204. Swing (R) 5300 Rechargeable Lithium- ion Cell (Boston Power). Available online: https://www.tme.eu/en/Document/87b5645f730e56cad507df13706ef5b9/Swing5300.pdf (accessed on 16 August 2018).
205. Chamberlain, R. Progress of xEV batteries with blended NCM/NCA cathode materials (Boston Power). In Proceedings of the Asia-Pacific Lithium Battery Congress 2014, Shenzhen, China, 26–28 March 2014; p. 28.
206. Specification of C12 Fe-Battery Cell Name: FP58146410A. Available online: https://cdn.shopify.com/s/files/1/0634/4605/files/BYD_ESS_and_DESS_Fe_battery_spec_201406.pdf (accessed on 21 July 2019).
207. Lu, H.-l. The Development Status & Trends of Main Materials for the WW Power LIB. In Proceedings of the 2016′ 第五届中国电池市场年会，暨第一届动力电池应用国际峰会，第二届中国电池行业智能制造研讨会 (The 5th China Battery Market Annual Conference, the 1st International Power Battery Application Summit, the 2nd China Battery Industry Intelligent Manufacturing Seminar) (CBEA), Beijing, China, 14–15 November 2016; p. 37.
208. Lima, P. BYD Blade Prismatic Battery Cell Specs and Possibilities (Update). Available online: https://pushevs.com/2020/05/26/byd-blade-prismatic-battery-cell-specs-possibilities/?source=content_type%3Areact%7Cfirst_level_url%3Aarticle%7Csection%3Amain_content%7Cbutton%3Abody_link (accessed on 29 January 2021).
209. Cadenza Innovation, Inc. Final Scientific/Technical Report Novel Low Cost and Safe Lithium-Ion Electric Vehicle Battery (DE-AR0000392). Available online: https://cadenzainnovation.com/wp-content/uploads/2018/05/2018-03-28-Cadenza-ARPA-e-DE-AR0000392-Final-Report-webversion-002.pdf (accessed on 29 November 2019).
210. CATL NCM 50AH Lithium Battery. Available online: http://www.evlithium.com/CATL-Battery.html (accessed on 23 July 2019).
211. Hummel, P.; Bush, T.; Gong, P.; Yasui, K.; Lee, T.; Radlinger, J.; Langan, C.; Lesne, D.; Takahashi, K.; Jung, E.; et al. *UBS Q-Series Tearing Down the Heart of an Electric Car: Can Batteries Provide an Edge, and Who Wins?* Report; UBS: New York, NY, USA, 2018.
212. 72AH & 96AH NCM Lithium Battery (CATL). Available online: http://www.evlithium.com/lifepo4-battery-news/558.html (accessed on 24 July 2019).
213. Pouch Cell Battery. Available online: http://www.dfdxny.com/en/list.php?id=46 (accessed on 14 November 2024).
214. Shanghai Electric Gotion New Energy Technology Co., Ltd. Battery Cells. Available online: https://www.shanghai-electric.com/group_en/c/2019-08-12/557924.shtml (accessed on 29 May 2021).
215. Ueki, K.; Kitano, S.; Toriyama, J.-I.; Seyama, Y.; Nishiyama, K. Development of Large-sized Long-life Type Lithium-ion Cells for Electric Vehicle. *GS Yuasa Tech. Rep.* **2012**, *9*, 26–29.

216. Kitano, S.; Nishiyama, K.; Toriyama, J.-I.; Sonoda, T. Development of Large-sized Lithium-ion Cell "LEV50" and Its Battery Module "LEV50-4" for Electric Vehicle. *GS Yuasa Tech. Rep.* **2008**, *5*, 21–26.
217. Baldwin, R. Inside the Factory Building GM's Game-Changing Bolt EV. Available online: https://www.engadget.com/2016/12/09/inside-the-factory-building-gm-s-game-changing-bolt-ev/?guccounter=1# (accessed on 24 July 2019).
218. Nisewanger, J. Jaguar and Chevy Have LG in Common. Available online: https://electricrevs.com/2018/03/09/jaguar-and-chevy-have-lg-in-common/ (accessed on 24 July 2019).
219. Microvast (gen4 HpCO). Available online: https://li-bat.ru/katalog/229-litievaya__batareya_microvast_hpco_3_7v_43_an__6000_tsiklov_1542748419.html (accessed on 24 July 2019).
220. Panasonic NCR18650B 3400mAh (Green). Available online: https://www.shoptronica.com/files/Panasonic-NCR18650.pdf (accessed on 14 November 2024).
221. Morris, C. What's Inside a Tesla Battery Cell [Video]. Available online: https://evannex.com/blogs/news/whats-inside-a-tesla-battery-cell (accessed on 31 October 2019).
222. CATL 的电芯成本被日韩两强吊打？对，也不对 (The cost of battery cells in China is beaten by Japan and South Korea? Yes and No). Available online: http://m.cbea.com/ldc/201812/532680.html (accessed on 13 February 2019).
223. Tesla Model 3 Battery Pack & Battery Cell Teardown Highlights Performance Improvements. Available online: https://cleantechnica.com/2019/01/28/tesla-model-3-battery-pack-cell-teardown-highlights-performance-improvements/ (accessed on 13 February 2019).
224. Lima, P. Samsung SDI 94 Ah Battery Cell Full Specifications. Available online: https://pushevs.com/2018/04/05/samsung-sdi-94-ah-battery-cell-full-specifications/ (accessed on 6 July 2018).
225. Honda, K. SCiB^{TM} Batteries Become a Vable Option for HEV/EV Manufacturers. Available online: https://wenku.baidu.com/view/ca025530ee06eff9aef807ce.html (accessed on 24 July 2019).
226. Lu, H.-L. 國際儲能系統應用概況與電力子整合設計方向分析 (Overview of international energy storage system applications and analysis of power sub-integration design directions) (ESS, Itri IEK). In Proceedings of the 電力子協會 呂學隆 (Electric Power Electronics Association) Seminar, Taipei, Taiwan, 14 December 2018; p. 49.
227. Takami, N.; Ise, K.; Harada, Y.; Iwasaki, T.; Kishi, T.; Hoshina, K. High-energy, fast-charging, long-life lithium-ion batteries using $TiNb_2O_7$ anodes for automotive applications. *J. Power Sources* **2018**, *396*, 429–436. [CrossRef]
228. Ise, K.; Morimoto, S.; Harada, Y.; Takami, N. Large lithium storage in highly crystalline $TiNb_2O_7$ nanoparticles synthesized by a hydrothermal method as anodes for lithium-ion batteries. *Solid State Ion.* **2018**, *320*, 7–15. [CrossRef]
229. High Energy EV Pouch Cells (Range EV Cells) Zenlabs. Available online: https://static1.squarespace.com/static/59dbcd906f4ca35190c9aeb4/t/5cac327223d25a000188813b/1554788979381/EPS+Range.pdf (accessed on 22 December 2019).

230. INL/EXT-14-32849. U.S. Department of Energy Vehicle Technologies Program. Battery Test Manual For Plug-In Hybrid Electric Vehicles (Revision 3). Available online: https://inldigitallibrary.inl.gov/sites/sti/sti/6308373.pdf (accessed on 16 March 2017).
231. Gao, J.-K. Positive material requirement of EDV Li-ion batteries from LFP to LMO/NCM to Li-rich Mn based material, etc. In Proceedings of the China Forum on LEMD, Tianjin, China, 8–10 April 2013; p. 40.
232. Lu, H.-l. An Update on the Chinese xEV Market, and a Technical Comparison of its Batteries (ITRI IEK). In Proceedings of the World Mobility Summit 2016, Munich, Germany, 17–19 October 2016.
233. IAA Commercial Vehicles Battery Technology (SBLimotive). Available online: https://www.vda.de/dam/vda/publications/Battery%20Technology/1286965063_en_1943346300.pdf (accessed on 12 December 2019).
234. ESS Batteries by Samsung SDI (2017.05). Available online: https://www.samsungsdi.com/upload/ess_brochure/201705SamsungSDI_ESS_EN.pdf (accessed on 6 August 2018).
235. Shen, X. The latest progress of the BYD lithium ion Battery Technology for xeV. In Proceedings of the 10th China International Battery Fair (CIBF), Shenzhen, China, 20–22 June 2012; p. 40.
236. World Smart Energy Week 2013 東京現場直擊系列報導三 (Tokyo Live Coverage Series Report Three). Available online: https://www.materialsnet.com.tw/material/DocView_MaterialNews.aspx?id=10933 (accessed on 6 December 2018).
237. Gandoman, F.H.; Jaguemont, J.; Goutam, S.; Gopalakrishnan, R.; Firouz, Y.; Kalogiannis, T.; Omar, N.; Van Mierlo, J. Concept of reliability and safety assessment of lithium-ion batteries in electric vehicles: Basics, progress, and challenges. *Appl. Energy* **2019**, *251*, 17. [CrossRef]
238. Holtstiege, F.; Kasnatscheew, J.; Wagner, R.; Placke, T.; Winter, M. Simple Experiments Giving Deep Insights into Capacity Fade and Capacity Loss Mechanisms of Li Battery Materials. In Proceedings of the Chemistry Symposium, AABC Europe, Mainz, Germany, 30 January–2 February 2017; p. 30.
239. Smith, K. High-Energy Long-Life Li-Ion (L3B) via Pre- and Continuous-Lithiation. In Proceedings of the Advanced Automotive Battery Conference USA, Virtual, 3–5 November 2020; p. 20.
240. Grolleau, S. Cycling Aging of Lithium-ion Batteries. In Proceedings of the Summer-School on Materials for Batteries (Mat4bat), La Roschelle, France, 2–4 June 2015; p. 44.
241. ＧＳユアサによる蓄電池の運用監視技術 の開発の歩み（その３） 蓄電池の残存性能の評価技術 (GS Yuasa's development of battery operation monitoring technology (part 3)—Technology for evaluating remaining battery performance). *GS Yuasa Tech. Rep.* **2018**, *15*, 29–30.
242. Origuchi, M. Impact of Charging Method on Battery ZE. In Proceedings of the Advanced Automotive Battery Conference, Strasbourg, France, 24–28 June 2013; p. 30.

243. Lithium Iron Phosphate (LFP or $LiFePO_4$). Available online: https://www.powertechsystems.eu/home/tech-corner/lithium-iron-phosphate-lifepo4/#prettyPhoto/ (accessed on 14 November 2024).
244. Wood, E.; Alexander, M.; Bradley, T.H. Investigation of battery end-of-life conditions for plug-in hybrid electric vehicles. *J. Power Sources* **2011**, *196*, 5147–5154. [CrossRef]
245. Wood, E. *Battery End-of-Life Considerations for Plug-In Hybrid Electric Vehicles*; Colorado State University: Fort Collins, CO, USA, 2011.
246. Walter, L.; Robertson, D.; Basco, J.; Prezas, P.; Bloom, I. Electrochemical Performance Testing (ES201). In Proceedings of the 2014 DOE Annual Merit Review, Washington, DC, USA, 16–20 June 2014; p. 22.
247. Perner, A.; Hockgeiger, E. Challenges of Modular Battery Design for Electric Vehicles. In Proceedings of the Advanced Automotive Batteries (AABC), Strasbourg, France, 24–28 June 2013; p. 16.
248. MOLICEL® IHR18650C 2 Ah Power Cell (30 January 2013). Available online: https://www.master-instruments.com.au/file/64178/1/Molicel-IHR18650C.pdf (accessed on 3 June 2019).
249. Warnecke, A. Ageing effects of Lithium-ion batteries. In Proceedings of the Workshop EPE Conference (Batteries 2020), Geneva, Switzerland, 8–10 September 2015; p. 20.
250. Schlasza, C.; Ostertag, P.; Chrenko, D.; Kriesten, R.; Bouquain, D. Review on the aging mechanisms in Li-ion batteries for electric vehicles based on the FMEA method. In Proceedings of the IEEE Transportation Electrification Conference and Expo (ITEC), Dearborn, MI, USA, 15–18 June 2014.
251. Vetter, J.; Novak, P.; Wagner, M.R.; Veit, C.; Moller, K.C.; Besenhard, J.O.; Winter, M.; Wohlfahrt-Mehrens, M.; Vogler, C.; Hammouche, A. Ageing mechanisms in lithium-ion batteries. *J. Power Sources* **2005**, *147*, 269–281. [CrossRef]
252. Waldmann, T.; Iturrondobeitia, A.; Kasper, M.; Ghanbari, N.; Aguesse, F.; Bekaert, E.; Daniel, L.; Genies, S.; Gordon, I.J.; Loble, M.W.; et al. Review-Post-Mortem Analysis of Aged Lithium-Ion Batteries: Disassembly Methodology and Physico-Chemical Analysis Techniques. *J. Electrochem. Soc.* **2016**, *163*, A2149–A2164. [CrossRef]
253. Lin, C.; Tang, A.H.; Mu, H.; Wang, W.W.; Wang, C. Aging Mechanisms of Electrode Materials in Lithium-Ion Batteries for Electric Vehicles. *J. Chem.* **2015**, *2015*, 104673. [CrossRef]
254. Ignatiev, E. *Performance Degradation Modelling and Technoeconomic Analysis of Lithium-Ion Battery Energy Storage Systems*; Lappeenranta University of Technology: Lappeenranta, Finland, 2016.
255. Merichko, T. *Battery Durability in Electrified Vehicle Applications: A Review of Degradation Mechanisms and Durability Testing (Final Report EP-C-12-014 WA 3-01)*; FEV North America, Inc.: Auburn Hills, MI, USA, 2016.
256. Wang, Q.; Wang, S.; Zhang, J.; Zheng, J.; Yu, X.; LI, H. Overview of the failure analysis of lithium ion batteries. *Energy Storage Sci. Technol.* **2017**, *6*, 1008–1025. [CrossRef]

257. Research, S. 배터리4 대주요소재시장동향및개발전략방향 (Battery 4 major material market trends and development strategy direction). In Proceedings of the 2016 7th Korea Advanced Battery Conference (KABC2016), Coex, Seul, Republic of Korea, 29 September 2016; p. 50.
258. Takeshita, H. *IIT LIB-Related Study Program11-12 (February 2012)*; Institute of Information Technology, Ltd.: Tokyo, Japan, 2012; Available online: https://wenku.baidu.com/view/f917455be45c3b3567ec8b8f.html?_wkts_=1716389557071&needWelcomeRecommand=1 (accessed on 22 May 2024).
259. Lu (Mark), H.-L. Commercial Technology & Product Development Trends of Cathode & Anode Materials for LIB in 2015. In Proceedings of the Second International Forum on Cathode & Anode Materials for Advanced Batteries, Hangzhou, China, 22 April 2015; p. 55.
260. *LIB 材料市場速報 (Materials and Market News) （15Q1)*; B3 Corporation: Tokyo, Japan, 2015; p. 30. Available online: https://www.docin.com/p-1282371064.html?docfrom=rrela (accessed on 22 May 2024).
261. Lu (Mark), H.-L. Present Technology & Market Development Trends of Electrolyte & Separator for LIB. In Proceedings of the 2015 the Second International Forum on Electrolyte & Separator for Advanced Batteries, Shenzhen, China, 11 November 2015; p. 33.
262. *B3 E13Q2 2013-2014 Chapter 2 LIB-Equipped Vehicle Market Bulletin (13Q2)*; Tokyo, Japan, 2013; p. 40. Available online: https://wenku.baidu.com/view/231faba4aeaad1f347933f12.html?re=view (accessed on 4 June 2024).
263. Maltsev, T. *Thermal Behaviour of Li-Ion Cells*; KTH: Stockholm, Sweden, 2012.
264. Yakushi, B. EV/HEV Safety (Nissan). Available online: https://www.slideshare.net/StphaneBARBUSSE/nissan-presentation-bobyakushi-ev-hev-safety (accessed on 13 November 2019).
265. Bugga, R.; Krause, C.; Billings, K.; Ruiz, J.P.; Brandon, E.; Darcy, E.; Iannello, C. Performance of Commercial High Energy and High Power Li-Ion Cells in Jovian Missions Encountering High Radiation Environments. In Proceedings of the NASA Battery Workshop, Huntsville, AL, USA, 19–21 November 2019; p. 28.
266. World Smart Energy Week 2014 東京現場直擊系列報導二 (Tokyo live coverage series two). Available online: http://www.materialsnet.com.tw/material/DocView_MaterialNews.aspx?id=11697 (accessed on 7 November 2018).
267. Lithium Ion Cell for Aerospace Applications LFC 40 (GS Yuasa). Available online: http://www.gsyuasa-lp.com/SpecSheets/LFC40_Spec.pdf (accessed on 13 November 2019).
268. 动力电池的创新与发展 (Innovation and development of power batteries) (Yinlong, Gree). In Proceedings of the First LTO Battery Industrial Forum (LBIF2017), Wuxi, China, 6–7 July 2017; p. 41.
269. 冲压件 (Stamping Parts). Available online: http://www.gdrunye.com/productinfo/1222036.html (accessed on 14 November 2024).

270. Коштял, Ю.М.; Жданов, В.В. Обзор современных материалов для производства литий-ионных аккумуляторов. In Proceedings of the 27-я Международная специализированная выставка "Автономные источники тока", Moscow, Russia, 21–23 March 2018; p. 17, (Koshtyal, Yu.M.; Zhdanov, V.V. Review of modern materials for the production of lithium-ion batteries. In Proceedings of the 27th International Specialized Exhibition "Autonomous Power Sources", Moscow, Russia, 21–23 March 2018; p. 17.).
271. Wachtler, M. *Li-Ion Batteries Lecture Winter Term 2016/17 Electrolytes*; Zentrum für Sonnenenergie- und Wasserstoff-Forschung Baden-Württemberg (ZSW): Ulm, Germany, 6 February 2017; p. 61. Available online: https://www.zsw-bw.de/fileadmin/user_upload/PDFs/Vorlesungen/lib/170206_Uni-Ulm_Lecture_LIB_Wachtler_Electrolytes.pdf (accessed on 9 October 2019).
272. Ohkubo, K.; Sukino, K.; Yamate, S.; Kozono, S.; Katayama, Y.; Nukuda, T. Enhancement of Low Temperature Power Performance for Lithium-ion Cells with Lithium Titanium Oxide Negative Electrode by New Functional Electrolyte: Application of Carboxylic Acid Ester as Solvent and Organic Titanium Compound as Additive. *GS Yuasa Tech. Rep.* **2009**, *6*, 14–19.
273. Yaakov, D.; Gofer, Y.; Aurbach, D.; Halalay, I.C. On the Study of Electrolyte Solutions for Li-Ion Batteries That Can Work Over a Wide Temperature Range. *J. Electrochem. Soc.* **2010**, *157*, A1383–A1391. [CrossRef]
274. Ren, Y.-H.; Wu, B.-R.; Mu, D.-B.; Yang, C.-W.; Zhang, C.-Z.; Wu, F. Optimization of Electrolyte Conductivity for Li-ion Batteries Based on Mass Triangle Model. *Chem. Res. Chin. Univ.* **2013**, *29*, 116–120. [CrossRef]
275. de Leon, S. LIB 18650 Cells New Replacement Cylindrical Cell Sizes (20650, 20700, 21700) Report. Available online: https://www.sdle.co.il/wp-content/uploads/2018/12/15-LIB-18650-Cells-New-Replacement-Cylindrical-Cell-Sizes-20650-20700-21700-Report-ver-3-presentation-for-conferences.pdf (accessed on 13 November 2019).
276. Toyal-Carbo. Available online: http://www.toyal.co.jp/eng/products/haku/hk_tc.html (accessed on 16 October 2018).
277. Friesen, A.; Schappacher, F.; Winter, M. Energy Density, Lifetime and Safety—Not Only an Issue of Lithium Ion Batteries. In Proceedings of the 81th Annual Meeting of DPG and Spring Meeting, Münster, Germany, 27–31 March 2017; p. 36.
278. Zhang, J.; Fang, W. Kinetics of the EDV Li-ion Cells on Electrode Thickness, Porosity and Tortuosity (Asahi Kasei). In Proceedings of the International Battery Seminar, Fort Lauderdale, FL, USA, 21–22 March 2017; p. 35.
279. LiB の性能を向上する高粘度CMC 第一工業製薬 社報 No.574 拓人2015 秋 (High viscosity CMC to improve the performance of LiB Daiichi Kogyo Seiyaku Co., Ltd. Company Report No. 574 Takuto 2015 Autumn). Available online: https://www.dks-web.co.jp/catalog_pdf/574_1.pdf (accessed on 29 October 2018).

280. 高性能动力三元材料技术开发进展 谭欣欣 (Advances in high-performance power ternary materials technology development, Hunan Shanshan Energy Technology co. Ltd.). In Proceedings of the 2016' 第五届中国电池市场年会, 暨第一届动力电池应用国际峰会, 第二届中国电池行业智能制造研讨会 (The 5th China Battery Market Annual Conference, the 1st International Power Battery Application Summit, the 2nd China Battery Industry Intelligent Manufacturing Seminar) (CBEA2016), Beijing, China, 14–15 November 2016; p. 35.
281. Nishida, T. The Limitations of Anode Material and Future Development Direction (Hitachi Chemical). In Proceedings of the Asia-Pacific Lithium Battery Congress 2014, Shenzhen, China, 26–28 March 2014; p. 38.
282. Liang, G. Highlights on Phostech's Recent Business Development and Advanced Life Power (R) Grade. Available online: https://www.yumpu.com/en/document/read/34821972/highlights-on-phostechs-recent-business-development-and (accessed on 14 November 2019).
283. Fetcenko, M. 适用于小型移动、固定以及交通运输的金属氢化物镍电池技术进展 (Nickel Metal Hydride Batteries for Portable, Stationary and Transportation Application). In Proceedings of the of 12th China International Battery Fair (CIBF2016), Shenzhen, China, 26 May 2016; p. 37.
284. Nagai, H. Lithium-Ion Secondary Battery. US8945768B2, 6 May 2011.
285. Zheng, Z.; Deng, T.; Nie, Y.; Zhou, X. Graphite Negative Material for Lithium-Ion battery, Method for Preparing the Same and Lithium-On Battery. US9281521 B2, 12 April 2013.
286. Introduction of Denka Black (Denki Kagaku Kogyo K.K.). Available online: http://utsrus.com/index.php?file_id=10768&Itemid=7&option=com_virtuemart&page=shop.getfile&product_id=5833 (accessed on 14 October 2018).
287. DENKA 锂电池导电剂 乙炔黑 (Lithium Battery Conductive Agent Acetylene Black) Li-400, Li-250. Available online: https://www.achatestrade.com/productinfo/782551.html (accessed on 14 November 2024).
288. DENKA BLACK Li のグレード (Grade). Available online: http://www.denka.co.jp/scm/product/scm/detail_003523.html (accessed on 26 February 2018).
289. *Lithium-Ion Batteries*; Springer: New York, NY, USA, 2009; p. 452. [CrossRef]
290. Lin, J. BAK xEV Battery R&D and Industrialization. In Proceedings of the 2014 U.S.-China Electric Vehicle and Battery Technology Workshop, Seattle, WA, USA, 18 August 2014; p. 53.
291. Lei, H.; Blizanac, B.; Atanassova, P.; DuPaskier, A.; Oljaca, M. Carbon materials to Advance Li-ion and Lead-acid Battery Performance. In Proceedings of the CIBF 2014, International Conference on the Frontier of Advanced Batteries, Shenzhen, China, 20–22 June 2014; p. 29.
292. Tatsuo, N. 高速充放電リチウムイオン二次電池の開発 (Express Charging/Discharging Lithium-Ion Secondary Batteries). *FB テクニカルニュース (Tech. News)* **2008**, *11*, 3–18.
293. Imoto, H. The development of non-graphitezable carbon "Carbotron P" for automotive lithium-ion batteries (KMBJ). In Proceedings of the Advanced Automotive Battery Conference EUROPE (AABC), Strasbourg, France, 24–28 June 2013; p. 19.

294. Yoshio, M.; Gunawardhana, N. Modern graphite, electrolyte and its additives (Saga University). In Proceedings of the Shenzhen International Li-Ion Battery Summit, Shenzhen, China, 3–6 December 2011; p. 91.
295. Wang, S. Development of Low Temperature & Fast Charging Anode Materials for xEV Power Battery. In Proceedings of the 2nd HEV Market & Advanced Battery Technology Development Seminar, Hangzhou, China, 13–14 October 2016; p. 33.
296. Lee, J.-J. "Soft Carbon" as a LIB anode material for xEV application. In Proceedings of the Korea Advanced Battery Conference, Coex, Seoul, Republic of Korea, 29 May 2013; p. 24.
297. Coclasure, A. Electrode Scale and Electrolyte Transport Effects on Extreme Fast Charging of Lithium-Ion Cells. In Proceedings of the 38th Annual International Battery Seminar & Exhibition, Virtual, 9–11 March 2021; p. 22.
298. Ruiz, V.; Pfrang, A.; Kriston, A.; Omar, N.; Van den Bossche, P.; Boon-Brett, L. A review of international abuse testing standards and regulations for lithium-ion batteries in electric and hybrid electric vehicles. *Renew. Sustain. Energy Rev.* **2018**, *81*, 1427–1452. [CrossRef]
299. Kong, L.X.; Li, C.; Jiang, J.C.; Pecht, M.G. Li-Ion Battery Fire Hazards and Safety Strategies. *Energies* **2018**, *11*, 2191. [CrossRef]
300. Battery Abuse Testing Laboratory. Available online: https://energy.sandia.gov/wp-content//gallery/uploads/BatLab-Final2-SAND-2012-0502Pno-marks.pdf (accessed on 31 July 2019).
301. Lamb, J.; Orendorff, C.J. Battery Safety R&D at Sandia National Laboratories. Available online: https://www.osti.gov/servlets/purl/1336278 (accessed on 31 July 2019).
302. Cell Safety Test (Espec). Available online: https://www.espec.co.jp/english/products/trustee/test/safety.html (accessed on 31 July 2019).
303. Battery Safety Testing. Development and Validation Testing to Current and Emerging Standards (TuV). Available online: https://www.tuvsud.com/en/industries/mobility-and-automotive/automotive-and-oem/automotive-testing-solutions/battery-safety-testing (accessed on 31 July 2019).
304. Bonoski, J. Fire Protection for Lithium-Ion Batteries. In Proceedings of the 34th International Battery Seminar & Exhibit 2017 (Battery Safety Session), Ft. Lauderdale, FL, USA, 21 March 2019; p. 34.
305. Kritzer, P. Safer Batteries! In Proceedings of the Advanced Automotive Battery Conference (AABC 2021 Europe), Virtual, 20 January 2021.
306. Sheet Materials for Lithium-Ion Battery Cases (LB Series). Available online: https://www.nikkeikin.com/products/board/p3.html (accessed on 14 November 2019).
307. Analysis Of Samsung SDI Power Battery Technology For Lithium Battery Giant. Available online: http://www.optimumnanolithiumbattery.com/news/analysis-of-samsung-sdi-power-battery-technolo-4920947.html (accessed on 6 August 2018).
308. Maehliss, J. Lithium-Ionen Batterietechnologie. Available online: https://www.hs-rm.de/fileadmin/persons/khofmann/Gastvortraege/Vortragsfolien/20160603-Maehliss-Lithium-Ionen-Batterietechnologie.pdf (accessed on 14 May 2018).

309. Anderson, N.; Tram, M.; Darcy, E. 18650 Cell Bottom Vent: Preliminary Evaluation into its Merits for Preventing Side Wall Rupture. In Proceedings of the S&T Meeting, San Diego, CA, USA, 7 December 2016; p. 53.
310. Kokam Li-ion Cell. Available online: https://kokam.com/cell (accessed on 23 November 2020).
311. Battery Safety 101: Anatomy—PTC vs. PCB vs. CID. Available online: https://batterybro.com/blogs/18650-wholesale-battery-reviews/18306003-battery-safety-101-anatomy-ptc-vs-pcb-vs-cid (accessed on 14 May 2018).
312. Kokam Cell Brochure 20160304. Available online: https://www.west-l.ru/uploads/tdpdf/sf_slpb-cell-brochure1.pdf (accessed on 6 July 2018).
313. Barreras, J.V.; Raj, T.; Howey, D.A.; Schaltz, E. Results of Screening over 200 Pristine Lithium-Ion Cells. In Proceedings of the 2017 IEEE Vehicle Power and Propulsion Conference (VPPC), Belfort, France, 11–14 December 2017; pp. 1–6.
314. BYD Blade Battery—Unsheathed to Safeguard the World. Available online: https://www.youtube.com/watch?v=dIt5z4wT9RE (accessed on 30 January 2021).
315. Lithium-Ion Battery Lids. Available online: https://www.schott.com/en-gb/products/battery-and-capacitor-lids-p1000270/product-variants?tab=lithium-ion-battery-lids (accessed on 14 November 2024).
316. What Can We Learn from the Samsung Galaxy Note7 Battery Safety Event. Available online: https://www.sdle.co.il/wp-content/uploads/2018/12/18-What-can-we-learn-from-the-Samsung-Galaxy-Note-7-battery-safety-event-ver-5.pdf (accessed on 14 November 2024).
317. Liu, X.Y. 电芯装配工序安全手册 (Battery Assembly Process Safety Manual). In Proceedings of the Amperex Technology Limited Seminar, Dongguan, China, 9 October 2007; p. 17.
318. 3M Marches Electroniques. Available online: http://www.docin.com/p-206678983.html (accessed on 17 February 2017).
319. Strapping Tape (200m L × 10mm W × 0.03mm Thickness) for Pouch/Cylinder Cell—EQ-LiB-ST. Available online: http://www.mtixtl.com/Li-ionBatteryStrappingTape200mLx10mmWx0.03mmThicknessforPouc.aspx (accessed on 3 April 2017).
320. Zhang, J.; Ramadass, P.; Fang, W. Separator for Li-ion. In Proceedings of the 中国锂电池电解质、隔膜材料技术与 市场发展论坛 (China Lithium Battery Electrolyte, Diaphragm Material Technology and Market Development Forum), Guangzhou, China, 14–16 August 2013; p. 44.
321. Patents by Inventor Tatsuhiro Yaegashi (UACJ CORPORATION). Available online: http://patents.justia.com/inventor/tatsuhiro-yaegashi (accessed on 28 February 2017).
322. Amionx SafeCore—Battery Safety from the Inside-Out. Available online: http://www.amionxsafecore.com/wp-content/uploads/2018/12/Amionx-SafeCore-White-Paper.pdf (accessed on 20 January 2021).
323. King, J. Enabling the Next Generation of Batteries through Safety. In Proceedings of the 37th International Battery Seminar&Exhibit, Vitual, 28 July 2020; p. 23.

324. ITRI※ Grants Taiwan Mitsui Chemicals Exclusive License to Manufacture and Sell STOBA™—STOBATM Curbs Thermal Runaway to Prevent Lithium-ion Battery Fires. Available online: https://jp.mitsuichemicals.com/en/release/2014/140929.htm (accessed on 22 February 2020).
325. 刘兴江 (Liu Xingjiang). 化学电源前沿技术概述 *(Introduction to Frontline Technology for Chemical Power)*; Tianjin, China, 1 February 2017; p. 67. Available online: https://wenku.baidu.com/view/76210e1ab80d6c85ec3a87c24028915f804d840d.html?from=search (accessed on 4 June 2024).
326. Doughty, D.H.; Roth, P.E. A General Discussion of Li Ion Battery Safety. *Electrochem. Soc. Interface* **2012**, 37–44. [CrossRef]
327. Doughty, D.H. *Vehicle Battery Safety Roadmap Guidance. Subcontract Report NREL/SR-5400-54404*; National Renewable Energy Laboratory: Golden, CO, USA, 2012; p. 131.
328. Roth, P.E.; Doughty, D.H. Proceedings of the Advanced Automotive Battery and Ultracapacitor Conference (AABC-06), Baltimore, MD, USA, 15–19 May 2006.
329. Hao, R. Discoveries and Understanding of Materials and Cell Chemistry for xEV Batteries. In Proceedings of the Advanced Automotive Battery Conference (AABC 2021 Europe), Virtual, 20 January 2021; p. 21.
330. Arora, P.; Zhang, Z.M. Battery separators. *Chem. Rev.* **2004**, *104*, 4419–4462. [CrossRef] [PubMed]
331. Lu, H.-S. Commercialized Technology and Future Market Development Trends of Electrolyte & Separator for LIB. In Proceedings of the Third International Forum on Electrolyte & Separator Materials for Advanced Batteries, Nanjing, China, 18–19 October 2018; p. 22.
332. リチウムイオン2次電池用セパレータ開発動向 (東レ株式会社, Setela). (Development Trends of Separators for Lithium-Ion Secondary Batteries (Toray Industries, Inc., Setela)). *TRC News*. 1 May 2017, pp. 1–6. Available online: https://www.toray-research.co.jp/service/trcnews/pdf/201705-01.pdf (accessed on 16 May 2024).
333. Lu, M. Present Technology & Market Development Trends of Electrolyte & Separator for LIB. In Proceedings of the 中国锂电池电解质、隔膜材料技术与 市场发展论坛 (China Lithium Battery Electrolyte, Diaphragm Material Technology and Market Development Forum), Guangzhou, China, 14–16 August 2013; p. 48.
334. Kim, J.H. LG Li-ion battery separator for safety of EV Li-ion battery. In Proceedings of the EV Li-ion battery forum, Shanghai, China, 2–3 September 2009; p. 18.
335. Nguyen, T.T.D.; Abada, S.; Lecocq, A.; Bernard, J.; Petit, M.; Marlair, G.; Grugeon, S.; Laruelle, S. Understanding the Thermal Runaway of Ni-Rich Lithium-Ion Batteries. *World Electr. Veh. J.* **2019**, *10*, 79. [CrossRef]
336. Cardillo, S. Технология изготовления сепаратора для литий-ионных батарей (Separator manufacturing technology for lithium-ion batteries) (Celgard). In Proceedings of the 27th International Specialized Exhibition "Autonomous Power Sources" Interbat, Moscow, Russia, 21–23 March 2018; p. 21.

337. Ouyang, D.X.; Chen, M.Y.; Huang, Q.; Weng, J.W.; Wang, Z.; Wang, J. A Review on the Thermal Hazards of the Lithium-Ion Battery and the Corresponding Countermeasures. *Appl. Sci.* **2019**, *9*, 2483. [CrossRef]
338. Lithium-ion Battery Safety—Assessment by Abuse Testing, Fluoride Gas Emissions and Fire Propagation. Available online: https://publications.lib.chalmers.se/records/fulltext/251352/251352.pdf (accessed on 14 November 2024).
339. Balakrishnan, P.G.; Ramesh, R.; Kumar, T.P. Safety mechanisms in lithium-ion batteries. *J. Power Sources* **2006**, *155*, 401–414. [CrossRef]
340. Brandt, K.; Garche, J. *Electrochemical Power Sources: Fundamentals, Systems, and Applications (Li-Battery Safety)*; Elsevier: Amsterdam, The Netherlands, 2018; p. 670. [CrossRef]

6 Lithium-Ion Electric Energy Storage for Stationary Applications

6.1. High-Power Electric Energy Storage

Industrial energy storage enhances energy efficiency during energy generation and transmission, including renewable sources. According to Sandia's laboratory database [1], pumped storage power plants have the most significant share ($\approx$97% installed capacity, Figure 56). But due to the requirements imposed on the terrain, this type of energy storage can only be applied in territories with a particular landscape.

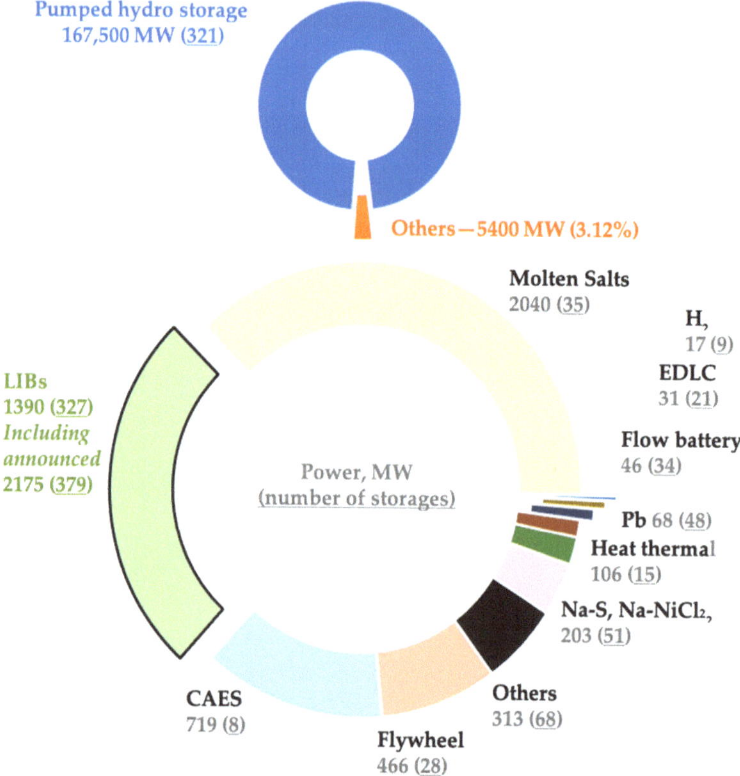

Figure 56. Contribution of the main types of energy storage to the total power balance (based on the Sandia Lab database: database version 09.2021 [1], rated power not less than 100 kW, status—operational). Source: Figure by authors.

The remaining $\approx$3% falls on electrochemical (such as lithium-ion (LIB), sodium–sulphur (Na-S), sodium–nickel chloride (Na-NiCl$_2$), lead–acid, and flow batteries), mechanical (compressed air, flywheels, etc.), and thermal (molten salts, heat thermal, etc.) storage. For the storage of energy obtained from renewable sources (solar), the largest share in terms of installed capacity and energy falls on thermal—molten salts (Figure 57). The thermal storages (molten salts) under construction have high power and energy capacity and are located near large systems to obtain energy from solar radiation. They are followed by LIBs (1.38 GW/2.4 GWh) and Na-S and Na-NiCl_2 (0.2 MW/1.28 GWh) stationary energy storages, which are smaller and can be used in distributed energy systems. The installed capacity and energy of flow batteries, electric double-layer capacitor (EDLC) batteries, and lead–acid (PbA, lead–acid–carbon—PbA/C, hybrid systems—PbA/EDLC) storage batteries are significantly lower in total (Figure 57).

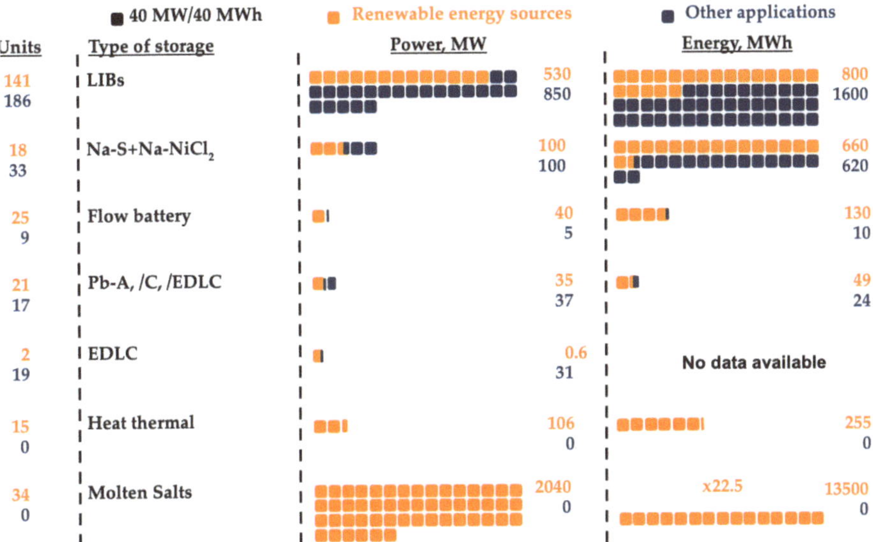

Figure 57. Installed energy and power capacity of stationary electrochemical and thermal energy storage used in the generation of energy from renewable sources [1]. Source: Figure by authors.

According to other data [2], in 2018 and 2019, the total supply of LIBs for stationary storage devices exceeded 10.5 GWh and 11.5 GWh. In 2018, about 87% and 10% of installed lithium-ion batteries for stationary applications were for large energy storage systems and home storage systems, respectively. More than half of the electric energy storage (EES) with more than 100 kWh capacity was installed in South Korea (the main applications are given in Table 37).

This chapter briefly describes large industrial energy storage systems (with power above 500 kW—ESSs) and uninterruptible power supplies (UPSs). Furthermore, it contains a list of lithium-ion modules and cells produced by the leading manufacturers used to design stationary ESS. Moreover, industrial lithium-ion cells with various pairs of active cathode and anode materials are compared.

Table 37. Applicability of lithium-ion EES [3,4].

Generation	T&D (Transmission and Distribution)	Demand
Ancillary Services	T&D Infrastructure Services	Customer Energy Management Services
- Regulation - Spinning reserve - Non-spinning reserve - Supplemental reserve - Voltage support - Black start - Load following/Ramping support for Renewables	- Frequency regulation - Transmission upgrade deferral - Transmission congestion relief - Distribution upgrade deferral - Voltage support	- Power quality - Power reliability - Retail electric energy time-shift - Demand charge management
Bulk Energy Services		
- Electric energy time-shift (Arbitrage) - Electric supply capacity		

Source: Authors' compilation based on data from references cited in the caption of the table.

Examples of ESSs created using LICs with various material chemistries are given in Table 38. The power-to-energy ratio varies between 0.25 and 4 and depends on ESS functions. For more details on the ESS operation during frequency regulation, levelling of power consumption and generation peaks, smoothing power profiles at energy generation based on renewable power sources, etc., refer to papers [4,5].

In general, an ESS consists of energy storage, a power conditioning system, a transformer, switchgear, and an energy management system [6,7]. In addition, an ESS can consist of five safety devices: a fuse, a contactor, a circuit breaker, a BMS system, and a fire suppression system.

Lithium-ion electric energy storage can comprise a 20-foot, 40-foot, or 53-foot container, or be designed as a group of racks with modules mounted in a protective case (Tesla powerpack) [8,9]. The voltage of container-type EESs may vary within the range of 600 ÷ 1100 V (depending on the model and manufacturer), and the discharge current can reach several thousands of amperes. The primary characteristics of an EES (a 20-foot container) manufactured by Saft can be found elsewhere [10]. The overall energy varies from 700 kWh (EES model—IM +20P2) to 1180 kWh (EES model—IM +20E). The nominal discharge/charge power varies from 2300/900 kW (IM +20E) to 2800/2900 kW (IM +20P2). The voltage range is 630 ÷ 867 V [10].

Table 38. Lithium-ion electric energy storage systems.

LIC Manufacturer	City	P, MW	E, MWh	P/E	LIC Material	Ref.
A123	Angamos	20	5	4.00	LFP/C	[11]
A123	Almacena	1	3	0.33	LFP/C	[12]
GS Yuasa	Kushiro Town	10	6.75	1.48	LMO/C	[13]
GS Yuasa	Everett	1	0.5	2.00	LMO/C	[14]
LG	Tehachapi	8	32	0.25	NCM/C	[15]
LG	Rutland	2	2	1.00	NCM/C	[16]
Saft	Anahola	6	4	1.50	NCA/C	[17]
Saft	Salinas	5	1.3	3.85	NCA/C	[17]
Toshiba	Minami-Soma	40	40	1.00	M.O./LTO	[18]
Toshiba	Helsinki	1.2	0.6	2.00	M.O./LTO	[19]

Source: Authors' compilation based on data from references cited in the table.

The container comprises several high-voltage racks connected in parallel to the common line (illustration is given elsewhere [6]). The racks accommodate series-connected modules, including lithium-ion cells connected in series and in parallel. The energy storage is controlled through a three-level battery management system. The top-level BMS interacts with the power conditioning system and energy management system (or is a part of these systems). Each rack can be disconnected to facilitate maintenance. The cooling system can be of air or liquid type. The air-cooling system can be used to cool the entire container space and functional cooling devices (rack with cells, control panel, etc.). In addition, the EES is equipped with an emergency fire suppression system. The main components of the container-type lithium-ion battery are given in Table 39.

Table 39. Basic components of EES.

Description	Function
Lithium-ion cell	Energy storage
Module	- Optimum heat dissipation is provided through ventilation holes made in heat-sinking plates. - Standard structure. - The option of proper configuration build-up by the creation of serial and parallel connections.
Module battery management system	See the Slave BMS below

Table 39. *Cont.*

Description	Function
Rack with modules	- Parallel connections are required for the build-up of a 1 ÷ 10 MWh energy storage. - Constructed as per the users' requirements for energy and voltage. - Equipped with several safety mechanisms. - Improved availability of the system ensured by the backup function.
Master BMS	- Operation support within the preset voltage range. - Overcurrent protection and alarm - Charge and discharge current measurement. - Overtemperature alarm. - Automatic shutdown and return to operation. - Magnetic switch.
Slave BMS	- Collection of cell parameters (temperature and other information). - LIC balanced by the state of charge (SOC). - Data transmission to Master BMS via the interface (for example, CAN bus).
On/Off control switch	- Facilitates maintenance (sectional shutdown). - Safe activation of the battery system through the pre-charging function.
Rack switch	- Current surge protection. - Isolation of battery module rack.
Protection door	- A barrier for protection in case of emergencies.
DC distribution panel	- Equipped with a main switch and central DC control unit
BMS of ESS	- Processes and displays information on the status of the battery system at the rack level and the level of modules and cells. - Records the readings of energy, voltage, current, state of health (SOH), and state of charge (SOC). - Data access via a user interface. - Detection of abnormal operation at the level of modules and LIC, quick diagnostics. - Activates the security systems in case of emergencies. - Data recording to electronic log and storage for three years. - Remote control via Ethernet.
Air conditioning system (HVAC)	- Heating, ventilation, and air conditioning in racks with modules are provided directly. - The ventilation system may be unavailable in some other LIC-based ESSs. For example, the Tesla powerpack ESS is also equipped with a liquid cooling system.
Fire suppression system	- Reduction in fire intensity or fire suppression in case of emergency.

Source: Table by authors.

6.2. Uninterruptible Power Supplies

The primary function of uninterruptible power supplies (UPSs) is to maintain the power supply of the facilities—data centres, factories, business centres, etc.—for a short period of 10 ÷ 15 min. As a rule, this time interval is sufficient to start the emergency independent power supply (diesel generator, etc.). Therefore, high-power LICs that can be discharged by 4 ÷ 6C currents are commonly used to manufacture UPSs. Such LICs are selected because though the cost of increased-power LICs is higher than that of the cells with high specific energy, it is necessary to design UPSs with more significant energy storage using high-energy cells to provide a preset power. Hence, it is reasonable to install one UPS with high-power cells instead of several UPSs with less powerful cells to reach the target power. Additionally, industrial buildings may have limited space for energy storage arrangements.

A rack of uninterruptible power supplies may include a cabinet cover, switched mode power supply, fire suppression system, fan, battery protection unit (BPU, battery disconnector, BMS, fuse, contactor, current sensor), and battery module (an illustration of the rack can be found elsewhere [6]).

The functional characteristics of UPSs can be illustrated using the example of Samsung products: DC UPS 0.5C (120 V), AC UPS 4C (600 V) and AC UPS 6C (600 V) with following energy (power) 23.8 kWh (11.9 kW), 35.7 kWh (143 kW) and 35.7 kWh (214 kW). More detailed information about these battery racks can be found elsewhere [3,20].

As the safety requirements are strict, each rack with modules can be equipped with a fire suppression system and battery protection unit.

Direct current-to-alternating current conversion is generally supported. For a detailed description of the UPS units and their functions, refer to the presentation by Saft (Flex'ion) [21].

6.3. Modules and Lithium-Ion Cells

Electric energy storage and uninterruptible power supplies based on lithium-ion cells consist of modules. The modules' design depends on their functional purpose and the types of LICs used.

The following components can be distinguished in the module structure: lithium-ion cells, battery management system, case parts, and insulating components. Cooling (temperature control) of the module can be passive (the construction of the module with passive cooling can be found elsewhere [22]) or active. Active cooling (temperature control) can be provided by air injection with fans (for example, Yuasa LIM30H-8A1 [22]) or liquid pumping (for example, Tesla powerpack [23]). In addition, when LICs have laminated foil (pouch) cases, aluminium plates (Enerdel, Moxie+ [24]) can be used to improve heat dissipation.

The characteristics of some modules produced by the major manufacturers are given in Table 40. The average discharge voltage and operating voltage range depending on a number of LICs connected in series (S) and on voltage (depending on LIC-active materials) applied in the manufacture of LICs. As a rule, the average voltage of the module varies from 22 V to 53 V. There are also some modules with a high average discharge voltage, for example, KBM 255 2P 20S (Kokam).

The operating voltage range is monitored by the battery management system and can be narrowed to extend the cycle life, but this reduces the stored energy and usable energy. The module capacity is a sum of the capacities of LICs connected in parallel (P). LICs with a high capacity of 20 Ah and above are often used for the manufacture of modules. Still, many companies (for example, Tesla, Valence, Sony) apply cells of lower capacity (form factors 21700, 18650 and 26650) to manufacture EES modules.

Usually, no more than three LICs are connected in parallel when manufacturing modules from medium- and high-capacity cells. Therefore, it becomes necessary to switch high currents due to a significant number of parallel connections inside a unit (a rack). Furthermore, suppose resistance rises because of ageing in one or more of the parallel-connected cells. In that case, a higher current will flow through the other cells, leading to the module's imbalance and failure.

The weight of the modules usually varies within 20 ÷ 35 kg, but there are some exceptions—modules 936C08 (Leclanche) and KBM 255 2P 20S (Kokam) with a weight above 90 kg. The specific energy of the modules depends on the specific energy of the LICs and the module design. The lowest specific energy is typical for the modules manufactured with cells made using lithium titanate anodes (42 ÷ 83 Wh/kg). A little higher specific energy is typical for the modules, including cells with positive lithium manganese oxide spinel cathodes ($LiMn_2O_4$, 44 ÷ 86 Wh/kg) or lithium iron phosphate cathodes (62 ÷ 97 Wh/kg). Finally, the maximum specific energy (80 ÷ 130 Wh/kg) is typical for the modules manufactured using cells with positive electrodes based on lithium nickel cobalt manganese oxide and lithium nickel cobalt aluminium oxide.

The specific power of the modules depends on the LIC power, the limits established for heat release, and the module design. LIC power depends on the applied materials, design, and manufacturing technology (for more details, refer to Chapter Lithium-Ion Cells and Batteries for Hybrid, Electric Passenger Vehicles, Section Technical and Design Parameters Affecting LICs Power). The power of LICs mounted in stationary EESs varies within a wide range of 60 ÷ 950 W/kg. The maximum values of power for the LFP/graphite, LMO/LTO, LMO/C, LMO/graphite, NCM/graphite, and NCA/graphite modules reviewed in this section are 570 W/kg, 294 W/kg (the modules built from SCiB^{TM} 10 Ah LICs have a higher power), 518 W/kg, 690 W/kg, and 950 W/kg, respectively.

Table 40. Lithium-ion cell modules for EES and UPS.

Manufacturer	Model	U, V	ΔU, V	C, Ah	m, kg	V, L	E, Wh/kg	P, W/kg	T, °C	LIC	Ref.
BYD	B-Plus 2.5	51.2	43.2 ÷ 57.6	50	34.8	30.8	70	72	−20 ÷ 50	FV50NP	[25]
GS Yuasa	LIM50E-7G	25.9	19.3 ÷ 29.05	47.5	15	10.0	86	518	−10 ÷ 45	LEV50	[26,27]
GS Yuasa	LIM30H-8	28.8	22 ÷ 33.2	30	19.5	14.7	44	860	−25 ÷ 45	LIM30H	[22,28]
Hitachi	CH75-6	22.2	16.2 ÷ 25.2	75	20	16.4	83	250	−20 ÷ 50	CH75	[29]
Leclanche	936C08 Titanate	46	34 ÷ 54	90	99	90.7	42	139	0 ÷ 40	LecCell 30Ah	
LG	M4863P3B	51.8	42 ÷ 58.8	63	25	16.6	132	132	–	JH3	[30,31]
LG	M4864P6B	51.5	42 ÷ 58.8	64	28	16.8	118	229	–	JP3	[30,31]
Litarion	Litastore 2.3	51.1	46.2 ÷ 57.7	44	22	15.0	104	77	−20 ÷ 50	Litacell 44	[32]
Kokam	KBM 255 2P 20S	74	60 ÷ 84	75	92.5	62.2	120	240	−20 ÷ 55	SLPB120255255	[33]
NEC (A123)	ALM 12V7 (1C)	52.8	– ÷ 57.6	100	–	–	–	–	−30 ÷ 55	AMP20	[34]
Saft	Synerion 48E	48	42 ÷ 56	45	19	17.2	116	132	−20 ÷ 60	VL 45E	[35]
Saft	Synerion 48M	48	42 ÷ 56	41	19	17.2	104	398	−20 ÷ 60	VL 41M	[36]
Saft	Synerion 48P	48	–	30	19	17.2	80	950	−20 ÷ 60	VL 30P	[37]
Saft	Flex'ion 46 M Fe	46	35 ÷ 53.2	39	18.5	17	97	300	−20 ÷ 40	VL 41M Fe	[38,39]
Saft	Flex'ion 46 P Fe	46	35 ÷ 53.2	28	18.5	17	70	568	−20 ÷ 40	VL 30P Fe	[38,39]
Samsung	M2994	29.6	25.6 ÷ 33.2	94	22	13.0	127	64	–	94 Ah	[3]
Samsung	U6-M020	29.2	24 ÷ 33.6	67	17	14.4	118	690	–	67 Ah	[20]
Sony	IJ1001M	51.2	– ÷ 57.6	24	17	14.5	71	z	−20 ÷ 60	Fortelion 26650	[40,41]
Toshiba	Type3-23 2P12S	27.6	18 ÷ 32.4	45	15	8.6	83	294	−30 ÷ 50	SCiB™ 23 Ah	[42,43]
Valence	U27-36XP	38.4	30 ÷ 40.8	46	19.6	11.8	91	211	–	18650/26650	[44]
Valence	P40-24	25.6	20 ÷ 27.2	40	16.5	11.0	62	274	–	18650/26650	[44]

Source: Authors' compilation based on data from references cited in the table.

The operational temperature range of the modules is preset not only based on LIC characteristics, but also considering the BMS design. For example, LecCell 30 Ah LICs with lithium titanate anodes can be charged and discharged at −20 °C. But the modules built using such cells have a higher temperature (0 ÷ 40 °C) at the lower performance limit. Similarly, the operational temperature range of the FM01202CCA01A module (Toshiba, 2011) outfitted with lithium titanate cells varies within −5 to 40 °C (the storage temperature range is −25 ÷ 55 °C). Refer to Table 40 for the temperature ranges allowed for module discharging. Lithium titanate LICs can be charged without a significant performance drop at low temperatures, as the lithium titanate ($Li_4Ti_5O_{12}$) potential is approximately at the level of 1.55 V. Therefore, no lithium plating (dendrite formation) occurs. Moreover, LICs with non-graphitised carbon (soft carbon, hard carbon) anodes can be charged at low temperatures (down to −10 °C, Lim30H-8) since their average potential is relatively high compared with the lithium potential. The potential of lithium graphite is slightly higher than that of lithium and, when charging at low temperatures, lithium plating may occur (if the excess of active graphite material does not sufficiently resist a drop in potential to a value close to zero vs. Li/Li^+). As a result, the internal resistance may rise sharply, which shortens the cycle life and increases the risk of emergency situations. However, some modules (LIM50e, GS Yuasa) where graphite functions as active anode material can be charged with the ≈C/5 current at a temperature of −20 °C [45]. Carbon coating can be applied to the LIC graphite to reduce the lower charging threshold, and/or the electrolyte may contain special additives required to maintain high conductivity of SEI film, and/or the anode capacity is much higher than the cathode capacity.

The average discharge voltage of the considered LICs for EESs lies within the range of 2.3 to 3.7 V, and it depends on active cathode and anode materials as well as the shape of the discharge curves. Figures 58–64 show the discharge curves for LICs with various pairs of active cathode and anode materials. All curves are plotted with the same scale to facilitate the comparison process. An excessive amount of active anode material should be applied during LIC manufacture (the negative electrode capacity is higher than that of the positive electrode).

The potential of graphite and lithium titanate varies slightly within quite a wide range of lithiation degrees. Thus, the actual shape of a discharge curve depends on the shape of the cathode material discharge curve. If non-graphitised materials (soft carbon, hard carbon) are applied, the potential relative to lithium gradually increases (decreases) when lithium is extracted (inserted) from (into) the structure. In this regard, the shape of a discharge curve for LICs with a pair of LMO/C active materials differs from discharge curves for the other LICs.

The maximum DC (continuous) discharge current depends on the materials used (size, particle size distribution, tap density, chemical nature), design (terminal construction, number of electrodes, etc.), and LIC manufacturing technology (thickness, porosity of electrodes, etc.). In addition, this value is limited by the safe operation conditions (the heating-up temperature at discharge).

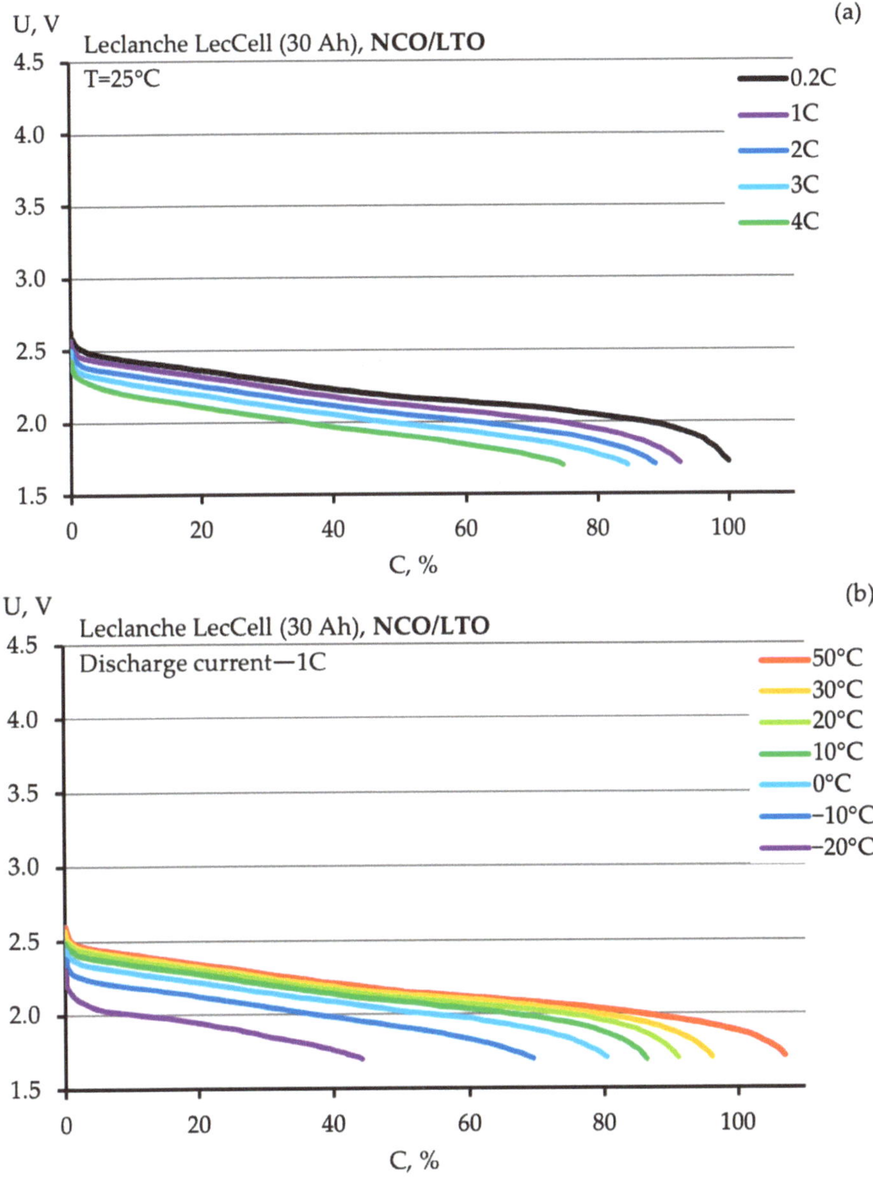

Figure 58. Discharge current (**a**) and temperature (**b**) effect on the shape of LecCell 30 Ah LICs (NCO/LTO, Leclanche) discharge curves. Source: Figure by authors.

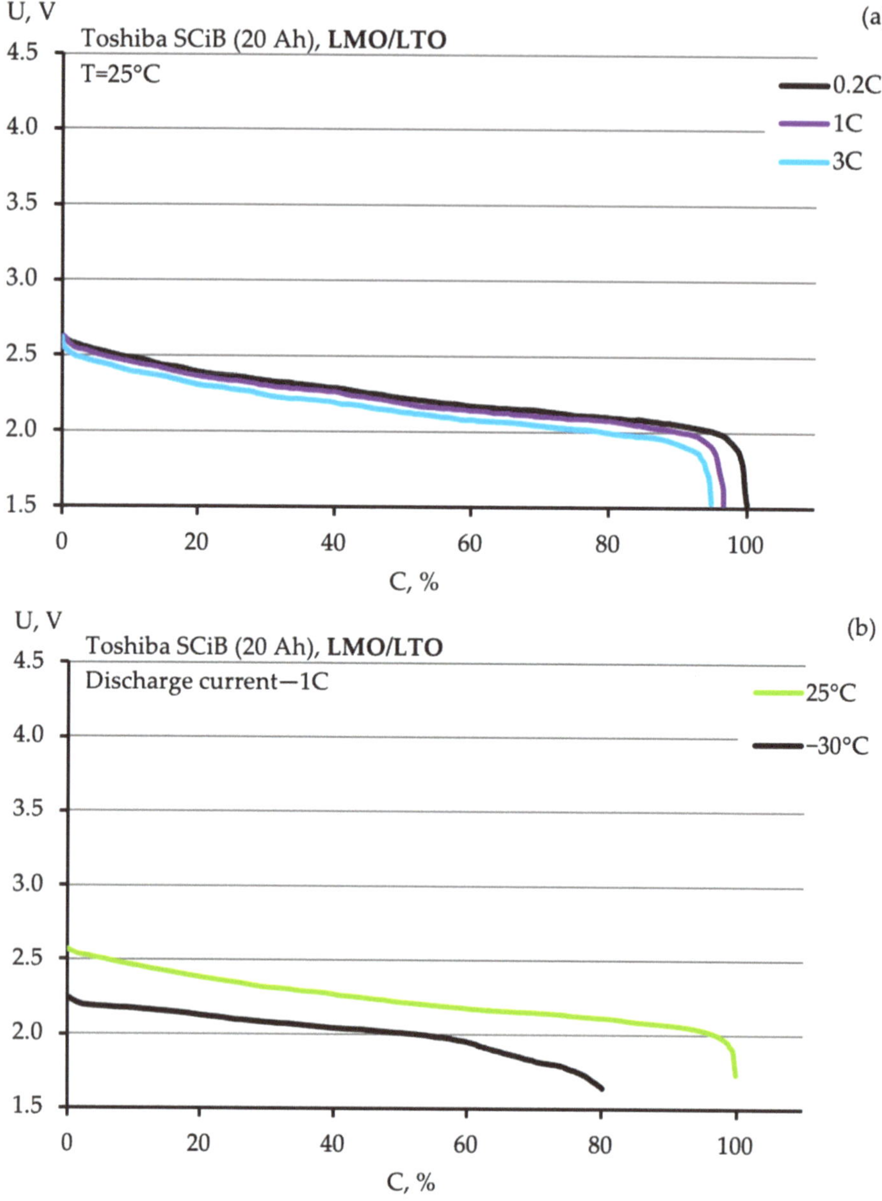

Figure 59. Discharge current (**a**) and temperature (**b**) effect on the shape of SCiB^{TM} 20 Ah (LMO/LTO, Toshiba) LICs; the graphs are plotted on the basis of data in references [43,47]. (**a**) The LIC discharge current in the module is limited by the 3C value. Source: Figure by authors.

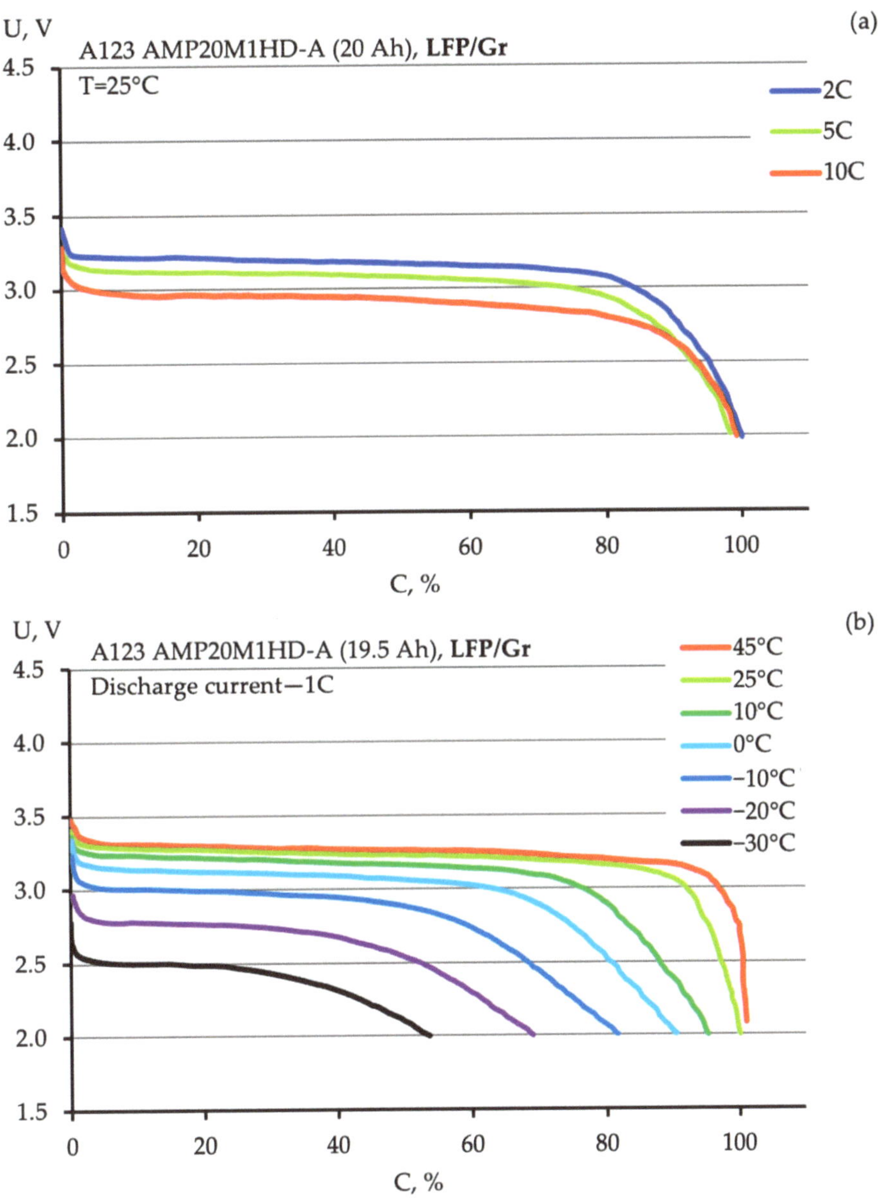

Figure 60. Discharge current (**a**) and temperature (**b**) effect on the shape of AMP20M1HD-A 20 Ah (LFP/graphite, A123) LIC discharge curves; the graphs are plotted on the basis of data in the technical manual [48]. Source: Figure by authors.

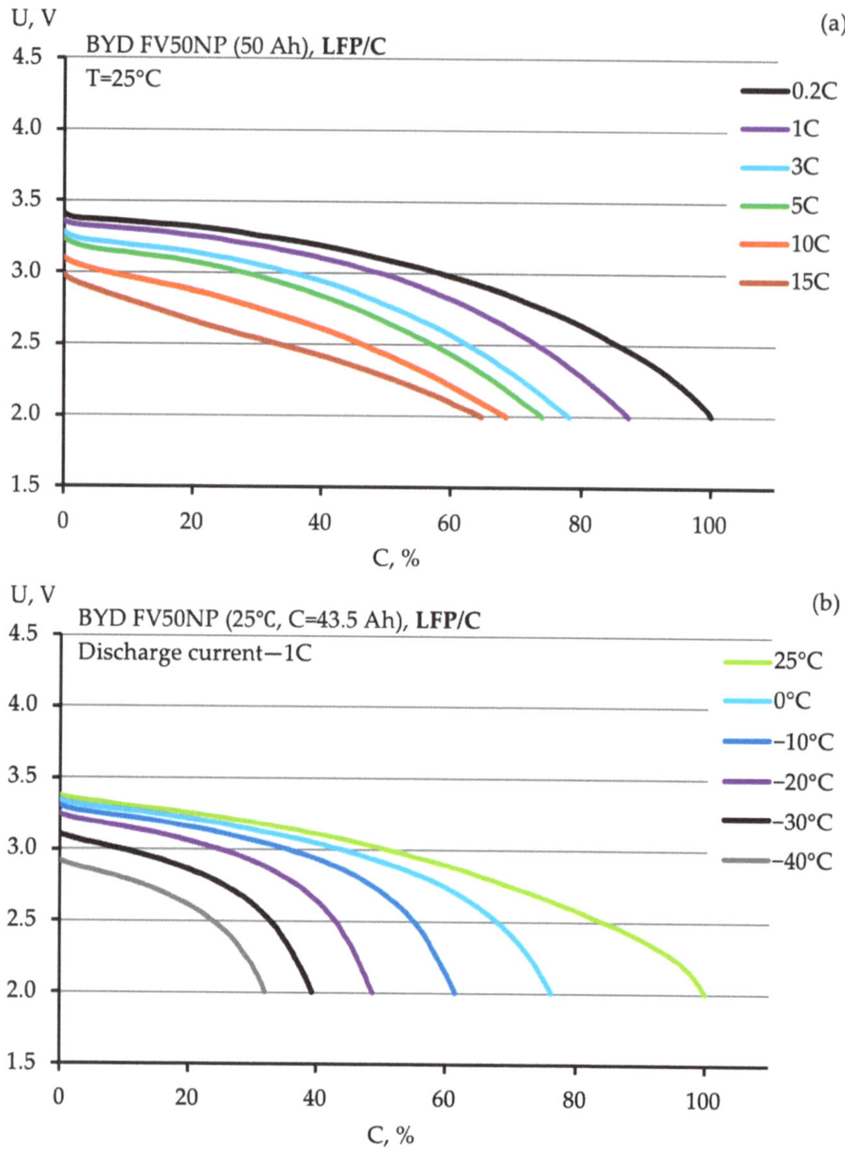

Figure 61. Discharge current (**a**) and temperature (**b**) effect on the shape of FV50NP (LFP/carbon (HC or SC), BYD) LIC discharge curves. Source: Figure by authors.

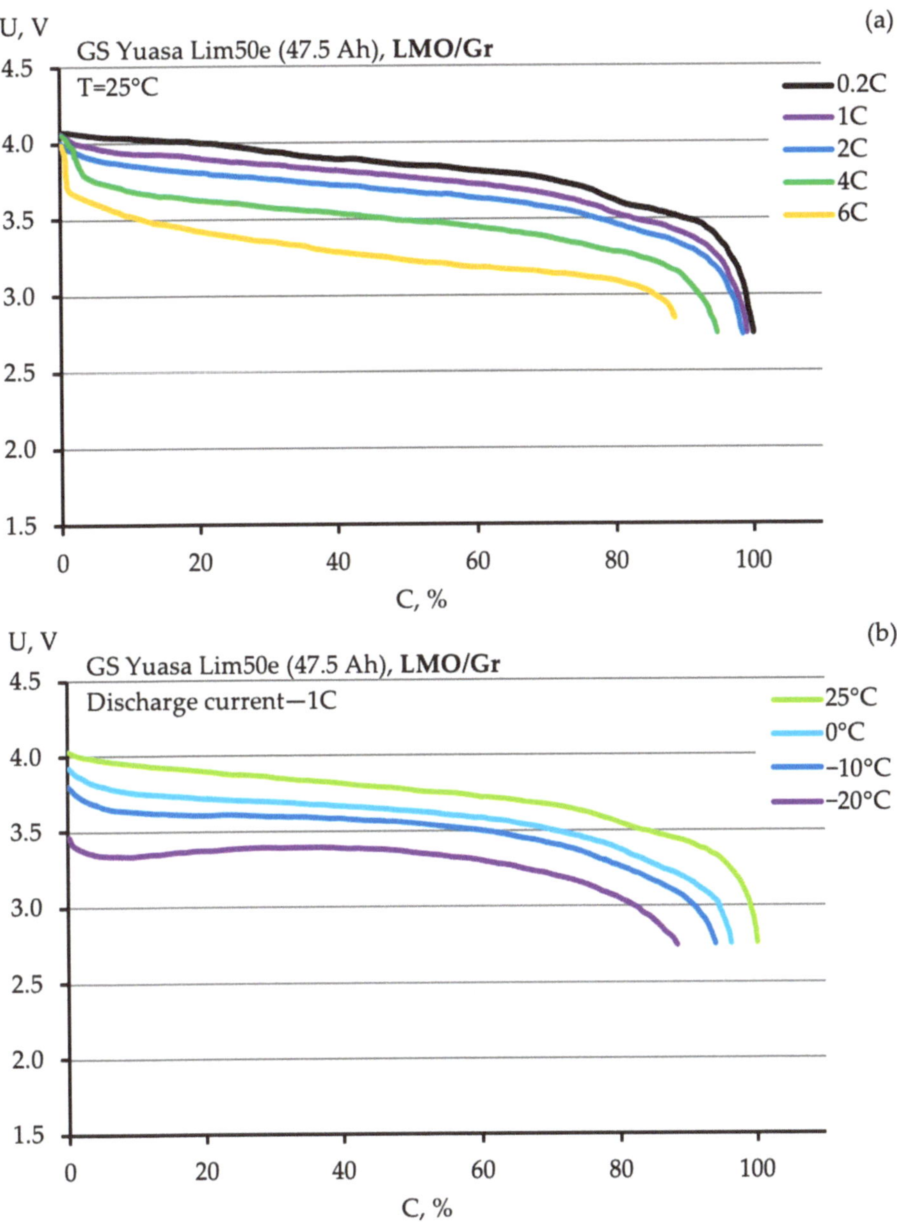

Figure 62. Discharge current (**a**) and temperature (**b**) effect on the shape of LIM50e 47.5 Ah (LMO/graphite, Yuasa) LIC discharge curves; the graphs are plotted on the basis of data in booklet [22]. Source: Figure by authors.

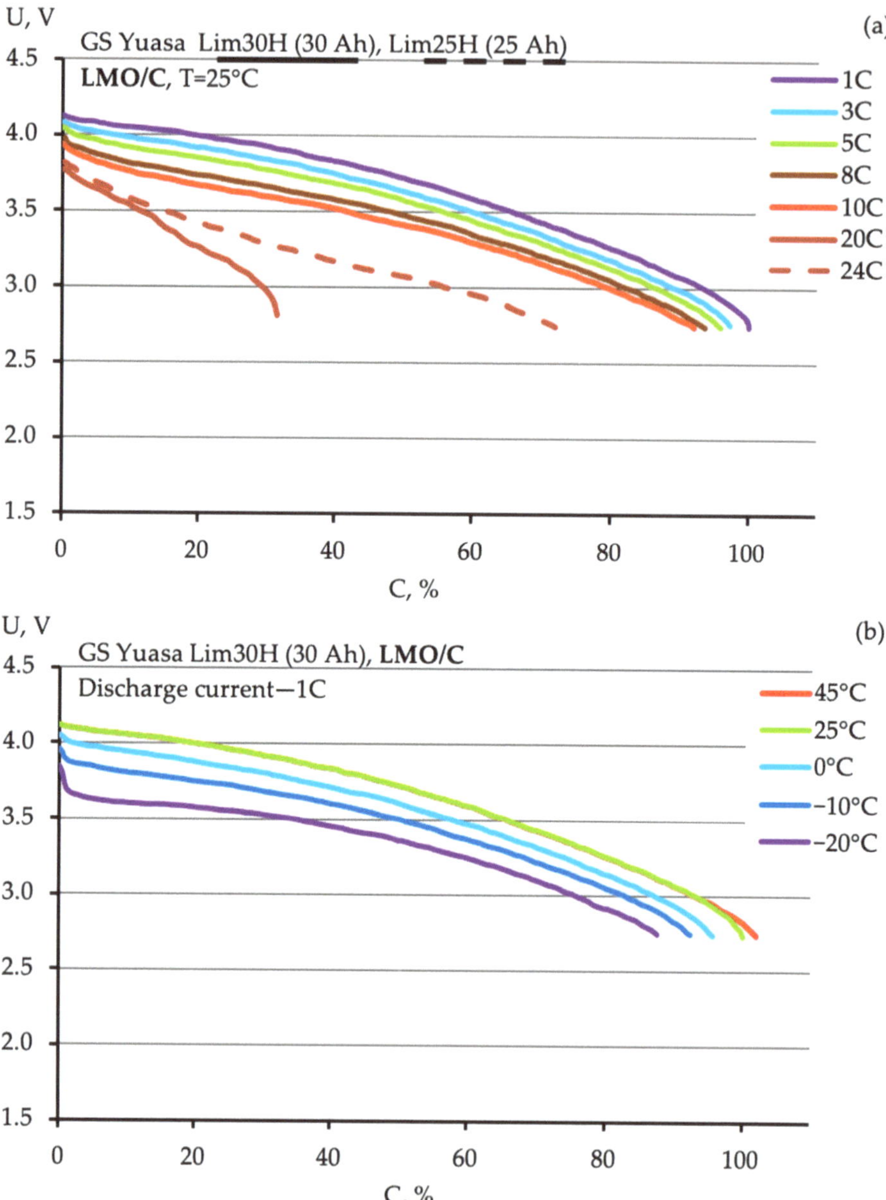

Figure 63. Discharge current (**a**) and temperature (**b**) effect on the shape of LIM30H 30 Ah (LMO/HC or SC, GS Yuasa) LIC discharge curves; the graphs are plotted on the basis of data in references [22,28]. Source: Figure by authors.

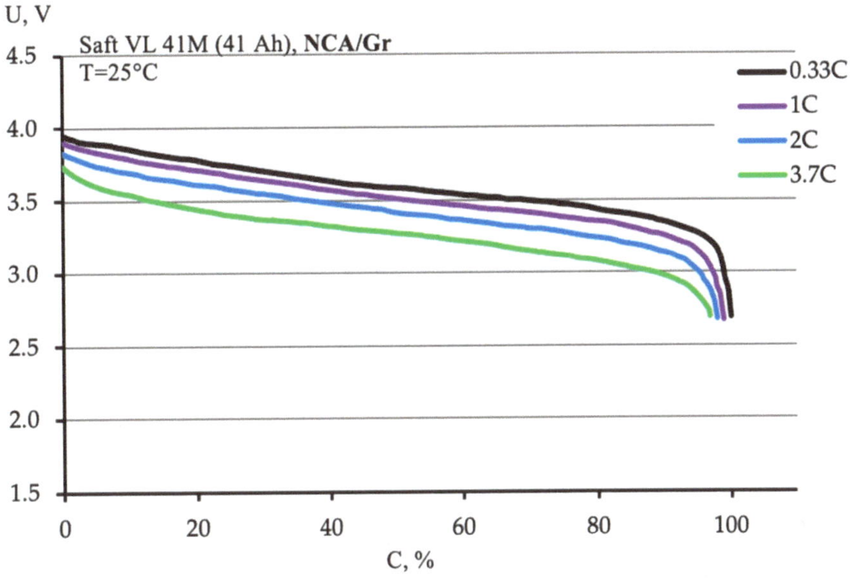

Figure 64. Discharge current effect on the shape of VL 41M (NCA/Gr, Saft) LIC discharge curves; the graphs are plotted on the basis of data in booklet [49]. Source: Figure by authors.

The operating time expressed in number of the charge/discharge cycles (cycle life) of industrial LICs, as a rule, is higher than that of LICs used in the portable equipment. Table 41 gives information on the number of charge/discharge cycles for various cells used in stationary electric energy storage applications. The actual values can be either higher or lower since they depend on the specific operating conditions and the manufacturing quality of the LIC under consideration. Since the test conditions (discharge current, depth of discharge) and shutdown conditions (80 or 70% of initial capacity) are different, it is rather difficult to compare LICs.

It is also necessary to consider the amount of energy transmitted for one charge/discharge cycle and its relation to the LIC full energy.

The number of charge/discharge cycles typical for lithium titanate cells may be 20,000 or higher. If the energy transmission per cycle relative to a unit of mass or volume is compared for LICs with an M.O./LTO electrochemical system and for LICs with an NCM/graphite electrochemical system, the number of charge/discharge cycles may be almost the same. Due to higher specific energy, the NCM/graphite LICs will be discharged less, which extends the cycle life considerably.

Table 41. Industrial lithium-ion cells for electric energy storage.

Company	Model	Materials	U, V	C, Ah	m, g	Sizes, mm	E, Wh/kg	Imax, C	nC	Ref.
BYD	FV50NP	LFP/Gr	3.2	50	1700	28 × 100 × 375	95	6	6000	[50]
GS Yuasa	Lim50E	LMO/Gr	3.7	47.5	1700	44 × 171 × 113	104	6	5000	[22,51,52]
GS Yuasa	LIM30H	LMO/C	3.6	30	2100	47 × 170 × 136	51	20	>3000	[22,28]
Hitachi	CH75	LMO/C	3.7	75	3000	d67 × 410	87	3	8000	[53–56]
Leclanche	LecCell 30 Ah	NCO/LTO	2.3	30	1100	12 × 178.5 × 285	61	4	15,000	[46,57]
LG	JH3	MO/Gr	3.7	63	1180	16 × 353 × 100	197	1	–	[58,59]
LG	JP3	MO/ Gr	3.7	64	1165	16 × 353 × 100	203	2	–	
Litarion	Litacell 44	NCM/Gr	3.65	44	990	12 × 155 × 249	162	3	6000	[60,61]
Kokam	SLPB120255255	NCM/Gr	3.7	75	1580	11.6 × 268 × 265	175	3	10,000	[62]
NEC (A123)	AMP20	LFP/Gr	3.3	20	496	7.25 × 160 × 227	131	10	6000	[48,63]
Saft	VL 45E	NCA/ Gr	3.6	45	1070	d54.3 × 222	149	2	>3000	[64]
Saft	VL 41M	NCA/ Gr	3.6	41	1070	d54.3 × 222	136	3.7	–	[49]
Saft	VL 30P	NCA/ Gr	3.6	30	1100	d54.3 × 222	97	10	–	[65]
Saft	VL 41M Fe	LFP/Gr	3.3	40	1030	d54.3 × 222	128	2.3	–	[66]
Saft	VL 30P Fe	LFP/Gr	3.3	29	1020	d54.3 × 222	94	10	–	[66,67]
Samsung	94 Ah	M.O./Gr	3.68	94	2010	45 × 173 × 125	174	–	6000	[68]
Samsung	2170033J	M.O./Gr	3.56	3.27	62	d21 × 70	188	1	2000	[69–71]
Sony	US26650 FTC1	LFP/Gr	3.2	3	86	d26 × 65	112	6.7	6000	[40,72]
Toshiba	SCiB™ 23 Ah	NMC/LTO	2.3	23	550	22 × 116 × 106	100	–	17,000	[73,74]
Toshiba	SCiB™ 20 Ah	NMC/LTO	2.3	20	515	22 × 116 × 106	93	–	17,000	[75–77]
Toshiba	SCiB™ 10 Ah	M.O./LTO	2.4	10	510	22 × 116 × 106	47	–	>20,000	[78,79]
Valence	APR18650M1B	LFP/Gr	3.3	1.1	39	d18 × 65	93	27	>2000	[80]
Valence	ANR26650M1B	LFP/Gr	3.3	2.5	76	d26 × 65	109	20	>2000	[80]

Note: C—HC or SC; Gr—graphite; LFP—$LiFePO_4/C$; LMO—$LiMn_2O_4$; LTO—$Li_4Ti_5O_{12}$; M.O.—component (components) from LMO, NCA, NCM; NCA—$LiNi_{0.85}Co_{0.1}Al_{0.05}O_2$; NCM—$LiNi_aCo_bMn_cO_2$ a + b + c = 1; NCO—$LiNi_dCo_eO_2$; d + e = 1 series; nC—number of cycles. Source: Authors' compilation based on data from references cited in the table.

The graphs in Figures 58a, 59a, 60a, 61a, 62a, 63a and 64a show that the average discharge voltage and discharge capacity reduce in proportion to the discharge current increase, and the curve shape varies. In other words, the LIC's usable energy drops. Ragone plots can be used to compare the LICs of various designs manufactured using different active materials. The mathematical processing (see Figure 12a, Chapter Lithium-Ion Cells with High Specific Energy) of the discharge curves for LICs used for stationary facilities demonstrates (Figure 65) that the nominal energy varies in the range of 50 ÷ 215 Wh/kg (110 ÷ 500 Wh/L). The power determined at continuous discharge does not exceed 1.2 kW/kg (2.2 kW/L, and for a prototype of Toshiba cell with niobium and titanium oxide as anode material—3.1 kW/L) for most LICs.

To describe the performance of LICs in a wide temperature range, manufacturers give the discharge curves obtained at 1C current (the LICs are charged at room temperature). After integration, the curves of the LIC specific energy vs. temperature are plotted. When the temperature drops, the specific energy reduces regardless of the LIC type and scope of application (Figure 66). At a temperature of minus 20 °C, the specific energy equals 20 ÷ 80 Wh/kg for LICs used for stationary facilities. As for LICs with high specific energy, the specific energy with a discharge current of 1C (−20 °C) may reach ≈200 Wh/kg. A relative drop in specific energy at −20 °C (relative to the specific energy at room temperature) is within the range of 20 ÷ 60%.

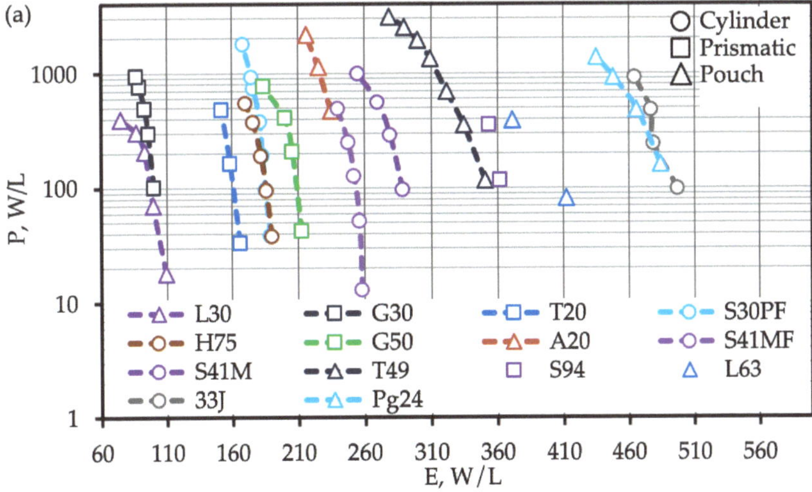

Figure 65. *Cont.*

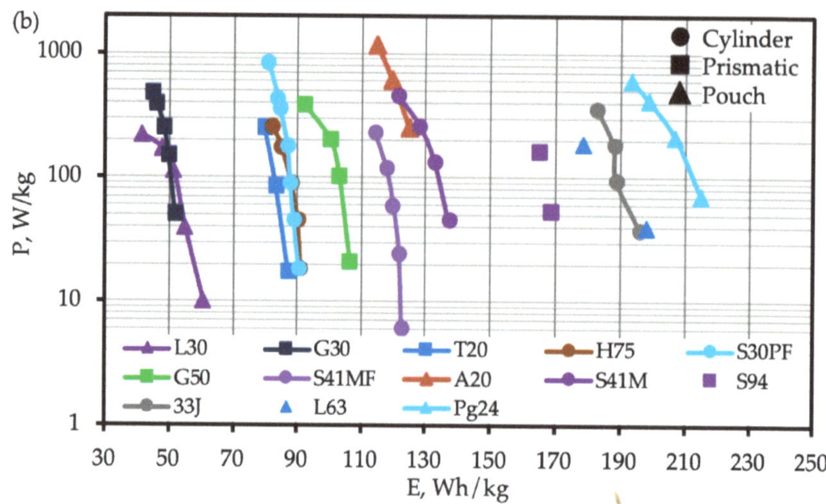

Figure 65. Energy (Wh/L, Wh/kg) vs. power (W/L, W/kg) of LICs used in stationary electric energy storage: L30—Leclanche 30 Ah [46], G30—GS Yuasa 30 Ah [22,28], T20—Toshiba 20 Ah [43], H75—Hitachi 75 Ah [56], S30PF—Saft VL30P Fe [67], G50—GS Yuasa 50 Ah [22], A20—A123 20 Ah [48], S41MF—Saft VL41M Fe [81], S41M—VL41M [49], T49 [82,83], S94—SDI 94 Ah [68], L63—LG 63 Ah [58], 33J—SDI 2170 3.3 Ah, Pg24—Prologium 24 Ah. (**a**) Energy density and power density are expressed in Wh/L and W/L. (**b**) Specific energy and specific power are expressed in Wh/kg and W/kg. For both (**a**,**b**) figures the form of a marker on the plot reveals the type of cell case: circle, square, and triangle corresponds to cylindrical, prismatic and pouch cell cases. Source: Figure by authors.

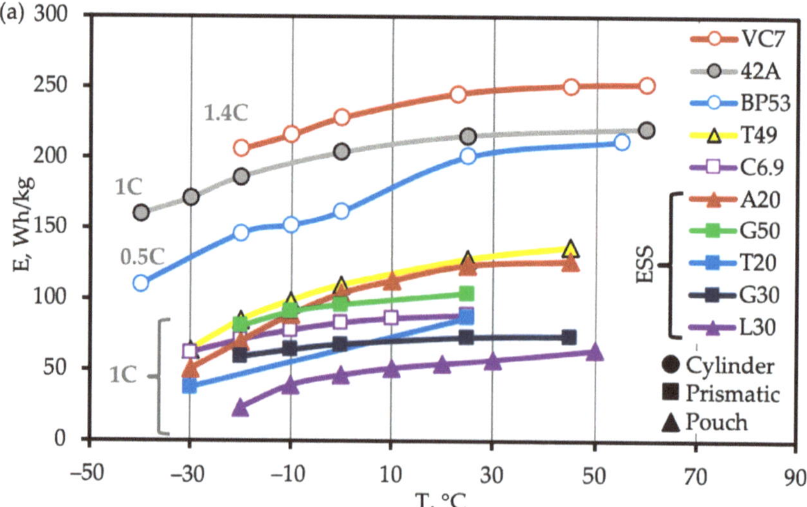

Figure 66. *Cont.*

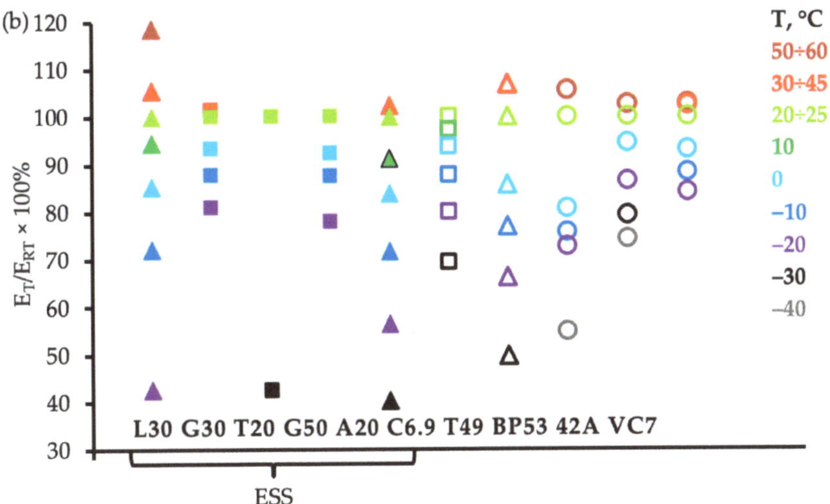

Figure 66. Discharge temperature (charging at room temperature) effect on variation of absolute (**a**) and relative (**b**) values of specific energy of L30 [57], G30 [22,28], T20 [47], G50 [22], A20 [48], C6.9 (CATL 6.9 Ah) [84], T49 (Toshiba 49 Ah) [82], BP53 (Boston Power 5.3 Ah) [85,86], 42A (Molicel 4.2 Ah) [87], and VC7 (Sony 3.5 Ah) [88] LICs. Source: Figure by authors.

6.4. Conclusions

The following topics were detailed in this chapter:

- Standard designs of electric energy storage and uninterruptible power supplies;
- Characteristics of modules and lithium-ion cells by major manufacturers;
- Comparison of results for LICs of various designs with different types of active materials under normal climatic conditions and within a wide temperature range.

Nominal specific energy varies in the range of 50 ÷ 215 Wh/kg (110 ÷ 500 Wh/L) under normal climatic conditions. The power determined at continuous discharge does not exceed 1.2 kW/kg (2.2 kW/L) for most LICs. When the temperature drops, the specific energy and power of LICs reduce to a different extent regardless of the active cathode and anode materials used during manufacture. The degree of reduction depends both on the used active materials and the design and manufacturing technology.

References

1. DOE Global Energy Storage Database. Available online: https://sandia.gov/ess-ssl/gesdb/public/projects.html (accessed on 3 January 2022).
2. Takeshita, H. Latest LIB Market Trends (The Automotive and Power Storage Markets Move into Full-scale Growth. In Proceedings of the 10th Int'l Rechargeable Battery Expo Battery Japan 2019, Tokyo, Japan, 28 February 2019; pp. 1–6.
3. Smart Battery Systems for Energy Storage. Available online: https://www.samsungsdi.com/upload/ess_brochure/SamsungSDI_ESS_201609EN.pdf (accessed on 14 November 2024).
4. Electricity Storage Services and Benefits. Available online: https://ez-pdh.com/course-material/EE1109-Electricity-Storage-Services.pdf (accessed on 24 July 2024).
5. Masgrangeas, D. Storage Energy System for On-grid and Off Grid application. Managing Massive Wind/PV Integration with li-ion Energy Storage. In Proceedings of the 27-я Международная специализированная выставка "Автономные источники тока", Москва (27th International Specialized Exhibition "Autonomous Power Sources", Moscow, Russia, 21–23 March 2018; p. 18.
6. Kokam ESS Brochure ver_5.0. 2018. Available online: http://kokam.com/data/2018_Kokam_ESS_Brochure_ver_5.0.pdf (accessed on 5 June 2018).
7. Plug&Save Kokam All in ONE Series Rev. 1.2. Available online: http://kokam.com/data/Kokam_All-in-One%20Brochure_rev_1_2.pdf (accessed on 5 June 2018).
8. Ozbek, M.; Pierquet, B.; Baglino, A.D. Scalable and Flexible Cell-Based Energy Storage System. US 2017/0093156A1, 30 March 2018.
9. Tesla's New Patent for Its Powerpack Explains How Its Energy Storage Stations Can Go Up to 1 GWh. Available online: https://electrek.co/2017/03/31/tesla-paten-powerpack-energy-storage-station/ (accessed on 5 July 2018).
10. Intensium Max. Containerized Energy Storage Systems (Saft). Available online: https://www.norwatt.es/archivos/productos/pdf/baterias_ion_litio_saft_intensium_plus_es.pdf (accessed on 5 July 2018).
11. AES Energy Storage Angamos Battery Energy Storage System (BESS). Available online: http://energystorage.org/energy-storage/case-studies/aes-energy-storage-angamos-battery-energy-storage-system-bess (accessed on 4 July 2018).
12. Almacena Lithium Ion Battery. Available online: https://gesdb.sandia.gov/projects.html#470 (accessed on 14 November 2024).
13. GS Yuasa Delivers 6750 kWh Lithium-Ion Battery System to Kushiro Town Toritoushi Wildland Solar Power Plant. Available online: https://www.yuasa.de/en/2017/08/gs-yuasa-delivers-6750kwh-lithium-ion-battery-system-to-kushiro-town/ (accessed on 4 July 2018).
14. Mitsubishi International Corporation and GS Yuasa International Ltd. Join MESA-1 Energy Storage Project. Available online: https://www.cleantechalliance.org/2014/04/30/mitsubishi-international-corporation-and-gs-yuasa-international-ltd-join-mesa-1-energy-storage-project/ (accessed on 4 July 2018).

15. North America's Largest Battery Energy Storage System Now Operational. Available online: http://www.lgchem.com/global/lg-chem-company/information-center/press-release/news-detail-635 (accessed on 16 May 2018).
16. Stafford Hill Solar Farm & Microgrid: Lithium Ion. Available online: https://gesdb.sandia.gov/projects.html#612 (accessed on 14 November 2024).
17. Sanchez, J. Almacenamiento Electroquímico con baterías de Li-ion (Saft) (Electrochemical Storage with Li-ion batteries (Saft)). In Proceedings of the Jornada "Presente y futuro de sistemas de almacenamiento eléctico con baterías" (Conference "Present and future of electrical storage systems with batteries") F2e, Castellon de la Plana, Spain, 30 September 2016; p. 30.
18. Tohoku Electric Power Co., Inc. Battery Energy Storage System (BESS) at Minamisoma Substation for Improving the Supply-demand Balance. Available online: http://www.scib.jp/en/applications/energy.htm (accessed on 4 July 2018).
19. Megawatt-Class Battery Energy Storage System Will Boost Use of Renewable Energy. Available online: https://www.toshiba.co.jp/about/press/2015_06/pr2301.htm (accessed on 4 July 2018).
20. Samsung SDI Energy Storage System Battery Business. Available online: https://www.samsungsdi.com/upload/ess_brochure/201809_SamsungSDI%20ESS_EN.pdf(accessed on 14 November 2024).
21. Мельников, В. (Melnikov, V.) Использование литий-ионных высокомощных батарей для центров обработки данных (Using Li-Ion High Power Batteries for Data Centers). In Proceedings of the 28-я международная специализированная выставка "Автономные источники тока" (28th international specialized exhibition "Autonomous power sources"), Moscow, Russia, 20 March 2019; p. 23.
22. Lithium Ion Battery for Industrial Use (Lim50E, Lim40, Lim30H), GS Yuasa. Available online: https://files.vogel.de/vogelonline/vogelonline/companyfiles/10901.pdf (accessed on 11 July 2018).
23. Thermal Control System (Tesla, Powerpack). Available online: https://www.tesla.com/powerpack?redirect=no (accessed on 18 July 2018).
24. Moxie+ Battery Module. Available online: http://enerdel.com/4708-2/ (accessed on 18 July 2018).
25. B-Box 2.5—10.0 B-Plus 2.5. Available online: https://imeon-energy.com/wp-content/uploads/byd-b-box-2.5-10.0-lithium-battery-48v.pdf?x77238 (accessed on 14 November 2024).
26. GS Yuasa Lithium Power Announces the New LIM50E Series of Lithium-Ion Battery Modules. Available online: http://www.s399157097.onlinehome.us/PDFS/LIM50E_Modules.pdf (accessed on 6 July 2018).
27. LIM50EN Series Long Life Batteries for Energy Storage. Available online: http://gsyuasa-es.com/Downloads/LIM50EN_SERIES_BROCHURE.pdf (accessed on 14 November 2024).
28. Uebo, Y.; Shimomura, K.; Nishie, K.; Nanamoto, K.; Matsubara, T.; Seike, H.; Kuzuhara, M. Development of High Power Li-ion Cell "LIM25H" for Industrial Applications. *GS Yuasa Tech. Rep.* **2015**, *12*, 12–17.

29. Nagaura, Y.; Oichi, R.; Shimada, M.; Kaneko, T. Battery-powered Drive Systems: Latest Technologies and Outlook. *Hitachi Rev.* **2017**, *66*, 54–60.
30. LG Ess Catalogue. 2016. Available online: https://eu.krannich-solar.com/fileadmin/content/data_sheets/storage_systems/italy/2016_LGChem_Catalog_Global_0_.pdf (accessed on 14 July 2018).
31. 170830_LG Chem Catalog(Global).indd. Available online: http://www.lgchem.com/upload/file/product/ESS_LGChem_Catalog_Global%20[0].pdf (accessed on 6 July 2018).
32. LITASTORE 2.3 Stationary Energy Storage System Powered by Lithium-ion Ceramic Technology. Available online: https://cdn.enfsolar.com/z/pp/m5cvaq144nr/5a3099ebe7a4b.pdf (accessed on 14 November 2024).
33. Kokam Battery Module Series. Available online: https://kokamdirect.com/wp-content/uploads/2017/02/Module_series_spec.pdf (accessed on 14 November 2024).
34. ALM 12V7 Characteristics (NEC). Available online: https://www.neces.com/products-services/battery-systems/battery-components-and-accessories/48v100/ (accessed on 6 July 2018).
35. Murray, D.B. Energy Storage Systems for Wave Energy Converters and Microgrids. Ph.D. Thesis, University College Cork, Cork, Ireland, 2013. Available online: https://core.ac.uk/download/pdf/61575028.pdf (accessed on 14 November 2024).
36. Synerion 48M Medium Power Capability Module. Available online: http://www.aegtranzcom.com/Repository/PDF-ENERGY/Saft_Com_Synerion48M_en_0512_Protg.pdf (accessed on 26 September 2016).
37. Saft's High Power Capability Synerion 48P. Available online: http://www.aegtranzcom.com/Repository/PDF-ENERGY/Saft_Com_Synerion48P_en_0512_Protege.pdf (accessed on 26 September 2016).
38. Flex'ion® Transportation, Storage, Installation and Operation Instructions (Saft). Available online: https://www.saftbatteries.com/download_file/6X7JMGAnv3Fm6HdmtEv%252B2gtlbZ1bRRVHkjS11M6md92GD2EF7vU%252F3Oybbz3WOlG%252BxR8srpA5iCdJ%252FV3IQzTVHQyiTucngZKEg9KkYCLkowAvgaG1huq0DbMmlt11iVLGUiN1CZrD8oxsKJTUOJRDsJeJ9K1uwrHik%252FpWlyyYRSmq19oPPw%253D%253D/FX2017-0066_Flex%2527ion+TSIO+Sheet_RevA_March+2018.pdf (accessed on 14 July 2018).
39. Flex'ion Li-Ion Battery Systems (Saft). Available online: https://www.saftbatteries.com/download_file/6X7JMGAnv3Fm6HdmtEv%252B2gtlbZ1bRRVHkjS11M6md92GD2EF7vU%252F3Oybbz3WOlG%252BxR8srpA5iCdJ%252FV3IQzTVHQyiTucngZKEg9KkYCLkowAvgaG1hurhEImefACH1rJZXa%252FBrUQ7O2fhjgwIFgdej4Cog5xFdaB%252FLV8dtgo5MDOzyxzEVA%253D%253D/ProductBrochureFlexion_electronic_0318_BD.pdf (accessed on 6 July 2018).
40. Sony Energy Storage System Using Olivine Type Battery 'FORTELION' 13.03.2012. Available online: https://www.eiseverywhere.com/file_uploads/89b02d8a4305f5ffe09d0c27691441af_O-302YasudaMasayuki.pdf (accessed on 6 July 2018).
41. Sony to Ship 1.2 kWh Energy Storage Modules Using Rechargeable Lithium-Ion Batteries Made from Olivine-Type Phosphate. Available online: http://www.sony.net/SonyInfo/News/Press/201104/11-053E/ (accessed on 26 September 2016).

42. Industrial Lithium-Ion Battery SCiB (Toshiba). Available online: http://www.scib.jp/en/download/ToshibaRechargeableBattery-en.pdf (accessed on 7 July 2018).
43. Toshiba SCiB™ Modules. Available online: http://www.scib.jp/en/product/module.htm (accessed on 26 September 2016).
44. Valence Modules with External BMS. Available online: https://previous.lithiumwerks.com/wp-content/uploads/2017/02/Valence-Module-Range-071717.pdf (accessed on 14 November 2024).
45. Lithium Power Cabinets Lithium-ion Energy Storage & Power Management (Gs Yuasa). Available online: https://www.beltrona.com/out/media/Yuasa_Lithium_Power_Cabinets_Web(1).pdf (accessed on 26 November 2018).
46. LecCell LTO (Leclanche). Available online: http://pdf.directindustry.com/pdf/leclanche/leccell-lto/20918-734815.html (accessed on 6 July 2018).
47. Miyamoto, H.; Enomoto, T.; Kosugi, H. SCiB™ Battery Modules for Electric Vehicles. Technology Administration & Planning Office, Corporate Technology Planning Division, Toshiba Corporation, 1-1, Shibaura 1-chome, Minato-ku, Tokyo 105-8001, Japan. 東芝レビュ (Toshiba Rev.) **2011**, *66*, 56–59.
48. Battery Pack Design, Validation, and Assembly Guide Using A123 Systems Battery Pack Design, Validation, and Assembly Guide Using A123 Systems AMP20m1HD-A Nanophosphate® Cells (493005-002, Rev. 2). Available online: https://www.buya123products.com/uploads/vipcase/b24d4f5b63934c59d43e93b3bb4db60a.pdf (accessed on 14 November 2024).
49. Medium Power Lithium-Ion Cells VL M (Saft). Doc № 54042-2-0305 (March 2005). Available online: http://www.sonicgrouphk.com/uploadfile/products/20080728124655188.pdf (accessed on 14 November 2024).
50. Electrical Energy Storage Systems and Automobile Applications. Available online: https://wenku.baidu.com/view/cc918f1da8114431b90dd8a0.html?from=search (accessed on 6 July 2018).
51. Yuasa Battery Sales (uk) Ltd Launches Lithium-Ion Cells for Industrial Power Applications. Available online: https://www.yuasa.co.uk/2013/10/yuasa-battery-sales-uk-ltd-launches-lithium-ion-cells-industrial-power-applications/ (accessed on 6 July 2018).
52. Product Safety Datasheet. Available online: http://www.gsyuasa-lp.com/SpecSheets/LIM50E-MSDS.pdf (accessed on 11 July 2018).
53. Technology and Its Advance of the Large Format Energy Storage Battery Systems (Hitachi). Available online: https://wenku.baidu.com/view/c40750c8aff8941ea76e58fafab069dc50224763 (accessed on 6 July 2018).
54. Nomura, T.; Myogadani, Y.; Ohno, M.; Emori, A.; Takeda, K. Container-type Energy Storage System with Grid Stabilization Capability. *Hitachi Rev.* **2014**, *63*, 432–437.
55. Hirota, S.; Hara, K.; Ochida, M.; Mishiro, Y. Energy Storage System with Cylindrical Large Formatted Lithium Ion Batteries for Industrial Applications. In Proceedings of the 2015 IEEE International Telecommunications Energy Conference (INTELEC), Osaka, Japan, 18–22 October 2015; p. 6.

56. Ito, T. Energy Storage System with Cylindrical Large Formatted Lithium Ion Batteries for Industrial Applications. In Proceedings of the 12th China International Battery Fair (CIBF2016), Shenzhen, China, 24–26 May 2016; p. 34.
57. LecCell 30Ah High Energy Lithium Nickel Cobalt Oxide (NCO)—Titanate (LTO) System. Available online: http://blizzard.cs.uwaterloo.ca/iss4e/wp-content/uploads/2019/04/leccell-30ah-high-energy.pdf (accessed on 14 June 2023).
58. Beltran, H.; Ayuso, P.; Vicente, N.; Beltrán-Pitarch, B.; García-Cañadas, J.; Pérez, E. Equivalent circuit definition and calendar aging analysis of commercial Li(NixMnyCoz)O2/graphite pouch cells. *J. Energy Storage* **2022**, *52*, 104747. [CrossRef]
59. smartguru0429. [인터배터리 2019] 국내최대 규모 전시회에 스마 ((InterBattery 2019) Smart at Korea's largest exhibition). Available online: https://m.blog.naver.com/smartguru0429/221681591643 (accessed on 18 January 2021).
60. Electrovaya Products (ex Litarion). Available online: http://electrovaya.com/battery-products/ (accessed on 12 July 2018).
61. LITACELL® LC-44, 44 Ah Lithium-ion Ceramic Cell. Available online: http://litarion.com/content/datasheets/LITACELL-LC44.pdf (accessed on 27 April 2018).
62. Kokam Cell Brochure 20160304. Available online: https://www.west-l.ru/uploads/tdpdf/sf_slpb-cell-brochure1.pdf (accessed on 6 July 2018).
63. Nanophosphate® Lithium Ion Prismatic Pouch Cell AMP20M1HD-A (A123). Available online: https://www.buya123products.com/uploads/vipcase/468623916e3ecc5b8a5f3d20825eb98d.pdf (accessed on 6 July 2018).
64. High Energy Lithium-Ion Cell VL 45 E (Saft). DOC № 54041-2-0305 (March 2005). Available online: http://www.sonicgrouphk.com/uploadfile/products/20080728124633172.pdf (accessed on 14 November 2024).
65. Kuempers, J. Neue Fahrzeugkonzepte mit dem Fokus Nachhaltigkeit: Lithium-Ionen-Batterien fur Elektrofahrzeuge (New vehicle concepts with a focus on sustainability: lithium-ion batteries for electric vehicles). In Proceedings of the 14. Zulieferforum der Arbeitsgemeinschaft Zulieferindustrie (14th Supplier Forum of the German Supplier Industry Association), Frankfurt, Germany, 26 January 2010; p. 31.
66. Saft Recent Progress and Products for Energy Storage Applications. In Proceedings of the 11th China International Battery Fair (CIBF 2014), Shenzhen, China, 20–22 June 2014; p. 53.
67. VL 30P Fe Super-Phosphate® Rechargeable LiFePO4 Cell (Datasheet). Document No 54087-2-0513, Edition: May 2013. Available online: https://saft.com/products-solutions/products/ (accessed on 13 October 2019).
68. Lima, P. Samsung SDI 94 Ah Battery Cell Full Specifications. Available online: https://pushevs.com/2018/04/05/samsung-sdi-94-ah-battery-cell-full-specifications/ (accessed on 6 July 2018).
69. SDI INR21700 33J Specification. Available online: https://www.facebook.com/ecoluxshop/photos/pcb.892652754215166/892652690881839/?type=3&theater (accessed on 7 July 2018).

70. Lu, M. Future Trends and Key Issues in the Global Lithium-ion Batteries Market and Related Technologies. In Proceedings of the 13th China International Battery Fair (CIBF 2018), Shenzhen, China, 22–24 May 2018; p. 30.
71. Samsung 33J—A 21700 Li-Ion Battery Designed for Tesla. Available online: https://endless-sphere.com/forums/viewtopic.php?t=98717 (accessed on 18 December 2020).
72. Sony Energy Storage System Using Olivine Type Battery. Available online: https://na.eventscloud.com/file_uploads/89b02d8a4305f5ffe09d0c27691441af_O-302YasudaMasayuki.pdf (accessed on 14 November 2024).
73. Characteristics of SCiB™ Cells. Available online: http://www.scib.jp/en/product/cell.htm (accessed on 30 August 2019).
74. Industrial Lithium-Ion Battery SCiB (Sip Series). Available online: https://www.tipsh.toshiba.com.cn/en/file/yangben_20190606_EN.pdf (accessed on 14 November 2024).
75. 东芝钛酸锂电池SCiB系列新增20Ah-HP电池 (Toshiba Adds 20Ah-HP Battery to SCiB Series of Lithium Titanate Batteries). Available online: https://libattery.ofweek.com/2022-01/ART-36001-8140-30546096.html (accessed on 14 November 2024).
76. Hashimoto, T.; Kawamata, T.; Shimada, K. Commencement of Operation of Large-Scale Battery Energy Storage System for Nishi-Sendai Substation of Tohoku Electric Power Co., Inc. Technology Administration & Planning Office, Corporate Technology Planning Division, Toshiba Corporation, 1-1, Shibaura 1-chome, Minato-ku, Tokyo 105-8001, Japan. *東芝レビュ (Toshiba Rev.)* **2015**, *70*, 45–48.
77. Hirota, K. Battery Energy Storage Systems for Rolling Stock Using SCiB™ Lithium-Ion Battery. Technology Administration & Planning Office, Corporate Technology Planning Division, Toshiba Corporation, 1-1, Shibaura 1-chome, Minato-ku, Tokyo 105-8001, Japan. *東芝レビュ (Toshiba Rev.)* **2016**, *71*, 16–19.
78. Saruwatari, H.; Yamamoto, D. 10 Ah-Class SCiB™ Lithium-Ion Battery for Idling Stop Systems and Micro Hybrid Vehicles. Technology Administration & Planning Office, Corporate Technology Planning Division, Toshiba Corporation, 1-1, Shibaura 1-chome, Minato-ku, Tokyo 105-8001, Japan. *東芝レビュ (Toshiba Rev.)* **2016**, *71*, 44–47.
79. Murashi, Y.; Yajima, A. High-Power Type 10 Ah SCiB™ Lithium-Ion Battery Contributing to Reduction of Carbon Dioxide Emissions. *東芝レビュ (Toshiba Rev.)* **2017**, *72*, 65–68.
80. Lithium Werks Power Cells. Available online: https://lithiumwerks.com/power-cells/ (accessed on 8 July 2018).
81. Saft VL 41M Fe Super-Phosphate® Rechargeable LiFePO4 Cell (Datasheet). Document No 31107-2-0214, Edition: February 2014. Available online: https://saft.com/products-solutions/products/ (accessed on 13 October 2019).
82. Takami, N.; Ise, K.; Harada, Y.; Iwasaki, T.; Kishi, T.; Hoshina, K. High-energy, fast-charging, long-life lithium-ion batteries using $TiNb_2O_7$ anodes for automotive applications. *J. Power Sources* **2018**, *396*, 429–436. [CrossRef]
83. Ise, K.; Morimoto, S.; Harada, Y.; Takami, N. Large lithium storage in highly crystalline $TiNb_2O_7$ nanoparticles synthesized by a hydrothermal method as anodes for lithium-ion batteries. *Solid State Ion.* **2018**, *320*, 7–15. [CrossRef]

84. Lu, H.-L. An Update on the Chinese xEV Market, and a Technical Comparison of Its Batteries (ITRI IEK). In Proceedings of the World Mobility Summit 2016, Munich, Germany, 17–19 October 2016.
85. Chamberlain, R. Progress of xEV batteries with blended NCM/NCA cathode materials (Boston Power). In Proceedings of the Asia-Pacific Lithium Battery Congress 2014, Shenzhen, China, 26–28 March 2014; p. 28.
86. Swing (R) 5300 Rechargeable Lithium Ion Cell (Boston Power). Available online: https://www.tme.eu/en/Document/87b5645f730e56cad507df13706ef5b9/Swing5300.pdf (accessed on 16 August 2018).
87. Product Data Sheet Model INR-21700-P42A. Available online: https://www.imrbatteries.com/content/molicel_p42a.pdf (accessed on 18 December 2020).
88. Sony VC7. Available online: http://queenbattery.com.cn/index.php?controller=attachment&id_attachment=83 (accessed on 4 May 2018).

7 Miniature Lithium-Ion Cells for the Internet of Things, Wearable Devices, and Medical Applications

Miniature lithium-ion cells are widely used in the following applications:

- Wearable devices (smart clocks, earphones, smart glasses, fitness sensors, styluses);
- Medical appliances (neurostimulators, pacemakers, hearing aid devices, implantable devices) with an approximate share of rechargeable batteries equal to 15% [1];
- Components of the Internet of Things, industrial sensors, and sensors for environmental and agricultural monitoring, etc.;
- Self-contained sensors that support energy harvesting from the environment (sunlight, vibration, heat, radio waves, etc. [2]).

Table 42 contains information on the lithium-ion cells that power some miniature devices. The functions of miniature LICs are the following [3]: storing energy harvested from the environment (energy harvesters) and supplying power to various devices, operating as uninterruptible power supplies (real-time clock), and equalising power supply parameters [4]. Currently, miniature LICs filled with liquid, gel, and solid-state electrolytes are produced.

7.1. Lithium-Ion Cells Filled with Liquid and Gel Electrolytes

Liquid and gel electrolytes are used in LICs of any shape: cylindrical, prismatic, coin, and button cells (Tables 43 and 44).

For LICs implanted into the human body and being in contact with the skin, the material of the case and tabs is critical. For example, an LIC case (battery case) can be made of titanium (Wyon [5], Quallion (Ti 6-4 [6], EaglePicher [7,8]), steel (EaglePicher [9,10]) or organic compatible plastic (Wyon [11]) depending on the specific medical application. Tabs can be made of platinum, iridium, titanium, or molybdenum [6]. A higher weight of case parts will reduce the specific energy and power of the LICs.

The miniature and small-sized LICs filled with liquid and gel electrolytes are similar in design. For example, the electrochemical block of a button-type LIC can be composed of two long electrodes isolated by a separator and coiled into a roll impregnated with electrolyte [12].

Another design of the electrochemical block is a stack of electrodes isolated by a separator and impregnated with liquid electrolyte (stacking technology—Wyon [13–15]). The electrochemical block shape can be adapted to the LIC case shape due to the said design.

A positive electrode and a negative electrode with an active layer applied to one side, isolated by a separator, and impregnated with electrolyte are commonly used to manufacture coin-type LICs. The active layer can be applied by the slurry coating process. There is no preferred orientation of electrode-active material in such a case. There are also coin-type LICs (produced by NGK) in which a solid-cast plate ($LiCoO_2$) functions as the active layer of the positive electrode. The feature of this cells is a cathode with preferred spatial orientation of the crystal lattice (the crystals may be grown on the case inner wall). The positive electrode designed in such a way that allows the specific energy of LICs to be increased due to the incredible amount of active material in the active layer (as opposed to an active layer produced by the slurry coating process).

The active materials applied for the manufacturing of miniature and small-sized LICs filled with liquid and gel electrolytes are almost the same. LICs are manufactured with LCO-based cathodes (including high-voltage ones), NCM- and NCA-based cathodes, and anodes made of graphite, a mixture of graphite, and silicon-containing substances such as lithium titanate (see Tables 43 and 44). In addition, some produced coin-type LICs have a pair of positive and negative active materials not available in larger LICs: $Li_xTi_yO_4$ (cathode)/C (anode)—TC 920S. Rechargeable lithium metal batteries are also manufactured. During the operation of these batteries, lithium ions do not intercalate into the active material structure on the anode. Rather, the following lithium alloys are formed instead: lithiated mixed manganese oxide/silicon monoxide (MS621FE), and manganese oxide/aluminium and lithium alloy (ML2032). Since the structure of active anode materials varies significantly at charging/discharging, the service life in cycles is highly dependent on a depth of discharge (active materials used in electrochemical reactions).

The current positive and negative electrode collectors are traditionally aluminium and copper, respectively. However, in the case of lithium titanate anodes, aluminium foil can be applied as a current collector of the negative electrode due to its high potential relative to lithium (approx. 1.55 V).

Table 42. Miniature lithium-ion cells for various applications.

Application /Product	Product Manufacturer (Partner)	Cell					Ref.
		Grade	Manufacturer	V, cm^3	C, mAh	E, mWh	
Wearable electronics							
Stylus pen	Samsung	SLB	Nichicon	0.05	0.35	1	[16]
Galaxy Buds	Samsung	CP1254 A3	Varta	0.62	60	233	[17]
Galaxy Watch Active 2 40 mm	Samsung	ICP52121	Samsung	1.94	240	920	[18]
GoPro Hero8	GoPro	A53129	ATL	6.5	1220	4700	[19,20]
Sensors and IoT							
Real-time clock (backup)	Microcrystal	Ceracharge 1812	TDK	0.016	0.1	0.117	[21]
Sub battery for smoothing voltage and current	–	Ceracharge 1812	TDK	0.016	0.1	0.117	[4]
Perpetual sensor for industrial IoT	–	P180	Ilika	0.36 (0.1)	0.18	0.612	[22,23]
Perpetual sensor for smart home	–	M250	Ilika	0.108	0.292	1.02	[22]
Condition monitoring of turbine blades Vibration harvester and sensor	–	M250	Ilika	0.108	0.292	1.02	[22]
Environmental sensor	Ricoh	SLB Series	Nichicon	0.05	0.35	1	[24]
Plant growth monitoring for AgriTech: Temperature, moisture, light level sensors	–	2x M250	Ilika	0.216	0.584	2.04	[22]
Environment sensor	IPS	Mec201	IPS	0.129	0.7	2.8	[25]
Agricultural sensor energy harvesting RE01 FDSoI	Renesas	ET2016C-R	NGK	0.63	25	57.5	[26–28]
Energy harvesting	Ricoh	Pouch	NGK	–	–	–	[26,28]
Smart card + fingerprint identification	Jinco	Pouch	NGK	–	–	–	[26,28]
Power source module for IoT	Torex Semiconductor	Pouch/ coin	NGK	–	–	–	[28,29]
Industry 4.0 Environmental sensor	Esima project	3P/CP1654 A3	Varta	3.3 (6.1)	360	1332	[30]

Table 42. *Cont.*

Application /Product	Product Manufacturer (Partner)	Cell					Ref.
		Grade	Manufacturer	V, cm^3	C, mAh	E, mWh	
		Medical/Health					
Implantable cell	–	QL0003B	Quallion	0.08	3	10.8	[31]
Neuromodulator Cochlear implants	–	D-00122	EaglePicher	0.08	3	10.8	[7]
Hearing aid	–	W101, etc.	Wyon	0.368	40	152	[32]
Neuromodulator Cochlear implants Sleep apnoea	–	D-00020	EaglePicher	0.88	50	182	[8]
Fitness sensor Cosinuss°One	Cosinuss GmbH	CP1254 A3	Varta	0.62	60	233	[33]
Neuromodulator Drug pumps Body fluid pumps Cardiac monitors	–	95-1399 95-1250	EaglePicher	3.86 5.97	225 325	810 1170	[9] [10]

Source: Authors' compilation based on data from references cited in the table.

During the deep discharge (a rise in the negative electrode potential) of LICs (for example, (NCA/Gr) voltage below 0.5 V [34]), the copper current collector will be dissolved, which will drastically reduce the cycle life later (due to the further deposition of copper particles on the positive electrode [35]). To minimise the effects of deep over-discharge, the Quallion company uses the patented Zero-Volt Technology to produce miniature LICs for medical applications. Furthermore, by using titanium foil instead of copper foil, selecting the ratios of electrode capacities in such a way that, in case of zero difference of potentials between the cathode and anode, the anode potential does not exceed the values, the result is that no dissolution of SEI film or current collector of the negative electrode is occurred [6,36].

When comparing LICs with equivalent pairs of active materials and similar parameters of the active layer (porosity, thickness, etc.), the specific energy (Wh/kg, Wh/L) and power (W/kg, W/L) increase with the rise in the volume and cell thickness (room temperature—Figure 67, low temperatures—Figure 68a). An increase in the specific energy can be explained by a more significant amount of active material and the decreased mass and volume percentage of case parts and electrochemically inactive components.

LICs with high specific energy and a capacity below 245 mAh have a much lower energy density (<500 Wh/L) than the superior small-sized LICs used for smartphone power supplies (3880 mAh, <760 Wh/L), and high internal resistance (the impedance value in some range is inversely proportional to the electrode area) [37]. An increased internal resistance (and a lower capacity) results in a more intensive drop of voltage (power) at impulse load (pulse current discharge) and, consequently, leads to a higher risk of battery disconnection by the battery management system (protection circuit module). Therefore, the conclusion is that volume is one of the critical parameters limiting the specific characteristics of miniature LICs.

The energy density (specific energy) of a 1.22 Ah cell is close to the energy density (specific energy) of LICs with high specific energy for smartphones. High energy characteristics can be achieved by using materials with high specific energy (high-voltage lithium cobalt oxide) and efficient use of the LIC case inner space, i.e., a thicker electrode block (approximately 1 cm) and an elliptical shape of the case made of laminated foil (pouch).

Both primary cells (non-rechargeable) and secondary lithium power supplies (LICs) are produced in coin-type cases. The specific energy (Figure 69) of primary coin-type cells (CR2032) at room temperature can reach 210 Wh/kg (620 Wh/L), but they can only be discharged with low currents, and, in general, their specific power does not exceed 1 W/kg (3 W/L). On the other hand, for primary cells with increased power, the specific power can reach 8 W/kg (26 W/L) at a specific energy of 58 Wh/kg (185 Wh/L).

Table 43. Miniature lithium-ion cells in prismatic (metal, pouch (laminated foil)) and cylindrical cases.

	Manufacturer	Model	Active Materials	U, V	C, mAh	m, g	Th/d	H mm	L	E, Wh/ kg	E, Wh/ L	Interval, T, °C	Cycle Life	Ref.	Designation
	EaglePicher	D-00122	-	3.6	2.8	0.25	2.91	12.25	-	40	126	37	>1000	[7]	-
	EaglePicher	D-00020	-	3.6 3.64	50	2.5	4.5	12.04	17.78	72 73	228 206	37	>1000	[8]	EP50
	Murata	CT04120	LCO/ LTO	2.3	3	-	4	12	-	-	46 47	−20 ÷ 70	5000	[38–40]	CT3
	Murata	Umal	LCO/ LTO	2.3	12	-	2	14	21	-	47 49	−20 ÷ 70	5000	[3]	U12
	Nichicon	SLB03070LR35	−/LTO	2.4 2.5	0.35	0.2	3	7	-	5 5	17 20	−30 ÷ 60	>5000	[41,42]	N035
	Quallion	QL0003B	NCA /Gr	3.6	3	0.2	2.9	11.8	-	54 53	139 136	15 ÷ 42	>500	[6,43]	Q3
	Panasonic	320B	LCO/Gr	3.8 3.89	16 16.6	0.5	3.65	20	-	122 128	291 306	−20 ÷ 60	>500	[44]	P16B
	Panasonic	420A	LCO/Gr	3.8	23	0.8	4.7	20	-	109 116	252 269	−20 ÷ 60	>900	[45]	P23A
	Samsung	EB-BR830ABY	LCO/ –	3.85	240	3.95	4.5	20	21 (23.5)	232	475 435	-	-	[18]	S240
	ATL	A53129	LCO/ –	3.85	1220	17.65 *22*	1 *1.1*	2.5 *3.2*	3 *3.6*	280 *225*	760	-	-	[19,20]	A1220

Note: The determined values based on an analysis of the discharge curves given in the LIC specification are underlined; the data corresponding to batteries are italicised. U—voltage, C—capacity, m—mass, Th—thickness, d—diameter, H—height, L—length, E—specific energy, T—temperature. Source: Authors' compilation based on data from references cited in the table.

Table 44. Miniature lithium-ion cells in coin-type and button-type cases.

	Manufacturer	Model	Active Materials	U, V	C, mAh	m, g	Th/d	H mm	L	E, Wh/ kg	E, Wh/ L	Interval, T, °C	Cycle Life	Ref.	Designation
✕	Maxell	CLB2016	LCO/Gr	3.7 3.83	30	2.1	20	2	-	53 54	177 182	−20 ÷ 60	250	[46,47]	M30
⊠	Maxell	CLB2032	LCO/Gr	3.8 3.82	70	3	20	3.45	-	89 89	245 245	−20 ÷ 60	>500	[48]	M70
⁘	Maxell	CLB937	LCO/Gr	3.7	18	0.7	9.5	3.9	-	95	241	−20 ÷ 60	>500	[46,49]	-
✚	Maxell	TC920S	$Li_xTi_yO_4/C$	1.5	3.5	0.43	9.5	2.05	-	12.2	36.1	−20 ÷ 60	>500	[50]	-
△	Maxell	ML2032	MnO_2/Li-Al	2.3	65 71	3	20	3.2	-	55	165	−20 ÷ 60	1000 (10%) 300 (20%)	[51]	M65
▲	Seiko	MS621FE	$Li_aMn_bO_c/SiO_x$	3 2.51	5.5	0.23	6.8	2.1	-	60	181	−20 ÷ 60	100 (100%) 1000 (20%)	[52–54]	S5.5
◇	NGK	ET1210C-H	LCO/LTO	2.3	4	0.5	12	1	-	18	81	−40 ÷ 85	-	[55,56]	-
◆	NGK	ET2016C-R	LCO/LTO	2.3	25	2	20	1.6	-	29	114	−40 ÷ 85	-	[55,56]	-
○	Varta	CP7840	NCM/Gr	3.7 3.71	16 17.2	0.7	7.8	4	-	85 90	310 332	−20 ÷ 60	>500	[55,56]	V17
●	Varta	CP1254 *	NCM/ Gr + Si	3.7 3.71	60 63	1.6	12.1	5.4	-	139 146	356 375	−20 ÷ 60	>500	[12,57–59]	V63
◉	Varta	CP1654	NCM/Gr	3.7 3.71	120 125	3.2	16.1	5.4	-	139 144	404 420	−20 ÷ 60	>500	[58]	V125
□	Wyon	-	LCO/Gr	3.7	0.16	0.016	9.15	5.6	-	169	420	0 ÷ 50	>2000	[5,60]	W0160
■	Wyon	W101	LCO/Gr	3.8	40	0.9	9.15	5.6	-	169	420	0 ÷ 50	>2000	[32]	-

Note: The determined values based on the discharge curves given in the LIC specification are underlined. U, C, m, Th, d, H, L, E, T—description is given in note to Table 43. * The cells [61] manufactured by EVE (Bean cells), MIC-power, ZeniPower, GreatPower, etc., have the same CP1254 case (and its equivalents) as the cells produced by Varta. EVE manufactures CP1254 cells of various power and specific energy with NCM and LCO cathodes [62]. Source: Authors' compilation based on data from references cited in the table.

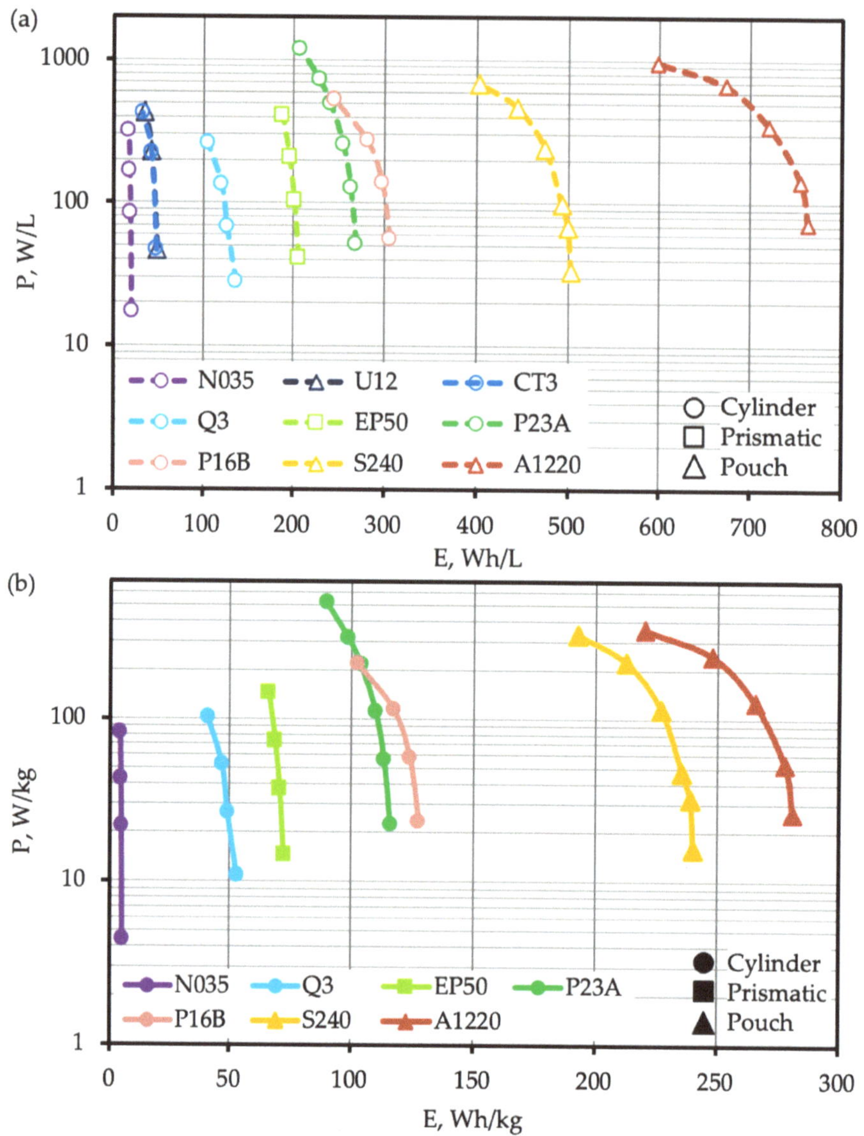

Figure 67. Interrelation of energy density (Wh/L) and power density (W/L) (**a**), specific energy (Wh/kg) and specific power (W/kg) (**b**) of lithium-ion cells N035 [41], Q3 [43], EP50 [8], P23A [45], P16B [44], S240, and A1220 (S240 and A1220 were tested in the Ioffe Institute). Source: Figure by authors.

The coin-type rechargeable LICs (CLB2032) have a much lower specific energy than primary cells, i.e., 90 Wh/kg (245 Wh/L). However, their power (16 W/kg, 49 W/L) at nominal discharge current exceeds the maximum power for high-power primary elements (Figure 69). The maximum power of LICs may reach 168 W/kg (467 W/L) at specific energy equal to 55 Wh/kg (168 Wh/L).

When operated at low temperatures, coin-type LICs also reach much higher power and lower energy density than primary cells mounted in the same case (Figure 68b).

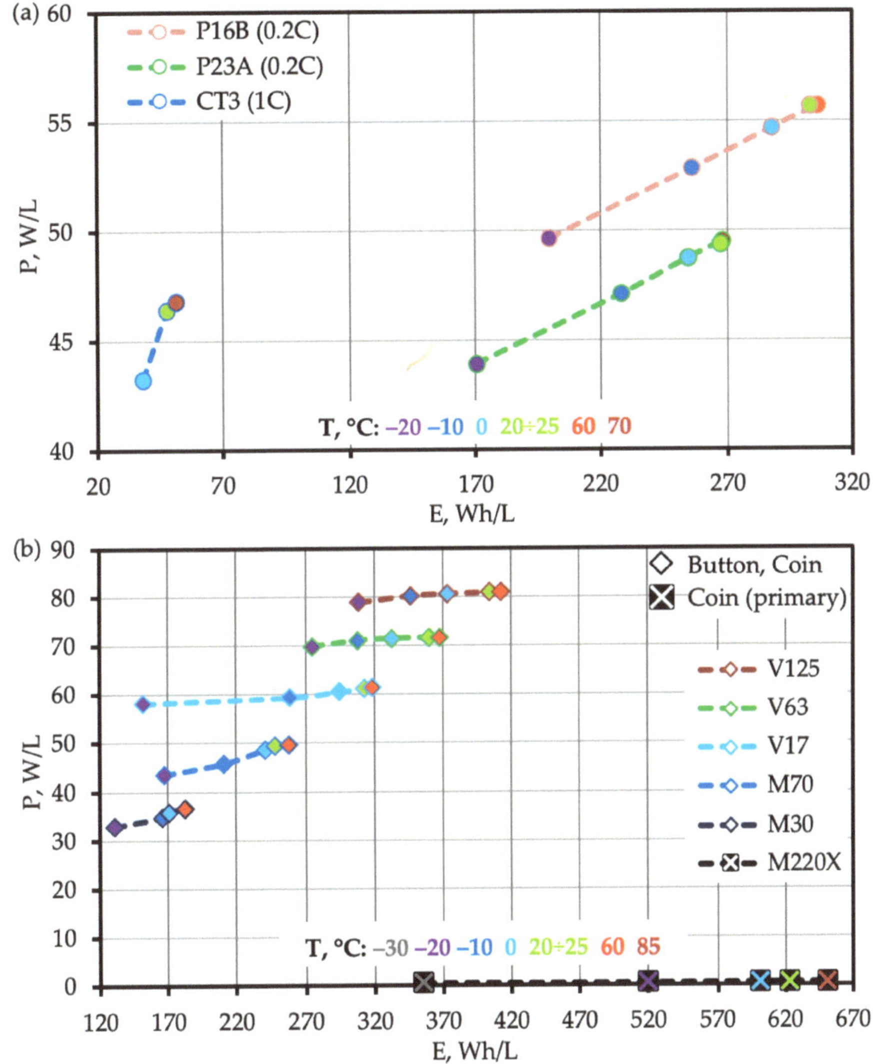

Figure 68. Temperature effect on energy density and power density of lithium-ion cells: (**a**) CT3 [39], P23A [45], P16B [44]; (**b**) M30 [47], M70 [48], V17 [58], V63 [58], V125 [58], and primary cell M220X [63]. Source: Figure by authors.

For coin- and button-type LICs, their cases' inner space is also a critical parameter. The greater height of the case allows the LIC's internal space to be used more effectively and increases the energy density (specific energy). As a result, the button-type LICs feature a higher energy density (specific energy) and power density than the coin-type cells, even at lower capacities (Figure 69). The same can be observed at low temperatures (Figure 68b).

For more information on lithium-ion cells (filled with liquid and gel electrolytes) and batteries for wearable devices, refer to presentation [37].

7.2. Miniature Solid-State Lithium-Ion Cells (mSSLICs)

There are some advantages to the development of miniature solid-state lithium-ion cells (mSSLICs):

- The risk of electrolyte leakage during operation and storage can be avoided due to the use of a solid-state electrolyte, which improves operational safety and extends calendar service life.
- Since only lithium ions move in solid-state electrolytes, the risk of side reactions is minimised; consequently, the service life can be extended (by at least a lifetime).
- Solid-state inorganic electrolytes can be used to increase the charging voltage and the voltage of the cell itself (bipolar batteries, voltage increase—impedance reduction—Prologium), which allows the energy density to be increased.
- Solid-state electrolytes allow the operating temperature range to be expanded. Low temperatures do not cause the solidification of the electrolyte. In contrast, at high temperatures, the risk of side reactions is reduced—damage to SEI film, solvent pressure build-up, swelling of the case, electrolyte decomposition, and inflammation.
- Solid-state inorganic electrolytes slow down combustion, which reduces negative consequences in the event of emergencies.
- There are manufacturing technologies appropriate for the mass production of miniature mSSLICs (in particular, the production of multilayer mSSLICs using multilayer ceramic capacitor manufacturing technology).

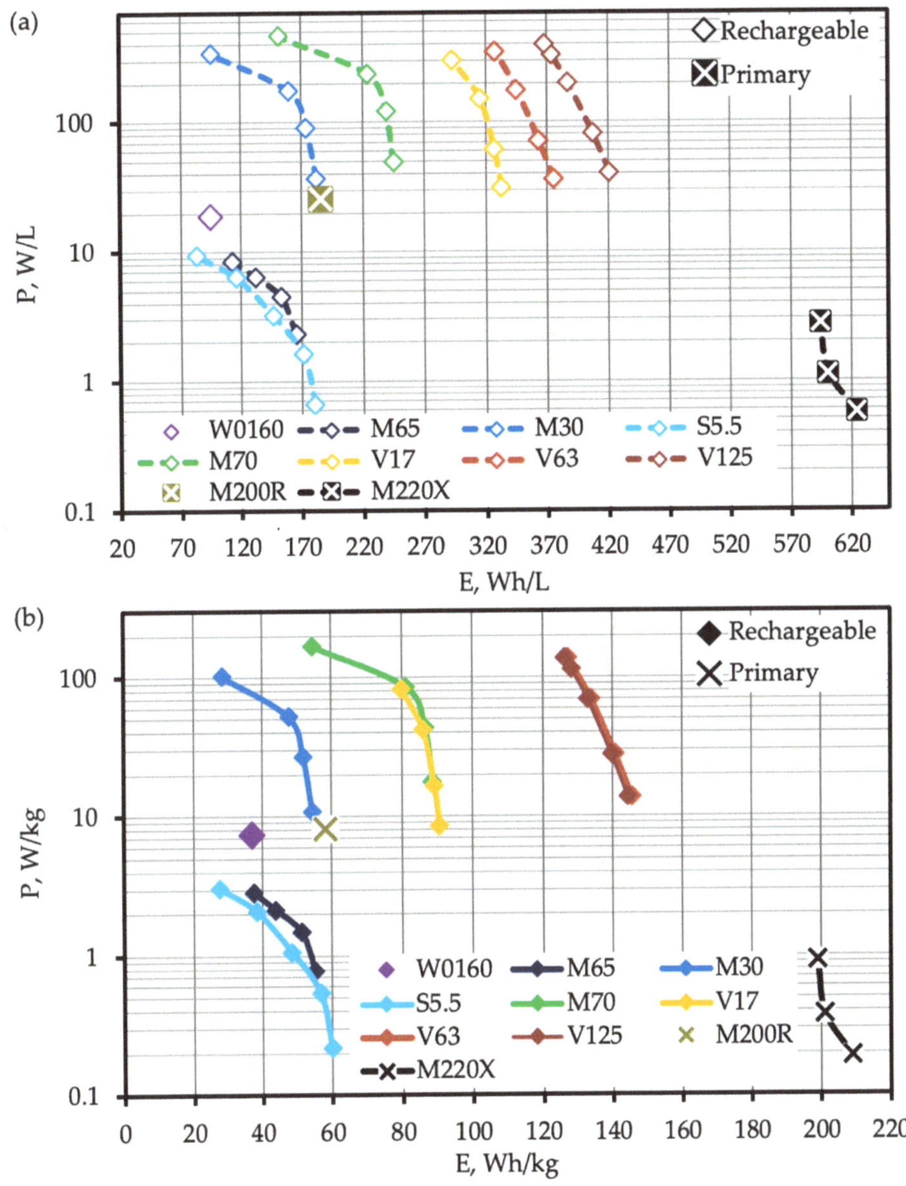

Figure 69. Interrelation of energy density (Wh/L) and power density (W/L) (**a**), specific energy (Wh/kg) and specific power (W/kg) (**b**) of lithium-ion cells (W0160 [5], M65 [51], M30 [47], S5.5 [53], M70 [48], V17 [58], V63 [58], V125 [58]) and primary cells (M200R [64] and M220X [63]). Source: Figure by authors.

Design versions of solid-state LICs are shown in Figure 70. The following types are available: bulk, multilayer (design is similar to design of multilayer ceramic capacitors MLCC) and thin film (manufactured based on the planar process). The LICs manufactured using the mixed method, i.e., the slurry coating process and integrated circuit manufacturing technology, have recently been presented [65].

The electrochemical block of bulk-type mSSLICs (Figure 70a) consists of a pair (or several pairs) of electrodes, including a current collector and an active layer isolated by a solid-state electrolyte (inorganic or polymer) that functions as a separator as well. Polymer electrolytes can be produced based on partially oxidised polyethylene (PEO) [66,67] and other organic compounds polymerised on exposure to temperatures [68].

Among the most widely used solid-state inorganic electrolytes (glass, glass-ceramic, crystalline) are oxygen-containing [69–71] ($Li_{3-x}PO_{4-y}N_z$, $Li_{1+x}Ge_{2-x}Al_x(PO_4)_3$, $Li_{1.3}Ti_{1.7}Al_{0.3}(PO_4)_3$, $La_{2/3-x}Li_{3x}TiO_3$, $Li_7La_3ZrO_{12}$, Li_3BO_3-Li_2SO_3 etc.) and sulphide substances [69,70] ($Li_7P_3S_{11}$, Li_3PS_4, $Li_{10}GeP_2S_{12}$ (LGPS), Li_6PS_5X (X—Cl, Br), and other types). Oxide electrolytes are easier to synthesise and handle and it is easier to manufacture cells with the use of such electrolytes. They form a more stable interface with the cathode and lithium, but they have a lower conductivity of lithium ions than sulphide electrolytes [69,72–74].

A significant amount of solid-state electrolyte is added to the electrode-active layer to provide ionic conductivity. For example, the analysis of electrode section images shows that the areas of active cathode material (NCA), two electrolyte components (LGPS, Li_6PS_5Cl), and pore space are as follows: 49.2%, 39.1% + 1.3%, and 10.4%, respectively [75]. The amount of active material can be increased by applying a solid-state electrolyte coating to particles of active cathode material [69]. Along with solid-state electrolyte, the active layer can also contain conductive additives [76] to provide proper electronic conductivity.

Bulk-type mSSLICs can be manufactured by the operations of dry [70] and liquid application followed by drying (including mSSLICs filled with polymer electrolyte), pressure moulding, sintering, etc. A review of the mSSLIC components' manufacturing methods is given elsewhere [77]. However, the complexity of arranging mass production [70] is one of the barriers to the widespread use of bulk-type mSSLICs.

For the positive electrodes of small-sized LICs, the fraction of the cross-sectional area of the particles of the positive-electrode-active material [78] is much higher

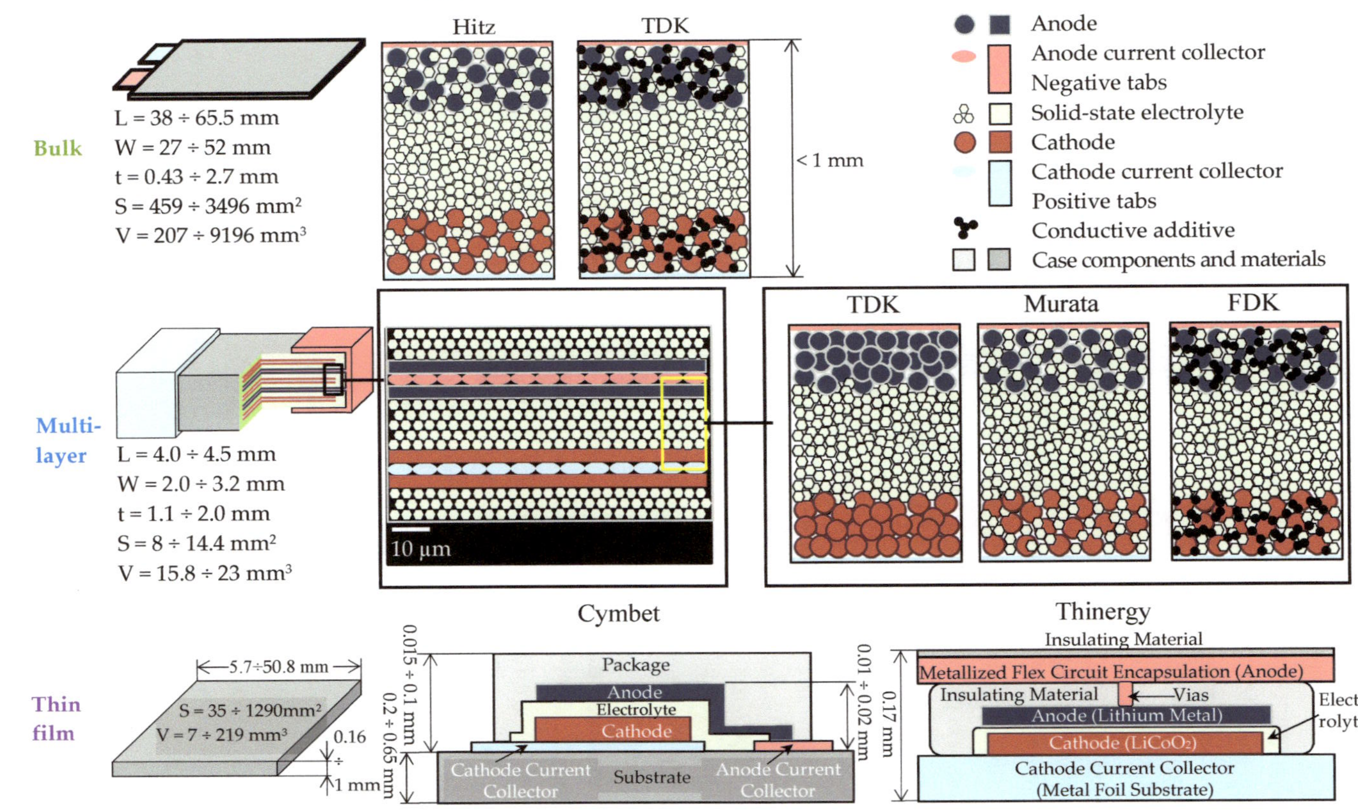

Figure 70. Schematic structure of miniature solid-state lithium-ion cells developed by Hitz [70], TDK [4], Murata [79], FDK [76], Cymbet [4,70,80–82], and Thinergy [83–85]. Source: Figure by authors.

Multilayer mSSLICs (Figure 70) consist of several double-sided positive and negative electrodes arranged in parallel, isolated by solid-state electrolyte, and with electrical contact with positive and negative tabs. The critical fabrication stage is the application of components using the screen-printing method, as the exact alignment of the layers being applied is required. A three-dimensional structure of many multilayer solid-state cells is formed on one plate after alignment and sequential layer-by-layer application. Next, the plate is split (cut) into many mSSLICs blanks, which are further annealed at high temperatures. After that, tabs are applied to both sides of two parallel end walls. This manufacturing method is similar to the one used for the production of multilayer ceramic capacitors (MLCCs) [86]; the latter approach is described in more detail in the documentation by Kemet [87].

The analysis of micrographs [4] of multilayer TDK mSSLICs sections (shown schematically in Figure 70) demonstrates that the approximate thickness values for the electrolyte layer, active layers (on both sides), and current collector are 13 μm, 4.6 μm, and 3.5 μm, respectively. In this regard, it can be concluded that the volume fraction of solid-state electrolyte in mSSLICs exceeds the fraction of active materials. Indeed, the safety data sheet states that the mass fraction of active material $Li_3V_2(PO_4)_3$ (anode and cathode) is 22 wt.%, while the mass fraction of electrolyte is 59.9 wt.% ($Li_{1.3}Al_{0.3}Ti_{1.7}(PO_4)_3$—97%, Li_3PO_4—3%). In contrast, the electrolyte weight in high-power LICs for hybrid vehicles is approx. $0.63 \div 0.92$ of the mass of cathode material [88].

Thin-film mSSLICs (Figure 70, thin film) consist of an electrochemical block (cathode, electrolyte, anode) arranged on a substrate and coated with a protective layer.

The literature describes several design types of thin-film mSSLICs [89,90], manufactured based on the planar process [91]. Figure 70 shows the devices similar in design to those manufactured by Cymbet [80] and Thinergy [85]. In the first case, layers of current collectors are applied to the substrate, and films of the cathode, electrolyte, and anode are deposited on top of the current collectors. The electrochemical block is closed with an insulating layer whose thickness varies from several microns [92] to hundreds of microns [22]. This type of substrate and the encapsulating layer occupy a significant part of the mSSLICs' weight and volume, the case thickness of mSSLICs is approx. 1 mm, and the thickness of the electrochemical block is $10 \div 20$ μm. It can be concluded that the energy density and power density, specific energy, and power in terms of weight and dimensional characteristics of the LICs, in general, will have low values. Several pairs of electrodes stacked on the same chip can increase the energy density and power density [93–95] and possibly reduce the cost by using active materials more efficiently.

The estimated [95] bill of materials (BOM) of thin-film cells with LiPON-based electrolyte equals 1.5 USD/Wh. However, as high-value equipment is used and production output is limited (as opposed to the slurry coating process), the price per Wh for a thin-film cell is significantly higher.

In the second design option (Thinergy), the cathode current collector serves as a substrate, and the thickness of the mSSLIC is much lower—approx. 170 μm [85]. Therefore, this type of mSSLIC possesses higher energy density and power density.

For more information on the design, manufacture methods, and characteristics of mSSLIC sample items, refer to papers [96,97].

Table 45 provides the characteristics of bulk-type mSSLICs filled with inorganic electrolyte manufactured by Hitz, NTK, and Prologium. As may be supposed, such cells are produced with the use of solid-state electrolytes made of sulphide (Hitz) and oxide (NTK, Prologium (the content of solid-state electrolyte is over 90% of overall electrolyte content [98])). At nominal current discharge, the energy density (specific energy) varies within 56 ÷ 222 Wh/L (20 ÷ 128 Wh/kg) depending on the active materials used. Table 45 also describes the characteristics of flexible LICs (Jenax, Lionrock, Panasonic) similar in dimensions, which may contain gel polymer or polymer electrolyte. The energy density (specific energy) of flexible LICs varies within 94 ÷ 184 Wh/L (55 ÷ 114 Wh/kg). Thus, it can be concluded that bulk-type mSSLICs and gel polymer (polymer) electrolyte LICs have similar energy density (specific energy) at nominal current discharge. The energy density (specific energy) of bulk-type mSSLICs filled with inorganic electrolyte (Figure 71) drops when the discharge current (power) rises, as well as at a maximum power of 200 ÷ 360 Wh/L (130 ÷ 200 Wh/kg).

If the temperature rises from 20 to 60 °C, the energy density and power density of the mSSLICs produced by Prologium (Figure 72) practically do not change at low discharge current (0.2C). A drop in the discharge temperature from 20 to −20 °C results in a reduction in the energy density and power density by approx. 40% and 14 ÷ 16%, respectively.

Table 45. Miniature laminated foil (pouch) case LICs filled with solid-state and gel (polymer) electrolyte.

	Manufacturer	Model	Active Materials	U, V	C, mAh	m, g	Th d	H mm	L	E, Wh/ kg	L	Interval, T, °C	Cycle Life	Ref.	Designation
✚	Hitz	AS LIB	–	3.65	140	25	2.7	52	65.5	20	56	−40 ÷ 120	–	[99]	–
△	Jenax	–	–	3.8	30	–	0.5	27	48	–	176	−20 ÷ 60	1000	[100,101]	–
◇	Lionrock	3X12019	–	3.6	80	5.2	1.35	19	120	55	94	−10 ÷ 60	–	[102]	–
▲	NGK	ET271704P-H	LCO/LTO	2.3	5	0.3	0.45	17	27	38	56	−40 ÷ 60	–	[55,56]	–
◆	NGK	EC382704P-C	LCO/Gr	3.8	27	0.8	0.45	27	38	128	222	0 ÷ 45	–	[27,55,56]	N27
○	Panasonic	CG-042839	–	3.8	18	0.7	0.45	28.5	39	98	137	−20 ÷ 60	>1000	[103,104]	–
□	Panasonic	CG-043555	–	3.8	42	1.4	0.45	35	55	114	184	−20 ÷ 60	–	[103,104]	–
●	Prologium	FLCB027038 AAAA	–	3.75 $\underline{3.87}$	16.5	0.8	0.43	27	38	77 $\underline{79}$	140 $\underline{144}$	−20 ÷ 60	>500		Pg16
■	Prologium	FLCB046046 AAAA	–	3.75 $\underline{3.86}$	45	1.6	0.43	46	46	105 $\underline{108}$	185 $\underline{190}$	−20 ÷ 60	>500		Pg45
⁚⁚	Toshiba	–	LMFP/LLZO/LTO *	12.5	102	14	1.8	70	110	90	91	−40 ÷ 80	>600	[105,106]	T102

Note: For specifications of the most advanced LIC models by Prologium, refer to FLCB027038AAAB, FLCB046046AAAB; the determined values based on the analysis of the discharge curves given in the LIC specification are underlined. U, C, m, Th, d, H, L, E, T—description is given in note to Table 43. * $LiMn_{0.85}Fe_{0.1}Mg_{0.05}PO_4$ (89 wt.%)/LLZO (96 wt.%) +4 wt.% PAN gel-polymer/$Li_4Ti_5O_{12}$ (94 wt.%). Source: Authors' compilation based on data from references cited in the table.

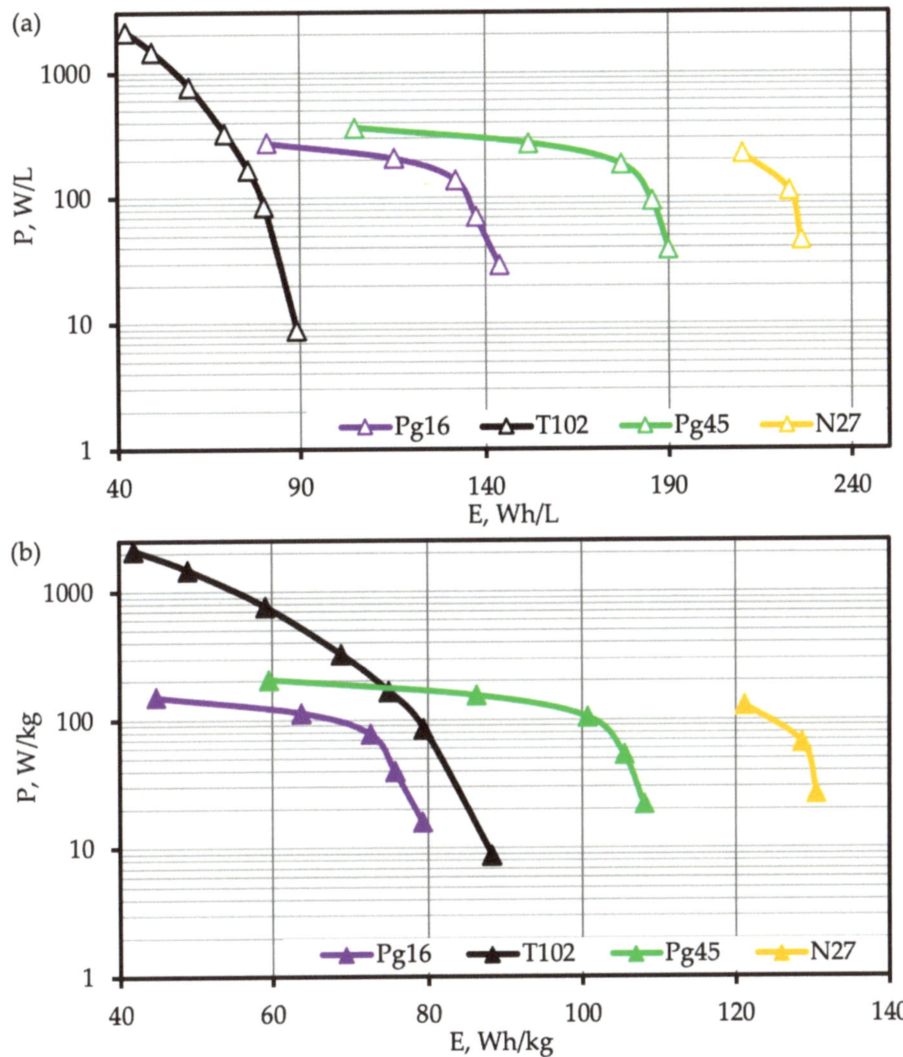

Figure 71. Variation in energy (Wh/L, Wh/kg) and power (W/L, W/kg) of mSSLIC T102 [105], Pg16, Pg45, and N27 [27] at room temperature ((**a**)—per unit of volume, (**b**)—per unit of mass). Source: Figure by authors.

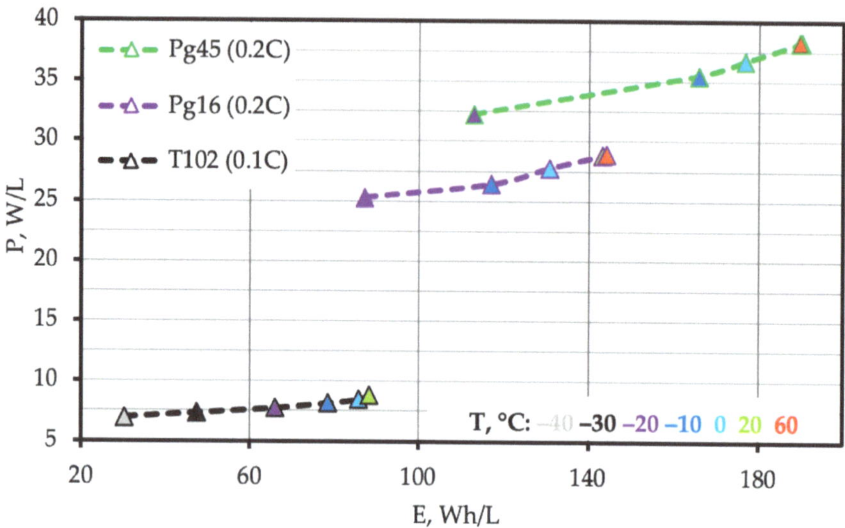

Figure 72. Temperature effect (at current discharge 0.1 ÷ 0.2C) on energy density and power density of mSSLIC T102 [105], Pg16, and Pg45. Source: Figure by authors.

Toshiba presented a bipolar solid-state cell with a nominal voltage of 12.5V [105, 106]. The solid-state electrolyte used is a 3 ÷ 8 micron layer of particles deposited from a suspension containing $Li_7La_3Zr_2O_{12}$, acrylic acid, and 1.2M $LiPF_6$ in solution PC:DEC 1:2 (volume ratio) [106]. $LiMn_{0.85}Fe_{0.1}Mg_{0.05}PO_4$ and $Li_4Ti_5O_{12}$ are used as active cathode materials. The cell capacity is 102 mAh, and the specific energy and energy density are almost the same, equalling 88 Wh/kg, 89 Wh/L. The cell can be discharged by high currents (30C) at room temperature. In this case, the specific power can reach 2 kW/kg (specific energy is 40 Wh/kg, Figure 71). The performance at low temperatures was demonstrated within the temperature range of −40 ÷ 25 °C and a discharge current of 0.1C.

Multilayer mSSLICs are manufactured by TDK and FDK (Table 46). Murata [79] and Taiyo Yuden [71,107] plan to arrange the mass production of their developments in the nearest future. TDK and FDK use oxygen-containing inorganic substances as a solid-state electrolyte. One can refer to the information given in the safety data sheet to conclude that TDK applies $Li_3V_2(PO_4)_3$ as the active cathode and anode materials, which can release and intercalate lithium ions depending on the potential. The voltage of mSSLICs by TDK varies in the range of 0 ÷ 1.6 V. The multilayer mSSLICs developed by FDK contain high-voltage cathode material $Li_2CoP_2O_7$ and an anode made of a material containing at least titanium and oxygen [76,108,109]. The average discharge voltage of 0.14 mAh mSSLICs produced by FDK is about 2.7 ÷ 3 V.

Table 46. Miniature multilayer (several pairs of electrodes) LICs filled with solid-state ceramic electrolyte.

	Manufacturer	Model	Active Materials	U, V	C, mAh	m, g	Th d	H mm	L	E, Wh/ kg	E, Wh/ L	Interval, T, °C	Cycle Life	Ref.	Designation
△	FDK	–	$Li_2CoP_2O_7$/oxide/Ti containing	3 2.7	0.14	0.05	2	2	4	8 7.4	26 23	−20 ÷ 105	–	[76,108, 109]	F0140
▲	FDK	–	$Li_2CoP_2O_7$/oxide/	3	0.5	0.08	1.6	3.2	4.5	19	65	−20 ÷ 105	–	[108]	–
–	Murata	–	/oxide/	3.8	10	–	4.5	5.6	9.6	–	157	−20 ÷ 125	–	[79,110]	–
●	TDK	Ceracharge 1812	$Li_3V_2(PO_4)_3$/ $Li_{1.3}Al_{0.3}Ti_{1.7}(PO_4)_3$ Li_3PO_4/$Li_3V_2(PO_4)_3$	1.4 1.18	0.1	0.045	1.1	3.2	4.5	4.5	9 8.9	−20 ÷ 80	1000 (DoD 100%)	[111,112]	T0100
○	Maxell *	PSB927L	/argyrodite/	2.3	8	–	0.95	2.65	–	–	101	−50 ÷ 125	–	[113,114]	M8

Note: The determined values based on the analysis of the discharge curves given in the LIC specification are underlined, * solid-state cell in a disc case, one or several pairs of electrodes. U, C, m, Th, d, H, L, E, T—description is given in note to Table 43. Source: Authors' compilation based on data from references cited in the table.

The energy density (specific energy) at a nominal discharge current of multilayer mSSLICs varies (Figure 73) from 7.4 Wh/L (2.9 Wh/kg) for TDK Ceracharge of 0.1 mAh to 65 Wh/L for FDK of 0.5 mAh. The maximum power density (specific power) for a TDK product (0.1 mAh) reaches 57 W/L (23 W/kg). The ratio of maximum power to nominal energy is approximately 7.7. As for the cells applied for the power supply of portable electronic devices and electric vehicles, this ratio is approximately 1 and less than 7, respectively. Thus, these mSSLICs are more like high specific energy devices than high-power devices in terms of the energy variation depending on the discharge current. (For PHEV and HEV LICs, the maximum power and nominal energy ratio is approximately 15 and 50, respectively).

If the temperature rises (20 ÷ 85 °C), the stored energy, usable energy, and power of the mSSLICs devised by TDK increase by less than 10% (Figure 74). On the other hand, FDK-produced mSSLICs (0.14 mAh) feature an increase in energy density of 57% (from 23 to 36 Wh/L) and power of 17% (from 2.33 to 2.73 W/L) as the temperature rises from 20 to 60 °C.

A temperature drop from room temperature to −20 °C is accompanied by a remarkable decrease in energy density (by 82 ÷ 87%) for both mSSLICs models under consideration, even at low discharge currents (0.1C—FDK and 0.2C—TDK). In contrast, the energy drop for LICs of various types filled with liquid (and gel polymer) electrolyte is usually below 60% when the discharge temperature drops from room temperature down to −20 °C, even at currents of 0.5 ÷ 1C (Figure 66b).

Thus, when the temperature drops from room temperature down to −20 °C, a bulk-type mSSLIC has a reduction in energy density (specific energy) close to the one recorded for LICs. The energy reduction of multilayer mSSLICs is higher than the temperature decrease. The significant decrease in energy and power is probably caused by an increase in the resistance to movement of lithium ions in the solid-state inorganic electrolyte and active cathode and anode materials.

The energy density (specific energy) of the considered thin-film mSSLICs (Figure 73) at nominal discharge current varies in the range of 2.5 ÷ 43.5 Wh/L (0.9 ÷ 17 Wh/kg) and overlaps with the range of 7.4 ÷ 65 Wh/L typical for multilayer mSSLICs. The energy density (specific energy) and power (specific power) of thin-film cells are low, as these factors significantly contribute to the volume and mass of mSSLICs in structural components. For example, the thickness of the electrochemical block in a 250 μAh Ilika mSSLIC is about 10 μm, the thickness of the encapsulation layer is 100 μm (or there may be an electrochemical block + the encapsulation layer), and the thickness of the encapsulated layer and substrate is 750 μm. According to reference [115], Ilika managed to master the technology of manufacturing thin-film solid-state LICs on silicon substrates measuring 100 μm thick, which is approximately 550 μm less than the value for the M250 model under consideration. The variations in energy and power reduced to the volume of the active-layer-encapsulated active layer and the cell assembly are shown in Figure 75. The energy density at the electrochemical block level reaches 700

Wh/L at nominal discharge current. The maximum power is approx. 5.6 kW/L. The energy density and power density at the mSSLIC level are about 70 times less.

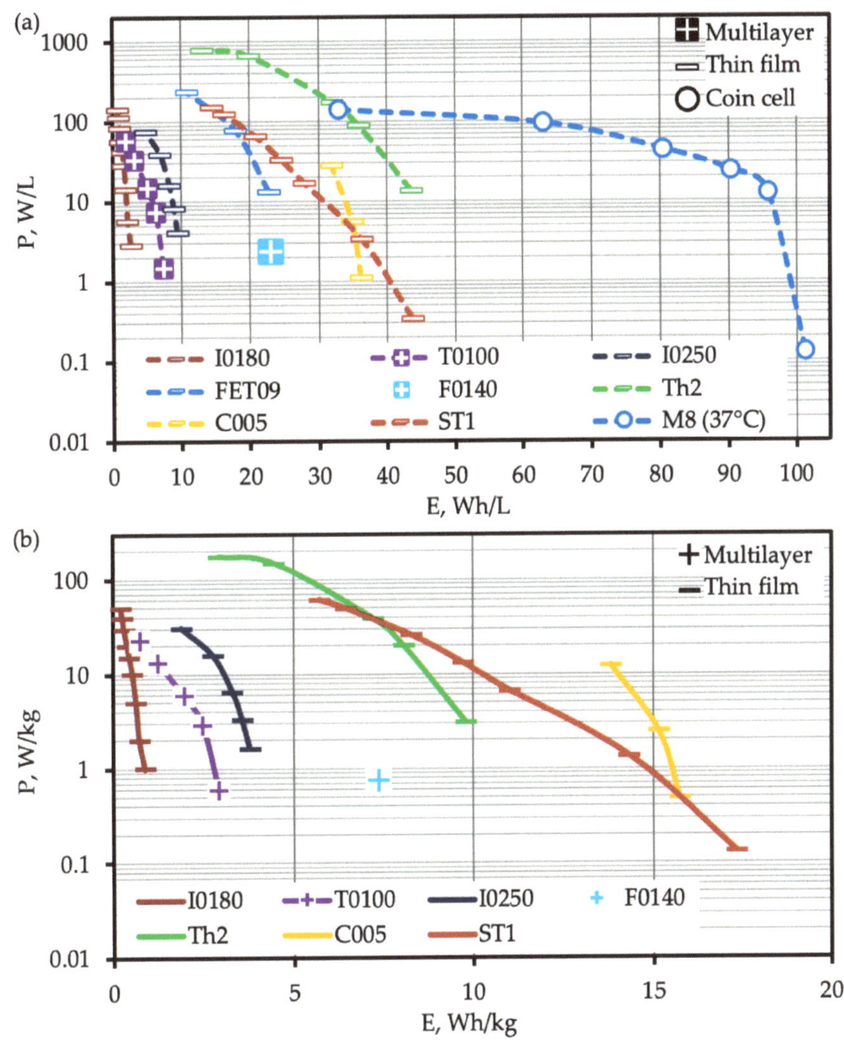

Figure 73. Variation in energy (Wh/L, Wh/kg) and power (W/L, W/kg) of mSSLIC I0180 (150 °C) [23], T0100 [112], I0250 [22,116], FET09 [117], F0140 [108], Th2 [83], C005 [81], ST1 [118], and M8 [113] (37 °C) at room temperature ((**a**)—per unit of volume, (**b**)—per unit of mass). Source: Figure by authors.

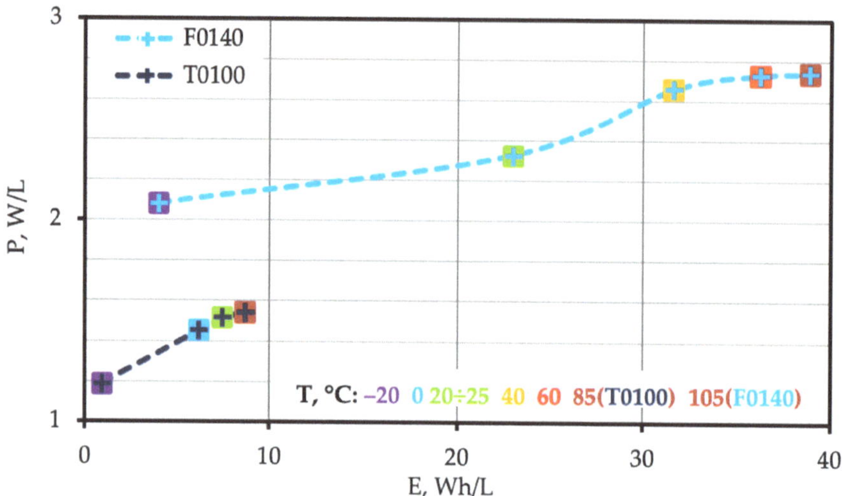

Figure 74. Temperature effect (at low current discharge) on energy density and power density of mSSLIC T0100 [112] and F0140 [76]. Source: Figure by authors.

The specific maximum power at a continuous discharge current of thin-film mSSLICs is much higher than the specific maximum power typical for multilayer mSSLICs. It varies in the range of 28 to 772 W/L depending on the design and materials. The mSSLIC manufactured by IPS is the only one with the maximum power density among all mSSLICs under consideration. This cell has a low thickness (0.17 mm) and a lithium anode. The ratio of maximum power (P_{max}) at direct discharge current and nominal energy (E_{nom}) varies from 0.8 (C005) to 17.7 (Th2). The P_{max}/E_{nom} ratio for I0250 and I0180 mSSLICs manufactured by Ilika equals 8 (20 °C) and 55 (150 °C, discharge current 50C), respectively. It is likely that the power rises proportionally to the temperature increase, but the total energy and power of I0180 cells are less than those of I0250 cells due to the greater volume and mass of the case parts.

The applied multilayer and thin-film mSSLICs can be mounted on printed circuit boards. In this regard, it is reasonable to compare mSSLICs of various types based on energy and power reduced to the area unit of the base (Figure 76). Since the area is small and the thickness (number of electrodes) is higher, the energy and power of multilayer T0100 mSSLICs with low energy density and power density reduced to the area unit approach (and in some cases exceed) the values characteristic of thin-film mSSLICs.

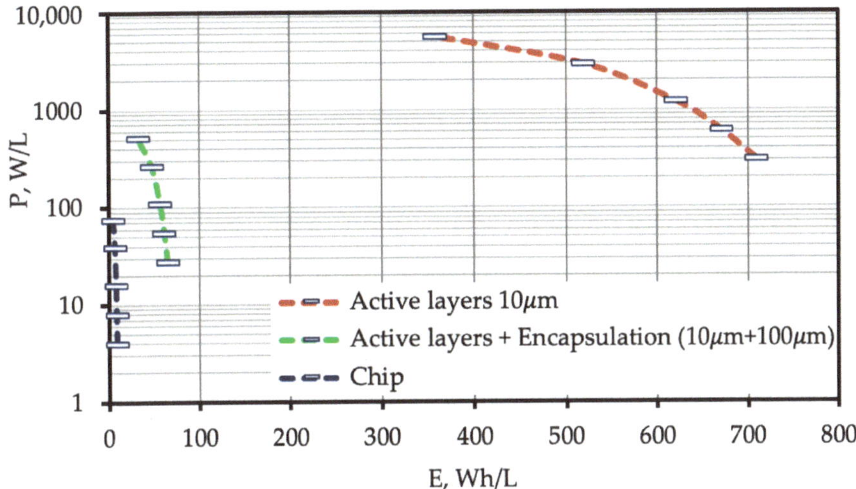

Figure 75. Energy and power reduced to the volume of active layers, encapsulated active layers, and the I0250 cell [22].

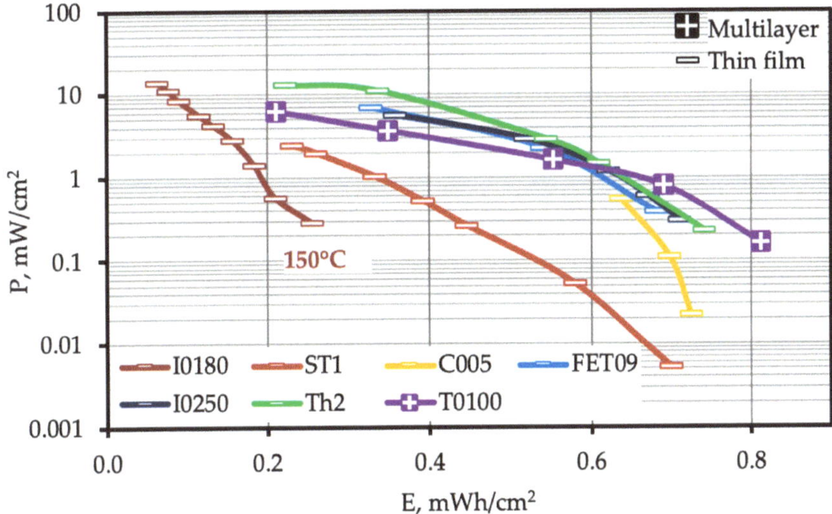

Figure 76. Specific energy (Wh/cm^2) and power (W/cm^2) at room temperature for the following types of mSSLICs: multilayer T0100 [112] and thin-film I0250 [22], I0180 (150 °C) [23], FET09 [117], F0140 [108], Th2 [83], C005 [81], and ST1 [118]. Source: Figure by authors.

7.3. Comparing Energy and Power for Miniature LICs

Miniature LICs can be installed on printed circuit boards of various devices, so the specific energy per unit of installation area can be a critical parameter. Figure 77 shows some LICs of different types in the ascending projection area of the case base. For example, the maximum nominal energy of button and coin LICs is related to the square of the installation area.

Some produced LICs have cases with a greater height, for example, CLB2032 and Varta 1254, which are characterised by an even higher nominal energy per unit of the installation area.

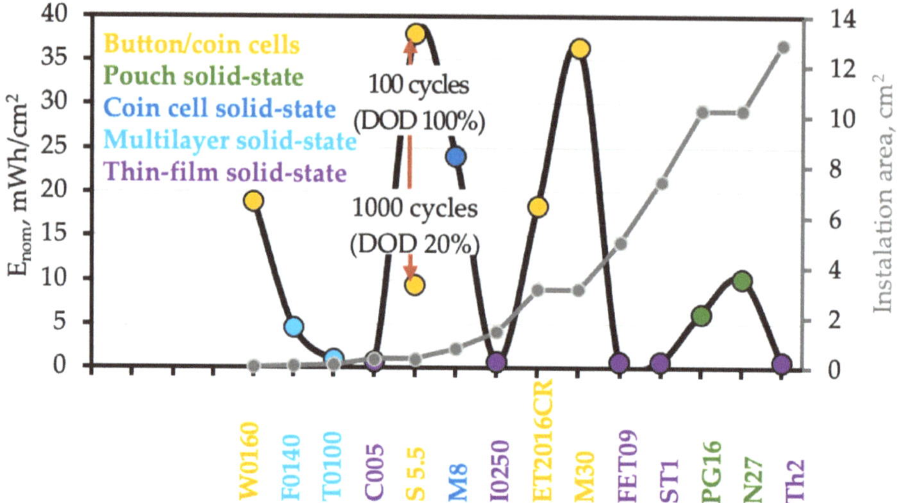

Figure 77. Nominal specific energy (mWh/cm^2) of various rechargeable lithium-ion cells (button/coin: W0160 [5], S 5.5 [52], ET2016CR [55], M30 [47]; bulk solid-state: PG16, N27 [27]; solid-state in the coin case: M8 [113]; multilayer solid-state: F0140 [108], T0100 [112]; thin-film solid-state: C005 [81], I0250 [22,116], FET09 [117], ST1 [118], Th2 [83]). Source: Figure by authors.

Usable energy and service life are expressed in cycles depending on the depth of discharge. For example, if the depth of discharge drops from 100% down to 20% for an MS621FE (S 5.5) rechargeable lithium-ion cell (the upper section of the discharge curve), the cycle life increases from 100 to 1000 cycles. At the same time, the energy used during discharge drops from 38 down to 9.4 mWh/cm^2. A similar interrelation between cycle life, depth of discharge, and the value of usable energy is typical for LICs, including the thin-film ones (Table 47).

Bulk-type mSSLICs have a small thickness (0.43 ÷ 0.45 mm). Their specific nominal energy relating to the maximum projection area (6 ÷ 10 mWh/cm^2) of LICs is somewhat below the typical values for miniature button-type and coin-type LICs. The minimum specific nominal energy is typical for multilayer mSSLICs (0.9 ÷ 4.6 mWh/cm^2) and thin-film mSSLICs (approximately 0.7 mWh/cm^2).

Table 47. Miniature thin-film solid-state LICs.

	Manufacturer	Model	Active Materials	U, V	C, mAh	m, g	Th d	H mm	L	E, Wh/ kg	E, Wh/ L	Interval, T, °C	Cycle Life	Ref.	Designation
△	Cymbet	CBC050 (Bare die)	$LiCoO_2$/ LiPON/	3.8 3.9	0.05 0.064	0.016	0.2	5.7	6.1	12 15.6	27 36	−40 ÷ 70	5000 (DoD, 10%)	[81,82]	C005
▲	Front Edge Technology	Nano-Energy	$LiCoO_2$/LiPON/Li	3.9 3.95	0.9 0.87	–	0.3	20	25	–	23	−40 ÷ 100 (170)	>1000	[117]	FET09
◇	Ilika	M250	–	3.5 3.52	0.25 0.29	0.27	0.75	12	12	3.2 3.8	8.1 10	−20 ÷ 100	5000 (DoD, 10%) 900 (100%)	[22,94, 116, 119]	I0250
◆	Ilika	P180	–	3.4	0.18	0.62	1 1	12 10	18 10	1	6	−40 ÷ 150	4000 (DoD, 5%)	[22,23, 120]	I0180
●	ST Micro-electronics	EFL1K0AF39	$LiCoO_2$/LiPON/ Li	3.9 3.99	1 1.3	0.3	0.16	25.8	28.8	13 17	33 44	−20 ÷ 60	4000 (SoC, 0 ÷ 75%)	[118]	ST1
■	IPS	Thinergy® MEC202	$LiCoO_2$/LiPON/ Li	3.9 3.96	2.2 2.4	0.975	0.17	25.4	50.8	8.6 10	39 44	−40 ÷ 85	5000 (100%)	[83,84]	Th2

Note: The determined values based on the discharge curves given in the LIC specification are underlined. U, C, m, Th, d, H, L, E, T—description is given in note to Table 43. DoD—depth of discharge, SoC—state of charge. Source: Authors' compilation based on data from references cited in the table.

Two graphs were plotted, i.e., energy density vs. capacity (Figure 78a) and energy density vs. cell volume (Figure 78b), to compare the energy characteristics of the LICs reviewed in this section (Tables 43–47).

Energy density rises in cells of the following types: thin-film, multilayer, bulk solid-state (and flexible cells filled with gel polymer electrolyte), coin and button types, and traditional (prismatic and cylindrical cells). For LICs with different form factors of their cases, there is a tendency for the energy density to rise proportionally to an increase in the thickness of the electrochemical block, volume, and nominal capacity of LICs.

Despite using materials similar in chemical nature, the energy density of button-type and prismatic LICs with high specific energy does not reach the values typical for small-sized LICs in cases 18650 and 21700 and in prismatic cases used for the power supply of smartphones. Therefore, at least for LICs with a volume and capacity less than S240 (cells for the power supply of smartwatches), the LIC (electrochemical block) volume is a critical parameter determining energy density.

The maximum energy density of miniature prismatic LICs for wearable devices is close to the values typical for the small-sized LICs (capacity below 6 Ah) with high specific energy used in quadcopters and the LICs with a capacity over 10 Ah used for the power supply of various types of electric transport (unmanned aerial vehicles, electric vehicles).

For small-sized unmanned aerial vehicles (quadcopters or multirotor aerial vehicles), the batteries are made of small-sized prismatic LICs with a lithium cobalt oxide (or NCM) positive electrode. These LICs have a volume and capacity almost equal to those of the prismatic cells used for the power supply of smartphones, but their energy density is lower, at the level of 475 ÷ 500 Wh/L (the values have been determined based on the volume calculated regarding the dimensions specified in the LIC marking—836678 (A4.28) and 914974 (A5.87); the measurable LIC dimensions are usually less than the said dimensions, and the energy density reaches 560 Wh/L for A4.28). The lower energy density is due to the need to manufacture electrodes with higher porosity (a smaller amount of active material) to provide sufficient power for quadcopter manoeuvres.

On the one hand, the prismatic LICs used in wearable devices have a small volume; on the other hand, lithium cobalt oxide is also used for cathode manufacture, making it possible to produce electrodes with a denser active layer (a greater mass and volume fraction of the active material). High-capacity cells with high specific energy are generally not manufactured using lithium cobalt oxide, probably due to safety and higher internal resistance. Instead, the cathodes of LICs with high specific energy for electric transport are manufactured using NCM and NCA active cathode materials with high nickel concentrations. These materials have a similar or higher capacity. However, the lower density of electrode-active layers and average discharge voltage negatively affects the energy density.

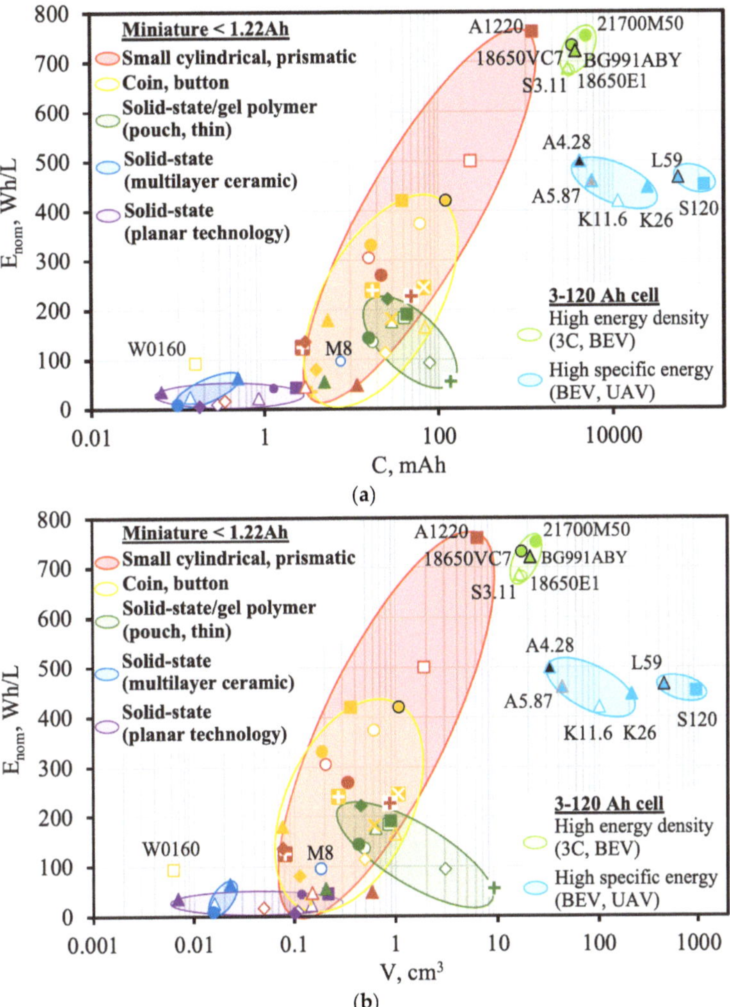

Figure 78. Energy density depending on nominal capacity (**a**) and volume (**b**) of miniature LICs (the symbols are interpreted in Tables 43–47), small-sized LICs with high specific energy (18650E1 [121], 18650VC7 [122], 21700M50 [123], S3.11 (with battery circuit board) [124], BG991ABY (with battery circuit board) [125,126]), and large-sized LICs used for the power supply of unmanned aerial vehicles (UAV)—A4.28 [127,128], A5.87 [129], K11.6 [130], K26 [130]) and battery electric vehicles (BEV)—L59 [131–134], S120 [135]. Source: Figure by authors.

7.4. Conclusions

In conclusion, this chapter presented Ragone plots showing the energy density vs. power density (Figure 79) and specific energy vs. specific power (Figure 80). The maximum power density (specific power) of most of the miniature LICs (C < 300 mAh) is within the range of 130 ÷ 500 W/L (55 ÷ 240 W/kg). Low power values are caused by the following factors:

- The relatively heavier weight of case parts that diminish power density (and especially specific power);
- For most miniature (wearable) products, the power density is not a critical parameter considered when developing miniature LICs, in contrast to the energy density and battery life.

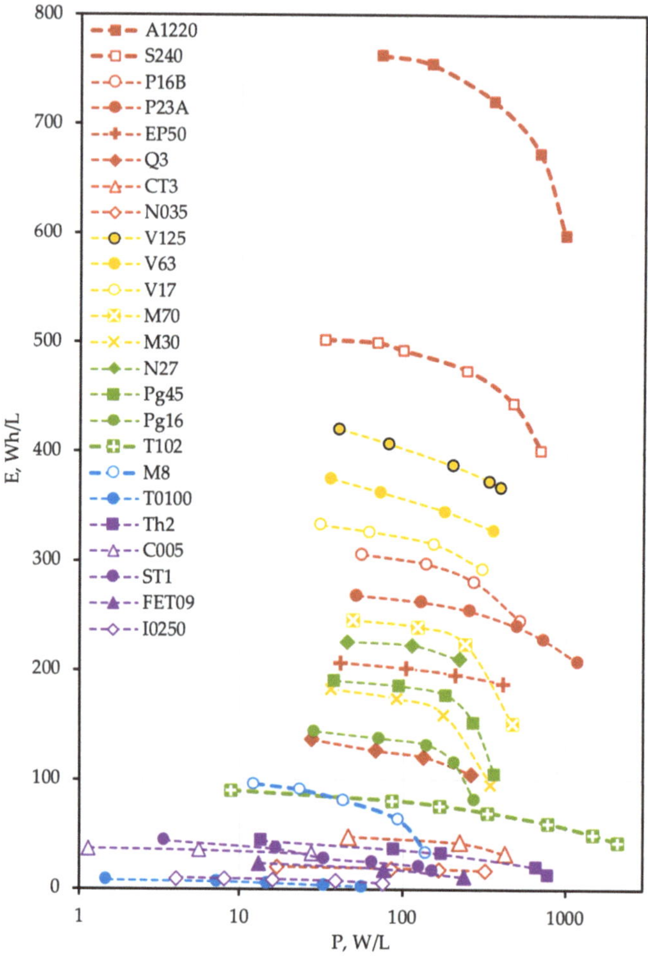

Figure 79. Energy density vs. power density of miniature (A1220, S240, P16B [44], P23A [45], EP50 [8], Q3 [43], CT3 [39], N035 [41]), button-type (V125 [58], V63 [58], V17 [58]), coin-type (M70 [48], M30 [47]), bulk solid-state (N27 [27], PG45, Pg16, T102 [105]), multilayer solid-state (T0100 [112]), coin solid-state (M8 [113]), and thin-film (Th2 [83], C005 [81], ST1 [118], FET09 [117], I0250 [22,116]) LICs. Source: Figure by authors.

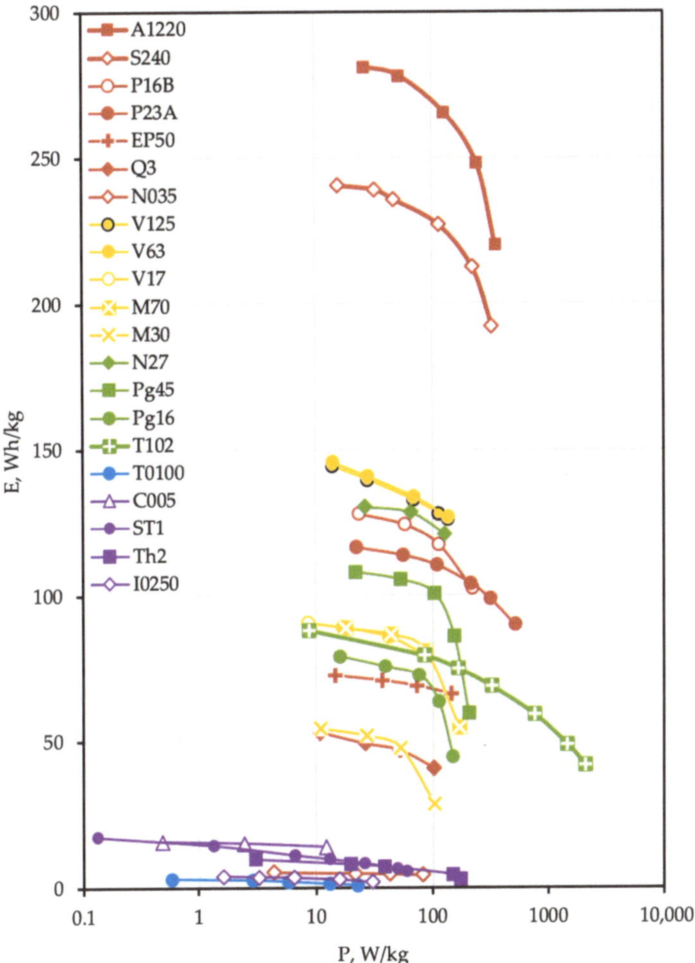

Figure 80. Specific energy vs. specific power of miniature (A1220, S240, P16B [44], P23A [45], EP50 [8], Q3 [43], CT3 [39], N035 [41]), button-type (V125 [58], V63 [58], V17 [58]), coin-type (M70 [48], M30 [47]), bulk solid-state (N27 [27], PG45, Pg16, T102 [105]), multilayer solid-state (T0100 [112]), and thin-film (Th2 [83], C005 [81], ST1 [118], FET09 [117], I0250 [22,116]) LICs. Source: Figure by authors.

References

1. Accountable Care and Battery Trends in Implantable Medical Devices. Available online: https://s3-eu-west-1.amazonaws.com/landingi-editor-uploads/9eh8WWiR/ESM009_EnerSys_Medical_eBook_A4_RGB_ENG_09_19.pdf (accessed on 11 August 2020).
2. Goldstein, M. IEEE Computer Society Phoenix Chapter—Internet of Things Innovations & Megatrends Update. Available online: https://www.slideshare.net/markgirc/ieee-computer-society-phoenix-chapter-internet-of-things-innovations-megatrends-update-121119 (accessed on 3 August 2020).
3. Small Energy Device Laminate Type (UMAL) Murata. Available online: https://www.murata.com/products/productdata/8799741476894/UMAL3DSE.pdf?1516851129000 (accessed on 29 March 2020).
4. CeraCharge™ World's First Rechargeable Solid-State SMD Battery August 2019. Available online: https://www.tdk-electronics.tdk.com/download/2427688/6fa95084f0dd78caac95dcb8f051082a/06-dl---ceracharge-presentation.pdf (accessed on 17 June 2020).
5. Implantable Batteries. Available online: https://www.wyon.ch/en/products#implantable-batteries (accessed on 8 June 2024).
6. Clarke, A.; Schrantz, K.; Macsalka, J. Benefits of Quallion's Zero-Volt™ Technology. In Proceedings of the 1st Medical Battery Conference, Dusseldorf, Germany, 16–17 November 2017; p. 22.
7. Contego® 3 mAh Battery. Available online: https://resonetics.com/wp-content/uploads/2023/03/Contego-3-010419-Rescopy23.pdf (accessed on 14 November 2024).
8. Contego® 50 mAh Battery. Available online: https://resonetics.com/wp-content/uploads/2022/08/resonetics-contego-50-mAh-Battery.pdf (accessed on 14 November 2024).
9. Contego® 225 mAh Battery. Available online: https://resonetics.com/wp-content/uploads/2022/08/resonetics-contego-225-mAh-Battery.pdf (accessed on 14 November 2024).
10. Contego® 325 mAh Battery. Available online: https://resonetics.com/wp-content/uploads/2022/08/resonetics-contego-325-mAh-Battery.pdf (accessed on 14 November 2024).
11. Prix SVC Ostschweiz 2016—Wyon AG Swiss Batteries. Available online: https://www.youtube.com/watch?v=TPq44GXO8tM&t=203s (accessed on 7 June 2024).
12. 4D Study of Silicon Anode Volumetric Changes in a Coin Cell Battery using X-ray Microscopy. Available online: https://asset-downloads.zeiss.com/catalogs/download/mic/6b928582-37b5-457d-bf5c-e26f4022c06b/EN_wp_4D_Study_Silicon_Anode_Coin_Cell_Battery_X-ray.pdf (accessed on 14 November 2024).
13. Wyser, P.; Sutter, R. Accumulator and Method for the Manufacture Thereof. US 2019/081313A1, 14 March 2019.
14. Wyser, P.J.; Sutter, R. Akkumulator und verfahren zur herstellung desselben. EP 3457453 A1, 13 September 2017.
15. Spreafico, C. Rechargeable Battery. EP 3457462, 13 September 2017.
16. Application Example: Stylus Pen (Nichicon). Available online: https://www.nichicon.co.jp/_assets/pdf/products/slb/stylus_pen_en.pdf (accessed on 17 June 2020).

17. Samsung Galaxy Buds Battery Replacement. Available online: https://ru.ifixit.com/Guide/Samsung+Galaxy+Buds+Battery+Replacement/127556 (accessed on 17 June 2020).
18. Samsung Galaxy Watch Active2 40 mm (SM-R830 SM-R835) Аккумуляторы EB-BR830ABY 240 mAh GH43-04968A. Available online: https://rounded.com/samsung-galaxy-watch-active2-40mm-sm-r830-sm-r835-battery-eb-br830aby-240mah-gh43-04968a.html (accessed on 8 August 2020).
19. Kirschenbaum, M. GoPro Hero8 Teardown. Available online: https://gethypoxic.com/blogs/technical/gopro-hero8-teardown (accessed on 8 July 2020).
20. GoPro AJBAT-001 Specs. Available online: https://www.bhphotovideo.com/c/product/1498986-REG/gopro_ajbat_001_rechargeable_li_ion_battery_for.html/specs (accessed on 25 January 2021).
21. CeraCharge™ Backup Battery for Real Time Clock (TDK Electronics AG Piezo & Protection Devices Business Group Multilayer Technology Munich, Germany 1 May 2020). Available online: https://www.tdk-electronics.tdk.com/download/2831894/a5932aa34a732bb2ad827025c862e61f/06-dl---real-time-clock.pdf (accessed on 17 June 2020).
22. Ming, L. Energy storage sources for autonomous IoT Sensing Devices. In Proceedings of the Integrated Power Conversion and Power Management (PwrSoC 2018), National Chiao Tung University (NCTU) in Hsinchu, Taiwan, 18 October 2018; p. 23.
23. Pasero, D. Extended Temperature Range Solid State Batteries for Industrial IoT (Sensor Solution International Online Webinar 2020). Available online: https://www.youtube.com/watch?v=9AyzJZuSBcM&feature=emb_logo (accessed on 16 July 2020).
24. Application Example: Environmental Sensor (Nichicon). Available online: https://www.nichicon.co.jp/_assets/pdf/products/slb/case2_en.pdf (accessed on 17 June 2020).
25. IPS-EVAL-EH-02 Wireless Environmental Sensor Energy Harvesting Evaluation Kit. Available online: https://www.digchip.com/datasheets/parts/datasheet/3430/IPS-EVAL-EH-02-pdf.php (accessed on 17 June 2020).
26. EnerCera—A New Li-Ion Battery Solution to Eliminate Bottlenecks in IoT Power Supply. Available online: https://iot.eetimes.com/enercera-a-new-li-ion-battery-solution-to-eliminate-bottlenecks-in-iot-power-supply/ (accessed on 17 June 2020).
27. Bush, S. More on: Rechargeable Cells for IoT. Available online: https://www.electronicsweekly.com/news/products/power-supplies/rechargeable-cells-iot-2020-03/ (accessed on 3 June 2020).
28. NGK202010 EnerCera English. Available online: https://www.youtube.com/watch?v=68Z5kMNgMKU&feature=emb_logo (accessed on 6 March 2021).
29. Chip-type Ceramic Secondary Battery "EnerCera" Series of New Power Sources for IoT Devices Exhibiting a Power Source Module Jointly with TOREX SEMICONDUCTOR LTD. in CES 2020. Available online: https://www.ngk-insulators.com/en/news/20191223_10646.html (accessed on 17 June 2020).
30. Energy Self-Sufficient Industrial Environmental Sensor (Varta). Available online: https://products.varta-microbattery.com/applications/mb_data/documents/application_note/IndustrialSensor_ApplicationNote_CoinPower.pdf (accessed on 17 June 2020).

31. Zhou, L.; Vest, A.N.; Peck, R.A.; Sredl, J.P.; Huang, X.C.; Bar-Cohen, Y.; Silka, M.J.; Pruetz, J.D.; Chmait, R.H.; Loeb, G.E. Minimally invasive implantable fetal micropacemaker: Mechanical testing and technical refinements. *Med. Biol. Eng. Comput.* **2016**, *54*, 1819–1830. [CrossRef] [PubMed]
32. Mikro-Batterein Ganz Gross (Micro Batteries Really Big). Available online: https://www.leaderdigital.ch/documents/magazine/special_2018_09_wyon_20_web.pdf (accessed on 8 June 2024).
33. Wearable Fitness Sensor (Varta). Available online: https://products.varta-microbattery.com/applications/mb_data/documents/application_note/WearableFitnessSensor_ApplicationNote_CoinPower.pdf (accessed on 17 June 2020).
34. Hendricks, C.E.; Mansour, A.N.; Fuentevilla, D.A.; Waller, G.H.; Ko, J.K.; Pecht, M.G. Copper Dissolution in Overdischarged Lithium-ion Cells: X-ray Photoelectron Spectroscopy and X-ray Absorption Fine Structure Analysis. *J. Electrochem. Soc.* **2020**, *167*, 9. [CrossRef]
35. Fear, C.; Juarez-Robles, D.; Jeevarajan, J.A.; Mukherjee, P.P. Elucidating Copper Dissolution Phenomenon in Li-Ion Cells under Overdischarge Extremes. *J. Electrochem. Soc.* **2018**, *165*, A1639–A1647. [CrossRef]
36. Fay, A. Zero-Volt: Medical and Satellite Battery Technology Can Help Improve Safety of Electric Vehicles. Available online: https://www.batterypoweronline.com/markets/batteries/zero-volt-medical-and-satellite-battery-technology-can-help-improve-safety-of-electric-vehicles/ (accessed on 2 August 2020).
37. Pan, L. Challenges of Wearable Batteries. In Proceedings of the 37th Annual International Battery Seminar & Exhibition, Virtual, 29–30 July 2020; p. 37.
38. Small Lithium-Ion Rechargeable Battery CT04120. Available online: https://www.murata.com/-/media/webrenewal/products/batteries/small/ct/pdf/ds-ct04120-001.ashx?la=en (accessed on 29 March 2020).
39. Small Lithium-Ion Secondary Battery (CT Series) Technical Note No.TCN-CT04120-001-J. Available online: https://www.murata.com/-/media/webrenewal/products/batteries/small/technology/technical_notes/tcn-ct04120-001.ashx?la=en (accessed on 30 May 2020).
40. Small Lithium-Ion Secondary Battery (CT04120) No. APN-SFT-001-E. Available online: https://www.murata.com/-/media/webrenewal/products/batteries/small/technology/technical_documents/apn-sft-001-safety-of-uma-series-against-overheat-ignition.ashx?la=en-gb (accessed on 3 June 2020).
41. Small Lithium-Ion Rechargeable Battery SLB03070LR35. Available online: https://www.nichicon.co.jp/_assets/pdf/products/slb/en_specification1.pdf (accessed on 29 March 2020).
42. Small Lithium-Ion Rechargeable Battery. Available online: https://www.nichicon.co.jp/_assets/pdf/products/slb/slben_slb_products_200605.pdf (accessed on 4 August 2020).
43. Micro3—QL0003B. Available online: https://s3-eu-west-1.amazonaws.com/landingi-editor-uploads/H6fKaBsv/Micro3_QL0003B.pdf (accessed on 3 June 2020).

44. CG-320B:Lithium-Ion Batteries. Available online: https://energy.panasonic.com/global/business/e/na/products/lithium-ion/models/CG-320B (accessed on 14 November 2024).
45. CG-420A:Lithium-Ion Batteries. Available online: https://energy.panasonic.com/global/business/e/na/products/lithium-ion/models/CG-420A (accessed on 14 November 2024).
46. Coin Type Lithium-ion Rechargeable Batteries. Safety Data Sheet. Reference No CLB20240101-01e. Available online: https://biz.maxell.com/en/rechargeable_batteries/sds/clb_battery_sdse.pdf (accessed on 14 November 2024).
47. Coin Type Lithium-ion Rechargeable Battery CLB2016 Data Sheet. Available online: https://biz.maxell.com/en/rechargeable_batteries/data_sheet/CLB2016_Data_sheet_e.pdf (accessed on 14 November 2024).
48. Coin Type Lithium-ion Rechargeable Battery CLB2032 Data Sheet. Available online: https://biz.maxell.com/en/rechargeable_batteries/data_sheet/CLB2032_Data_sheet_e.pdf (accessed on 14 November 2024).
49. Coin Type Lithium-ion Rechargeable Battery CLB937A Data Sheet . Available online: https://biz.maxell.com/en/rechargeable_batteries/data_sheet/CLB937A_Data_sheet_e.pdf (accessed on 14 November 2024).
50. Titanium Carbon Lithium Rechargeable Battery. Available online: https://www.yumpu.com/en/document/view/33392345/titanium-carbon-lithium-rechargeable-battery-maxell (accessed on 14 November 2024).
51. Data Sheet Lithium Manganese Dioxide Rechargeable Battery ML2032. Available online: https://biz.maxell.com/en/rechargeable_batteries/ML2032_DataSheet_17e.pdf (accessed on 2 August 2020).
52. MS Lithium Rechargeable Battery MS621FE. Available online: https://www.sii.co.jp/en/me/datasheets/ms-rechargeable/ms621fe/ (accessed on 27 June 2020).
53. Seiko Micro Battery. Available online: https://seiko-instruments.de/wp-content/uploads/2023/04/MicroBatteryCatalogue_E_2020_04_web.pdf (accessed on 14 November 2024).
54. Sugano, Y.; Xian, Z.P.; Tamachi, T.; Suzuki, T.; Shinoda, I. Compact Nonaqueous Electrolyte Secondary Battery and Manufacturing Method Therefor. JP 2013-101770 A, 23 May 2013.
55. Chip-Type Ceramic Secondary Battery EnerCera® Series. Available online: https://www.ngk-insulators.com/en/ces2020/pdf/Enercera.pdf (accessed on 3 June 2020).
56. New Product Information Session for FY2018 (Ended March 2019) NGK. Available online: https://www.ngk-insulators.com/en/ir/library/pdf/pre_FY2018_new_en.pdf (accessed on 29 March 2020).
57. Warta, A. High Energy Coin Cells for Wearable Products. In Proceedings of the 1st Medical Battery Conference, Dusseldorf, Germany, 16–17 November 2017; p. 20.
58. CoinPower Rechargeable Li-Ion Button Cells (Varta, Technical Book). Available online: https://products.varta-microbattery.com/applications/mb_data/documents/sales_literature_varta/2018_05_HANDBOOK_CoinPower_en.pdf (accessed on 30 May 2020).

59. Data Sheet—CP 1254 A3 (CoinPower®). Available online: https://products.varta-microbattery.com/applications/mb_data/documents/data_sheets/DS63125_3.pdf (accessed on 3 June 2020).
60. de Leon, S. Miniature Cylindrical Medical Batteries for Small Devices. Available online: https://www.sdle.co.il/wp-content/uploads/2018/12/20-medical-batteries-for-small-devices-v2-.pdf (accessed on 3 June 2020).
61. Varta AG: Strong Sell As Monopoly Story Shattered By Competition In Premium TWS Devices. Available online: https://newsfilter.io/articles/varta-ag-strong-sell-as-monopoly-story-shattered-by-competition-in-premium-tws-devices-ba071d2a863bf65a0f31c843412daa77 (accessed on 13 January 2021).
62. Hoying, T. Progress of Micro Electrochemical Power Technology for Wearable Devices. In Proceedings of the 37th Annual International Battery Seminar & Exhibition, Virtual, 28–29 July 2020; p. 12.
63. CR2032X Coin Manganese Dioxide Lithium Batteries. Available online: https://www.murata.com/products/productdata/8802803744798/CR2032X-DATASHEET.pdf?1577455212000 (accessed on 8 June 2020).
64. CR2032R. Available online: https://www.murata.com/zh-cn/products/productdetail?cate=cgsubMicroBatteries&partno=CR2032R (accessed on 8 June 2020).
65. Hahn, R.; Hoeppner, K.; Ferch, M.; Elia, G.; Kyeremateng, A.; Zoschke, K.; Oppermann, H. Batteries for novel medical applications based on wafer-level processing. In Proceedings of the 36th Annual International Battery Seminar & Exhibit, Fort Lauderdale, FL, USA, 26–27 March 2019; p. 33.
66. Mauger, A.; Julien, C.M.; Paolella, A.; Armand, M.; Zaghib, K. Building Better Batteries in the Solid State: A Review. *Materials* **2019**, *12*, 86. [CrossRef] [PubMed]
67. Xue, Z.G.; He, D.; Xie, X.L. Poly(ethylene oxide)-based electrolytes for lithium-ion batteries. *J. Mater. Chem. A* **2015**, *3*, 19218–19253. [CrossRef]
68. Kim, C.H.; Shin, E.J.; Shin, L.H. Flexible Secondary Battery and Method for Manufacturing Same. US 10497963B2, 22 July 2015.
69. 無機固体電解質を用いた全固体リチウム二次電池の開発 (Development of all-solid-state lithium secondary batteries using inorganic solid electrolytes) (The TRC News). 1 June 2018. Available online: https://www.toray-research.co.jp/service/trcnews/pdf/201806-01.pdf (accessed on 16 May 2024).
70. Sunayama, S. Development of All-solid-state Battery for Commercialization (Hitachi Zosen). In Proceedings of the 10th Battery Japan Technical Conference (Battery Japan 2019), Tokyo, Japan, 27 February 2019; pp. 19–31.
71. Masuda, H.; Matsushita, K.; Ito, D.; Fujita, D.; Ishida, N. Dynamically visualizing battery reactions by operando Kelvin probe force microscopy. *Commun. Chem.* **2019**, *2*, 6. [CrossRef]
72. Taylor, N.J.; Sakamoto, J. Solid-state batteries: Unlocking lithium's potential with ceramic solid electrolytes. *Am. Ceram. Soc. Bull.* **2019**, *98*, 26–31.
73. Lau, J.; DeBlock, R.H.; Butts, D.M.; Ashby, D.S.; Choi, C.S.; Dunn, B.S. Sulfide Solid Electrolytes for Lithium Battery Applications. *Adv. Energy Mater.* **2018**, *8*, 24. [CrossRef]

74. Ilika. Ilika (IKA)—Capital Markets Day December 2019: Solid-State Battery Technology. (28 min). Available online: https://www.youtube.com/watch?v=u9_6uFBhxrc (accessed on 21 December 2020).
75. Evaluation Items for All Solid State Battery. Available online: https://www.toray-research.co.jp/cn/technicaldata/pdf/TechData_P01825E.pdf (accessed on 3 August 2020).
76. FDKの全固体電池の製造プロセス (The Manufacturing Process of Solid-State Batteries). Available online: https://business.nikkei.com/atcl/seminar/19/00113/00005/?SS=imgview&FD=-1039926986 (accessed on 7 June 2020).
77. Schnell, J.; Gunther, T.; Knoche, T.; Vieider, C.; Kohler, L.; Just, A.; Keller, M.; Passerini, S.; Reinhart, G. All-solid-state lithium-ion and lithium metal batteries—Paving the way to large-scale production. *J. Power Sources* **2018**, *382*, 160–175. [CrossRef]
78. リチウムイオン二次電池 電極観察 (Lithium-Ion Secondary Battery Electrode Observation). Available online: https://www.mcanac.co.jp/db/technical-note/tec-9027.php (accessed on 14 November 2024).
79. CEATEC 2019: 村田製作所とソニーの技術を融合、容量25 mAhの全固体電池として昇華 (Murata Manufacturing and Sony combine their technologies to create a 25 mAh all-solid-state battery). Available online: https://monoist.atmarkit.co.jp/mn/articles/1910/15/news048.html (accessed on 3 August 2020).
80. Advanced Power Solutions for Wearable Technology and Internet of Everything Sensors Webinar Cymbet Corporation. Available online: https://docplayer.net/76394731-Advanced-power-solutions-for-wearable-technology-and-internet-of-everything-sensors-webinar-cymbet-corporation-all-rights-reserved.html (accessed on 3 August 2020).
81. EnerChip™ Bare Die. Available online: https://www.cymbet.com/wp-content/uploads/2019/02/DS-72-41-v6.pdf (accessed on 5 June 2020).
82. PI-72-04 Product Information Sheet. Available online: https://www.cymbet.com/wp-content/uploads/2019/02/PI-72-04.pdf (accessed on 5 June 2020).
83. Thinergy® MEC202 Solid-State, Flexible, Rechargeable Thin-Film Micro-Energy Cell. Available online: https://media.digikey.com/pdf/Data%20Sheets/Infinite%20Power%20Solutions%20PDFs/MEC202.pdf (accessed on 6 May 2020).
84. THINERGY® Micro-Energy Cell (MEC) Product Overview. Available online: https://datasheet.octopart.com/MEC201-10S-Infinite-Power-Solutions-datasheet-17049067.pdf (accessed on 6 June 2020).
85. THINERGY®Micro-Energy Cell (MEC) Product Overview (Infinite Power Solutions). Available online: https://studylib.net/doc/18415438/infinite-power-solutions (accessed on 3 August 2020).
86. MLCC (Multi-Layered Ceramic Capacitor). Available online: https://www.e-microtec.co.jp/en/search_en/mlcc_en.html (accessed on 14 November 2024).
87. KEMET Ceramic Capacitor Manufacturing. Available online: https://www.youtube.com/watch?v=gFEYuaY35Vo (accessed on 3 August 2020).
88. Takeshita, H. IIT LIB-Related Study Program 11-12 (February 2012). 2012. Available online: https://wenku.baidu.com/view/f917455be45c3b3567ec8b8f.html?_wkts_=1716389557071&needWelcomeRecommand=1 (accessed on 22 May 2024).

89. Oudenhoven, J.F.M.; Baggetto, L.; Notten, P.H.L. All-Solid-State Lithium-Ion Microbatteries: A Review of Various Three-Dimensional Concepts. *Adv. Energy Mater.* **2011**, *1*, 10–33. [CrossRef]
90. Salot, R. Thin Film Solid State Batteries. Available online: https://www.nipslab.org/wp-content/uploads/2019/09/NiPS-SS2019-Raphael-Salot.pdf (accessed on 14 March 2021).
91. Liang, X.P.; Tan, F.H.; Wei, F.; Du, J. Research progress of all solid-state thin film lithium Battery. *IOP Conf. Ser. Earth Environ. Sci.* **2019**, *218*, 012138. [CrossRef]
92. Dudney, N.J. Thin Film Batteries for Energy Harvesting. In *Energy Harvesting Technologies*; Priya, S., Inman, D.J., Eds.; Springer US: Boston, MA, USA, 2009; pp. 355–363.
93. Allen, D. Batteries Are Charging Forward to Support Medtech Miniaturization, Connectivity. Available online: https://www.mddionline.com/digital-health/batteries-are-charging-forward-support-medtech-miniaturization-connectivity (accessed on 3 August 2020).
94. Pasero, D. Integrated Perpetual Energy Source for Autonomous Sensing Devices. In Proceedings of the Integrated Power Conversion and Power Management (PwrSoC 2016), Universidad Politécnica de Madrid, Madrid, Spain, 4 October 2016; p. 18.
95. Suu, K. Manufacturing Technology of All-Solid-State Thin-Film Lithium Secondary Battery for IoT Applications (Ulvac). In Proceedings of the 34th International Battery Seminar & Exhibit 2017, Fort Lauderdale, FL, USA, 20–23 March 2017; p. 25.
96. Sun, C.W.; Liu, J.; Gong, Y.D.; Wilkinson, D.P.; Zhang, J.J. Recent advances in all-solid-state rechargeable lithium batteries. *Nano Energy* **2017**, *33*, 363–386. [CrossRef]
97. Dudney, N.J.; West, W.C.; Nanda, J. *Handbook Of Solid State Batteries (Second Edition) (Materials And Energy 6)*, 2nd revised ed.; World Scientific: Singapore, 2015.
98. Hsu, L.; Chang, S. Better understanding of solid state battery and highlights of ProLogium. In Proceedings of the 38th Annual International Battery Seminar & Exhibition, Virtual, 9–11 March 2021; p. 15.
99. Hitz (AS-LIB(R)). Available online: https://textream-cimg.west.edge.storage-yahoo.jp/cd/ae/1007004-ffcna9ba4aa5/532/188c3ad551fe5c44b6546c3db471019f.jpg (accessed on 7 June 2020).
100. Ward, C. CES 2019 | Jenax Promotes the Benefits of Their Flexible Battery to the Medical Wearable Industry. Available online: https://www.notebookcheck.net/Jenax-promotes-the-benefits-of-their-flexible-battery-to-the-medical-Wearable-industry.387881.0.html (accessed on 4 June 2020).
101. Flexible Lithium Polymer Batteryjflex Performance. Available online: https://jenaxinc.com/products/batteries/ (accessed on 4 June 2020).
102. Lionrock Batteries, 3X12019 Specification. Available online: https://lionrockbatteries.com/product/ (accessed on 4 June 2020).
103. Kolodzie, T. Pin-type and Flexible Lithium-ion Battery for Small Medical Devices. In Proceedings of the 1st Medical Battery Conference, Dusseldorf, Germany, 16 November 2017; p. 23.

104. 次世代社会を創るIoT ・ロボット・AI が総結集した (IoT, Robots, and AI Come Together to Create the Next Generation of Society) 「CEATEC JAPAN 2017. Available online: https://sgforum.impress.co.jp/article/4323?page=0,4 (accessed on 3 June 2020).
105. Takami, N.; Yoshima, K.; Harada, Y. 12 V-Class Bipolar Lithium-Ion Batteries Using $Li_4Ti_5O_{12}$ Anode for Low-Voltage System Applications. *J. Electrochem. Soc.* **2017**, *164*, A6254–A6259. [CrossRef]
106. Yoshima, K.; Harada, Y.; Takami, N. Thin hybrid electrolyte based on garnet-type lithium-ion conductor $Li_7La_3Zr_2O_{12}$ for 12 V-class bipolar batteries. *J. Power Sources* **2016**, *302*, 283–290. [CrossRef]
107. TAIYO YUDEN Develops All-Solid-State Lithium-Ion Secondary Batteries. Available online: https://www.yuden.co.jp/cms/wp-content/uploads/2019/12/Develops-All-Solid-State-Lithium-Ion-Secondary-Batteries_HP.pdf (accessed on 8 June 2020).
108. FDK. Developed High Capacity Model of Small All-Solid-State SMD Battery in this April, the World's Highest Level. Available online: http://www.fdk.com/whatsnew-e/release 20190509-e.html (accessed on 11 May 2019).
109. FDK SMD Battery (Li2CoP2O7 Cathode). Available online: https://www.marklines.com/statics/exhibitions/img/sew2018/images/fdk_DSC00908.jpg (accessed on 8 June 2020).
110. 全固体電池 (Solid-state Batteries). Available online: https://www.murata.com/-/media/webrenewal/campaign/events/japan/ceatec/9bey5lkh/ceatec-dl2020au/ce17-batteries-solid-state.ashx?la=ja-JP&cvid=20201013054941000000 (accessed on 14 November 2024).
111. Material Data Sheet CeraCharge B73xxxA. Available online: https://www.tdk-electronics.tdk.com/download/2657250/f1238f6244d4ddc9986bfc289b24a2ab/u-spec-tdk-ceracharge-b73xxxa.pdf (accessed on 8 June 2020).
112. Rechargeable Solid-State SMD Battery for IoT Applications. Available online: https://www.tdk-electronics.tdk.com/en/2471330/tech-library/articles/applications---cases---video/rechargeable-solid-state-smd-battery-for-iot-applications/2431020 (accessed on 8 June 2020).
113. Coin Type All-Solid-State Battery (Under Development) Maxell. Available online: https://biz.maxell.com/en/rechargeable_batteries/allsolidstate.html (accessed on 28 February 2020).
114. Installation of Production Equipment for Coin-Type All-Solid-State Batteries Using Sulfide-Based Solid-State Electrolytes. Available online: https://www2.maxell.co.jp/csr/pdf/CR20e_segment-energy.pdf (accessed on 15 March 2021).
115. Ilika plc Increased Energy Density of Stereax M250. Available online: https://www.directorstalk.net/ilika-plc-increased-energy-density-of-stereax-m250/ (accessed on 21 December 2020).
116. M250 Rechargeable Solid State Battery: 250 uAh, 3.5 V. Available online: https://www.ilika.com/images/uploads/general/Stereax_M250_Spec_Sheet_V1.01.pdf (accessed on 6 June 2020).

117. Ultrathin Lithium Rechargeable Battery. Available online: https://www.printedelectronicsworld.com/articles/1219/ultrathin-lithium-rechargeable-battery (accessed on 14 November 2024).
118. EnFilm™—Rechargeable Solid State Lithium Thin Film Battery (EFL1K0AF39). Available online: https://nl.mouser.com/datasheet/2/389/en.DM00408165-1148630.pdf (accessed on 3 June 2020).
119. Autonomous Self Powered Miniaturized Intelligent Sensor for Environmental Sensing and Asset Tracking in Smart IoT Environments. AMANDA Grant Agreement No: 825464 [H2020-ICT-2018-2020] Autonomous Self Powered Miniaturized Intelligent Sensor for Environmental Sensing and Asset Tracking in Smart IoT Environments. D1.1 SoA and Gap Analysis/Recommendations on ESS Features Report. Available online: https://amanda-project.eu/documents/public-deliverables/send/6-public-deliverables/7-amanda-d1-1 (accessed on 5 June 2020).
120. Stereax® P180 Extended Temperature Range Solid State Battery: -40° to + 150 °C. Available online: https://www.ilika.com/images/uploads/downloads/STEREAX-P180-SPECIFICATION-V3-6.pdf (accessed on 6 June 2020).
121. LG 18650 E1 3200mAh (Green). Available online: https://lygte-info.dk/review/batteries2012/LG%2018650%20E1%203200mAh%20(Green)%20UK.html (accessed on 14 November 2024).
122. Sony VC7. Available online: http://queenbattery.com.cn/index.php?controller=attachment&id_attachment=83 (accessed on 4 May 2018).
123. 【高工锂电•总工札记】LG和三星圆柱21700电芯和去钴化 ([Gaogong Lithium Battery • Chief Engineer's Notes] LG and Samsung Cylindrical 21700 Battery Cells and decobaltization). Available online: http://www.sohu.com/a/255303019_740349 (accessed on 13 February 2019).
124. Whitson, G. iPhone 11 Teardown. Available online: https://ru.ifixit.com/News/33016/iphone-11-teardown (accessed on 2 January 2021).
125. Samsung G975 Galaxy S10+ 4100 mAh (GH82-18827A). Available online: https://telefon-service.ru/part/gh82-18827a/ (accessed on 17 December 2020).
126. SAMSUNG Original Battery EB-BG973ABE EB-BG973ABU For Samsung GALAXY S10 Galaxy S10 X S10X SM-G973F SM-G9730 G9730 G973 3400mAh. Available online: https://www.aliexpress.com/item/33004882016.html (accessed on 2 January 2021).
127. MSDS 914974(TB50-4280mAh-22.8V). Available online: https://droneshopperth.com.au/wp-content/uploads/2017/01/ATL-TB50-4280mAh-22.8V-International-MSDS-report.pdf (accessed on 13 February 2019).
128. Inspire 2 Batteries - Part 1 Of Our In Depth Series - Heliguy™. Available online: https://www.heliguy.com/blogs/posts/inspire-2-batteries-part-1-of-our-in-depth-series/?srsltid=AfmBOopxQxhA5s8h-BEJv7vPyJvmexV-xDKOdh3xQ9igocuRy40agjTI (accessed on 14 November 2024).
129. MSDS 836678N(PH4-5870mAh -15.2V). Available online: https://droneshopperth.com.au/wp-content/uploads/2017/01/P4P-high-capacity-battery.pdf (accessed on 13 February 2019).

130. Superior Lithium Polymer Battery (SLPB) KOKAM Li-ion/Polymer Cell. Available online: https://moodle.utc.fr/pluginfile.php/202328/mod_resource/content/1/2019_Kokam_Cell_ver_4.1-compressed.pdf (accessed on 14 November 2024).
131. リチウムイオン電池応用・実用化先端技術開発事業」 事後評価）分科会 資料 7-1 ("Lithium-Ion Battery Application and Practical Use Advanced Technology Development Project" Post-Evaluation) Subcommittee Materials 7-1). Available online: https://www.nedo.go.jp/content/100873514.pdf (accessed on 13 February 2019).
132. Hummel, P.; Bush, T.; Gong, P.; Yasui, K.; Lee, T.; Radlinger, J.; Langan, C.; Lesne, D.; Takahashi, K.; Jung, E.; et al. *UBS Q-Series Tearing Down the Heart of an Electric Car: Can Batteries Provide an Edge, and Who Wins?* Report; UBS: New York, NY, USA, 2018.
133. Baldwin, R. Inside the Factory Building GM's Game-Changing Bolt EV. Available online: https://www.engadget.com/2016/12/09/inside-the-factory-building-gm-s-game-changing-bolt-ev/?guccounter=1# (accessed on 24 July 2019).
134. Nisewanger, J. Jaguar and Chevy Have LG in Common. Available online: https://electricrevs.com/2018/03/09/jaguar-and-chevy-have-lg-in-common/ (accessed on 24 July 2019).
135. Application of Samsung SDIs 120 Ah Cells in Current Evs. Available online: https://gosavetime.com/specifications-of-the-samsung-sdi-120-ah-battery-cell-for-electric-vehicles/ (accessed on 14 November 2024).

8 Conclusions

In conclusion, we summarise the specific energy and power—the main parameters of lithium-ion cells—considered when developing independent power sources. Figures 81 and 82 show the Ragone plots of LICs tested at room temperature and used to manufacture batteries for powering equipment (Ragone plots of miniature LICs, including solid-state LICs and LICs for medical applications, are shown in Figures 79 and 80). According to a literature review, the specific energy of LICs at nominal discharge currents varies from 50 to ≈300 Wh/kg (100 ÷ 760 Wh/L). For the developed, promising LIC prototypes and rechargeable lithium metal cells (discussed in Section Rechargeable Cells with Lithium Anodes in more detail), a more significant specific energy of 300 ÷ 400 Wh/kg (800 ÷ 1200 Wh/L) and 400 ÷ 700 Wh/kg (800 ÷ 1400 Wh/L), respectively, is observed. However, since their other characteristics do not meet the requirements (cycle life, price, temperature range, etc.), these cells are not yet widely used.

The maximum specific power of produced LICs, measured at continuous current, generally does not exceed 4.1 kW/kg (8.3 kW/L). There are also LICs—VL10V (Saft, LFP/Gr system) and VL5U (Saft, cathode material—NCA)—whose specific power (power density) reaches 4.7 kW/kg (10 kW/L) and 14 kW/kg (36 kW/L). However, they are not used in civil applications.

The general trend is a decrease in nominal specific energy (and provided energy at given power) with an increase in the allowed average power. Depending on the application, the balance between power and energy shifts to one side or another. For example, to manufacture independent power sources for unmanned aerial vehicles (quadcopters), LICs (for instance, A4.28; Figures 81 and 82, Table 48) with a high specific energy and power (energy and power density) are used. For high energy density small pouch cells used for this application one of the critical materials allowing high performance is lithiated cobalt oxide cathode, which is stable at high potentials relative to lithium.

The development of solid-state LICs (SSLICs, without a lithium anode on the negative electrode and with a solid electrolyte) is a promising direction. This type of lithium-ion cell is an alternative for LICs and batteries used in electric transport, portable electronics, etc. SSLICs are characterised by increased safety due to the absence of liquid, combustible electrolyte. It is also assumed that this type of LIC will provide high specific energy and power. However, due to their insufficiently high characteristics, SSLICs have not yet been widely used. For example, the energy density of Hitachi Zosen prototypes with a capacity of 0.14 Ah and 1 Ah (2021) is 55.6 Wh/L (20.4 Wh/kg) [1–3] and 91 Wh/L [4], respectively. On the other hand, the SSLICs produced by Prologium have higher specific energy and power. The nominal specific energy of the PCLB reaches 140 Wh/kg (350 Wh/L). At the maximum power of 250 W/kg (625 W/L), the specific energy decreases to 120 Wh/kg (300 Wh/L).

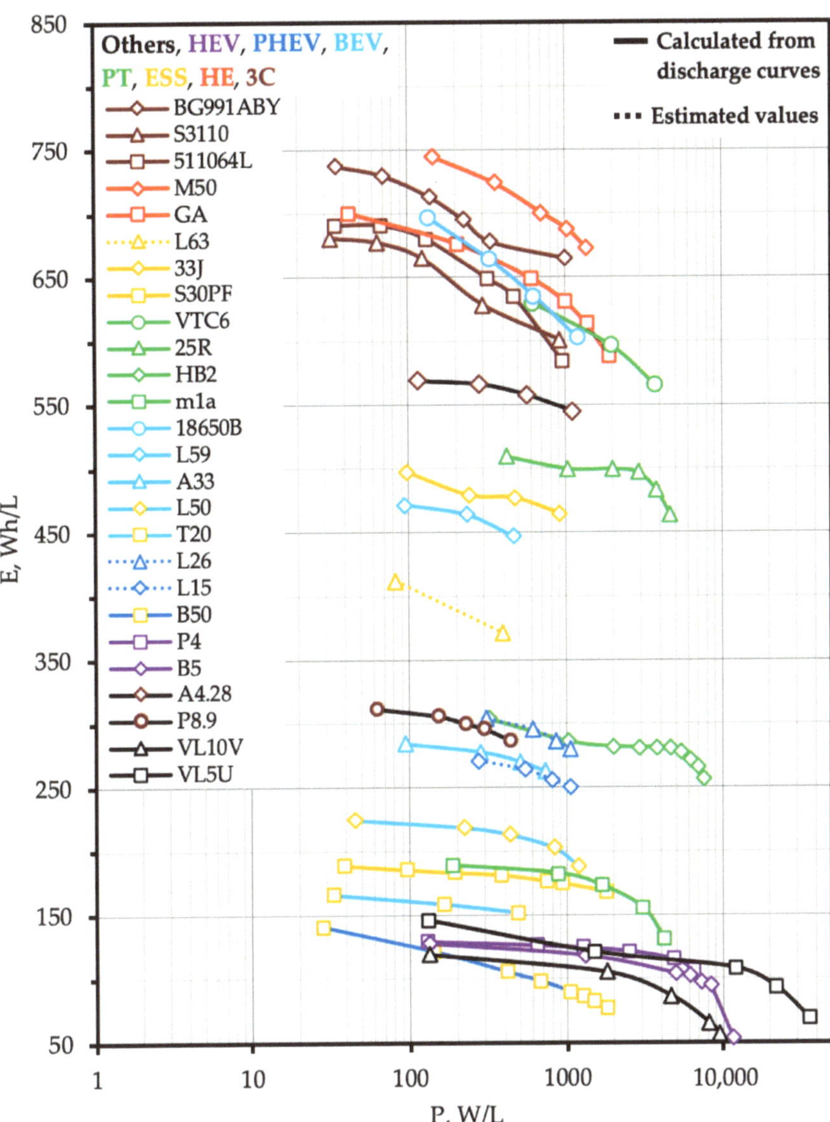

Figure 81. Ragone plots (energy density vs. power density) of LICs for various applications (3C—consumer electronics, BEV—battery electric vehicle, ESS—energy storage systems (stationary application), HE—other application which needs power source with high specific energy, HEV—hybrid electric vehicle, PHEV—plug-in hybrid electric vehicle, PT—power tools; explanation of designations is provided in Table 48). Source: Figure by authors.

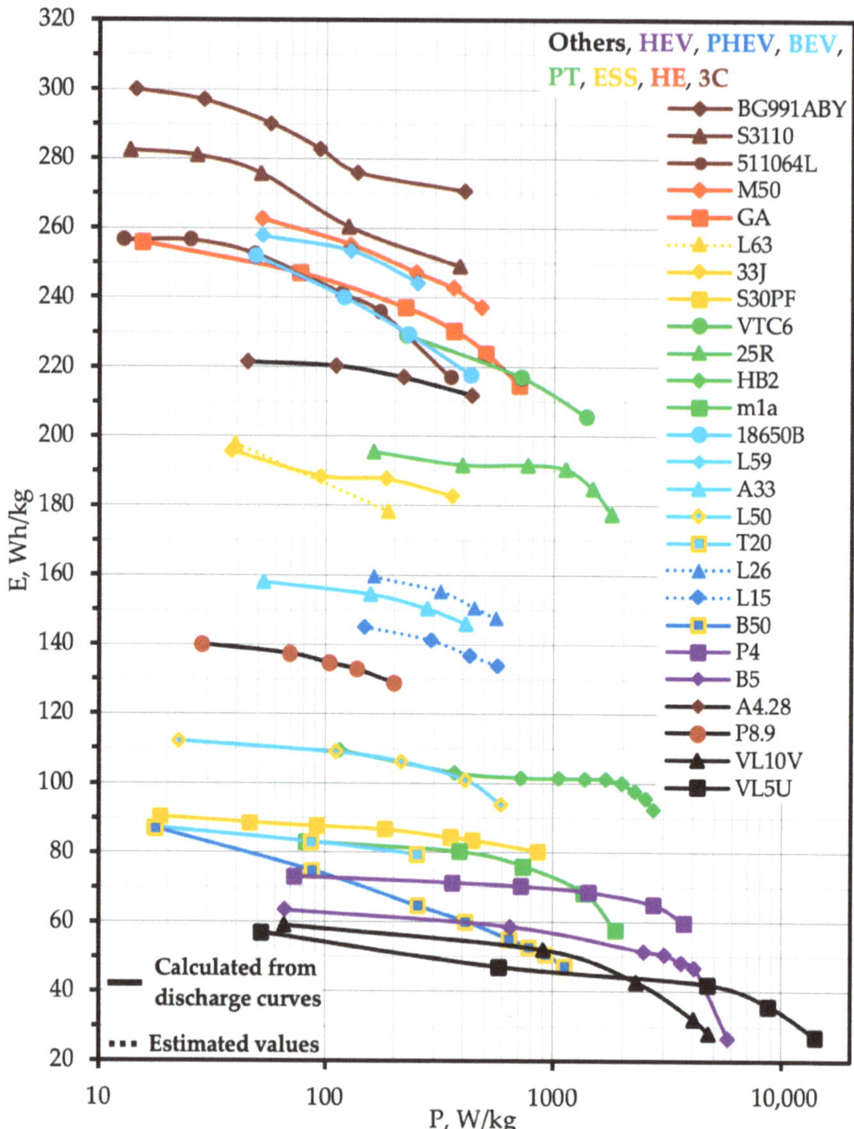

Figure 82. Ragone plots (specific energy vs. specific power) of LICs for various applications (description of 3C, BEV, ESS, HE, HEV, PHEV, PT abbreviations is given in the caption of Figure 81, explanation of designations is provided in Table 48). Source: Figure by authors.

As the temperature goes down, the performance of the LICs decreases. For comparison, dependencies that show the relation between the energy and power of LICs with various active materials were plotted (Figure 83). Figure 83 shows that the specific energy and power of LICs (Moli IHR [5], Saft MP176065xc [6]) with graphite-based anodes are higher than those of LICs (Lishen LTO 18Ah [7]) with lithium-titanate-based anodes. However, due to the higher potential of lithium titanate relative to lithium (≈1.55 V), the formation of lithium dendrides on the anode during charging is unlikely, even at low temperatures. In this regard, if a charge at low temperatures is required, the use of an LIC with a negative electrode based on lithium titanate is preferable.

Thus, we reviewed the materials used to make lithium-ion cells (Materials for Lithium-Ion Cells) and demonstrated the main characteristics of various types of lithium-ion cells and batteries used to power portable electronics (Lithium-Ion Cells with High Specific Energy), power tools (Advanced High-Power Lithium-Ion Cells for Electric Hand Tools), hybrid and electric vehicles (Lithium-Ion Cells and Batteries for Hybrid, Electric Passenger Vehicles), energy storage systems and uninterruptible power supplies (Lithium-Ion Electric Energy Storage for Stationary Applications), wearable devices (including implantable devices for medical applications), and IoT devices (Miniature Lithium-ion Cells for the Internet of Things, Wearable Devices, and Medical Applications). We do not claim to have provided an exhaustive presentation of current achievements in lithium-ion cells and materials for their production. Nevertheless, we hope that this book will be helpful for researchers and engineers engaged in the development of materials, cells, batteries, and devices that work autonomously.

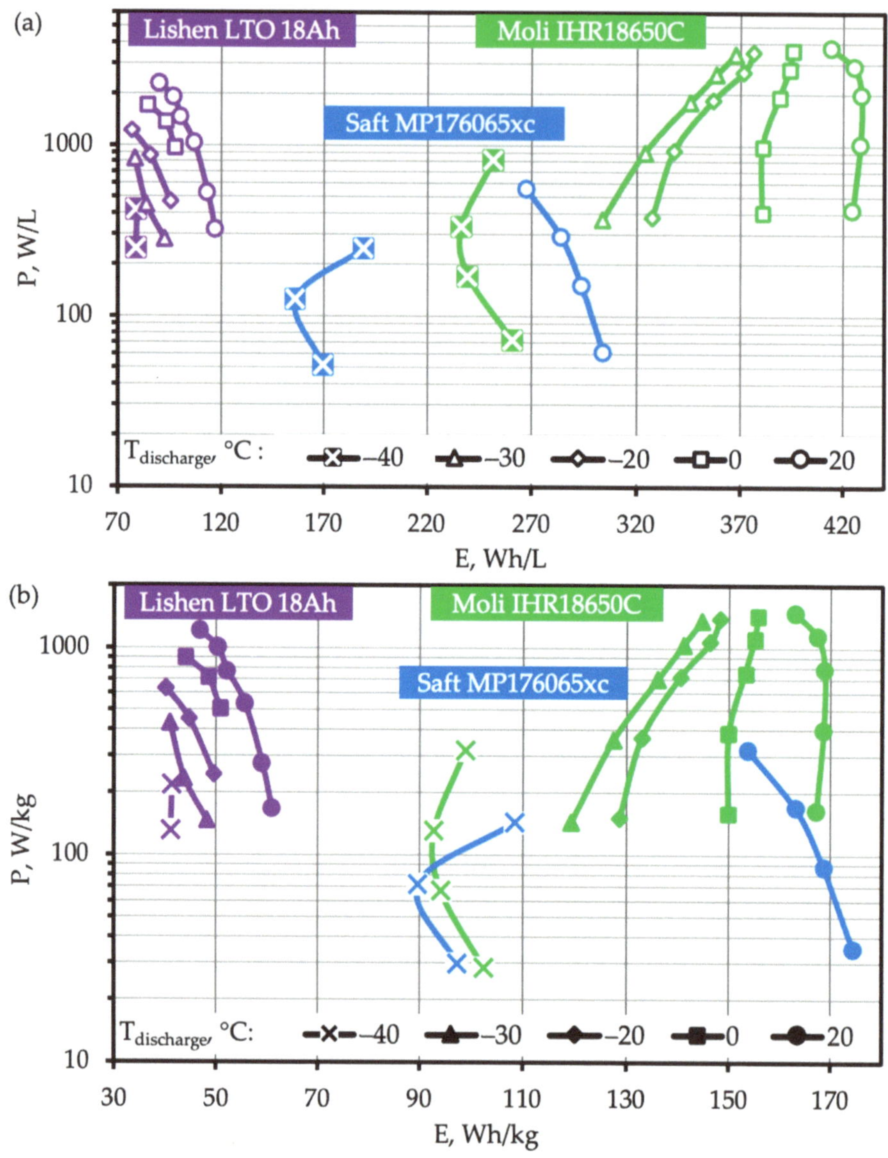

Figure 83. Change in energy (Wh/L, Wh/kg) and power (W/L, W/kg) of Lishen 18 Ah [7], Moli IHR18650C [5], and Saft M176065xc [6] discharged at room and lower temperatures. Energy and power related to volume (**a**) and weight (**b**) of the cells, correspondingly Source: Figure by authors.

Table 48. Explanation of LICs' designations given in Figures 81 and 82.

Applications	Designation	Manufacturer	Model	C, Ah	Initial Data
3C Smartphone	BG991ABY	ATL/Samsung	BG991ABY	3.88	Lab. testing
	S3110	Sunwoda	–	3.11	Lab. testing
	511064L	Sunwoda	ICP5/110/64	3.97	Lab. testing
High Energy	M50	LG	INR 21700M50	5	[8]
	GA	Panasonic	NCR18650GA	3.35	[9]
Energy Storage Systems	L63	LG	JH3	63	[10]
	33J	SDI	INR21700-33J	3.25	Lab. testing [11,12]
	S30PF	Saft	VL30P Fe	30	[13]
Power Tools	VTC6	Sony	US18650VTC6	3.1	[14]
	25R	SDI	INR18650-25R	2.5	[15]
	HB2	LG	LG 18650 HB2	1.5	[16]
	m1a	A123	18650 M1A	1.1	[17,18]
Battery Electric Vehicle	18650B	Panasonic	NCR18650B	3.35	[19]
	L59	LG	–	59	Lab. testing
	A33	AESC	33 Ah	33	[20]
	L50	LEJ	LEV50N	50	[21,22]
	T20	Toshiba	SCiB™	20	[23]
Plug-in Hybrid EV	L26	LG	–	26	[12,24]
	L15	LG	–	15	[12,25,26]
	B50	BYD	FV50NP	50	
Hybrid Electric Vehicle	P4	PEVE	–	4.1(3.6)	Lab. testing
	B5	BEC	EH5	5.0	Lab. testing
Others	A4.28	ATL	914974	4.28	Lab. testing
	P8.9	Prologium	PLCB	8.9	
	VL5U	Saft	VL5U	5	[27,28]
	VL10V	Saft	VL10V Fe	10	[28–31]

Source: Authors' compilation based on data from references cited in the table.

References

1. Nozawa, T. Hitachi Zosen to Roll Out Sulfide-Based All-Solid Battery in FY2019 (1). Available online: https://tech.nikkeibp.co.jp/dm/atclen/news_en/15mk/040102773/ (accessed on 2 December 2019).
2. Nozawa, T. Hitachi Zosen to Roll Out Sulfide-Based All-Solid Battery in FY2019 (2). Available online: https://tech.nikkeibp.co.jp/dm/atclen/news_en/15mk/040102774/ (accessed on 2 December 2019).
3. Sunayama, S. Development of All-solid-state Battery for Commercialization (Hitachi Zosen). In Proceedings of the 1Oth Battery Japan Technical Conference (Battery Japan 2019), Tokyo, Japan, 27 February 2019; pp. 19–31.
4. 日立造船が1Ahの全固体LIBを開発　宇宙利用に (Hitachi Zosen Develops 1 Ah Solid-State LIB for Space Use). Available online: https://xtech.nikkei.com/atcl/nxt/news/18/09822/ (accessed on 9 March 2019).
5. MOLICEL® IHR18650C 2 Ah Power Cell (2013-01-30). Available online: https://www.master-instruments.com.au/file/64178/1/Molicel-IHR18650C.pdf (accessed on 3 June 2019).
6. MP 176065 XC Rechargeable Lithium-Ion Cell 3.65 V High Energy Lithium-Ion Cell for Extemely Cold Conditions (Saft). Available online: https://buster-spb.ru/files//SAFT//MP//mp176065xcen.pdf (accessed on 2 December 2019).
7. Zhang, N. Lishen High Power Battery Technology for Start-stop. In Proceedings of the 2nd HEV Market & Advanced Battery Technology Development Seminar, Binjiang, China, 13–14 October 2016; p. 30.
8. 【高工锂电•总工札记】LG和三星圆柱21700电芯和去钴化 ([Gaogong Lithium Battery • Chief Engineer's Notes] LG and Samsung cylindrical 21700 battery cells and decobaltization). Available online: http://www.sohu.com/a/255303019_740349 (accessed on 13 February 2019).
9. Sanyo/Panasonic NCR18650GA 3500mAh (Red). Available online: https://lygte-info.dk/review/batteries2012/Sanyo%20NCR18650GA%203500mAh%20(Red)%20UK.html (accessed on 10 May 2018).
10. Beltran, H.; Ayuso, P.; Vicente, N.; Beltrán-Pitarch, B.; García-Cañadas, J.; Pérez, E. Equivalent circuit definition and calendar aging analysis of commercial Li(NixMnyCoz)O2/graphite pouch cells. *J. Energy Storage* **2022**, *52*, 104747. [CrossRef]
11. SDI INR21700 33J Specification. Available online: https://www.facebook.com/ecoluxshop/photos/pcb.892652754215166/892652690881839/?type=3&theater (accessed on 7 July 2018).
12. Lu, M. Future Trends and Key Issues in the Global Lithium-ion Batteries Market and Related Technologies. In Proceedings of the 13th China International Battery Fair (CIBF 2018), Shenzhen, China, 22–24 May 2018; p. 30.
13. VL 30P Fe Super-Phosphate® Rechargeable LiFePO4 Cell (Datasheet) Document No 54087-2-0513, Edition: May 2013. Available online: https://saft.com/products-solutions/products/ (accessed on 13 October 2019).

14. Lithium Ion Rechargeable Battery Technical Information, Revision 0.2. 30 June 2015. Available online: https://www.kupifonar.kz/upload/manuals/batteries/sony-us18650vtc6-techinfo.pdf (accessed on 13 February 2019).
15. Introduction of INR18650-25R Oct. 2013 Energy Business Division. Available online: https://www.powerstream.com/p/INR18650-25R-datasheet.pdf (accessed on 14 November 2024).
16. Technical Information of LG 18650HB2 (1.5 Ah). Available online: https://www.dnkpower.com/wp-content/uploads/2022/11/LGINR18650-HB2.pdf (accessed on 14 November 2024).
17. High Power Lithium Ion APR18650M1A. Available online: http://www.batteryspace.com/prod-specs/6612.pdf (accessed on 6 June 2016).
18. Jeevarajan, J. Performance and Safety Evaluation of High-Rate 18650 Lithium IronPhosphate Cells. In Proceedings of the NASA Battery Workshop, Huntsville, AL, USA, 17–19 November 2009; p. 36.
19. Panasonic NCR18650B 3400mAh (Green). Available online: https://www.shoptronica.com/files/Panasonic-NCR18650.pdf (accessed on 14 November 2024).
20. Ikezoe, M.; Hirata, N.; Amemiya, C.; Miyamoto, T.; Watanabe, Y.; Hirai, T.; Sasaki, T. *Development of High Capacity Lithium-Ion Battery for NISSAN LEAF*; SAE International 2012-01-0664; SAE International: Warrendale, PA, USA, 2012; p. 7. [CrossRef]
21. Ueki, K.; Kitano, S.; Toriyama, J.-I.; Seyama, Y.; Nishiyama, K. Development of Large-sized Long-life Type Lithium-ion Cells for Electric Vehicle. *GS Yuasa Tech. Rep.* **2012**, *9*, 26–29.
22. Kitano, S.; Nishiyama, K.; Toriyama, J.-I.; Sonoda, T. Development of Large-sized Lithium-ion Cell "LEV50" and Its Battery Module "LEV50-4" for Electric Vehicle. *GS Yuasa Tech. Rep.* **2008**, *5*, 21–26.
23. Toshiba SCiB™ Modules. Available online: http://www.scib.jp/en/product/module.htm (accessed on 26 September 2016).
24. Herrmann, M.; Matthe, R. Progress of battery systems at General Motors. In Proceedings of the World Mobility Summit, Munich, Germany, 18–20 October 2016; p. 24.
25. Darcy, E. Li-ion Pouch Cell Designs; Are They Ready for Space Applications? In Proceedings of the Large Li-ion Battery Technology and Application Symposium (AABC12), Orlando, FL, USA, 8 February 2012; p. 20.
26. IEK, I. 鑑往知來: 由全球鈕電池産黨與市場發展動向 看中國産黨轉型與升級之路 (Learn from the past and learn from the future: Looking at the transformation and upgrading of the Chinese Communist Party from the global button battery industry and market development trends). In Proceedings of the 2012 中國化學與物理電源行業協會理事會議 (China Chemical and Physical Power Supply Industry Association Board of Directors Meeting), Beijing, China, 14–15 December 2012; p. 126.
27. Allen, J.L.; Wolfenstine, J.; Xu, K.; Porschet, D.; Salem, T.; Tipton, W.; Behl, W.; Read, J.; Jow, T.R.; Gargies, S. *Evaluation of Saft Ultra High Power Lithium Ion Cells (VL5U)*; U.S. Army Research Laboratory: Adelphi, MD, USA, 2009; p. 24. Available online: https://apps.dtic.mil/sti/tr/pdf/ADA494956.pdf (accessed on 3 June 2024).

28. Wetz, D.A.; Shrestha, B.; Novak, P.M. Pulsed Evaluation of High Power Electrochemical Energy Storage Devices. *IEEE Trns. Dielectr. Electr. Insul.* **2013**, *20*, 1040–1048. [CrossRef]
29. Li-Ion Energy Storage Systems by Saft Batteries. 2008. Available online: http://www.who-sells-it.com/cy/saft-batteries-3792/li-ion-energy-storage-systems-18811.html (accessed on 21 November 2019).
30. Wetz, D.A.; Shrestha, B.; Novak, P.M. Pulsed Current Limitations of High Power Electrochemical Energy Storage Devices. In Proceedings of the IEEE International Power Modulator and High Voltage Conference (IPMHVC), San Diego, CA, USA, 3–7 June 2012; pp. 288–291.
31. Sud Chemie Creating Performance Technology. Available online: http://wenku.baidu.com/view/ec72b73b3968011ca3009148.html (accessed on 8 August 2016).

Appendix A. Lithium-Ion Cells Mentioned in Figures and Tables

Table A1. Lithium-ion cells sorted by company name.

№	Manufacturer	Designation in Text	Designation in Figures	Figures	Tables
1	–	433381	433381	11	–
2	–	476790	476790	13	–
3	–	606080	606080	11	–
4	–	3768110	3768110	11	–
5	A123	18650 M1A	M1A	24, 26, 27, 28, 81, 82	20, 21, 22, 24, 48
6	A123	AMP20m1HD	A20	60, 65, 66	30, 43
7	A123	UltraPhosphate 20 Ah	–	–	30
8	A123	UltraPhosphate 8 Ah	–	–	30
9	AESC	33 Ah (BEV)	A33	42, 81, 82	31, 48
10	AESC	4.3 Ah	–	–	31
11	AESC	56 Ah (BEV)	A56	42	31
12	Amprius	–	–	–	19
13	Amprius	–	–	–	19
14	Amprius	–	–	–	19
15	Amprius	–	–	–	19
16	Amprius	–	–	–	19
17	Amprius	–	–	22	–
18	Amprius	ANW2.6-405056	–	–	19
19	Amprius	ANW3.6-405056	–	–	19
20	Amprius	ANW4.0-455056	–	–	19
21	ATL	425882	–	–	–
22	ATL	6860C5	–	–	–
23	ATL	914974	A4.28	78, 81, 82	48
24	ATL	A53129	A1220	67, 78, 79, 80	42, 43
25	ATL	PH4 cell 5.87 Ah	A5.87	78	–
26	BEC	EH5	B5	39, 42, 81, 82	31, 48
27	BEC	EHW5	–	–	31
28	BMZ	21700 40PS1	40PS	26	–
29	BMZ	21700 50EL	50EL	18	–
30	BMZ	21700 52EM	52EM	18	–
31	Boston Power	Swing 5300	BP53	42, 43, 44, 66	31
32	BYD	50Ah	B50	42, 61, 81, 82	31, 41, 48

Table A1. *Cont.*

№	Manufacturer	Designation in Text	Designation in Figures	Figures	Tables
33	BYD	Blade battery (200 Ah)	–	–	31
34	BYD	C12 FP58146410A	–	–	31
35	Cadenza	Supercell	–	–	31
36	CATL	10 Ah (for 48 V LIB)	–	–	30
37	CATL	50 Ah	–	–	31
38	CATL	6.9 Ah (HEV)	–	44	31
39	CATL	70 Ah	–	–	31
40	Cymbet	CBC050 (Bare die)	C005	70, 73, 76, 77, 78, 79, 80	47
41	DFD	PSP12161227 55 Ah	–	–	31
42	E-one Moli	18650 BC	IBR	24, 26, 29, 30	20, 21, 24, 25
43	E-one Moli	18650 C	IHR	24, 26, 29, 30, 83	20, 21, 23, 24, 25
44	E-one Moli	18650 e	IMR	24, 26, 29, 30	20, 21, 22, 23, 24, 25
45	Eagle Picher	95-1399, 95-1250	–	–	42
46	Eagle Picher	D-00020	EP50	67, 78, 79, 80	42, 43
47	Eagle Picher	D-00122	–	78	42, 43
48	EIG	C020	E20	42	31
49	Embatt	3.0	–	–	19
50	Enevate	EN9072120	–	–	19
51	Envia	ENV 234426-XP	–	–	19
52	Farasis	–	–	22	19
53	FDK	0.14 mAh	F0140	70, 73, 74, 77, 78	46
54	FDK	0.5 mAh	–	78	46
55	Front Edge Technology	NanoEnergy	FET09	73, 76, 77, 78, 79	47
56	Ganfeng	5889C05	–	–	19
57	Ganfeng	7289C05	–	–	19
58	GS Yuasa	13Ah (PHEV)	G13	42	31
59	GS Yuasa	25 Ah (PHEV)	–	–	31
60	GS Yuasa	EX25A	–	–	31
61	GS Yuasa	LEV60F	–	–	30
62	GS Yuasa	LIM25H	–	63	–

Table A1. *Cont.*

№	Manufacturer	Designation in Text	Designation in Figures	Figures	Tables
63	GS Yuasa	LIM30H	G30	63, 65, 66	41
64	GS Yuasa	Lim50E	G50	62, 65, 66	41
65	Guoxuan	105 Ah	–	–	31
66	High Power	404798AC	404798AC	13	–
67	High Power	404798AD	404798AD	13	–
68	Hitachi	28 Ah (PHEV)	–	–	31
69	Hitachi	4.4 Ah (HEV)	–	–	31
70	Hitachi	40 Ah	–	–	31
71	Hitachi	5.2 Ah (HEV)	–	–	31
72	Hitachi	5.3 Ah (HEV)	–	–	31
73	Hitachi	5.5 Ah (HEV)	–	–	31
74	Hitachi	CH75	H75	65	41
75	Hitz	AS LIB	–	70, 78	45
76	Huahui new energy	HTC1865	HTC	24, 26	20, 21, 25
77	Hydro Quebec	Gen2 30Ah	–	–	19
78	Ilika	Goliath A6	–	–	19
79	Ilika	M250	I0250	73, 75, 76, 77, 78, 79, 80	42, 47
80	Ilika	P180	I0180	73, 76, 78	42, 47
81	IPS	Mec201	–	–	42
82	IPS	Thinergy® MEC202	Th2	70, 73, 76, 77, 78, 79, 80	47
83	Jenax	30 mAh	–	78	45
84	Kokam	SLPB065070180	K11.6	78	–
85	Kokam	SLPB080085270	K26	78	–
86	Kokam	SLPB120255255	–	–	41
87	Leclanche	LecCell30Ah	L30	58, 65, 66	41
88	LEJ	LEV50N	L50	42, 81, 82	31, 48
89	LG	18650 E1	E1	14, 15, 18, 78	15, 16
90	LG	18650 HB2	HB2	24, 25, 26, 81, 82	20, 21, 22, 23, 24, 48
91	LG	18650 HD2	HD2	24, 26	20, 21, 22, 23, 24
92	LG	18650 HG2	HG2	24, 26, 29, 30	20, 21, 23, 24, 25
93	LG	18650 MJ1	MJ1	15, 18, 19	15, 16

Table A1. *Cont.*

№	Manufacturer	Designation in Text	Designation in Figures	Figures	Tables
94	LG	21700 M50	M50	18, 78, 81, 82	15, 48
95	LG	26 Ah (Volt gen2)	L26	42, 81, 82	31, 48
96	LG	59 Ah (BEV)	L59	42, 78, 81, 82	31, 48
97	LG	CPI (15 Ah)	L15	42, 81, 82	31, 48
98	LG	JH3	L63	65, 81, 82	41, 48
99	LG	JP3	–	–	41
100	Lionrock	3X12019	–	78	45
101	Lishen	405582	405582	13	–
102	Lishen	10 Ah (for 48 V LIB)	–	43	30
103	Lishen	25 Ah (PHEV)	–	41, 42	31
104	Lishen	LB1212075 5.5 Ah	–	38	31
105	Lishen	Lishen 4.45	Lishen 4.45	12	–
106	Lishen	LP2270112-7.5 Ah	–	38	31
107	Lishen	Super Battery	L18	42, 45, 83	30
108	Litarion	Litacell 44	–	–	41
109	Maxell	CLB2016	M30	68, 69, 77, 78, 79, 80	44
110	Maxell	CLB2032	M70	68, 69, 78, 79, 80	44
111	Maxell	CLB937	–	78	44
112	Maxell	ML2032	M65	69, 78	44
113	Maxell	PSB927L	M8	73, 77, 78, 79	46
114	Maxell	TC 920S	–	78	44
115	Microvast	HpCO (Gen 4)	–	–	31
116	Microvast	LpCO (Gen 2)	–	42	31
117	Microvast	MpCO (Gen 3)	–	–	31
118	Molicel	21700 42A	42A	26, 66	–
119	Murata	10 mAh	–	70	46
120	Murata	CR2032R (primary cell)	M200R	69	–
121	Murata	CR2032X (primary cell)	M220X	68, 69	–
122	Murata	CT04120	CT3	67, 68, 78, 79	43
123	Murata	Umal	U12	67, 78	43
124	Murata/Sony	18650 VC7	VC7	16, 17, 18, 19, 66	16

Table A1. *Cont.*

№	Manufacturer	Designation in Text	Designation in Figures	Figures	Tables
125	Murata/Sony	18650 VTC6	VTC6	24, 26, 29, 30, 81, 82	20, 21, 22, 23, 24, 25, 48
126	NGK	EC382704P-C	N27	71, 77, 78, 79, 80	45
127	NGK	ET1210C-H	–	78	44
128	NGK	ET2016C-R	–	77	42, 44
129	NGK	ET271704P-H	–	78	45
130	Nichicon	SLB	–	–	42
131	Nichicon	SLB Series	–	–	42
132	Nichicon	SLB03070LR35	N035	67, 78, 79, 80	43
133	Panasonic	20 Ah (4.35 V)	–	–	31
134	Panasonic	320B	P16B	67, 68, 78, 79, 80	43
135	Panasonic	420A	P23A	67, 68, 78, 79, 80	43
136	Panasonic	CG-042839	–	78	45
137	Panasonic	CG-043555	–	78	45
138	Panasonic	18650 GA	GA	16, 18, 19, 81, 82	16, 48
139	Panasonic	18650 B	P3.3	42, 81, 82	31, 48
140	Panasonic (Tesla)	21700 4.85 Ah	–	–	31
141	PEVE	3.6 Ah (4.1) 2nd gen HEV	P4	39, 42, 81, 82	31, 48
142	PEVE	5 Ah 1st gen (HEV)	–	–	31
143	Prologium	32D3L8ABKA	P11	23	19
144	Prologium	36D3L8AAJA	P8.9	23, 81, 82	19, 48
145	Prologium	4360A5AAMA	P1.95	23	19
146	Prologium	59D3L8ABXA	Pg24	23, 65	19
147	Prologium	FLCB027038 AAAA	Pg16	71, 72, 77, 78, 79, 80	45
148	Prologium	FLCB046046 AAAA	Pg45	71, 72, 78, 79, 80	45
149	Quallion	QL0003B	Q3	67, 79, 80	42, 43
150	Saft	MP176065xc	–	83	–
151	Saft	VL 10V Fe	VL10V	81, 82	48
152	Saft	VL 30P	–	–	41
153	Saft	VL 30P Fe	S30PF	65, 81, 82	41, 48
154	Saft	VL 41M Fe	S41MF	65	41

Table A1. *Cont.*

№	Manufacturer	Designation in Text	Designation in Figures	Figures	Tables
155	Saft	VL 41M	S41M	64, 65	41
156	Saft	VL 45E	–	–	41
157	Saft	VL 5U	VL5U	81, 82	48
158	SDI	120 Ah	S120	78	31
159	SDI	18650 25R	25R	24, 26, 29, 30, 81, 82	20, 21, 22, 23, 24, 25, 48
160	SDI	18650 26F	–	20	–
161	SDI	18650 32A	–	–	15
162	SDI	18650 35E	35E	16, 18, 19	16
163	SDI	21700 30T	30T	26	–
164	SDI	21700 33J	33J	65, 81, 82	41, 48
165	SDI	21700 48G	48G	18	–
166	SDI	94 Ah	S94	42, 65	31, 41
167	SDI	BG900BBE	–	–	14
168	SDI	BG920ABE	–	–	14
169	SDI	BG930ABA	–	–	14
170	SDI	BG950ABA	–	–	14
171	SDI	BG960ABA	–	–	14
172	SDI	BG973ABU	BG973ABU	13	14
173	SDI	BG975ABU	BG975ABU	13	14
174	SDI	BG980ABY	–	–	14
175	SDI	BG991ABY	BG991ABY	13, 78, 81, 82	14, 48
176	SDI	BS901ABY	–	–	14
177	SDI	EB-1A2GBU	–	–	14
178	SDI	EB-B600	–	–	14
179	SDI	EB-BR830ABY	S240	67, 78, 79, 80	43
180	SDI	EB-l1g6	–	–	14
181	SDI	EB57515ZVU	–	–	14
182	SDI	ICP52121	–	–	42
183	SDI (SBL)	20.5 Ah (PHEV)	–	–	31
184	SDI (SBL)	24.5 Ah (PHEV)	S24	42	31
185	SDI (SBL)	5.2 Ah (HEV)	–	38, 46	31
186	SDI (SBL)	60 Ah (BEV)	–	–	31
187	SEEO	L13-001-001	–	–	19

Table A1. *Cont.*

№	Manufacturer	Designation in Text	Designation in Figures	Figures	Tables
188	Seiko	MS621FE	S5.5	69, 77, 78	44
189	Sion Power	Licerion	–	–	19
190	Sion Power	Licerion	–	–	19
191	Sion Power	Licerion	–	22	–
192	Sion power	Licerion HE	–	–	19
193	Sion Power	Licerion HP	–	–	19
194	Solid Energy	Hermes	–	22	19
195	Sony (Murata)	18650 VTC3	VTC3	–	20, 21, 23, 24
196	Sony (Murata)	18650 WH1	–	–	15
197	Sony (Murata)	26650 FTC1	–	–	41
198	ST Microelectronics	EFL1K0AF39	ST1	73, 76, 77, 78, 79, 80	47
199	Sunwoda	3.110 Ah	S3110, S3.11	13, 78, 81, 82	14, 48
200	Sunwoda	511064L	511064L	13, 81, 82	14, 48
201	TDK	Ceracharge 1812	T0100	70, 73, 74, 76, 77, 78, 79, 80	42, 46
202	Toshiba	102 mAh	T102	71, 72, 78, 79, 80	45
203	Toshiba	20 Ah (BEV)	–	59, 65, 66, 81, 82	31, 41, 48
204	Toshiba	20 Ah (for 48 V LIB)	–	–	30
205	Toshiba	49 Ah ($TiNb_2O_7$)	T49	42, 65, 66	31
206	Toshiba	5 Ah (for 48 V LIB)	–	–	30
207	Toshiba	SCiB™ 10	–	–	30, 41
208	Toshiba	SCiB™ 2.9	–	–	30
209	Toshiba	SCiB™ 23 Ah	–	–	41
210	Valence	18650 M1B	–	–	41
211	Valence	26650 M1B	–	–	41
212	Varta	CP1254 A3	V63	68, 69, 79, 80	42, 44
213	Varta	CP1654 A3	V125	68, 69, 79, 80	42, 44
214	Varta	CP7840	V17	68, 69, 78, 79, 80	44
215	Wyon	0.16 mAh	W0160	77, 78	44
216	Wyon	W101, etc.	–	–	42, 44
217	Zenlabs	51 Ah	–	–	31
218	Zenlabs	Glide drone	–	22	19
219	Zenlabs	Range EV	–	22	19

Table A2. Lithium-ion cells mentioned in figures and tables sorted by designation in the text.

Designation in Text	Manufacturer	Designation in Figures	Figure	Table	№ (A.1)
3.0	Embatt	–	–	19	49
405582	Lishen	405582	13	–	101
425882	ATL	–	–	–	21
433381	–	433381	11	–	1
476790	–	476790	13	–	2
606080	–	606080	11	–	3
914974	ATL	A4.28	78, 81, 82	48	23
3768110	–	3768110	11	–	4
–	Amprius	–	–	19	12
–	Amprius	–	–	19	13
–	Amprius	–	–	19	14
–	Amprius	–	–	19	15
–	Amprius	–	–	19	16
–	Amprius	–	22	–	17
–	Farasis	–	22	19	52
0.14 mAh	FDK	F0140	70, 73, 74, 77, 78	46	53
0.16 mAh	Wyon	W0160	77, 78	44	215
0.5 mAh	FDK	–	78	46	54
10 mAh	Murata	–	70	46	119
102 mAh	Toshiba	T102	71, 72, 78, 79, 80	45	202
105 Ah	Guoxuan	–	–	31	65
10 Ah (for 48 V LIB)	CATL	–	–	30	36
10 Ah (for 48 V LIB)	Lishen	–	43	30	102
120 Ah	SDI	S120	78	31	158
13 Ah (PHEV)	GS Yuasa	G13	42	31	58
18650 25R	SDI	25R	24, 26, 29, 30, 81, 82	20, 21, 22, 23, 24, 25, 48	159
18650 26F	SDI	–	20	–	160
18650 32A	SDI	–	–	15	161
18650 35E	SDI	35E	16, 18, 19	16	162
18650 B	Panasonic	P3.3	42, 81, 82	31, 48	139
18650 BC	E-one Moli	IBR	24, 26, 29, 30	20, 21, 24, 25	42

Table A2. *Cont.*

Designation in Text	Manufacturer	Designation in Figures	Figure	Table	№ (A.1)
18650 C	E-one Moli	IHR	24, 26, 29, 30, 83	20, 21, 23, 24, 25	43
18650 e	E-one Moli	IMR	24, 26, 29, 30	20, 21, 22, 23, 24, 25	44
18650 E1	LG	E1	14, 15, 18, 78	15, 16	89
18650 GA	Panasonic	GA	16, 18, 19, 81, 82	16, 48	138
18650 HB2	LG	HB2	24, 25, 26, 81, 82	20, 21, 22, 23, 24, 48	90
18650 HD2	LG	HD2	24, 26	20, 21, 22, 23, 24	91
18650 HG2	LG	HG2	24, 26, 29, 30	20, 21, 23, 24, 25	92
18650 M1A	A123	M1A	24, 26, 27, 28, 81, 82	20, 21, 22, 24, 48	5
18650 M1B	Valence	–	–	41	210
18650 MJ1	LG	MJ1	15, 18, 19	15, 16	93
18650 VC7	Murata/Sony	VC7	16, 17, 18, 19, 66	16	124
18650 VTC3	Sony (Murata)	VTC3	–	20, 21, 23, 24	195
18650 VTC6	Murata/Sony	VTC6	24, 26, 29, 30, 81, 82	20, 21, 22, 23, 24, 25, 48	125
18650 WH1	Sony (Murata)	–	–	15	196
20.5 Ah (PHEV)	SDI (SBL)	–	–	31	183
20 Ah (4.35 V)	Panasonic	–	–	31	133
20 Ah (BEV)	Toshiba	–	59, 65, 66, 81, 82	31, 41, 48	203
20 Ah (for 48 V LIB)	Toshiba	–	–	30	204
21700 30T	SDI	30T	26	–	163
21700 33J	SDI	33J	65, 81, 82	41, 48	164
21700 4.85 Ah	Panasonic (Tesla)	–	–	31	140
21700 40PS1	BMZ	40PS	26	–	28
21700 42A	Molicel	42A	26, 66	–	118
21700 48G	SDI	48G	18	–	165
21700 50EL	BMZ	50EL	18	–	29

Table A2. *Cont.*

Designation in Text	Manufacturer	Designation in Figures	Figure	Table	№ (A.1)
21700 52EM	BMZ	52EM	18	–	30
21700 M50	LG	M50	18, 78, 81, 82	15, 48	94
24.5 Ah (PHEV)	SDI (SBL)	S24	42	31	184
25 Ah (PHEV)	GS Yuasa	–	–	31	59
25 Ah (PHEV)	Lishen	–	41, 42	31	103
26650 FTC1	Sony (Murata)	–	–	41	197
26650 M1B	Valence	–	–	41	211
26 Ah (Volt gen2)	LG	L26	42, 81, 82	31, 48	95
28 Ah (PHEV)	Hitachi	–	–	31	68
3.110 Ah	Sunwoda	S3110, S3.11	13, 78, 81, 82	14, 48	199
3.6 Ah (4.1) 2nd gen HEV	PEVE	P4	39, 42, 81, 82	31, 48	141
30 mAh	Jenax	–	78	45	83
320B	Panasonic	P16B	67, 68, 78, 79, 80	43	134
32D3L8ABKA	Prologium	P11	23	19	143
33 Ah (BEV)	AESC	A33	42, 81, 82	31, 48	9
36D3L8AAJA	Prologium	P8.9	23, 81, 82	19, 48	144
3X12019	Lionrock	–	78	45	100
4.3 Ah	AESC	–	–	31	10
4.4 Ah (HEV)	Hitachi	–	–	31	69
404798AC	High Power	404798AC	13	–	66
404798AD	High Power	404798AD	13	–	67
40 Ah	Hitachi	–	–	31	70
420A	Panasonic	P23A	67, 68, 78, 79, 80	43	135
4360A5AAMA	Prologium	P1.95	23	19	145
49 Ah ($TiNb_2O_7$)	Toshiba	T49	42, 65, 66	31	205
5.2 Ah (HEV)	Hitachi	–	–	31	71
5.2 Ah (HEV)	SDI (SBL)	–	38, 46	31	185
5.3 Ah (HEV)	Hitachi	–	–	31	72
5.5 Ah (HEV)	Hitachi	–	–	31	73
50 Ah	BYD	B50	42, 61, 81, 82	31, 41, 48	32
50 Ah	CATL	–	–	31	37
511064L	Sunwoda	511064L	13, 81, 82	14, 48	200

Table A2. *Cont.*

Designation in Text	Manufacturer	Designation in Figures	Figure	Table	№ (A.1)
51 Ah	Zenlabs	–	–	31	217
56 Ah (BEV)	AESC	A56	42	31	11
5889C05	Ganfeng	–	–	19	56
59 Ah (BEV)	LG	L59	42, 78, 81, 82	31, 48	96
59D3L8ABXA	Prologium	Pg24	23, 65	19	146
5 Ah (for 48 V LIB)	Toshiba	–	–	30	206
5 Ah 1st gen (HEV)	PEVE	–	–	31	142
6.9 Ah (HEV)	CATL	–	44	31	38
60 Ah (BEV)	SDI (SBL)	–	–	31	186
6860C5	ATL	–	–	–	22
70 Ah	CATL	–	–	31	39
7289C05	Ganfeng	–	–	19	57
94 Ah	SDI	S94	42, 65	31, 41	166
95-1399, 95-1250	Eagle Picher	–	–	42	45
A53129	ATL	A1220	67, 78, 79, 80	42, 43	24
AMP20m1HD	A123	A20	60, 65, 66	30, 41	6
ANW2.6-405056	Amprius	–	–	19	18
ANW3.6-405056	Amprius	–	–	19	19
ANW4.0-455056	Amprius	–	–	19	20
AS LIB	Hitz	–	70, 78	45	75
BG900BBE	SDI	–	–	14	167
BG920ABE	SDI	–	–	14	168
BG930ABA	SDI	–	–	14	169
BG950ABA	SDI	–	–	14	170
BG960ABA	SDI	–	–	14	171
BG973ABU	SDI	BG973ABU	13	14	172
BG975ABU	SDI	BG975ABU	13	14	173
BG980ABY	SDI	–	–	14	174
BG991ABY	SDI	BG991ABY	13, 78, 81, 82	14, 48	175
Blade battery (200 Ah)	BYD	–	–	31	33
BS901ABY	SDI	–	–	14	176
C020	EIG	E20	42	31	48
C12 FP58146410A	BYD	–	–	31	34

Table A2. *Cont.*

Designation in Text	Manufacturer	Designation in Figures	Figure	Table	№ (A.1)
CBC050 (Bare die)	Cymbet	C005	70, 73, 76, 77, 78, 79, 80	47	40
Ceracharge 1812	TDK	T0100	70, 73, 74, 76, 77, 78, 79, 80	42, 46	201
CG-042839	Panasonic	–	78	45	136
CG-043555	Panasonic	–	78	45	137
CH75	Hitachi	H75	65	41	74
CLB2016	Maxell	M30	68, 69, 77, 78, 79, 80	44	109
CLB2032	Maxell	M70	68, 69, 78, 79, 80	44	110
CLB937	Maxell	–	78	44	111
CP1254 A3	Varta	V63	68, 69, 79, 80	42, 44	212
CP1654 A3	Varta	V125	68, 69, 79, 80	42, 44	213
CP7840	Varta	V17	68, 69, 78, 79, 80	44	214
CPI (15 Ah)	LG	L15	42, 81, 82	31, 48	97
CR2032R (primary cell)	Murata	M200R	69	–	120
CR2032X (primary cell)	Murata	M220X	68, 69	–	121
CT04120	Murata	CT3	67, 68, 78, 79	43	122
D-00020	Eagle Picher	EP50	67, 78, 79, 80	42, 43	46
D-00122	Eagle Picher	–	78	42, 43	47
EB-1A2GBU	SDI	–	–	14	177
EB-B600	SDI	–	–	14	178
EB-BR830ABY	SDI	S240	67, 78, 79, 80	43	179
EB-l1g6	SDI	–	–	14	180
EB57515ZVU	SDI	–	–	14	181
EC382704P-C	NGK	N27	71, 77, 78, 79, 80	45	126
EFL1K0AF39	ST Microelectronics	ST1	73, 76, 77, 78, 79, 80	47	198
EH5	BEC	B5	39, 42, 81, 82	31, 48	26
EHW5	BEC	–	–	31	27
EN9072120	Enevate	–	–	19	50
ENV 234426-XP	Envia	–	–	19	51

Table A2. *Cont.*

Designation in Text	Manufacturer	Designation in Figures	Figure	Table	№ (A.1)
ET1210C-H	NGK	–	78	44	127
ET2016C-R	NGK	–	77	42, 44	128
ET271704P-H	NGK	–	78	45	129
EX25A	GS Yuasa	–	–	31	60
FLCB027038 AAAA	Prologium	Pg16	71, 72, 77, 78, 79, 80	45	147
FLCB046046 AAAA	Prologium	Pg45	71, 72, 78, 79, 80	45	148
Gen2 30 Ah	Hydro Quebec	–	–	19	77
Glide drone	Zenlabs	–	22	19	218
Goliath A6	Ilika	–	–	19	78
Hermes	Solid Energy	–	22	19	194
HpCO (Gen 4)	Microvast	–	–	31	115
HTC1865	Huahui new energy	HTC	24, 26	20, 21, 25	76
ICP52121	SDI	–	–	42	182
JH3	LG	L63	65, 81, 82	41, 48	98
JP3	LG	–	–	41	99
L13-001-001	SEEO	–	–	19	187
LB1212075 5.5 Ah	Lishen	–	38	31	104
LecCell30 Ah	Leclanche	L30	58, 65, 66	41	87
LEV50N	LEJ	L50	42, 81, 82	31, 48	88
LEV60F	GS Yuasa	–	–	30	61
Licerion	Sion Power	–	–	19	189
Licerion	Sion Power	–	–	19	190
Licerion	Sion Power	–	22	–	191
Licerion HE	Sion power	–	–	19	192
Licerion HP	Sion Power	–	–	19	193
LIM25H	GS Yuasa	–	63	–	62
LIM30H	GS Yuasa	G30	63, 65, 66	41	63
Lim50E	GS Yuasa	G50	62, 65, 66	41	64
Lishen 4.45	Lishen	Lishen 4.45	12	–	105
Litacell 44	Litarion	–	–	41	108
LP2270112-7.5 Ah	Lishen	–	38	31	106
LpCO (Gen 2)	Microvast	–	42	31	116

Table A2. *Cont.*

Designation in Text	Manufacturer	Designation in Figures	Figure	Table	№ (A.1)
M250	Ilika	I0250	73, 75, 76, 77, 78, 79, 80	42, 47	79
Mec201	IPS	–	–	42	81
ML2032	Maxell	M65	69, 78	44	112
MP176065xc	Saft	–	83	–	150
MpCO (Gen 3)	Microvast	–	–	31	117
MS621FE	Seiko	S5.5	69, 77, 78	44	188
NanoEnergy	Front Edge Technology	FET09	73, 76, 77, 78, 79	47	55
P180	Ilika	I0180	73, 76, 78	42, 47	80
PH4 cell 5.87 Ah	ATL	A5.87	78	–	25
PSB927L	Maxell	M8	73, 77, 78, 79	46	113
PSP12161227 55 Ah	DFD	–	–	31	41
QL0003B	Quallion	Q3	67, 79, 80	42, 43	149
Range EV	Zenlabs	–	22	19	219
SCiB™ 10	Toshiba	–	–	30, 41	207
SCiB™ 2.9	Toshiba	–	–	30	208
SCiB™ 23 Ah	Toshiba	–	–	41	209
SLB	Nichicon	–	–	42	130
SLB Series	Nichicon	–	–	42	131
SLB03070LR35	Nichicon	N035	67, 78, 79, 80	43	132
SLPB065070180	Kokam	K11.6	78	–	84
SLPB080085270	Kokam	K26	78	–	85
SLPB120255255	Kokam	–	–	41	86
Super Battery	Lishen	L18	42, 45, 83	30	107
Supercell	Cadenza	–	–	31	35
Swing 5300	Boston Power	BP53	42, 43, 44, 66	31	31
TC 920S	Maxell	–	78	44	114
Thinergy® MEC202	IPS	Th2	70, 73, 76, 77, 78, 79, 80	47	82
UltraPhosphate 20 Ah	A123	–	–	30	7
UltraPhosphate 8 Ah	A123	–	–	30	8
Umal	Murata	U12	67, 78	43	123
VL 10V Fe	Saft	VL10V	81, 82	48	151
VL 30P	Saft	–	–	41	152
VL 30P Fe	Saft	S30PF	65, 81, 82	41, 48	153

Table A2. *Cont.*

Designation in Text	Manufacturer	Designation in Figures	Figure	Table	№ (A.1)
VL 41M Fe	Saft	S41MF	65	41	154
VL 41M	Saft	S41M	64, 65	41	155
VL 45E	Saft	–	–	41	156
VL 5U	Saft	VL5U	81, 82	48	157
W101	Wyon	–	–	42, 44	216

MDPI AG
Grosspeteranlage 5
4052 Basel
Switzerland
Tel.: +41 61 683 77 34

MDPI Books Editorial Office
E-mail: books@mdpi.com
www.mdpi.com/books

www.ingramcontent.com/pod-product-compliance
Lightning Source LLC
LaVergne TN
LVHW071926160726
843515LV00010B/2540

* 9 7 8 3 0 3 6 5 8 2 6 7 2 *